TODAY'S TECHNICIAN™

SHOP MANUAL

For Automotive Suspension & Steering Systems

SEVENTH EDITION

Mark Schnubel

Cengage

Australia • Brazil • Canada • Mexico • Singapore • United Kingdom • United States

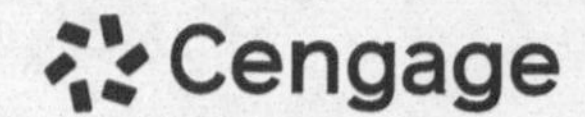

***Today's Technician: Automotive Suspension and Steering Systems,* Seventh Edition**
Mark Schnubel

SVP, GM Skills & Global Product Management: Jonathan Lau

Product Director: Matthew Seeley

Senior Product Manager: Katie McGuire

Product Assistant: Kimberly Klotz

Executive Director, Content Design: Marah Bellegarde

Learning Design Director: Juliet Steiner

Learning Designer: Mary Clyne

Vice President, Strategic Marketing Services: Jennifer Ann Baker

Marketing Director: Shawn Chamberland

Marketing Manager: Andrew Ouimet

Director, Content Delivery: Wendy Troeger

Manager, Content Deliver: Alexis Ferraro

Senior Content Manager: Meaghan Tomaso

Senior Digital Delivery Lead: Amanda Ryan

Senior Designer: Angela Sheehan

Service Provider/Compositor: SPi Global

Cover image(s): Photographicss/ShutterStock.com

Library of Congress Control Number: 2018960546

Book Only ISBN: 978-1-337-56735-0
Package ISBN: 978-1-337-56733-6

Cengage
200 Pier 4 Boulevard
Boston, MA 02210
USA

Notice to the Reader
Publisher does not warrant or guarantee any of the products described herein or perform any independent analysis in connection with any of the product information contained herein. Publisher does not assume, and expressly disclaims, any obligation to obtain and include information other than that provided to it by the manufacturer. The reader is expressly warned to consider and adopt all safety precautions that might be indicated by the activities described herein and to avoid all potential hazards. By following the instructions contained herein, the reader willingly assumes all risks in connection with such instructions. The publisher makes no representations or warranties of any kind, including but not limited to, the warranties of fitness for particular purpose or merchantability, nor are any such representations implied with respect to the material set forth herein, and the publisher takes no responsibility with respect to such material. The publisher shall not be liable for any special, consequential, or exemplary damages resulting, in whole or part, from the readers' use of, or reliance upon, this material.

Printed Number: 6 Print Year: 2024
Printed in Mexico

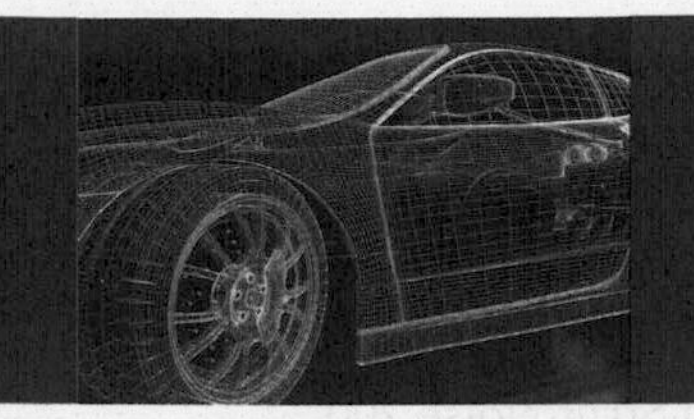

CONTENTS

PHOTO SEQUENCES

JOB SHEETS

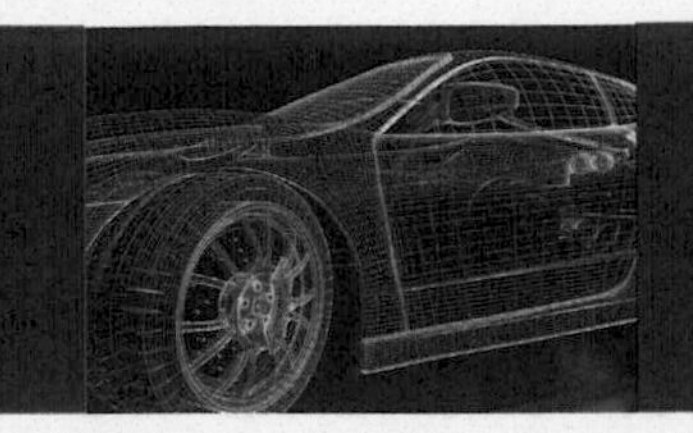

PREFACE

Thanks to the support the *Today's Technician* series has received from those who teach automotive technology, Cengage Learning, the leader in automotive-related textbooks, is able to live up to its promise to regularly provide new editions of texts of this series. We have listened and responded to our critics and our fans and present this new updated and revised seventh edition. By revising this series on a regular basis, we can respond to changes in the industry, changes in technology, changes in the accreditation process, and to the ever-changing needs of those who teach automotive technology.

The *Today's Technician* series features textbooks that cover all mechanical and electrical systems of automobiles and light trucks. Principally, the individual titles correspond to the areas for National Institute for Automotive Service Excellence (ASE) certification.

All titles in the *Today's Technician* series include remedial skills and theories common to all of the certification areas and advanced or specific subject areas that reflect the latest technological trends.

Today's Technician: Automotive Suspension & Steering Systems, 7e is designed to give students a chance to develop the same skills and gain the same knowledge that today's successful technician has. This edition also reflects the changes in the guidelines established by the ASE Education Foundation in 2017.

The purpose of the ASE Education Foundation is to evaluate technician training programs against standards developed by the automotive industry and recommend qualifying programs for accreditation by ASE. Programs can earn ASE accreditation upon the recommendation of the ASE Education Foundation. ASE Education Foundation's national standards reflect the skills that students must master. ASE Education Foundation accreditation ensures that the training programs meet or exceed industry-recognized, uniform standards of excellence.

The technician of today and for the future must know the underlying theory of all automotive systems and be able to service and maintain those systems. Dividing the material into two volumes, a Classroom Manual and a Shop Manual, provides the reader with the information needed to begin a successful career as an automotive technician without interrupting the learning process by mixing cognitive and performance learning objectives into one volume.

The design of Cengage Learning's *Today's Technician* series was based on features that are known to promote improved student learning. The design was further enhanced by a careful study of survey results, in which the respondents were asked to value particular features. Some of these features can be found in other textbooks, whereas others are unique to this series.

Each Classroom Manual contains the principles of operation for each system and subsystem. The Classroom Manual also discusses design variations in key components used by the different vehicle manufacturers, and considers emerging technologies that will be standard or optional features in the near future. This volume is organized to build upon basic facts and theories. Its primary objective is to help the reader gain an understanding of how each system and subsystem operates. This understanding is necessary to diagnose the complex automobiles of today and tomorrow. Although the basics contained in the Classroom Manual provide the knowledge needed for diagnostics, diagnostic procedures appear only in the Shop Manual. An understanding of the underlying theories is also a requirement for competence in the skill areas covered in the Shop Manual.

A spiral-bound Shop Manual delivers hands-on learning experiences with step-by-step instructions for diagnostic and repair procedures. Photo Sequences are used to illustrate some of the common service procedures. Other common procedures are listed and are accompanied with fine-line drawings and photos that let the reader visualize and conceptualize the finest details of the procedure. This volume explains the reasons for performing the procedures, as well as the circumstances when each particular service is appropriate.

The two volumes are designed to be used together and are arranged in corresponding chapters. Not only are the chapters in the volumes linked together, the contents of the chapters are also linked. The linked content is indicated by marginal callouts that refer the reader to the chapter and page where the same topic is addressed in the companion volume. This valuable feature saves users the time and trouble of searching the index or table of contents to locate supporting information in the other volume. Instructors will find this feature especially helpful when planning the presentation of material and when making reading assignments.

Both volumes contain clear and thoughtfully selected illustrations. Many of which are original drawings or photos specially prepared for inclusion in this series. This means that the art is a vital part of each textbook and not merely inserted to increase the number of illustrations.

The page design of this series uses available margin space to deliver helpful information efficiently without interrupting the pedagogical lesson material. This information includes examples of concepts just introduced in the text, explanations or definitions of terms that are not defined in the text, examples of common trade jargon used to describe a part or an operation, and unique applications of the system or service described in the text. Many textbooks also include this information but insert it in the main body of text; this tends to interrupt the reader's thought process. By placing this information to the side of the main text, students can read through the text uninterrupted and refer to the additional information when it is best for them.

Jack Erjavec
Series Editor

HIGHLIGHTS OF THIS EDITION—CLASSROOM MANUAL

The text was updated to include the latest technology in suspension and steering systems. Updated and expanded coverage of basic electrical theory and hybrid electric vehicle safety, hybrid vehicle steering systems, driver assist systems, active steering systems, rear active steering (RAS), CAN bus networking, computer-controlled suspension systems, and adaptive cruise control systems. Expanded coverage of tires and wheels including tire plus sizing and rim offset considerations. As well as expanded alignment theory coverage and sequencing of information. The text also includes the latest technology in driver assist systems including but not limited to vehicle stability control systems, traction control systems, active roll control, lane departure warning (LDW) systems, active cruise control, collision mitigation systems, telematics, and tire pressure monitoring systems (TPMS).

The first chapter explains the design and purpose of basic suspension and steering systems. This chapter provides students with the necessary basic understanding of suspension and steering systems. The other chapters in the book allow the students to build upon their understanding of these basic systems.

The second chapter explains all the basic theories required to understand the latest suspension and steering systems described in the other chapters. Students must understand these basic theories to comprehend the complex systems explained later in the text.

The other chapters in the book are designed to be stand alone to allow reordering of topics covered to fit individual program needs and explain all the current model systems and components such as wheel bearings, tires and wheels, shock absorbers and struts, front and rear suspension systems, computer-controlled suspension systems, steering columns and linkages, power steering pumps, steering gears and systems, updated information on four-wheel steering systems currently in production, frames, and four-wheel alignment. All of the art pieces have been replaced with color photos and color diagrams throughout the text to improve visual concepts of suspension and steering systems and components.

HIGHLIGHTS OF THIS EDITION—SHOP MANUAL

The chapters in the Shop Manual have been updated to explain the diagnostic and service procedures for the latest systems and components described in the Classroom Manual. Diagnostics is a very important part of an automotive technician's job. Therefore, proper diagnostic procedures are emphasized in the Shop Manual.

All Photo Sequences are in color with several new sequences added and existing sequences updated. These Photo Sequences illustrate the correct diagnostic or service procedure for a specific system or component. These Photo Sequences allow the students to visualize the diagnostic or service procedure. Visualization of these diagnostic and service procedures helps students to remember the procedures, and perform them more accurately and efficiently. The text covers the information required for the ASE test in Suspension and Steering Systems. New and updated Job Sheets have been created to meet current ASE Education Foundation tasks.

Chapter 1 explains the necessary safety precautions and procedures in an automotive repair shop. General shop safety and the required shop safety equipment are explained in the text. The text describes safety procedures when operating vehicles and various types of automotive service equipment. Correct procedures for handling hazardous waste materials are detailed in the text.

Chapter 2 describes suspension and steering diagnostic and service equipment and the use of service manuals. This chapter also explains employer and employee obligations, and ASE certification requirements.

The other chapters in the text have been updated to explain the diagnostic and service procedures for the latest suspension and steering systems explained in the Classroom Manual. New job sheets related to the new systems and components have been added in the text. All of the art pieces have been replaced with color photos and color diagrams throughout the text to improve the student's visualization of diagnostic and service procedures.

Mark Schnubel

CLASSROOM MANUAL

Features of the Classroom Manual include the following:

Cognitive Objectives

These objectives outline the chapter's contents and identify what students should know and be able to do upon completion of the chapter. Each topic is divided into small units to promote easier understanding and learning.

Terms To Know List

A list of key terms appears immediately after the Objectives. Students will see these terms discussed in the chapter. Definitions can also be found in the Glossary at the end of the manual.

Cross-References to the Shop Manual

References to the appropriate page in the Shop Manual appear whenever necessary. Although the chapters of the two manuals are synchronized, material covered in other chapters of the Shop Manual may be fundamental to the topic discussed in the Classroom Manual.

Margin Notes

The most important terms to know are highlighted and defined in the margin. Common trade jargon also appears in the margin and gives some of the common terms used for components. This helps students understand and speak the language of the trade, especially when conversing with an experienced technician.

CHAPTER 1

SUSPENSION AND STEERING SYSTEMS

Upon completion and review of this chapter, you should be able to understand and describe:

- How strength and rigidity are designed into a unitized body.
- The advantages of reduced vehicle weight.
- The design of a short-and-long arm (SLA) front suspension system.
- How limited independent rear wheel movement is provided in a semi-independent rear suspension system.
- The advantage of an independent rear suspension system.
- The purposes of vehicle tires.
- The terms *positive* and *negative offset* as they relate to vehicle wheel rims.
- Three different loads that are applied to wheel bearings.
- The purposes of shock absorbers.
- The difference between a shock absorber and a strut.
- Two different types of computer-controlled shock absorbers.
- The advantages of computer-controlled suspension systems.
- Two types of steering linkages.
- How the rack is moved in a rack and pinion steering gear.
- How a power steering pump develops hydraulic pressure.
- The result of incorrect rear wheel toe.
- The front wheel caster.
- The results of excessive negative camber.

Terms to Know

Angular bearing loads	Negative offset	Thrust bearing loads
Carbon dioxide (CO_2)	Positive camber	Thrust line
Greenhouse gas	Positive caster	Toe-out
Jounce travel	Positive offset	Wheel alignment
Negative camber	Radial bearing load	Wheel offset
Negative caster	Rebound travel	Wheel shimmy

INTRODUCTION

The suspension system must provide proper steering control and ride quality. Performing these functions is extremely important to maintaining vehicle safety and customer satisfaction. For example, if the suspension system allows excessive vertical wheel oscillations, the driver may lose control of the steering when driving on an irregular road surface. This loss of steering control can result in vehicle collisions and personal in... vertical wheel oscillations transfer undesirable vibrations... passenger compartment, which causes custom...

The suspension system and f... provide normal tire...

104 Chapter 4

rotate or move the component. Bearings are precision-machined assemblies, which provide smooth operation and long life. When bearings are properly installed and maintained, bearing failure is rare.

AUTHOR'S NOTE Only 21 percent of the power developed by a vehicle engine actually gets to the drive wheels. Much of the engine energy is lost in overcoming friction and wind resistance. The U.S. Department of Energy (DOE) is working with a major bearing manufacturer to explore the use and advantage of roller bearings to reduce friction in vehicle engines. The DOE estimates that if all vehicle engines used roller-bearing, low-friction technology, 100 million barrels of oil could be saved in a year.

BEARING LOADS

Shop Manual Chapter 4, page 148

A **thrust bearing load** may be referred to as an axial load.

When a bearing load is applied in a vertical direction on a horizontal shaft, it is called a **radial bearing load**. If the vehicle weight is applied straight downward on a bearing, this weight is a radial load on the bearing. A **thrust bearing load** is applied in a horizontal direction (**Figure 4-1**). For example, while a vehicle is turning a corner, horizontal force is applied to the front wheel bearings. When an **angular bearing load** is applied, the angle of the applied load is somewhere between the horizontal and vertical positions. This can also be referred to as a combination load.

BALL BEARINGS

Ball bearings have round steel balls between the inner and outer races.

Front and rear wheel bearings may be **ball bearings** or roller bearings. Either type of bearing contains these basic parts:

1. Inner race, or cone
2. Rolling elements, balls, or rollers
3. Separator, also called a cage or retainer
4. Outer race, or cup

The **inner race** supports the inner side of the rolling elements in a bearing.

The **rolling elements** are the precision-machined balls or rollers between the inner and outer bearing races.

The **inner race** is an accurately machined component. The inner surface of the race is mounted on the shaft with a precision fit. The **rolling elements** are mounted on a very

Radial load

Angular contact load

Thrust load

Figure 4-1 Types of bearing loads.

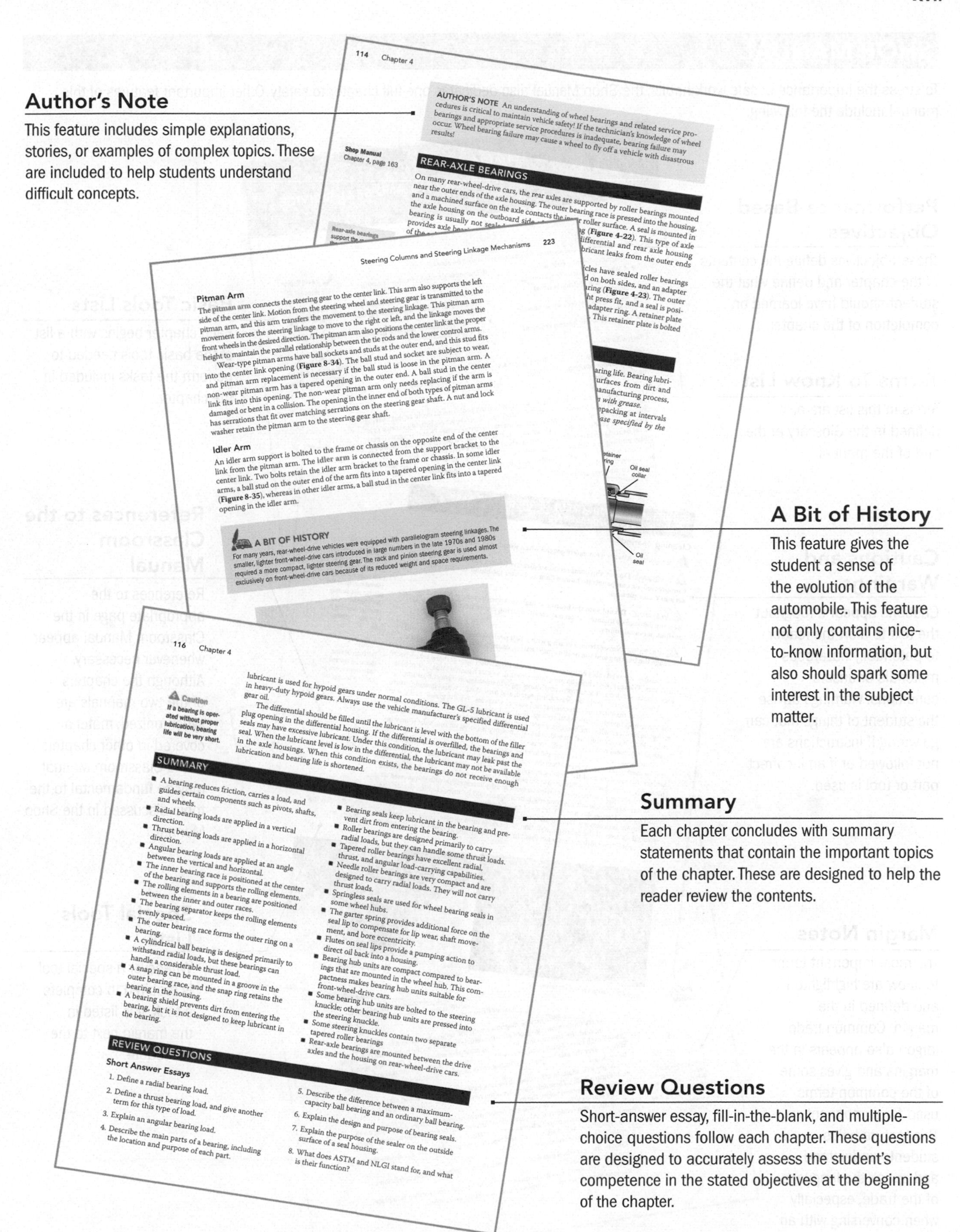

Author's Note

This feature includes simple explanations, stories, or examples of complex topics. These are included to help students understand difficult concepts.

A Bit of History

This feature gives the student a sense of the evolution of the automobile. This feature not only contains nice-to-know information, but also should spark some interest in the subject matter.

Summary

Each chapter concludes with summary statements that contain the important topics of the chapter. These are designed to help the reader review the contents.

Review Questions

Short-answer essay, fill-in-the-blank, and multiple-choice questions follow each chapter. These questions are designed to accurately assess the student's competence in the stated objectives at the beginning of the chapter.

SHOP MANUAL

To stress the importance of safe work habits, the Shop Manual also dedicates one full chapter to safety. Other important features of this manual include the following:

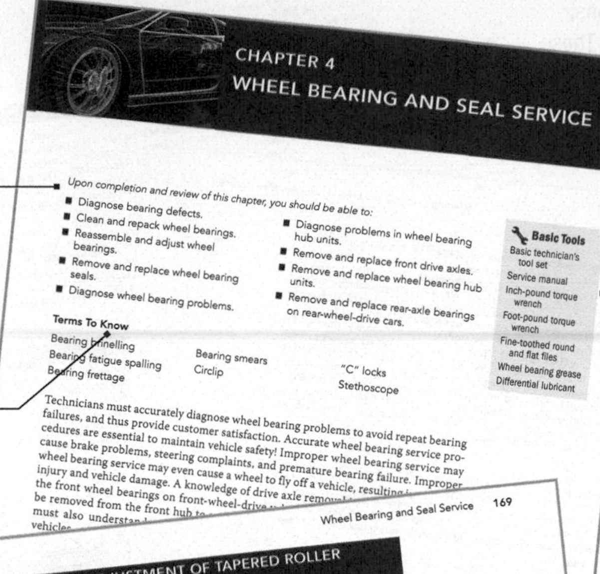

Performance-Based Objectives

These objectives define the contents of the chapter and define what the student should have learned on completion of the chapter.

Terms To Know List

Terms in this list are also defined in the Glossary at the end of the manual.

Basic Tools Lists

Each chapter begins with a list of the basic tools needed to perform the tasks included in the chapter.

Cautions and Warnings

Cautions appear throughout the text to alert the reader to potentially hazardous materials or unsafe conditions. Warnings advise the student of things that can go wrong if instructions are not followed or if an incorrect part or tool is used.

References to the Classroom Manual

References to the appropriate page in the Classroom Manual appear whenever necessary. Although the chapters of the two manuals are synchronized, material covered in other chapters of the Classroom Manual may be fundamental to the topic discussed in the Shop Manual.

Figure 4-4 Wrapping a bearing in waterproof paper.

Margin Notes

The most important terms to know are highlighted and defined in the margin. Common trade jargon also appears in the margins and gives some of the common terms used for components. This feature helps students understand and speak the language of the trade, especially when conversing with an experienced technician.

Special Tools Lists

Whenever a special tool is required to complete a task, it is listed in the margin next to the procedure.

Case Studies

Each chapter ends with a Case Study describing a particular vehicle problem and the logical steps a technician might use to solve the problem. These studies focus on system diagnosis skills and help students gain familiarity with the process.

Service Tips

Whenever a shortcut or special procedure is appropriate, it is described in the text. Generally, these tips describe common procedures used by experienced technicians.

ASE-Style Review Questions

Each chapter contains ASE-style review questions that reflect the performance objectives listed at the beginning of the chapter. These questions can be used to review the chapter as well as to prepare for the ASE certification exam.

Customer Care

This feature highlights those little things a technician can do or say to enhance customer relations.

Photo Sequences

Many procedures are illustrated in detailed Photo Sequences. These photographs show the students what to expect when they perform particular procedures. They also familiarize students with a system or type of equipment that the school might not have.

---

214 Chapter 5

9. Connect the ohmmeter leads from the signal return terminal in the wiring harness side of the actuator connector to a chassis ground. The ohmmeter should indicate less than 10 ohms. If the ohmmeter reading is higher than specified, check the signal return wire and the programmed ride control (PRC) module.

SERVICE TIP Wiring harness colors vary depending on the vehicle make and model year. The wiring colors in the following steps are based on Ford vehicles. If you are working on a different make of vehicle, refer to the wire colors in the vehicle manufacturer's service information.

CASE STUDY

A customer complained about a squeaking noise in the rear suspension of a 2009 Dodge Caravan. The customer said the noise occurred during normal driving at lower speeds. The technician lifted the car on a hoist and made a check of all rear suspension bushings in shock absorbers, track bar, and trailing arms. The spring insulators and all suspension bolts were checked visually. All of these items were in good condition, and there was no evidence of a squeaking noise as the chassis was bounced gently. Because the exact source of chassis noise can sometimes be difficult to locate, the technician performed a visual check of bushings, insulators, and fasteners in the front suspension. No problems were found in the front suspension.

One of the first requirements for successful automotive diagnosis is to obtain as much information as possible from the customer. The customer with the squeaking rear suspension in a Dodge Caravan lived in a part of the country where cold temperatures occur in the winter. This complaint occurred in January. The technician questioned the customer further about the conditions when the squeaking suspension noise was heard, and the customer revealed that the noise occurred when the temperature was severely cold. The customer also indicated that the noise disappeared at warmer temperatures.

The technician informed the customer that the only way to find the exact cause of this annoying squeak was to leave the car on the lot at the shop all night and check it first thing in the morning when it was colder. The customer complied with this suggestion, and the technician drove the car into the shop immediately the next morning. The squeaking noise occurred as the car was driven across the parking lot. The technician lifted the car on a hoist and listened with a stethoscope at each rear suspension bushing as a coworker gently bounced the rear suspension. No squeaking noise was heard at any of the rear suspension bushings. However, when the stethoscope pickup was placed on the left rear shock absorber, the squeaking noise was loud and clear. The shock absorber was quickly removed, and the squeaking noise was gone when the rear chassis was bounced gently. Replacement of the left rear shock absorber corrected this complaint and made the customer happy.

ASE-STYLE REVIEW QUESTIONS

1. A slight oil film appears on the lower shock absorber oil chamber, and the shock absorber performs satisfactorily on a road test:
   Technician A says the shock absorber is satisfactory.
   Technician B says the shock absorber may contain excessive pressure.
   Who is correct?
   A. A only
   B. B only
   C. Both A and B
   D. Neither A nor B

2. While discussing a shock absorber and strut bounce test:
   Technician A says the shock absorber is satisfactory if the bumper makes two free upward bounces.
   Technician B says the bumper must be pushed downward with considerable force.
   Who is correct?
   A. A only
   B. B only
   C. Both A and B
   D. Neither A nor B

---

8. [illegible]

The rear coil springs [illegible] cedure for spring removal from the [illegible]

INSTALLING STRUT CARTRIDGE, OFF-CAR

CUSTOMER CARE Check the cost of the strut cartridges versus the price of new struts. Give customers the best value for their repair dollar!

Many struts are a sealed unit, and thus rebuilding is impossible. However, some manufacturers supply a replacement cartridge that may be installed in the strut housing after the strut has been removed from the vehicle. Always follow the strut cartridge manufacturer's recommended installation procedure.

The following is a typical off-car strut cartridge installation procedure:

1. Install a bolt and two nuts in the upper strut-to-knuckle mounting bolt hole. Place a nut on the inside and outside of the strut flange.
2. Clamp this bolt in a vise to hold the strut.
3. Locate the line groove near the top of the strut body, and use a pipe cutter installed in this groove to cut the top of the strut body.
4. After the cutting procedure, remove the strut piston assembly from the strut (**Figure 5-26**).
5. Remove the strut from the vise and dump the oil from the strut.
6. Place the special tool supplied by the vehicle manufacturer or cartridge manufacturer on top of the strut body. Strike the tool with a plastic hammer until the tool shoulder contacts the top of the strut body. This action removes burrs from the strut body and places a slight flare on the body.
7. Remove the tool from the strut body.
8. Install the required amount of oil in the strut, place the new cartridge in the strut body, and turn the cartridge until it settles into indentations in the bottom of the strut body.
9. Place the new nut over the cartridge.
10. Using a special tool supplied by the vehicle or cartridge manufacturer, tighten the nut to the specified torque.
11. Move the strut piston rod in and out several times to check for proper strut operation.

A strut cartridge contains the inner working part of the strut, which may be installed in the outer housing of the old strut.

Special Tool
Pipe cutter

INSTALLING STRUT CARTRIDGE, ON-CAR

WARNING If a vehicle is hoisted or lifted in any way during an on-car strut cartridge replacement, the coil spring may fly off the strut, causing vehicle damage and personal injury.

On some vehicles, the front strut cartridge may be removed and replaced with the strut

---

Figure 7-2 Electronic stethoscope.

Riding Height Measurement

Regular inspection and proper main[illegible] tant to maintain vehicle safety. The [illegible] tion. Other suspension components, [illegible] they are worn. *Since incorrect riding [illegible] measurement is critical.* Reduced re[illegible] causes rapid steering wheel return a[illegible] riding height is less than specified. Th[illegible] manufacturer's specified location, [illegible] suspension system. On some vehicle[illegible] sured from the floor to the center of [illegible] level floor or an alignment rack (**Fig**[illegible] cedure for measuring front and rear curb riding height.

PHOTO SEQUENCE 11
Typical Procedure for Measuring Front and Rear Riding Height

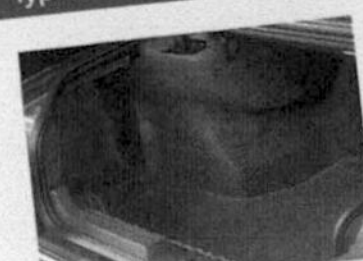

P11-1 Check the trunk for extra weight.

P11-2 Check the tires for normal inflation pressure.

P11-3 Park the car on a level shop floor or an alignment rack.

Figure 11-81 The lane departure warning system will stay engaged even while cornering if the cornering maneuver is completed in less than 100 seconds.

Some LDW systems have lane-departure prevention capabilities. If the vehicle is moving toward the lane markings on the left, the ABS and stability control computer lightly pulse the brakes on the right side of the vehicle. When the wheels on the right side slow down, the wheels on the left side turn faster than the wheels on the right side, and this action steers the car away from the highway markers on the left. Other LDW systems activate the electronic (electromechanical) steering on the vehicle to steer the vehicle away from the lane markers if a lane departure is starting to occur. The LDW system will engage the electromechanical power steering system to first cause a slight vibration of the steering wheel to warn driver to take corrective action to maintain lane position and then take corrective action to maintain lane position. If the driver exerts more than 2.21 ft.-lbs (3 Nm) of pressure on the steering wheel, the LDW system will disengage to allow a lane change without using turn signal. The LDW system will continue to function even while corning as long as the corning maneuver is completed in less than 100 seconds (**Figure 11-81**). The lane departure system can be adversely affect by weather and road surfaces. While the system is effective at detecting lane lines on asphalt, it can have issues detecting lane lines on concrete roads on bright sunny days. The system can also be affected by some wet surfaces due to reflection or snow-covered surfaces that cover lane lines. A dirty windshield that affects camera image will also cause lane departure system issues and will set a diagnostic trouble code "Lane Assist – no sensor visibility present".

AUTHOR'S NOTE Statistics compiled by one of the major car manufacturers indicate that 30 percent of all vehicle accidents in the United States involve rear-end collisions. In half of these collisions, the driver did not apply the brakes before collision occurred. The statistics also indicate that 75 percent of these rear-end collisions occur below 19 mph (30 km/h).

COLLISION MITIGATION SYSTEMS

Some vehicles are equipped with a collision mitigation system that uses radar and camera information to detect a vehicle in front. The collision mitigation computer is in constant communication with the PCM and ABS computers via the vehicle network. If the radar and camera signals indicate a vehicle in front is too close and a collision is imminent, the

Author's Note

This feature includes simple explanations, stories, or examples of complex topics. These are included to help students understand difficult concepts.

Rear Suspension Service 293

10. While diagnosing improper rear wheel tracking:
    A. Improper rear wheel tracking may be caused when both rear springs are sagged the same amount.
    B. A bent rear suspension tie rod does not affect rear wheel tracking.
    C. Improper rear wheel tracking may result in steering pull when driving straight ahead.
    D. Improper rear wheel tracking may cause front wheel shimmy.

ASE CHALLENGE QUESTIONS

1. A squeak in the rear suspension could be caused by all of the following EXCEPT:
   A. The suspension bushing.
   B. Weak spring leaves.
   C. Worn spring antifriction pads.
   D. A defective shock absorber.
2. A vehicle with a live axle coil spring rear suspension has become hard to steer with harsh ride quality. Which of the following could be the cause of this problem?
   A. Worn lateral link bushings
   B. Weak rear coil springs
   C. Bent rear shock rod
   D. Worn stabilizer bar bushings
3. A truck with a solid axle and parallel leaf spring rear suspension darts and acts erratic during turns. Which of the following is the most likely cause?
   A. Worn rear sway bar bushings
   B. Incorrect driveline angle
   C. Incorrect ride height
   D. Loose rear axle U-bolts
4. A car with independent rear suspension has excessive rear tire wear. An inspection of the rear tires shows they are worn on the inside edge and the tread is feathered.
   Technician A says the problem could be the tires are toeing out during acceleration.
   Technician B says worn control arm bushings could be the cause of the problem.
   Who is correct?
   A. A only
   B. B only
   C. Both A and B
   D. Neither A nor B
5. The steering on a front-wheel-drive car pulls to the right.
   Technician A says the strut on the right rear suspension assembly could be the problem.
   Technician B says worn bushings of the left rear lower arm assembly could be the problem.
   Who is correct?
   A. A only
   B. B only
   C. Both A and B
   D. Neither A nor B

ASE Challenge Questions

Each technical chapter ends with five ASE challenge questions. These are not more review questions; rather, they test the students' ability to apply general knowledge to the contents of the chapter.

ASE Practice Examination

A 50-question ASE practice exam, located in the Appendix, is included to test students on the content of the complete Shop Manual.

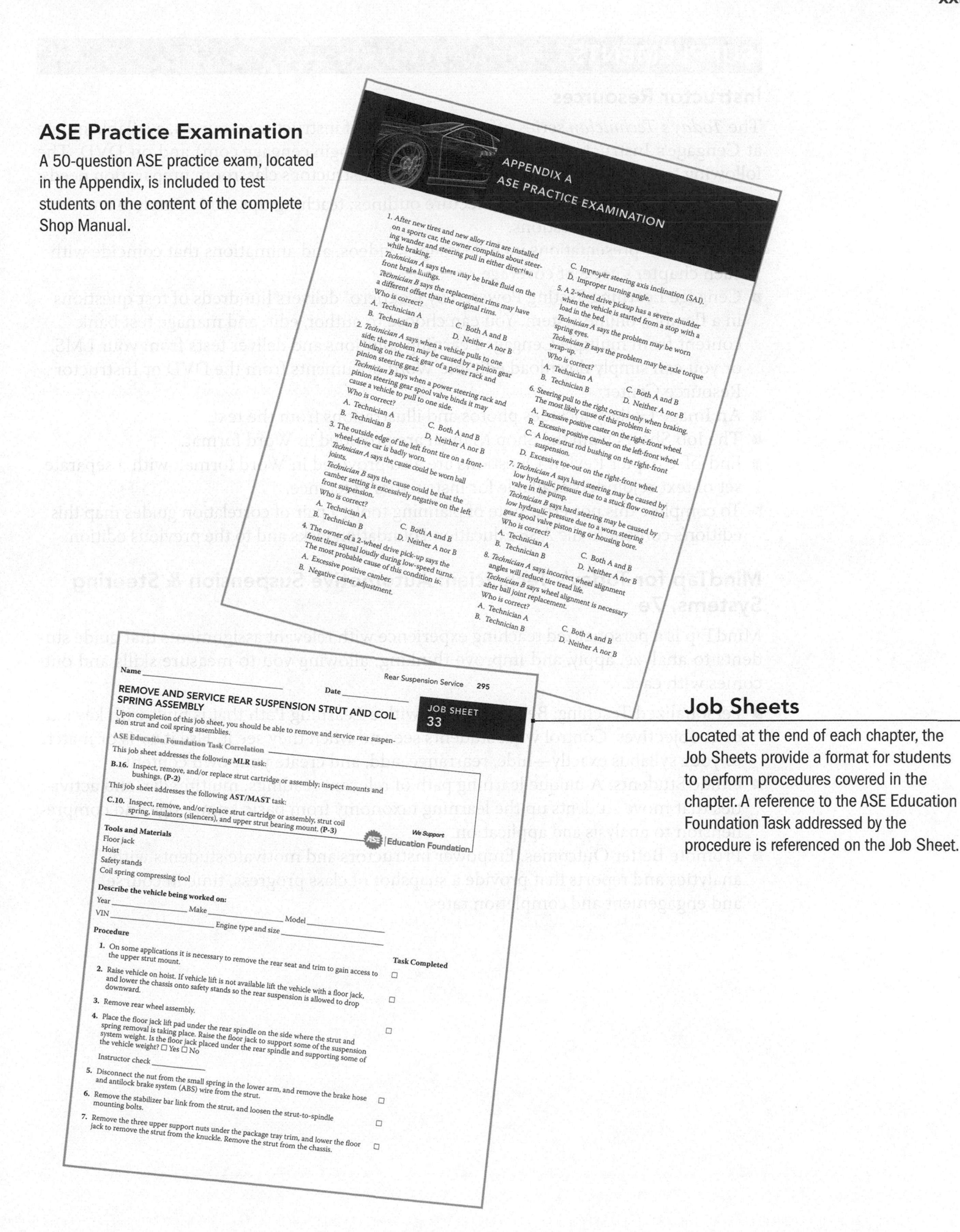

APPENDIX A
ASE PRACTICE EXAMINATION

1. After new tires and new alloy rims are installed on a sports car, the owner complains about steering wander and steering pull in either direction while braking.
Technician A says there may be brake fluid on the front brake linings.
Technician B says the replacement rims may have a different offset than the original rims.
Who is correct?
A. Technician A
B. Technician B
C. Both A and B
D. Neither A nor B

2. Technician A says when a vehicle pulls to one side; the problem may be caused by a pinion gear binding on the rack gear of a power rack and pinion steering gear.
Technician B says when a power steering rack and pinion steering gear spool valve binds it may cause a vehicle to pull to one side.
Who is correct?
A. Technician A
B. Technician B
C. Both A and B
D. Neither A nor B

3. The outside edge of the left front tire on a front-wheel-drive car is badly worn.
Technician A says the cause could be worn ball joints.
Technician B says the cause could be that the camber setting is excessively negative on the left front suspension.
Who is correct?
A. Technician A
B. Technician B
C. Both A and B
D. Neither A nor B

4. The owner of a 2-wheel drive pick-up says the front tires squeal loudly during low-speed turns. The most probable cause of this condition is:
A. Excessive positive camber.
B. Negative caster adjustment.
C. Improper steering axis inclination (SAI).
D. Improper turning angle.

5. A 2-wheel drive pickup has a severe shudder when the vehicle is started from a stop with a load in the bed.
Technician A says the problem may be worn spring eyes.
Technician B says the problem may be axle torque wrap-up.
Who is correct?
A. Technician A
B. Technician B
C. Both A and B
D. Neither A nor B

6. Steering pull to the right occurs only when braking. The most likely cause of this problem is:
A. Excessive positive caster on the right-front wheel.
B. Excessive positive camber on the left-front wheel.
C. A loose strut rod bushing on the right-front suspension.
D. Excessive toe-out on the right-front wheel.

7. Technician A says hard steering may be caused by low hydraulic pressure due to a stuck flow control valve in the pump.
Technician B says hard steering may be caused by low hydraulic pressure due to a worn steering gear spool valve piston ring or housing bore.
Who is correct?
A. Technician A
B. Technician B
C. Both A and B
D. Neither A nor B

8. Technician A says incorrect wheel alignment angles will reduce tire tread life.
Technician B says wheel alignment is necessary after ball joint replacement.
Who is correct?
A. Technician A
B. Technician B
C. Both A and B
D. Neither A nor B

Rear Suspension Service 295

Name ______________________ Date ______________

JOB SHEET 33

REMOVE AND SERVICE REAR SUSPENSION STRUT AND COIL SPRING ASSEMBLY

Upon completion of this job sheet, you should be able to remove and service rear suspension strut and coil spring assemblies.

ASE Education Foundation Task Correlation

This job sheet addresses the following **MLR** task:

B.16. Inspect, remove, and/or replace strut cartridge or assembly; inspect mounts and bushings. **(P-2)**

This job sheet addresses the following **AST/MAST** task:

C.10. Inspect, remove, and/or replace strut cartridge or assembly, strut coil spring, insulators (silencers), and upper strut bearing mount. **(P-3)**

We Support ASE Education Foundation

Tools and Materials

Floor jack
Hoist
Safety stands
Coil spring compressing tool

Describe the vehicle being worked on:

Year ________ Make ______________ Model ______________
VIN ______________ Engine type and size ______________

Procedure — **Task Completed**

1. On some applications it is necessary to remove the rear seat and trim to gain access to the upper strut mount. ☐
2. Raise vehicle on hoist. If vehicle lift is not available lift the vehicle with a floor jack, and lower the chassis onto safety stands so the rear suspension is allowed to drop downward. ☐
3. Remove rear wheel assembly. ☐
4. Place the floor jack lift pad under the rear spindle on the side where the strut and spring removal is taking place. Raise the floor jack to support some of the suspension system weight. Is the floor jack placed under the rear spindle and supporting some of the vehicle weight? ☐ Yes ☐ No ☐
Instructor check ________
5. Disconnect the nut from the small spring in the lower arm, and remove the brake hose and antilock brake system (ABS) wire from the strut. ☐
6. Remove the stabilizer bar link from the strut, and loosen the strut-to-spindle mounting bolts. ☐
7. Remove the three upper support nuts under the package tray trim, and lower the floor jack to remove the strut from the knuckle. Remove the strut from the chassis. ☐

Job Sheets

Located at the end of each chapter, the Job Sheets provide a format for students to perform procedures covered in the chapter. A reference to the ASE Education Foundation Task addressed by the procedure is referenced on the Job Sheet.

SUPPLEMENTS

Instructor Resources

The *Today's Technician* series offers a robust set of instructor resources, available online at Cengage's Instructor Resource Center (http://login.cengage.com) and on DVD. The following tools have been provided to meet any instructor's classroom preparation needs:

- An Instructor's Guide provides lecture outlines, teaching tips, and complete answers to end-of-chapter questions.
- PowerPoint presentations include images, videos, and animations that coincide with each chapter's content coverage.
- Cengage Learning Testing Powered by Cognero® delivers hundreds of test questions in a flexible, online system. You can choose to author, edit, and manage test bank content from multiple Cengage Learning solutions and deliver tests from your LMS, or you can simply download editable Word documents from the DVD or Instructor Resource Center.
- An Image Gallery includes photos and illustrations from the text.
- The Job Sheets from the Shop Manual are provided in Word format.
- End-of-Chapter Review Questions are also provided in Word format, with a separate set of text rejoinders available for instructors' reference.
- To complete this powerful suite of planning tools, a pair of correlation guides map this edition's content to the ASE Education Foundation tasks and to the previous edition.

MindTap for Today's Technician: Automotive Suspension & Steering Systems, 7e

MindTap is a personalized teaching experience with relevant assignments that guide students to analyze, apply, and improve thinking, allowing you to measure skills and outcomes with ease.

- Personalized Teaching: Becomes yours with a Learning Path that is built with key student objectives. Control what students see and when they see it. Use it as-is or match to your syllabus exactly—hide, rearrange, add, and create your own content.
- Guide Students: A unique learning path of relevant readings, multimedia, and activities that move students up the learning taxonomy from basic knowledge and comprehension to analysis and application.
- Promote Better Outcomes: Empower instructors and motivate students with analytics and reports that provide a snapshot of class progress, time in course, and engagement and completion rates.

REVIEWERS

The author and publisher would like to extend a special thanks to the instructors who reviewed this text and offered invaluable feedback:

Rodney Batch
University of Northwestern Ohio
Lima, OH

Tim LeVan
University of Northwestern Ohio
Lima, OH

Cory Peck
Vatterott College
St. Louis, MO

Steve Roessner
University of Northwestern Ohio
Lima, OH

Ronald Strzalkowski Flint
Baker College Flint, MI

Ronald Strzalkowski
Baker College
Flint, MI

Claude F. Townsend III
Oakland Community College
Bloomfield Hills, MI

CHAPTER 1
SAFETY

Upon completion and review of this chapter, you should be able to:

- Recognize shop hazards and take the necessary steps to avoid personal injury or property damage.
- Describe the general shop rules that must be followed in an automotive shop.
- Explain the purposes of the Occupational Safety and Health Act (OSHA).
- Explain two reasons why batteries are a shop hazard.
- Describe how long hair may be a shop hazard.
- Explain how incandescent trouble lights may be a shop hazard.
- Describe the harmful effects of carbon monoxide on the human body.
- Explain why asbestos dust in the shop air is a hazard.
- Describe the requirements for the location of safety equipment in the shop.
- Explain four different types of fires, and the fire extinguisher required for each type of fire.
- Describe three pieces of shop safety equipment other than fire extinguishers.
- Describe the main purposes of the right-to-know laws.
- Describe the purpose of safety data sheets (SDSs) and list the information that must be contained on these sheets.
- Explain the employer's responsibility regarding hazardous materials in the shop.
- Observe all general shop safety precautions.
- Demonstrate all precautions related to personal safety when working in the shop.
- Explain why smoking is dangerous in the shop.
- Explain why drug or alcohol use is dangerous in the shop.
- Describe basic electrical safety precautions.
- Explain gasoline safety precautions.
- Describe housekeeping safety precautions.
- Explain fire safety precautions.
- Describe the proper procedure for using a fire extinguisher.
- Describe the necessary precautions to maintain airbag safety.
- Describe the precautions to be observed when driving vehicles in the shop.
- Explain the proper procedure for lifting heavy objects.
- Describe the necessary steps for hand tool safety.
- Explain the necessary precautions when operating a vehicle lift.
- Describe hydraulic jack and safety stand safety precautions.
- Explain the necessary safety precautions when using power tools.
- Describe the precautions required to maintain compressed-air equipment safety.
- Explain the precautions and environmental concerns related to cleaning equipment safety.
- Describe the safety precautions related to working on a hybrid electric vehicle.

Terms To Know

Air bag system
Approved gasoline storage cans
Aqueous parts cleaning tank
Brake parts washer
Cold parts washer
Corrosive
Environmental Protection Agency (EPA)
Fluorescent trouble lights
Hazard Communication Standard
Hot cleaning tank
Hybrid vehicle
Ignitable
Incandescent bulb-type trouble lights
Multipurpose dry chemical fire extinguishers
Occupational Safety and Health Act (OSHA)
Reactive
Resource Conservation and Recovery Act (RCRA)
Right-to-know laws
Safety data sheets (SDSs)
Toxic
Workplace hazardous materials information system (WHMIS)

INTRODUCTION

Safety is extremely important in the automotive shop! The knowledge and practice of safety precautions prevent serious personal injury and expensive property damage. Automotive students and technicians must be familiar with shop hazards and shop safety rules. The first step in providing a safe shop is learning about shop hazards and safety rules. The second, and most important, step in this process is applying your knowledge of shop hazards and safety rules while working in the shop. In other words, you must develop safe working habits in the shop through your understanding of shop hazards and safety rules. When shop employees have a careless attitude toward safety, accidents are more likely to occur; therefore, all shop personnel must develop a serious attitude toward safety. The result of this attitude is serious shop personnel who will learn and adopt all shop safety rules.

Shop personnel must be familiar with their rights regarding hazardous waste disposal. These rights are explained in the **right-to-know laws**. Shop personnel must also be familiar with the types of hazardous materials in the automotive shop and the proper disposal methods for these materials according to state and federal regulations.

OCCUPATIONAL SAFETY AND HEALTH ACT

The **Occupational Safety and Health Act (OSHA)** was passed by the U.S. government in 1970. The purposes of this legislation are:

1. To assist and encourage the citizens of the United States in their efforts to ensure safe and healthful working conditions by providing research, information, education, and training in the field of occupational safety and health.
2. To ensure safe and healthful working conditions for working men and women by authorizing enforcement of the standards developed under the Act.

Because approximately 25 percent of workers are exposed to health and safety hazards on the job, the OSHA is necessary to monitor, control, and educate workers regarding health and safety in the workplace.

SHOP HAZARDS

Service technicians and students encounter many hazards in an automotive shop. When these hazards are known, basic shop safety rules and procedures must be followed to avoid personal injury. Some of the hazards in an automotive shop include the following:

1. Flammable liquids, such as gasoline and paint. These must be handled and stored properly in approved, closed containers to comply with safety regulations.
2. Flammable materials, such as oily rags. These must be stored properly in closed containers to avoid a fire hazard.
3. Batteries contain a corrosive sulfuric acid solution and produce explosive hydrogen gas while charging.
4. Loose sewer and drain covers may cause foot or toe injuries.
5. Caustic liquids, such as those in hot cleaning tanks, are harmful to skin and eyes.
6. High-pressure air in the shop's compressed-air system can be very dangerous or fatal if it penetrates the skin and enters the bloodstream. High-pressure air released near the eyes may cause eye injury.
7. Frayed cords on electrical equipment and lights may result in severe electrical shock.
8. Hazardous waste material, such as batteries and caustic cleaning solutions. These must be handled with adequate personal protection to avoid injury (**Figure 1-1**).
9. Carbon monoxide from vehicle exhaust is poisonous and potentially fatal.
10. Loose clothing or long hair may become entangled in rotating parts on equipment or vehicles, resulting in serious injury.
11. Dust and vapors generated during some repair jobs are harmful. Asbestos dust, which may be released during brake lining service and clutch service, is a contributor to lung cancer.
12. High noise levels from shop equipment such as an air chisel may be harmful to the ears.
13. Oil, grease, water, or parts cleaning solutions on shop floors may cause someone to slip and fall, resulting in serious injury.
14. The incandescent bulbs used in some trouble lights may shatter if the light is dropped, igniting flammable materials in the area and causing a fire. Many insurance companies now require the use of trouble lights with fluorescent bulbs in the shop.

Figure 1-1 Always wear recommended safety clothing and equipment when handling hazardous materials.

SHOP SAFETY RULES

Applying basic shop rules helps prevent serious, expensive accidents. Failure to comply with shop rules may cause personal injury or expensive damage to vehicles and shop facilities. It is the responsibility of the employer and all shop employees to make sure that shop rules are understood and followed until these rules become automatic habits. The following basic shop rules should be observed:

1. Always wear safety glasses and other protective equipment required by a service procedure (**Figure 1-2**). For example, a brake parts washer must be used to avoid breathing asbestos dust into the lungs. Asbestos dust is a known cause of lung cancer. This dust is encountered in manual transmission clutch facings and brake linings.
2. Tie long hair securely behind your head, and do not wear loose or torn clothing.
3. Do not wear rings, watches, or loose hanging jewelry. If jewelry such as a ring, metal watchband, or chain makes contact between an electrical terminal and ground, the jewelry will become extremely hot, resulting in severe burns.
4. Do not work in the shop while under the influence of alcohol or drugs.
5. Set the parking brake when working on a vehicle. If the vehicle has an automatic transmission, place the gear selector in park unless a service procedure requires another selector position. When the vehicle is equipped with a manual transmission, position the gear selector in neutral with the engine running or in reverse with the engine stopped.
6. Always connect a shop exhaust hose to the vehicle tailpipe, and be sure the shop exhaust fan is running. If it is absolutely necessary to operate a vehicle without a shop exhaust pipe connected to the tailpipe, open the large shop door to provide adequate ventilation. Carbon monoxide in the vehicle exhaust may cause severe headaches and other medical problems. High concentrations of carbon monoxide may result in death!
7. Keep hands, clothing, and wrenches away from rotating parts such as cooling fans. Remember that electric-drive fans may start rotating at any time, even with the ignition off.

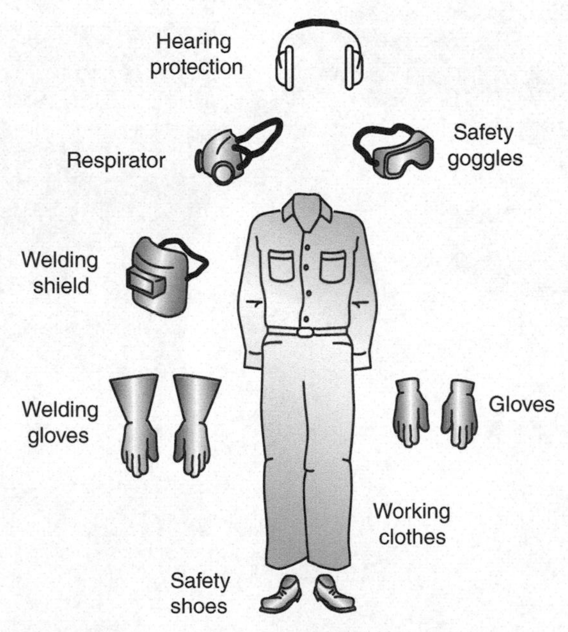

Figure 1-2 Shop safety clothing and equipment, including safety goggles, respirator, welding shield, proper work clothes, ear protection, welding gloves, work gloves, and safety shoes.

A hybrid vehicle has a power train with two power sources. The most common type of hybrid vehicle has a gasoline engine and an electric-drive motor(s).

8. Always leave the ignition switch off unless a service procedure requires another switch position.
9. Always follow the vehicle manufacturer's recommended procedure to disable the high-voltage (HV) electrical system and wait for 5 minutes before working on a **hybrid vehicle**. A switch under the steering column is pushed to the off position to disable the HV system on some hybrid vehicles. On other hybrid vehicles, the HV disconnect switch is mounted at or near the HV battery pack. On some hybrid vehicles, the HV system retains voltage for 5 minutes after the disable switch is turned off.
10. Do not smoke in the shop. If the shop has designated smoking areas, smoke only in those areas.
11. Store oily rags and other discarded combustibles in covered metal containers designed for this purpose.
12. Always use the wrench or socket that fits properly on the bolt. Do not substitute metric for English wrenches, or vice versa.
13. Keep tools in good condition. For example, do not use a punch or a chisel with a mushroomed end because a piece of the mushroomed metal could break off when struck with a hammer, resulting in severe eye or other injury.
14. Do not leave power tools running and unattended.
15. Serious burns may be prevented by avoiding contact with hot metal components, such as exhaust manifolds and other exhaust system components, radiators, and some air-conditioning hoses.
16. When a lubricant such as engine oil is drained, always wear heavy plastic gloves because the oil could be hot enough to cause burns.
17. Prior to getting under a vehicle, be sure the vehicle is placed securely on safety stands.
18. Operate all shop equipment, including lifts, according to the equipment manufacturer's recommended procedure. Do not operate equipment unless you are familiar with the correct operating procedure.
19. Do not run or engage in horseplay in the shop.
20. Obey all state and federal fire, safety, and environmental regulations.
21. Do not stand in front of or behind vehicles.
22. Always place fender, seat, and floor mat covers on a customer's vehicle before working on the car.
23. Inform the shop foreman of any safety dangers, as well as suggestions for safety improvement.
24. Do not direct high-pressure air from an air gun toward human skin or near the eyes. High-pressure air may penetrate the skin and enter the bloodstream. Air in the bloodstream may be fatal! High-pressure air discharged near the eyes may cause serious eye damage.

AIR QUALITY

Vehicle exhaust contains small amounts of carbon monoxide, which is a poisonous gas. Weak concentrations of carbon monoxide in the shop air may cause nausea and headaches; strong concentrations may be fatal. All shop personnel are responsible for air quality in the shop. Shop management is responsible for providing an adequate exhaust system that can remove exhaust fumes from the maximum allowable number of vehicles that may be running in the shop at one time. Technicians should never run a vehicle in the shop unless a shop exhaust hose is installed on the tailpipe of the vehicle. The exhaust fan must be switched on to remove exhaust fumes.

If shop heaters or furnaces have restricted chimneys, they release carbon monoxide into the shop air. Therefore, chimneys should be checked periodically for restriction and proper ventilation. Diesel exhaust contains some carbon monoxide, but particulate emissions are also present in the exhaust from these engines. Particulates are small carbon particles that can be harmful to the lungs.

Monitors are available to measure the level of carbon monoxide in the shop. Some of these monitors read the amount of carbon monoxide present in the shop air; others provide an audible alarm if the concentration of carbon monoxide exceeds the danger level.

The sulfuric acid solution in car batteries is a corrosive, poisonous liquid. If a battery is charged with a fast charger at a high rate for a period of time, the battery becomes hot, and the sulfuric acid solution begins to boil. Under these conditions, the battery may emit strong sulfuric acid fumes that may be harmful to the lungs. If this happens, the battery charger should be turned off or the charging rate should be reduced considerably.

Some automotive clutch facings and brake linings contain asbestos. Never use an air hose to blow dirt from these components, because this action disperses asbestos dust into the shop where it may be inhaled by technicians and other people in the shop. A brake parts washer or a vacuum cleaner with special attachments must be used to clean the dust from these components.

Even though technicians take every precaution to maintain air quality in the shop, some undesirable gases may still get into the air. For example, exhaust manifolds may get oil on them during an engine overhaul. When the engine is started and these manifolds get hot, the oil burns off the manifolds and pollutes the shop air with oil smoke. Adequate shop ventilation must be provided to take care of this type of air contamination.

SHOP SAFETY EQUIPMENT

Fire Extinguishers

Fire extinguishers are one of the most important pieces of safety equipment. All shop personnel must know the location of each fire extinguisher in the shop. If you have to waste time looking for an extinguisher after a fire starts, the fire could get out of control before you get the extinguisher into operation. Fire extinguishers should be located where they are easily accessible at all times. A decal on each fire extinguisher identifies the type of chemical in the extinguisher and provides operating information (**Figure 1-3**). Shop personnel should be familiar with the following types of fires and fire extinguishers:

1. *Class A fires* are those involving ordinary combustible materials such as paper, wood, clothing, and textiles. **Multipurpose dry chemical fire extinguishers** are used on these fires.
2. *Class B fires* involve the burning of flammable liquids such as gasoline, oil, paint, solvents, and greases. These fires can also be extinguished with multipurpose dry chemical fire extinguishers. In addition, fire extinguishers containing halogen, or halon, may be used to extinguish class B fires. The chemicals in this type of extinguisher attach to the hydrogen, hydroxide, and oxygen molecules to stop the combustion process almost instantly. However, the resultant gases from the use of halogen-type extinguishers are very toxic and harmful to the operator of the extinguisher.
3. *Class C fires* involve the burning of electrical equipment such as wires, motors, and switches. These fires can be extinguished with multipurpose dry chemical fire extinguishers.
4. *Class D fires* involve the combustion of metal chips, turnings, and shavings. Dry chemical fire extinguishers are the only type of extinguisher recommended for these fires.

Figure 1-3 Types of fire extinguishers.

Some multipurpose dry chemical fire extinguishers may be used on Class A, B, C, or D fires. Additional information regarding which types of extinguishers are used for various types of fires is provided in **Table 1-1**.

Causes of Eye Injuries

Eye injuries can occur in various ways in the automotive shop. Some of the more common eye accidents are:

1. Thermal burns from excessive heat
2. Irradiation burns from excessive light such as from an arc welder
3. Chemical burns from strong liquids such as battery electrolyte
4. Foreign material in the eye
5. Penetration of the eye by a sharp object
6. A blow from a blunt object

Wearing safety glasses and observing shop safety rules will prevent most eye accidents.

Eyewash Fountains

If a chemical gets in your eyes, it must be washed out immediately to prevent a chemical burn. An eyewash fountain is the most effective way to wash the eyes. An eyewash fountain is similar to a drinking water fountain, but the eyewash fountain has water jets placed throughout the fountain top. Every shop should be equipped with some type of eyewash facility (**Figure 1-4**). Be sure you know the location of the eyewash fountain in the shop.

Safety Glasses and Face Shields

The mandatory use of eye protection, either safety glasses or a face shield, is one of the most important safety rules in an automotive shop. Face shields protect the face; safety glasses protect the eyes. When grinding, safety glasses must be worn, and a face shield can be worn. Many shop insurance policies require the use of eye protection in the shop. Some automotive technicians have been blinded in one or both eyes because they did not wear safety glasses. All safety glasses must be equipped with safety glass and should provide some type of side protection (**Figure 1-5**). When selecting a pair of safety glasses, they should feel comfortable on your face; if they are uncomfortable, you may remove

TABLE 1-1 GUIDE TO FIRE EXTINGUISHER SELECTION

	Class of fire	Typical fuel involved	Type of extinguisher
Class A Fires (green)	**For ordinary combustibles** Put out a class A fire by blowing its temperature or by coating the burning combustibles.	Wood Paper Cloth Rubber Plastics Rubbish Upholstery	Water* Foam* Multipurpose dry chemical
Class B Fires (red)	**For flammable liquids** Put out a class B fire by smothering it. Use an extinguisher that gives a blanketing, flame-interrupting effect; cover whole flaming liquid surface.	Gasoline Oil Grease Paint Lighter fluid	Foam* Carbon dioxide Halogenated agent Standard dry chemical Purple K dry chemical Multipurpose dry chemical
Class C Fires (blue)	**For electrical equipment** Put out a class C fire by shutting off power as quickly as possible and by always using a nonconducting extinguisher agent to prevent electric shock.	Motors Appliances Wiring Fuse boxes Switchboards	Carbon dioxide Halogenated agent Standard dry chemical Purple K dry chemical Multipurpose dry chemical
Class D Fires (yellow)	**For combustible metals** Put out a class D fire of metal chips, turnings, or shavings by smothering or coating with a specially designed extinguisher agent.	Aluminum Magnesium Potassium Sodium Titanium Zirconium	Dry power extinguisher and agents only

*Cartridge-operated water, foam, and soda-acid types of extinguishers are no longer manufactured. These extinguishers should be removed from service when they are due for their next hydrostatic pressure test.

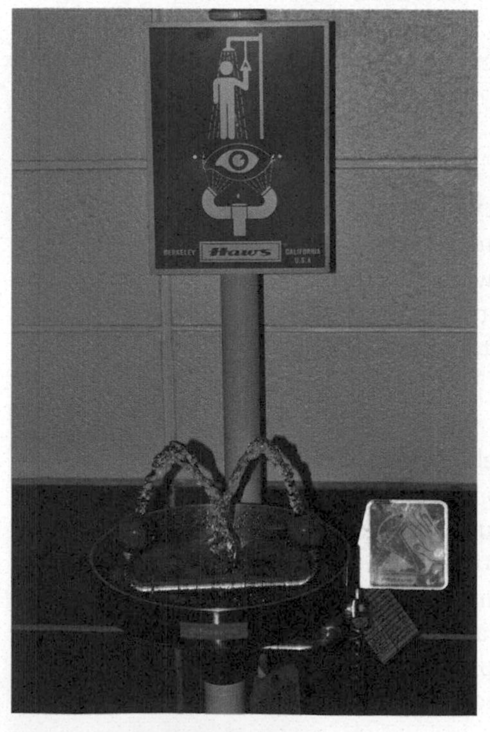

Figure 1-4 Eyewash fountain.

Figure 1-5 Safety glasses with side protection must be worn in the automotive shop.

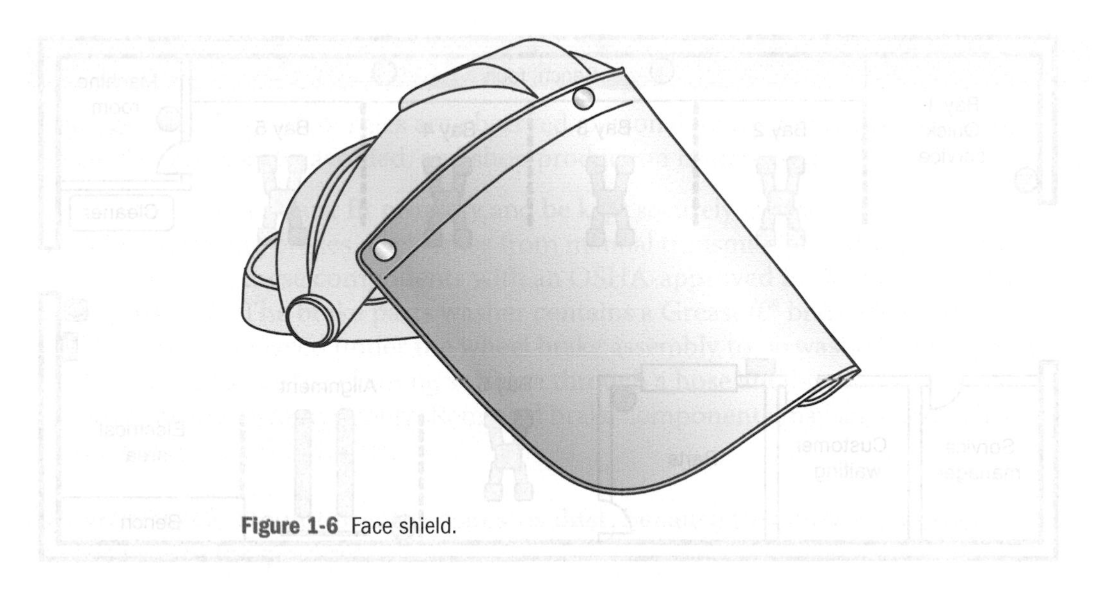

Figure 1-6 Face shield.

them, leaving your eyes unprotected. A face shield should be worn when handling hazardous chemicals or when using an electric grinder or buffer (**Figure 1-6**).

First-Aid Kits

First-aid kits should be clearly identified and conveniently located (**Figure 1-7**). These kits contain such items as bandages in a variety of sizes and ointment required for minor cuts. All shop personnel must be familiar with the location of first-aid kits. At least one of the shop personnel should have basic first-aid training. This person should be in charge of administering first aid and keeping first-aid kits filled.

SHOP LAYOUT

There are many different types of shops in the automotive service industry, including:

- New car dealers
- Independent repair shops
- Specialty shops
- Service stations
- Fleet shops

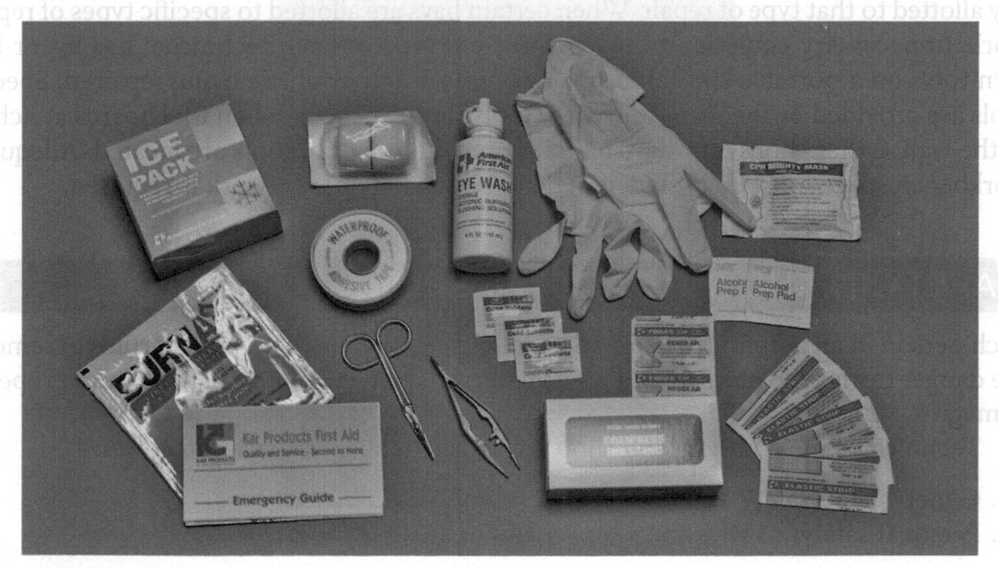

Figure 1-7 First-aid kit.

ELECTRICAL SAFETY

In the automotive shop you will be using electric drills, shop lights, wheel balancers, and wheel aligners.

Observe the following electrical safety precautions on this equipment:

1. Frayed cords on electrical equipment must be replaced or repaired immediately.
2. All electrical cords from lights and electrical equipment must have a ground connection. The ground connector is the round terminal in a three-pronged electrical plug. Do not use a two-pronged adapter to plug in a three-pronged electrical cord. Three-pronged electrical outlets should be mandatory in all shops.
3. Do not leave electrical equipment running and unattended.

GASOLINE SAFETY

Gasoline is a very explosive liquid! One exploding gallon of gasoline has a force equal to 14 sticks of dynamite. The expanding vapors from gasoline are extremely dangerous. These vapors are present even in cold temperatures. Vapors formed in gasoline tanks on cars are controlled, but vapors from a gasoline storage can may escape, resulting in a hazardous situation. Therefore, gasoline storage containers must be placed in a well-ventilated space.

A gasoline can may be called a jerry can.

Approved gasoline storage cans have a flash-arresting screen at the outlet (**Figure 1-11**). This screen prevents external ignition sources from igniting the gasoline within the can while the gasoline is being poured.

Follow these safety precautions regarding gasoline containers:

1. Always use approved gasoline containers that are painted red for proper identification.
2. Do not fill gasoline containers completely full. Always leave the level of gasoline at least one inch from the top of the container. This allows for expansion of the gasoline at higher temperatures. If gasoline containers are completely full, the gasoline will expand when the temperature increases. This expansion forces gasoline from the can and creates a dangerous spill.
3. If gasoline containers must be stored, place them in a well-ventilated area such as a storage shed. Do not store gasoline containers in your home or in the trunk of a vehicle.
4. When a gasoline container must be transported, be sure it is secured against upsets.
5. Do not store a partially filled gasoline container for long periods of time, because it may give off potentially dangerous vapors.

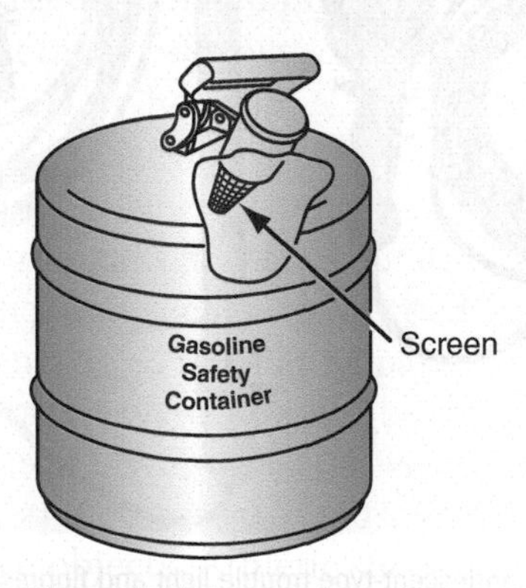

Figure 1-11 Approved gasoline container.

6. Never leave gasoline containers open except while filling or pouring gasoline from the container.
7. Do not prime an engine with gasoline while cranking the engine.
8. Never use gasoline as a cleaning agent.

FIRE SAFETY

When fire safety rules are observed, personal injury and expensive fire damage to vehicles and property may be avoided. Follow these fire safety rules:

1. Familiarize yourself with the location and operation of all shop fire extinguishers.
2. If a fire extinguisher is used, report it to management so the extinguisher can be recharged.
3. Do not use any type of open-flame heater to heat the work area.
4. Do not turn on the ignition switch or crank the engine with a gasoline line disconnected.
5. Store all combustible materials such as gasoline, paint, and oily rags in approved safety containers.
6. Clean up gasoline, oil, or grease spills immediately.
7. Always wear clean shop clothes. Do not wear oil-soaked clothes.
8. Do not allow sparks and flames near batteries.
9. Be sure that welding tanks are securely fastened in an upright position.
10. Do not block doors, stairways, or exits.
11. Do not smoke when working on vehicles.
12. Do not smoke or create sparks near flammable materials or liquids.
13. Store combustible shop supplies such as paint in a closed steel cabinet.
14. Store gasoline in approved safety containers.
15. If a gasoline tank is removed from a vehicle, do not drag the tank on the shop floor.
16. Know the approved fire escape route from your classroom or shop to the outside of the building.
17. If a fire occurs, do not open doors or windows. This action creates extra draft, which makes the fire worse.
18. Do not pour water on a gasoline fire, because the water will make the fire worse.
19. Call the fire department as soon as a fire begins, and then attempt to extinguish the fire.
20. If possible, stand 6 to 10 feet from the fire and aim the fire extinguisher nozzle at the base of the fire with a sweeping action.
21. If a fire produces a lot of smoke in the room, remain close to the floor to obtain oxygen and avoid breathing smoke.
22. If the fire is too hot or the smoke makes breathing difficult, get out of the building.
23. Do not re-enter a burning building.
24. Keep solvent containers covered except when pouring from one container to another. When flammable liquids are transferred from bulk storage, the bulk container should be grounded to a permanent shop fixture, such as a metal pipe. During this transfer process, the bulk container should be grounded to the portable container (**Figure 1-12**). These ground wires prevent the buildup of a static electric charge, which could cause a spark and a disastrous explosion. Always discard or clean empty solvent containers, because fumes in these containers are a fire hazard.
25. Familiarize yourself with different types of fires and fire extinguishers, and know the type of extinguisher to use on each type of fire.

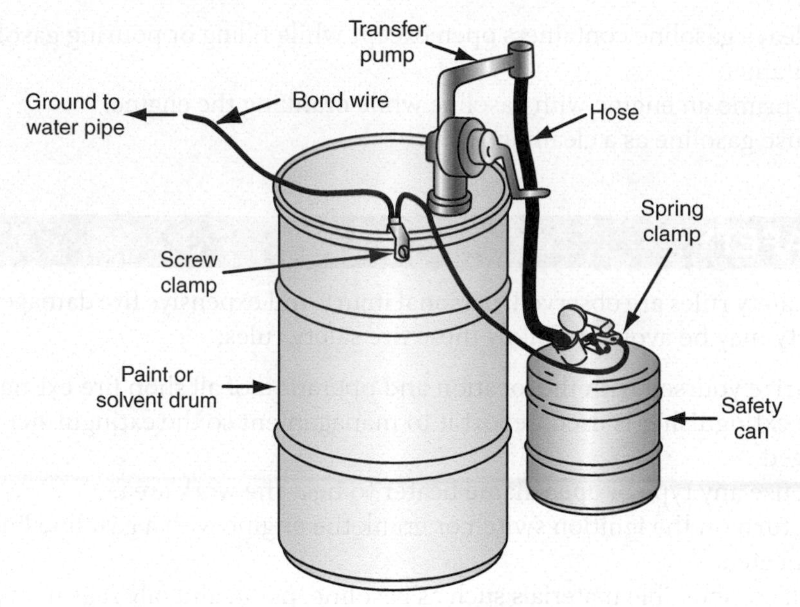

Figure 1-12 Safe procedures for flammable liquid transfer.

Using a Fire Extinguisher

Everyone working in the shop must know how to operate the fire extinguishers.

There are several different types of fire extinguishers, but their operation usually involves the following steps:

1. Get as close as possible to the fire without jeopardizing your safety.
2. Grasp the extinguisher firmly and aim the extinguisher at the fire.
3. Pull the pin from the extinguisher handle.
4. Squeeze the handle to dispense the contents of the extinguisher.
5. Direct the fire extinguisher nozzle at the base of the fire, and dispense the contents of the extinguisher with a sweeping action back and forth across the fire. Most extinguishers discharge their contents in 8 to 25 seconds.
6. Always be sure the fire is extinguished.
7. Always keep an escape route open behind you so a quick exit is possible if the fire gets out of control.

VEHICLE OPERATION

When driving a customer's vehicle, observe the following precautions to prevent accidents and maintain good customer relations:

1. Prior to driving a vehicle, make sure the brakes are operating and fasten the safety belt.
2. Before you start the engine, check to be sure there is no person or object under the car.
3. If the vehicle is parked on a lift, be sure the lift is fully down and that the lift arms or components are not in contact with the vehicle chassis.
4. Before driving away, check to see if any objects are directly in front of or behind the vehicle.
5. Always drive slowly in the shop, and watch carefully for personnel and other moving vehicles.

6. Make sure the shop door is up high enough so there is plenty of clearance between the top of the vehicle and the door.
7. Watch the shop door to be certain that it is not coming down as you attempt to drive under the door.
8. If a road test is necessary, wear your seat belt, obey all traffic laws, and never drive in a reckless manner.
9. Do not squeal tires when accelerating or turning corners.

If customers observe that service personnel take good care of their car by driving carefully and by installing fender, seat, and floor mat covers, the service department's image is greatly enhanced in their eyes. These procedures impress upon the customers that shop personnel respect their car. Conversely, if grease spots are found on the upholstery or fenders after service work is completed, the customers will probably think the shop is careless, not only in car care but also in service work quality.

HOUSEKEEPING SAFETY

CUSTOMER CARE When customers see that you are concerned about their vehicle and that you operate a shop with excellent housekeeping habits, they will be impressed and will likely keep returning for service.

Careful housekeeping habits prevent accidents and increase worker efficiency. Good housekeeping also helps impress upon the customer that quality work is a priority in your shop.

Follow these housekeeping rules:

1. Keep aisles and walkways clear of tools, equipment, and other items.
2. Be sure all sewer covers are securely in place.
3. Keep floor surfaces free of oil, grease, water, and loose material.
4. Sweep up under a vehicle before lowering the vehicle on the lift.
5. Proper trash containers must be conveniently located, and these containers should be emptied regularly.
6. Access to fire extinguishers must be unobstructed at all times, and fire extinguishers should be checked for proper charge at regular intervals.
7. Tools must be kept clean and in good condition.
8. When not in use, tools must be stored in their proper location.
9. Oily rags must be stored in approved, covered containers (**Figure 1-13**). A slow generation of heat occurs from the oxidation of oil on these rags. Heat may continue to be generated until the ignition temperature is reached. The oil and rags then begin to burn, causing a fire. This action is called spontaneous combustion. However, if the oily rags are in an airtight, approved container, there is not enough oxygen to cause burning.
10. Store paint, gasoline, and other flammable liquids in a closed steel cabinet (**Figure 1-14**).
11. Rotating components on equipment and machinery must have guards, and all shop equipment should have regular service and adjustment schedules.
12. Keep the workbenches clean. Do not leave heavy objects, such as used parts, on the bench after you are finished with them.
13. Keep parts and materials in their proper location.
14. When not in use, creepers must not be left on the shop floor. Creepers should be stored in a specific location.

Figure 1-13 Dirty shop towels or rags must be kept in an approved, closed container.

Figure 1-14 Store combustible materials in an approved safety cabinet.

An air bag system is designed to protect the driver and/or passengers in a vehicle collision.

15. The shop should be well lit, and all lights should be in working order.
16. Frayed electrical cords on lights or equipment must be replaced.
17. Walls and windows should be cleaned regularly.
18. Stairs must be clean, well lit, and free of loose material.

SERVICE TIP When you are finished with a tool, never set it on the customer's car. After using a tool, the best place for it is in your toolbox or on the workbench. Many tools have been lost by leaving them on customers' vehicles.

AIR BAG SAFETY

1. When service is performed on any air bag system component, always disconnect the negative battery cable, isolate the cable end, and wait for the amount of time specified by the vehicle manufacturer before proceeding with the necessary diagnosis or service. The average waiting period is 2 minutes, but some vehicle manufacturers specify up to 10 minutes. **Photo Sequence 1** shows a typical procedure for removing an air bag module.

   On some recent-model vehicles, the air bag system is divided into different disabling and enabling zones for diagnostic and service purposes. When performing vehicle service on or near air bag system components, the vehicle manufacturer recommends disabling the air bag components, in the zone where the air bag components are located rather than disconnecting the vehicle battery (**Figure 1-15**).

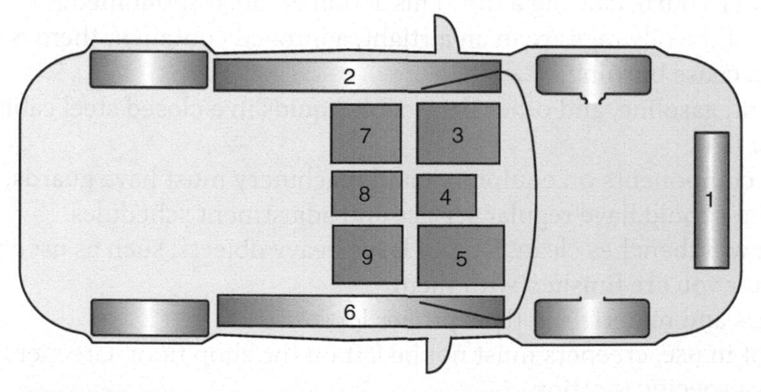

Figure 1-15 Air bag system disabling and enabling zones.

PHOTO SEQUENCE 1

Typical Procedure for Removing Air Bag Module

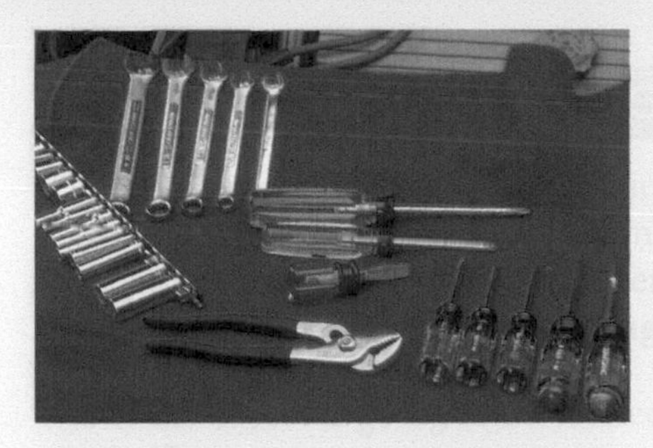

P1-1 Tools required to remove the air bag module: safety glasses, seat covers, screwdriver set, torx driver set, battery terminal pullers, battery pliers, assorted wrenches, ratchet and socket set, and service manual.

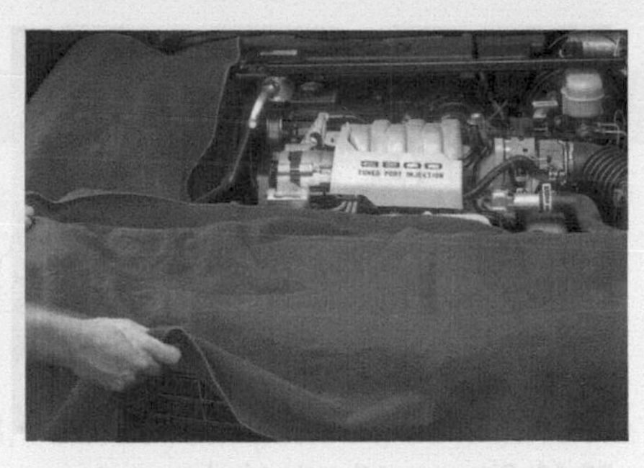

P1-2 Place the seat and fender covers on the vehicle.

P1-3 Place the front wheels in the straight-ahead position, and turn the ignition switch to the lock position.

P1-4 Disconnect the negative battery cable.

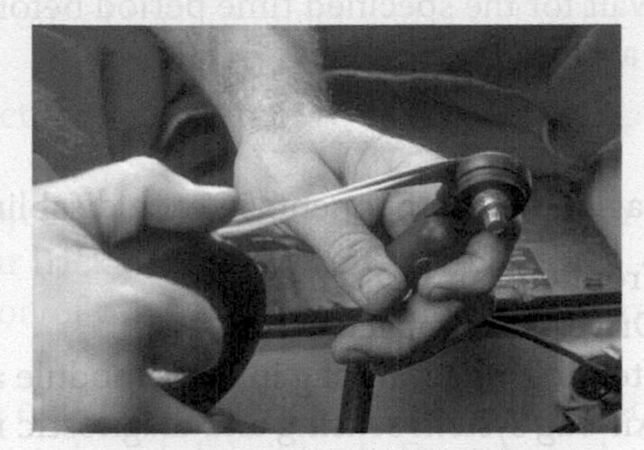

P1-5 Tape the cable terminal to prevent accidental connection with the battery post. Note: A piece of rubber hose can be substituted for the tape.

P1-6 Remove the SIR fuse from the fuse box. Wait 10 minutes to allow the reserve energy to dissipate.

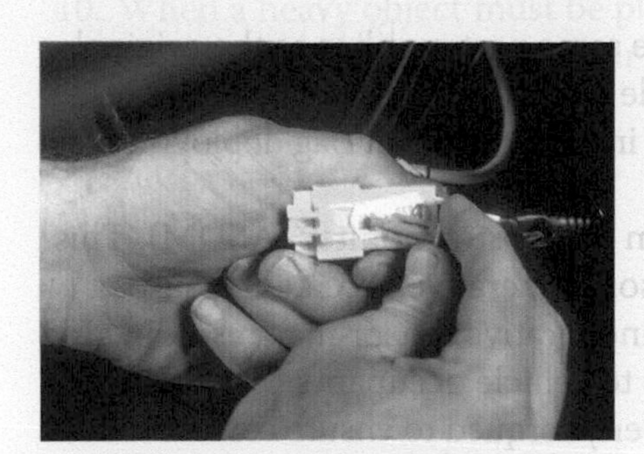

P1-7 Remove the connector position assurance (CPA) from the yellow electrical connector at the base of the steering column.

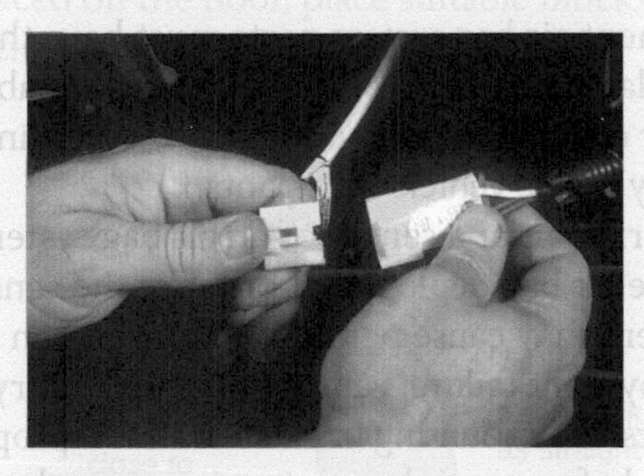

P1-8 Disconnect the yellow two-way electrical connector.

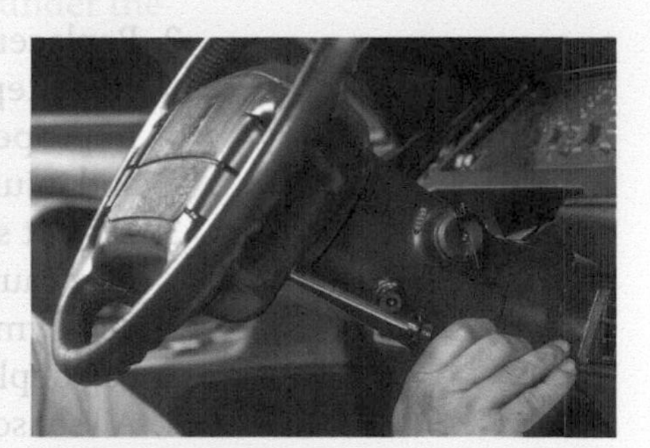

P1-9 Remove the four bolts that secure the module from the rear of the steering wheel.

HAND TOOL SAFETY

Many shop accidents are caused by improper use and care of hand tools.
Follow these safety steps when working with hand tools:

1. Maintain tools in good condition and keep them clean. Worn tools may slip and result in hand injury. If a hammer with a loose head is used, the head may fly off and cause personal injury or vehicle damage. If your hand slips off a greasy tool, it may cause some part of your body to hit the vehicle, causing injury.
2. Using the wrong tool for the job may damage the tool, fastener, or your hand if the tool slips. If you use a screwdriver as a chisel or pry bar, the blade may shatter, causing serious personal injury.
3. Use sharp-pointed tools with caution. Always check your pockets before sitting on the vehicle seat. A screwdriver, punch, or chisel in the back pocket may put an expensive tear in the upholstery. Do not lean over fenders with sharp tools in your pockets.
4. Tools that are intended to be sharp should be kept sharp. A sharp chisel, for example, will do the job faster with less effort.

LIFT SAFETY

Special precautions and procedures must be followed when a vehicle is raised on a lift.
Follow these steps for safe lift operation:

When a vehicle is raised on a lift, the vehicle must be raised high enough to allow engagement of the lift locking mechanism.

1. Always be sure the lift is completely lowered before driving a vehicle on or off the lift.
2. Do not hit or run over lift arms and adaptors when driving a vehicle on or off the lift. Have a coworker guide you when driving a vehicle onto the lift. Do not stand in front of a lift with the car coming toward you.
3. Be sure the lift pads contact the car manufacturer's recommended lifting points shown in the service manual. If the proper lifting points are not used, components under the vehicle such as brake lines or body parts may be damaged. Failure to use the recommended lifting points may cause the vehicle to slip off the lift, resulting in severe vehicle damage and personal injury.
4. Before a vehicle is raised or lowered, close the doors, hood, and trunk lid.
5. When a vehicle has been lifted a short distance off the floor, stop the lift and check the contact between the hoist lift pads and the vehicle to be sure the lift pads are still on the recommended lifting points.
6. When a vehicle has been raised, be sure the safety mechanism is in place to prevent the lift from dropping accidentally.
7. Prior to lowering a vehicle, always make sure there are no objects, tools, or people under the vehicle.
8. Do not rock a vehicle on a lift during a service job.
9. When a vehicle is raised, removal of some heavy components may cause vehicle imbalance. For example, because front-wheel-drive cars have the engine and transaxle at the front of the vehicle, these cars have most of their weight on the front end. Removing a heavy rear-end component on these cars may cause the back end of the car to rise off the lift. If this happens, the vehicle could fall off the lift!
10. Do not raise a vehicle on a lift with people in the vehicle.
11. When raising pickup trucks and vans on a lift, remember these vehicles are higher than a passenger car. Be sure there is adequate clearance between the top of the vehicle and the shop ceiling or components under the ceiling.

12. Do not raise a four-wheel-drive vehicle with a frame contact lift unless proper adaptors are used. Lifting a vehicle on a frame contact lift without the proper adaptors may damage axle joints.
13. Do not operate a front-wheel-drive vehicle that is raised on a frame contact lift. This may damage the front-drive axles.

HYDRAULIC JACK AND SAFETY STAND SAFETY

WARNING Always make sure the safety stand weight capacity rating exceeds the vehicle weight that you are planning to raise. A safety stand with insufficient weight capacity may collapse, resulting in vehicle damage and/or personal injury.

WARNING Never lift a vehicle with a floor jack if the weight of the vehicle exceeds the rated capacity of the jack. If the vehicle weight exceeds the weight rating of the floor jack, the vehicle weight may collapse the floor jack, resulting in jack and vehicle damage and/or personal injury.

Accidents involving the use of floor jacks and safety stands may be avoided if these safety precautions are followed:

1. Never work under a vehicle unless safety stands are placed securely under the vehicle chassis and the vehicle is resting on these stands (**Figure 1-17**).
2. Prior to lifting a vehicle with a floor jack, be sure that the jack lift pad is positioned securely under a recommended lifting point on the vehicle. Lifting the front end of a vehicle with the jack placed under a radiator support may cause severe damage to the radiator and support.
3. Position the safety stands under a strong chassis member such as the frame or axle housing. The safety stands must contact the vehicle manufacturer's recommended lifting points.
4. Because the floor jack is on wheels, the vehicle and safety stands tend to move as the vehicle is lowered from a floor jack onto the safety stands. Always be sure the safety stands remain under the chassis member during this operation, and be sure the safety stands do not tip. All the safety stand legs must remain in contact with the shop floor.
5. When the vehicle is lowered from the floor jack onto the safety stands, remove the floor jack from under the vehicle. Never leave a jack handle sticking out from under a vehicle, which may trip someone and cause an injury.

Figure 1-17 Safety stands.

POWER TOOL SAFETY

Power tools use electricity, shop air, or hydraulic pressure as a power source. Careless operation of power tools may cause personal injury or vehicle damage.

Follow these steps for safe power tool operation:

1. Do not operate power tools with frayed electrical cords.
2. Be sure the power tool cord has a proper ground connection.
3. Do not stand on a wet floor while operating an electric power tool.
4. Always unplug an electric power tool before servicing the tool.
5. Do not leave a power tool running and unattended.
6. When using a power tool on small parts, do not hold the part in your hand. The part must be secured in a bench vise or with locking pliers.
7. Do not use a power tool on a job where the maximum capacity of the tool is exceeded.
8. Be sure that all power tools are in good condition.
9. Always operate these tools according to the tool manufacturer's recommended procedure.
10. Make sure all protective shields and guards are in position.
11. Maintain proper body balance while using a power tool.
12. Always wear safety glasses or a face shield.
13. Wear ear protection.
14. Follow the equipment manufacturer's recommended maintenance schedule for all shop equipment.
15. Never operate a power tool unless you are familiar with the tool manufacturer's recommended operating procedure, because serious accidents can occur.
16. Always make sure that the wheels are securely attached and in good condition on the electric grinder.
17. Keep fingers and clothing away from grinding and buffing wheels. When grinding or buffing a small part, hold the part with a pair of locking pliers.
18. Always make sure the sanding or buffing disc is securely attached to the sander pad.
19. Special heavy-duty sockets must be used on impact wrenches. If ordinary sockets are used on an impact wrench, they may break and cause serious personal injury.

COMPRESSED-AIR EQUIPMENT SAFETY

The shop air supply contains high-pressure air in the shop compressor and air lines. Serious injury or property damage may result from careless operation of compressed-air equipment.

Follow these steps to improve safety:

1. Never operate an air chisel unless the tool is securely connected to the chisel with the proper retaining device.
2. Never direct a blast of air from an air gun toward any part of your body. If air penetrates the skin and enters the bloodstream, it may cause very serious health problems and even death.
3. Safety glasses or a face shield should be worn for all shop tasks, including those tasks involving the use of compressed-air equipment.
4. Wear ear protection when using compressed-air equipment.
5. Always maintain air hoses and fittings in good condition. If an end suddenly blows off an air hose, the hose will whip around, causing possible personal injury.

6. Do not direct compressed air against the skin, because it may penetrate the skin, especially through small cuts or scratches, and enter the bloodstream, causing death or serious health complications. Use only OSHA-approved air gun nozzles.
7. Do not use an air gun to blow debris off clothing or hair.
8. Do not clean the workbench or floor with compressed air. This action may blow very small parts toward your skin or into your eye. Small parts blown by compressed air may also cause vehicle damage. For example, if the car in the next stall has the air cleaner removed, a small part may find its way into the carburetor or throttle body. When the engine is started, this part will likely be pulled into a cylinder by engine vacuum, and the part will penetrate through the top of a piston.
9. Never spin bearings with compressed air because the bearing will rotate at extremely high speed. This may damage the bearing or cause it to disintegrate, causing personal injury.
10. All pneumatic tools must be operated according to the tool manufacturer's recommended operating procedure.
11. Follow the equipment manufacturer's recommended maintenance schedule for all compressed-air equipment.

CLEANING EQUIPMENT SAFETY AND ENVIRONMENTAL CONSIDERATIONS

Cleaning Equipment Safety

All technicians are required to clean parts during their normal work routines. Face shields and protective gloves must be worn while operating cleaning equipment. In most states, environmental regulations require that the runoff from steam cleaning must be contained in the steam cleaning system. This runoff cannot be dumped into the sewer system. Because it is expensive to contain this runoff in the steam cleaning system, the popularity of steam cleaning has decreased. The solution in hot and cold cleaning tanks may be caustic, and contact between this solution and skin or eyes must be avoided. Parts cleaning often creates a slippery floor, and care must be taken when walking in the parts cleaning area. The floor in this area should be cleaned frequently. When the cleaning solution in hot or cold cleaning tanks is replaced, environmental regulations require that the old solution be handled as hazardous waste. Use caution when placing aluminum or aluminum alloy parts in a cleaning solution. Some cleaning solutions will damage these components. Always follow the cleaning equipment manufacturer's recommendations.

Parts Washers with Electromechanical Agitation

Some parts washers provide electromechanical agitation of the parts to provide improved cleaning action (**Figure 1-18**). These parts washers may be heated with gas or electricity. Various water-based **hot cleaning tank** solutions are available, depending on the type of metals being cleaned. For example, Kleer-Flo Greasoff® No. 1 powdered detergent is available for cleaning iron and steel. Nonheated electromechanical parts washers are also available, and these washers use cold cleaning solutions such as Kleer-Flo Degreasol® formulas.

A **hot cleaning tank** uses a heated solution to clean metal parts.

Many cleaning solutions, such as Kleer-Flo Degreasol® 99R, contain no ingredients listed as hazardous by the Environmental Protection Agency's Resource Conservation and Recovery Act (RCRA). This cleaning solution is a blend of sulfur-free hydrocarbons, wetting agents, and detergents. Degreasol® 99R does not contain aromatic or chlorinated solvents, and it conforms to California's Rule 66 for clean air. Always use the cleaning solution recommended by the equipment manufacturer.

Refrigerants

When servicing automotive air-conditioning systems, it is illegal to vent refrigerants to the atmosphere. Certified equipment must be used to recover and recycle the refrigerant and to recharge air-conditioning systems. This service work must be performed by an EPA-certified technician.

The EPA approves a Refrigerant Recovery and Recycling Review and Quiz supplied by the National Institute for Automotive Service Excellence (ASE). Upon successful completion of the ASE review and quiz, a technician is certified to service mobile air-conditioning refrigerant systems. The ASE review and quiz meets Section 609 regulations in the Clean Air Act Amendments.

HAZARDOUS WASTE DISPOSAL

Hazardous waste materials in automotive shops are chemicals or components no longer needed by the shop. These materials pose a danger to the environment and to people if they are disposed of in ordinary trash cans or sewers. However, no material is considered hazardous waste until the shop has finished using it and is ready to dispose of it. The **Environmental Protection Agency (EPA)** publishes a list of hazardous materials, which is included in the Code of Federal Regulations. Waste is considered hazardous if it is included in the EPA list of hazardous materials, or if it has one or more of these characteristics:

The Environmental Protection Agency (EPA) is the agency responsible for air and water quality in the United States.

A material that is **reactive** reacts with some other chemicals and gives off gas(es) during the reaction.

A material that is **corrosive** causes another material to be gradually worn away by chemical action.

A **toxic** substance is poisonous to animal or human life.

A substance that is **ignitable** can be ignited spontaneously or by another source of heat or flame.

1. **Reactive.** Any material that reacts violently with water or other chemicals is considered hazardous. If a material releases cyanide gas, hydrogen sulfide gas, or similar gases when exposed to low-pH acid solutions, it is hazardous.
2. **Corrosive.** If a material burns the skin or dissolves metals and other materials, it is considered hazardous.
3. **Toxic.** Materials are hazardous if they leach one or more of eight heavy metals in concentrations greater than 100 times the primary drinking water standard.
4. **Ignitable.** A liquid is hazardous if it has a flashpoint below 140°F (60°C). A solid is hazardous if it ignites spontaneously.

Federal and state laws control the disposal of hazardous waste materials. Every shop employee must be familiar with these laws. Hazardous waste disposal laws include the **Resource Conservation and Recovery Act (RCRA)**. This law states that hazardous material users are responsible for hazardous materials from the time they become waste until the proper waste disposal is completed. Many automotive shops hire an independent hazardous waste hauler to dispose of hazardous waste material (**Figure 1-21**). The shop owner or manager should have a written contract with the hazardous waste hauler. Rather than hauling hazardous waste material to an approved hazardous waste disposal site, a shop may choose to recycle the material in the shop; therefore, hazardous waste material must be properly and safely stored. The user is responsible for the transportation of this material until it arrives at an approved hazardous waste disposal site and is processed according to the law.

The RCRA controls these types of automotive waste:

1. Paint and body repair products waste
2. Solvents for parts and equipment cleaning
3. Batteries and battery acid
4. Mild acids used for metal cleaning and preparation
5. Waste oil, engine coolants, and antifreeze
6. Air-conditioning refrigerants
7. Engine oil filters

Figure 1-21 Hazardous waste hauler.

Never, under any circumstances, use these methods to dispose of hazardous waste material:

1. Pour hazardous wastes on weeds to kill them.
2. Pour hazardous wastes on gravel streets to prevent dust.
3. Throw hazardous wastes in a dumpster.
4. Dispose of hazardous wastes anywhere but an approved disposal site.
5. Pour hazardous wastes down sewers, toilets, sinks, or floor drains.

The right-to-know laws state that employees have a right to know when the materials they use at work are hazardous. The right-to-know laws started with the **Hazard Communication Standard** published by the OSHA in 1983. This document was originally intended for chemical companies and manufacturers that required employees to handle hazardous materials in their work situation. Currently, most states have established their own right-to-know laws. Meanwhile, federal courts have decided to apply these laws to all companies, including automotive service shops. Under the right-to-know laws, the employer has three responsibilities regarding the handling of hazardous materials by its employees.

First, all employees must be trained about the types of hazardous materials they will encounter in the workplace. Employees must be informed about their rights under legislation regarding the handling of hazardous materials. All hazardous materials must be properly labeled, and information about each hazardous material must be posted on **safety data sheets (SDSs)**, which are available from the manufacturer (**Figure 1-22**). A major change that occurred under OSHA's HazCom 2012 regulations has to do with SDSs. OSHA's current requirement aligns itself with the United Nations' Globally Harmonized System of Classification and Labeling of Chemicals (GHS) Safety Data Sheet (SDS) preparation requirements. What has historically been called a Material Safety Data Sheet, or MSDS, will now be referred to as a Safety Data Sheet, or SDS. In Canada, SDS may be called **workplace hazardous materials information systems (WHMIS)**.

Safety data sheets (SDSs) provide all the necessary data about hazardous materials.

The employer has a responsibility to place SDSs where they are easily accessible by all employees. The SDSs provide extensive information about hazardous materials such as:

1. Chemical name
2. Physical characteristics

WARNING When the vehicle has been left unattended, recheck that the service disconnect has not been reinstalled by a well-meaning associate.

When working on an HEV, always assume the HV system is live until you have proven otherwise—you can never be too safe. Your first mistake may be your last! If the vehicle has been driven into the service department, you know that the HV system was energized because most HEVs do not move without the HV system operating.

It is critical that the proper tools be used when working on the HV system. These include protective hand tools and a digital multimeter (DMM) with an insulation test function. The meter must be capable of checking for insulation up to 1000 V and measuring resistance at over 1.1 mega ohms. In addition, the DMM insulation test function is used to confirm proper insulation of the HV system components after a repair is performed.

Whenever possible use the one-hand rule when servicing the HV system (**Figure 1-23**). The one-hand rule means working with only one hand while servicing the HV systems so that in the event of an electric shock the HV will not pass through your body. It is important to follow this rule when performing the HV checkout procedure because confirmation of HV system power down has not been proven yet.

Insulated Glove Integrity Test

The rubber insulating gloves (lineman gloves), which the technician must wear for protection while serving the HV system, are your first line of defense when it comes to preventing contact with energized electrical components and must be tested for integrity before they are used. Also, pay attention to the date code on the gloves; gloves have an expiration date, and they do not last forever. In addition, for heavy services, use the leather protective glove on top of the isolating gloves.

Not just any gloves will do: lineman gloves must meet current ASTM D120 specifications and NFPA 70E standards. These requirements are enforced by OSHA as part of their CFR 1910.137 regulation. These standards dictate testing, retesting, and manufacturing criteria for lineman gloves. For HV vehicles the lineman gloves must meet Class "0" requirements of a rating of 1000 V AC (**Figure 1-24**). Electrical protective gloves are categorized by the amount of voltage, both AC and DC, to which they have been

Figure 1-23 It is recommended to follow the one-hand rule when testing any high-voltage system.

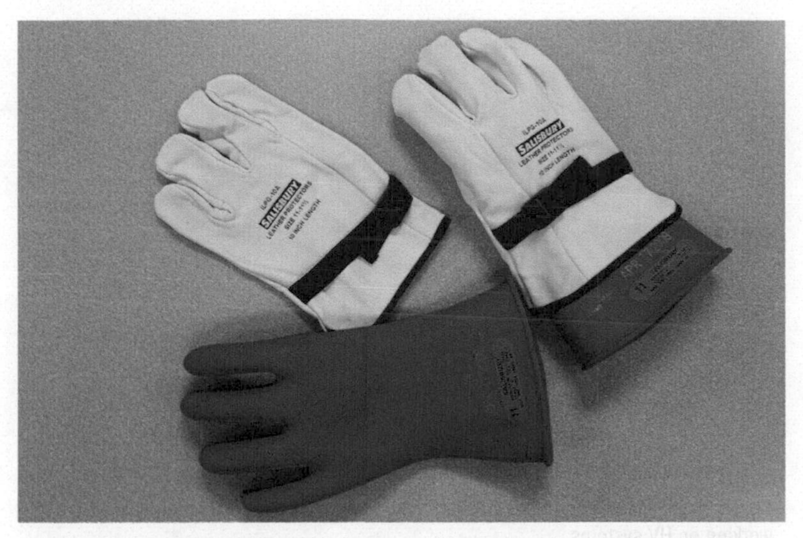

Figure 1-24 Insulated rubber isolation gloves rated Class 0 must be worn when working on high-voltage system.

proof-tested. In addition, the technician should wear rubber-soled shoes, cotton clothing, and safety glasses with side shields as part of their personnel protective equipment. Remove all jewelry and make sure metal zippers are not exposed. Always have a second set of insulating gloves available and let someone in the shop know their location. When preparing to work on any HV vehicle, let associates know in the event they must come to your aid.

OSHA regulations require that all insulating gloves must be electrically tested before first issue and retested every 6 months thereafter by a test laboratory. For this reason, most shops will discard and replace gloves after 6 months. Any gloves unused after 12 months must be retested or discarded. The manufacturers and/or suppliers of insulating gloves can assist in providing test laboratory locations for retest certification if you prefer that over replacement of the glove. It is recommended that gloves be stored out of direct sunlight and away from sources of ozone (i.e., electric motors). They should be stored in a glove bag flat, never folded, and hung up rather than laid down on a flat surface.

WARNING **Do not use insulating gloves that have been used for over 6 months, or gloves that are new in the package but older than 12 months (unless they have been recertified by a licensed test laboratory). Perform a daily or prior-to-use safety inspection of insulating gloves before working on any HV vehicle.**

Daily or prior-to-use safety inspection of rubber insulating gloves procedure:

1. Visually inspect rubber gloves prior to use for cracks, tears, holes, signs of ozone damage, possible chemical contact, and signs of abrasion or after any situation that may have caused damage to the gloves.
2. OSHA requires a glove air inflator test.
3. Blow air into glove to inflate them and seal the opening by folding the base of the glove.
4. Slowly role the base of the glove toward the fingers to increase the pressure (**Figure 1-25**).
5. Look and feel for pin holes on all surface sides.
6. If a leak or damage is detected, discard the glove. As an extra precaution, the glove should be rendered unusable (i.e., cut in half).

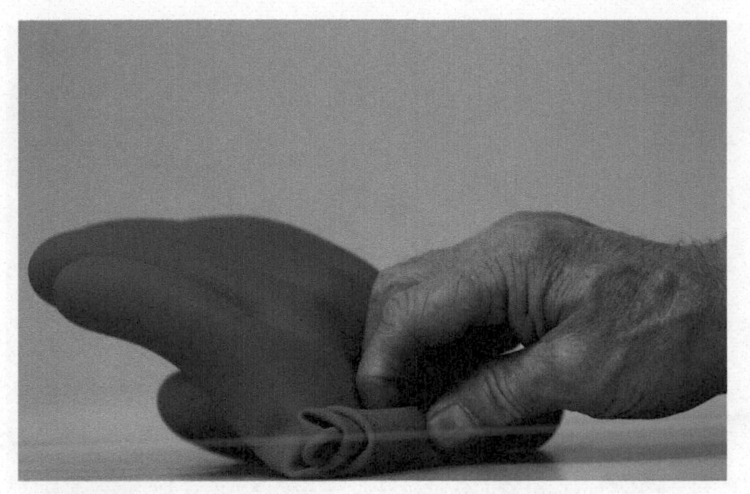

Figure 1-25 Insulated rubber isolation gloves must be tested for leaks before working on HV systems.

High-Voltage Service Plug

The HEV is equipped with an HV service plug that disconnects the HV battery from the system. Usually this plug is located near the battery (**Figure 1-26**). Prior to disconnecting the HV service plug, the vehicle must be turned off. Some manufactures also require the negative terminal of the auxiliary battery be disconnected. Once the HV service plug is removed, the HV circuit is shut off at the intermediate position of the HV battery.

The HV service plug assembly contains a safety interlock reed switch. The reed switch is opened when the clip on the HV service plug is lifted. The open reed switch turns off power to the service main relay (SMR). The main fuse for the HV circuit is inside the HV service plug assembly.

However, never assume that the HV circuit is off. The removal of the HV service plug does not disable the individual HV batteries. Use a DMM to verify that 0 V are in the system before beginning service. When testing the circuit for voltage, set the voltmeter to the 400 VDC scale.

After the HV service plug is removed, a minimum of 5 minutes must pass before beginning service on the system. This is required to discharge the high voltage from the condenser in the inverter circuit.

Figure 1-26 The HV service plug is generally located near the HV battery.

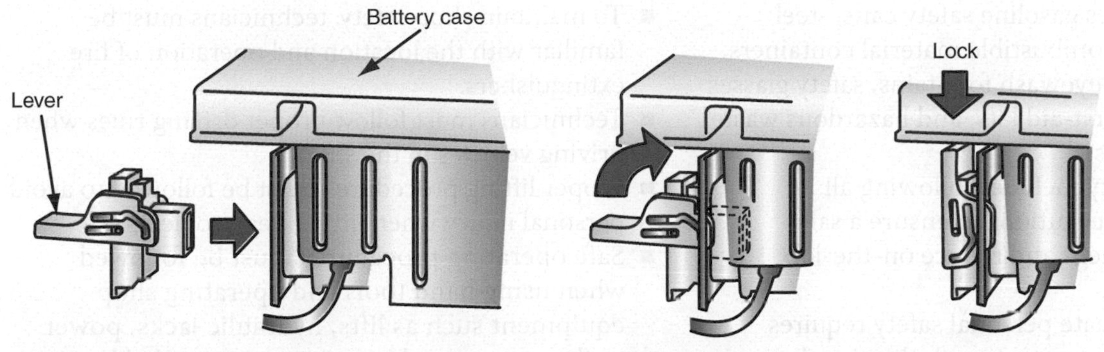

Figure 1-27 The HV service plug lever must be placed in the down and locked position after installation.

To install the HV service plug, make sure the lever is locked in the DOWN position (**Figure 1-27**). Slide the plug into the receptacle, and lock it in place by lifting the lever upward. Once it is locked in place, it closes the reed switch returning power to the system.

CASE STUDY

A technician raised a Grand Marquis on a lift to perform an oil and filter change and a chassis lubrication. This lift was a twin-post-type with separate front and rear lift posts. On this type of lift, the rear wheels must be positioned in depressions in the floor to position the rear axle above the rear lift arm. Then the front lift post and arms must be moved forward or rearward to position the front lift arms under the front suspension. The front lift arms must also be moved inward or outward so they are lifting on the vehicle manufacturer's specified lifting points.

The technician carefully positioned the front lift post and arms properly, but forgot to check the position of the rear tires in the floor depressions. The car was raised on the lift, and the technician proceeded with the service work. Suddenly there was a loud thump, and the rear of the car bounced up and down! The rear lift arms were positioned against the floor of the trunk rather than on the rear axle, and the lift arms punched through the floor of the trunk, narrowly missing the fuel tank. The technician was extremely fortunate the car did not fall off the lift, which would have resulted in more severe damage. If the rear lift arms had punctured the fuel tank, a disastrous fire could have occurred! Luckily, these things did not happen.

The technician learned a very important lesson about lift operation. Always follow all the recommended procedures in the lift operator's manual! The trunk floor was repaired at no cost to the customer, and fortunately, the shop and the vehicle escaped without major damage.

SUMMARY

- The U.S. Occupational Safety and Health Act (OSHA) of 1970 ensures safe and healthful working conditions and authorizes enforcement of safety standards.
- Many hazardous materials and conditions can exist in an automotive shop, including flammable liquids and materials, corrosive acid solutions, loose sewer covers, caustic liquids, high-pressure air, frayed electrical cords, hazardous waste materials, carbon monoxide, improper clothing, harmful vapors, high noise levels, and spills on shop floors.
- Safety data sheets (SDSs) provide information regarding hazardous materials, labeling, and handling.
- The danger regarding hazardous conditions and materials may be avoided by eliminating shop hazards and applying the necessary shop rules and safety precautions.
- The automotive shop owner/management must supply the necessary shop safety equipment, and all shop personnel must be familiar with the location and operation of this equipment. Shop safety

equipment includes gasoline safety cans, steel storage cabinets, combustible material containers, fire extinguishers, eyewash fountains, safety glasses and face shields, first-aid kits, and hazardous waste disposal containers.

- General shop safety includes following all safety rules and precautions to ensure a safe working environment and reduce on-the-job injuries.
- Maintaining adequate personal safety requires technicians to wear protective clothing such as proper footwear with steel toe caps, and safety glasses or a face shield. Technicians must avoid wearing loose fitting clothing that may become entangled in rotating equipment or components.
- To provide adequate personal safety, technicians must avoid wearing watches, jewelry, or rings.
- Shop safety requires a properly ventilated shop with an adequate exhaust removal system to avoid carbon monoxide gas in the shop.
- Shop safety equipment such as first-aid kits, eyewash fountains, and fire extinguishers must be clearly marked and easily accessible.
- Electrical cords must have a ground connection, and frayed electrical cords must be repaired or replaced immediately.
- Gasoline must be stored in clearly marked, approved gasoline containers.
- Fire safety in the shop includes storing combustible materials in approved, covered safety containers, avoiding sparks and flames near combustible materials, and quickly cleaning up gasoline, oil, or grease spills.
- To maintain shop safety, technicians must be familiar with the location and operation of fire extinguishers.
- Technicians must follow proper driving rules when driving vehicles in the shop.
- Proper lifting procedures must be followed to avoid personal injury when lifting heavy objects.
- Safe operating procedures must be followed when using hand tools and operating shop equipment such as lifts, hydraulic jacks, power tools, compressed-air equipment, and cleaning equipment.
- Hybrid system can use voltage in excess of 300 V, both AC and DC; it is critical that you as a service technician be familiar with and follow all safety and service precautions.
- Always remove the HV service plug prior to servicing any HV system and place the plug, such as on your toolbox, where it cannot be accidentally reinstalled by someone else.
- Do not attempt to test or service the HV system of an HEV for 5 minutes after the HV service plug is removed.
- Confirm that the HV circuits have 0 V using a DMM before performing any service procedure on the HV system of an HEV. Never assume that the HV circuit is off.
- Test the integrity of your lineman gloves (insulating gloves) prior to servicing any HV system. Do not use linemen gloves that fail a leak test, visual inspection, or are over 6 months old.

ASE-STYLE REVIEW QUESTIONS

1. Breathing carbon monoxide may cause:

   A. Arthritis
   B. Cancer
   C. Impaired vision
   D. Headaches

2. While discussing shop rules:

   Technician A says the wire harness, terminals, and connectors of the HV system are identified by red.

   Technician B says to remove the service plug prior to disconnecting or reconnecting any HV connections or components.

   Who is correct?

   A. A only
   B. B only
   C. Both A and B
   D. Neither A nor B

3. *Technician A* says that OSHA is a state agency responsible for Occupational Standards for Heating and Air-Conditioning.

   Technician B says that an OSHA poster must be displayed in the employees' common area, such as the break room.

   Who is correct?

   A. A only
   B. B only
   C. Both A and B
   D. Neither A nor B

4. While discussing personal safety:

   Technician A says rings and jewelry may be worn in the automotive shop.

   Technician B says some electric-drive cooling fans may start rotating at any time.

Who is correct?

A. A only
B. B only
C. Both A and B
D. Neither A nor B

5. While lifting heavy objects in the automotive shop:
   A. Bend your back to pick up the heavy object.
   B. Place your feet as far as possible from the object.
   C. Bend forward to place the object on the workbench.
   D. Straighten your legs to lift an object off the floor.

6. All of the following statements are true concerning HV system safety, EXCEPT:
   A. Turn the power switch to the off position prior to performing a resistance check.
   B. Do not attempt to test or service the system for 5 minutes after the HV service plug is removed.
   C. Test lineman gloves for damage and leaks prior to use.
   D. Disconnect the motor generators prior to turning the ignition off.

7. While operating hydraulic equipment safely in the automotive shop remember that:
   A. Safety stands have a maximum weight capacity.
   B. The driver's door should be open when raising a vehicle on a lift.
   C. A lift does not require a safety mechanism to prevent lift failure.
   D. Four-wheel-drive vehicles should be lifted on a frame contact lift.

8. While discussing shop hazards:

   Technician A says high-pressure air from an air gun may penetrate the skin.

   Technician B says air in the bloodstream may be fatal.

   Who is correct?

   A. A only
   B. B only
   C. Both A and B
   D. Neither A nor B

9. While discussing hazardous material disposal:

   Technician A says certain types of hazardous waste material can be poured down a floor drain.

   Technician B says a shop is responsible for hazardous waste materials from the time they become waste until the proper waste disposal is completed.

   Who is correct?

   A. A only
   B. B only
   C. Both A and B
   D. Neither A nor B

10. When working in an automotive shop, hazardous conditions must be avoided by:
    A. Storing oily rags in open containers.
    B. Storing flammable liquids such as paint in metal safety cabinet.
    C. Leaving sewer covers loose for quick access.
    D. Cleaning brake parts with an air gun.

Name ______________________________ **Date** ____________

JOB SHEET
1

DEMONSTRATE PROPER LIFTING PROCEDURES

Upon completion of this job sheet, you should be able to follow the proper procedure when lifting heavy objects.

ASE Education Foundation Correlation

This task applies to the following **RST** tasks:

1-1. Identify general shop safety rules and procedures.

1-2. Utilize safe procedures for handling of tools and equipment.

We Support ASE | Education Foundation

Tools and Materials

A heavy object with a weight that is within the weight-lifting capability of the technician.

Procedure

Task Completed

Have various members of the class demonstrate proper weight-lifting procedures by lifting an object off the shop floor and placing it on the workbench. Other class members are to observe and record any improper weight-lifting procedures.

1. If the object is to be carried, be sure your path is free from loose parts or tools. ☐
2. Position your feet close to the object, and position your back reasonably straight for proper balance. ☐
3. Your back and elbows should be kept as straight as possible. Continue to bend your knees until your hands reach the best lifting location on the object to be lifted. ☐
4. Be certain the container is in good condition. If a container falls apart during the lifting operation, parts may drop out of the container, resulting in foot injury or part damage. ☐
5. Maintain a firm grip on the object; do not attempt to change your grip while lifting is in progress. ☐
6. Straighten your legs to lift the object, keeping the object close to your body. Use leg muscles rather than back muscles. ☐
7. If you have to change direction of travel, turn your whole body. Do not twist. ☐
8. Do not bend forward to place an object on a workbench or table. Position the object on the front surface of the workbench and slide it back, watching your fingers under the object to avoid pinching them. ☐
9. If the object must be placed on the floor or a low surface, bend your legs to lower the object. Do not bend your back forward because this movement strains back muscles. ☐
10. When a heavy object must be placed on the floor, put suitable blocks under the object to prevent jamming your fingers under the object. ☐

11. List any improper weight-lifting procedures observed during the weight-lifting demonstrations.

1. ____________________
2. ____________________
3. ____________________
4. ____________________
5. ____________________
6. ____________________

Instructor's Response

____

____

____

Name ______________________________ Date ______________

JOB SHEET
2

LOCATE AND INSPECT SHOP SAFETY EQUIPMENT

Upon completion of this job sheet, you should be familiar with the location of shop safety equipment and know if this equipment is serviced properly.

ASE Education Foundation Correlation

This task applies to the following **RST** tasks:

1-7. Identify the location and the types of fire extinguishers and other fire safety equipment; demonstrate knowledge of the procedures for using fire extinguishers and other fire safety equipment.

1-8. Identify the location and use of eye wash stations.

We Support ASE Education Foundation

Procedure **Task Completed**

1. Fire extinguishers: Are the fire extinguishers tagged to indicate they have been checked or serviced recently? ☐ Yes ☐ No
   Draw a basic diagram of the shop layout in the space below, and mark the locations of the fire extinguishers and fire exits.

2. Eyewash fountain or shower: Is the eyewash fountain or shower operating properly?
   ☐ Yes ☐ No
   Mark the location of the eyewash fountain or shower on the shop layout diagram in Step 1.
3. First-aid kits: Are the first-aid kits properly stocked with supplies? ☐ Yes ☐ No
   Mark the location of the first-aid kits on the shop layout diagram in Step 1.
4. Electrical shut-off box: Mark the location of the shop electrical shut-off box on the shop layout diagram in Step 1. ☐
5. Are the trash containers equipped with proper covers? ☐ Yes ☐ No
   Mark the location of the trash containers on the shop layout diagram in Step 1.
6. Metal storage cabinet: Mark the location of metal storage cabinet(s) for combustible materials on the shop layout diagram in Step 1. ☐

Instructor's Response

____

____

____

Name ______________________________ Date ______________

JOB SHEET 3

SHOP HOUSEKEEPING INSPECTION

Upon completion of this job sheet, you should be able to apply shop housekeeping rules in your shop.

ASE Education Foundation Correlation

This task applies to the following **RST** tasks:

1-1. Identify general shop safety rules and procedures.

1-2. Utilize safe procedures for handling of tools and equipment.

We Support ASE Education Foundation

Procedure **Task Completed**

When students from another class of Automotive Technology are working in the shop, evaluate their shop housekeeping procedures using the 16 shop housekeeping procedures. List all the improper shop housekeeping procedures that you observed in the space provided at the end of the job sheet.

1. Keep aisles and walkways clear of tools, equipment, and other items. ☐
2. Be sure all sewer covers are securely in place. ☐
3. Keep floor surfaces free of oil, grease, water, and loose material. ☐
4. Proper trash containers must be conveniently located, and these containers should be emptied regularly. ☐
5. Access to fire extinguishers must be unobstructed at all times; fire extinguishers should be checked for proper charge at regular intervals. ☐
6. Tools must be kept clean and in good condition. ☐
7. When not in use, tools must be stored in their proper location. ☐
8. Oily rags and other combustibles must be placed in proper, covered metal containers. ☐
9. Rotating components on equipment and machinery must have guards. All shop equipment should have regular service and adjustment schedules. ☐
10. Benches and seats must be kept clean. ☐
11. Keep parts and materials in their proper location. ☐
12. When not in use, creepers must not be left on the shop floor. Creepers should be stored in a specific location. ☐
13. The shop should be well-lit, and all lights should be in working order. ☐
14. Frayed electrical cords on lights or equipment must be replaced. ☐
15. Walls and windows should be cleaned regularly. ☐

16. Stairs must be clean, well lit, and free of loose material. Observed improper shop housekeeping procedures:

1. ____
2. ____
3. ____
4. ____
5. ____
6. ____

Instructor's Response

____

____

____

Name ______________________ Date __________

JOB SHEET 4

PERSONNEL AND SHOP SAFETY ASSESSMENT

As a service professional, one of your first concerns should be safety. Upon completion of this job sheet you should have an increased awareness of personnel and shop safety item. As you take these personnel assessment and survey your shop answering the following questions, you will learn to evaluate your workplace and personnel safety.

ASE Education Foundation Correlation

This task applies to the following **RST** tasks:

1-1. Identify general shop safety rules and procedures.

1-10. Comply with the required use of safety glasses, ear protection, gloves, and shoes during lab/shop activities.

1-11. Identify and wear appropriate clothing for lab/shop activities.

We Support ASE Education Foundation

Procedure

Give a brief description following each step. Always wear eye protection when working or walking around a service facility.

1. Before you evaluate your work place you must first evaluate yourself. Are you dressed for work? ☐ Yes ☐ No

a. If yes, why do you believe your attire is appropriate?

____

____

b. If no, what must you correct to be properly attired?

____

____

Are your safety glasses OSHA approved? ☐ Yes ☐ No

c. Do you have side shields? ☐ Yes ☐ No

Are you wearing leather boots or shoes with oil-resistant soles? ☐ Yes ☐ No

d. Does your footwear have steel toes? ☐ Yes ☐ No

Is your shirt tucked into your pants? ☐ Yes ☐ No

If you have long hair, is it tied back or under a hat? ☐ Yes ☐ No

Next carefully inspect your shop, noting any potential hazards.

Note: A hazard is not necessarily a safety violation but in an area of which you must be aware (i.e., pothole in parking lot).

Are there safety areas marked around grinders and other machinery?
☐ Yes ☐ No

What is the air pressure in the shop set at? ______________________

Where are the tools stored in the shop? ______________________

____

____

What recommendations would you make to improve the tool storage?

____

____

Have you been instructed on proper use of shop vehicle hoist (lift)?
☐ Yes ☐ No

e. If not, ask your instructor to demonstrate vehicle lift use.

Where is the first-aid kit located? ______________________

Where is the eyewash station located? ______________________

List the location of the exits. ______________________

What is the emergency evacuation plan and where are you to assemble once you evacuate the building?

____

____

____

____

____

Where are the SDSs located? ______________________

2. Does the shop have a hazardous spill response kit? ☐ Yes ☐ No

List any additional questions your instructor would like you to answer:

____

____

____

____

____

____

____

____

____

____

____

Instructor's Response

____

____

____

CHAPTER 2
TOOLS AND SHOP PROCEDURES

Upon completion and review of this chapter, you should be able to:

- Diagnose steering and suspension problems using a basic diagnostic procedure.
- Raise a vehicle with a floor jack and lower the vehicle so it is supported on safety stands.
- Raise a vehicle with a lift.
- Test power steering pump pressure with an appropriate pressure gauge and valve assembly.
- Measure ball joint wear with a dial indicator designed for this purpose.
- Remove and replace a coil spring on a strut using a coil spring compressor tool.
- Use a tire changer to dismount and mount a tire.
- Use a scan tool to diagnose a computer-controlled suspension system.
- Balance tire-and-wheel assemblies with an electronic wheel balancer.
- Perform a four wheel alignment with a computer wheel aligner.
- Fulfill employee obligations when working in the shop.
- Accept job responsibility for each job completed in the shop.
- Describe the ASE automotive technician testing and certification process including the eight areas of certification.

Terms To Know

ASE blue seal of excellence
Axle pullers
Ball joint removal and installation tools
Bearing pullers
Brake pedal jack
Coil spring compressor tool
Computer four wheel aligner
Control arm bushing tools
Diagnostic procedure charts
Dial indicators
Electronic wheel balancers
Floor jack
Hydraulic press
International system (IS)
Machinist's rule
Magnetic wheel alignment gauge
Pitman arm puller
Plumb bob
Power steering pressure gauge
Rim clamp
Safety stands
Scan tools
Seal drivers
Steering wheel locking tool
Stethoscope
Tie rod sleeve adjusting tool
Tie rod end and ball joint puller
Tire tread depth gauge
Toe gauge
Track gauges
Tram gauge
Turning radius gauge turntables
U.S. customary (USC) system
Vacuum hand pump
Vehicle lift

USING SUSPENSION AND STEERING EQUIPMENT

Learn to use equipment properly the first time. When improper procedures are learned and become a habit, it is more difficult to break these wrong habits. Using service and test equipment will help you to work more safely and efficiently, and in the long term, this should improve your income. We will discuss the proper use of some of the most common service and test equipment used by suspension and steering technicians.

MEASURING SYSTEMS

Two systems of weights and measures are commonly used in the United States. One system is the **U.S. customary (USC) system**, which is commonly referred to as the English system. Well-known measurements for length in the USC system are the inch, foot, yard, and mile. In this system, the quart and gallon are measurements for volume; ounce, pound, and ton are measurements for weight. A second system of weights and measures is called the **international system (IS)**, also known as the metric system.

In the USC system, the basic linear measurement is the yard; in the metric system, it is the meter. Each unit of measurement in the metric system is related to the other metric units by a factor of 10. Thus, every metric unit can be multiplied or divided by 10 (or 100 or 1000) to obtain larger units (multiples) or smaller units (submultiples). For example, the meter maybe divided by 100 to obtain centimeters (1/100 meter) or by 1000 to obtain millimeters (1/1000 meter).

The U.S. government passed the Metric Conversion Act in 1975 in an attempt to move American industry and the general public to accept and adopt the metric system. The automotive industry has adopted the metric system, and in recent years most bolts, nuts, and fittings on vehicles have been changed to metric. During the early 1980s, some vehicles had a mix of English and metric bolts. Imported vehicles have used the metric system for many years. Although the automotive industry has changed to the metric system, the general public in the United States has been slow to convert from the USC system to the metric system. One of the factors involved in this change is cost. What would it cost to change every highway distance and speed sign in the United States to read kilometers? Probably hundreds of millions, or even billions, of dollars.

Service technicians must be able to work with both the USC and the metric system. One meter (m) in the metric system is equal to 39.37 inches (in.) in the USC system. A metric tape measure may be graduated in millimeters, and 10 millimeters = 1 centimeter (**Figure 2-1**).

Some common equivalents between the metric and USC systems are:

1 meter (m) = 39.378 inches
1 centimeter (cm) = 0.3937 inch
1 millimeter (mm) = 0.03937 inch
1 inch (in.) = 2.54 centimeters
1 inch = 25.4 millimeters

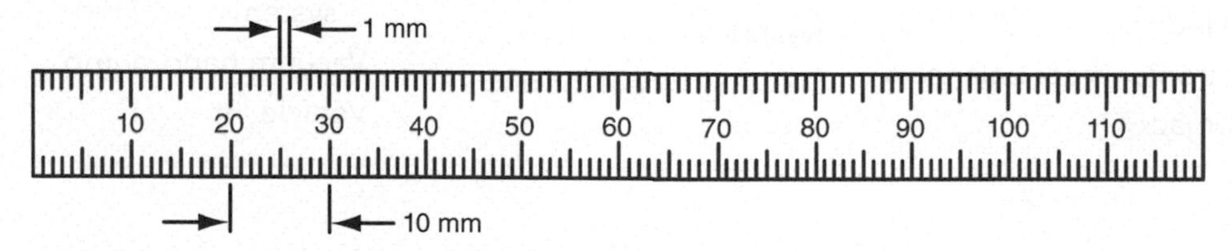

Figure 2-1 Metric tape graduated in millimeters.

In the USC system, phrases such as 1/8 of an inch are used for measurements. The metric system uses a set of prefixes. For example, in the word kilometer the prefix kilo means 1000, indicating that there are 1000 meters in a kilometer. Common prefixes in the metric system include:

Name	Symbol	Meaning
mega	M	one million
kilo	k	one thousand
hecto	h	one hundred
deca	da	ten
deci	d	one tenth of
centi	c	one hundredth of
milli	m	one thousandth of
micro	F	one millionth of

Measurement of Mass

In the metric system, mass is measured in grams, kilograms, or tons: 1000 grams (g) = kilogram (kg). In the USC system, mass is measured in ounces, pounds, or tons. When converting pounds to kilograms, 1 pound = 0.453 kilograms.

Measurement of Length

In the metric system, length is measured in millimeters, centimeters, meters, or kilometers: 10 millimeters (mm) = 1 centimeter (cm). In the USC system, length is measured in inches, feet, yards, or miles.

When distance conversions are made between the two systems, some of the following conversion factors are used:

1 inch = 25.4 millimeters
1 foot = 30.48 centimeters
1 yard = 0.91 meters
1 mile = 1.60 kilometers

Measurement of Volume

In the metric system, volume is measured in milliliters, cubic centimeters, and liters: 1 cubic centimeter = 1 milliliter. If a cube has a length, depth, and height of 10 centimeters (cm), the volume of the cube is 10 cm × 10 cm × 10 cm = 1000 cm^3 = 1 liter. When volume conversions are made between the two systems, 1 cubic inch = 16.38 cubic centimeters. If an engine has a displacement of 350 cubic inches, 350 × 16.38 = 5733 cubic centimeters, and 5733 ÷ 1000 = 5.7 liters.

BASIC DIAGNOSTIC PROCEDURE

One of the most important parts of a technician's job is diagnosing automotive problems. Each year, more vehicle systems are controlled by increasingly sophisticated electronic systems, and this makes accurate diagnosis even more important.

The following basic diagnostic procedure may be used to diagnose various automotive problems:

1. Be absolutely sure the problem is identified. Obtain all the information you can from the customer; for example, politely ask the customer to describe the exact symptoms of the problem. Ask the customer at what vehicle speed, engine temperature, and atmospheric temperature the problem occurs. From your discussion, find out when, where, and how the problem occurs.
2. Verify the customer complaint. If necessary, road test the vehicle under the same conditions as the customer described to verify the problem.
3. Think of the possible causes of the problem based on your own experience in diagnosing similar problems on other vehicles.
4. Consult original equipment manufacturer (OEM) or generic technical service bulletin (TSB) information regarding service procedures, parts replacement, or computer reprogramming that are designed to correct the problem.
5. Perform the necessary diagnostic tests to locate the exact cause of the complaint. Always begin the diagnostic tests with the quickest, easiest test, and if the problem is not located, proceed with the more complicated, time-consuming tests. During your diagnosis, watch for other problems with the vehicle that may give trouble in the near future. If there are any other potential problems such as worn belts and radiator hoses, advise the customer about them and attempt to obtain the customer's approval to repair these problems. If the customer does not approve the repair of these potential problems, describe them on the work order. If a car is repaired based on a customer's complaint, and two weeks later the customer brings the vehicle back with another serious complaint, the customer may assume that the repairs were not completed thoroughly.
6. After the cause of the complaint is definitely located, advise the customer about the extent and cost of the repairs. After the customer approves the necessary repairs, perform the appropriate service work on the vehicle.
7. Road test the vehicle if necessary to be sure the customer's complaint is eliminated.

Caution

Do not hit the seal case with a hammer. This action will damage the seal.

SUSPENSION AND STEERING TOOLS

CUSTOMER CARE Some automotive service centers have a policy of performing some minor service as an indication of their appreciation to the customer. This service may include cleaning all the windows and/or vacuuming the floors before the car is returned to the customer.

Although this service involves more labor costs for the shop, it may actually improve profits over a period of time. When customers find their windows cleaned and/or the floors vacuumed, it impresses them with the quality of work you do and the fact you care about their vehicle. They will likely return for service, and tell their friends about the quality of service your shop performs.

Caution

Always be sure the seal goes into the housing squarely and evenly. If the seal does not go squarely into the housing, the outer seal case may be distorted, and this condition may cause an oil leak around the seal housing.

Seal Drivers

WARNING **When using any suspension and steering tools, the vehicle manufacturer's recommended procedure in the service manual must be followed. Improper use of tools may lead to personal injury.**

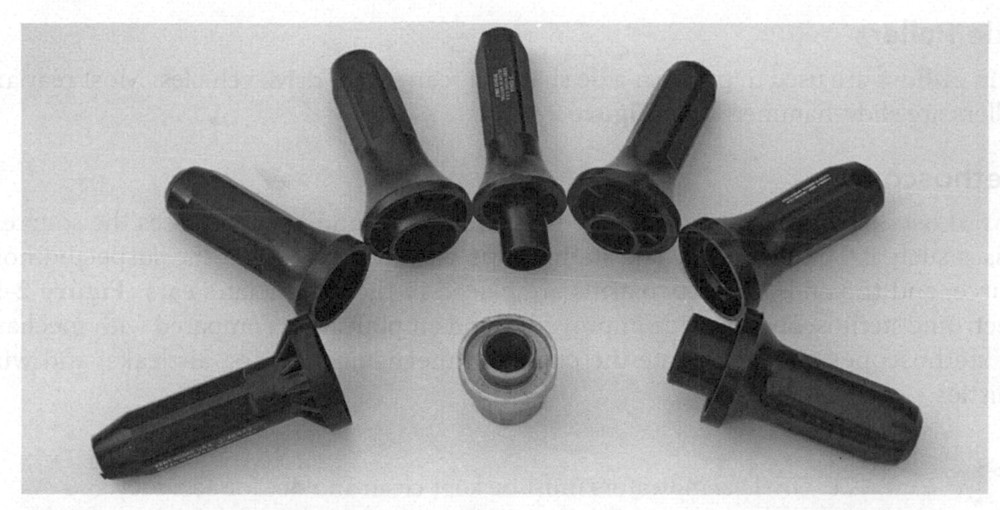

Figure 2-2 Seal drivers.

Seal drivers are designed to fit squarely against the seal case and inside the seal lip. A soft hammer is used to tap the seal driver and drive the seal straight into the housing. Some tool manufacturers market a seal driver kit with drivers to fit many common seals (**Figure 2-2**).

Seal drivers are available in various diameters to fit squarely against the outside edge of different-size seals. The seal driver handle is tapped with a soft hammer to install the seal.

Bearing Pullers

A variety of **bearing pullers** are available to pull different sizes of bearings in various locations (**Figure 2-3**). Some bearing pullers are slide-hammer-type, whereas others are screw-type.

Bearing pullers are designed to fit over the outer diameter of the bearing or through the center opening in the bearing, pulling the bearing from its mounting shaft.

SERVICE TIP When installing a seal, the garter spring must face toward the flow of lubricant.

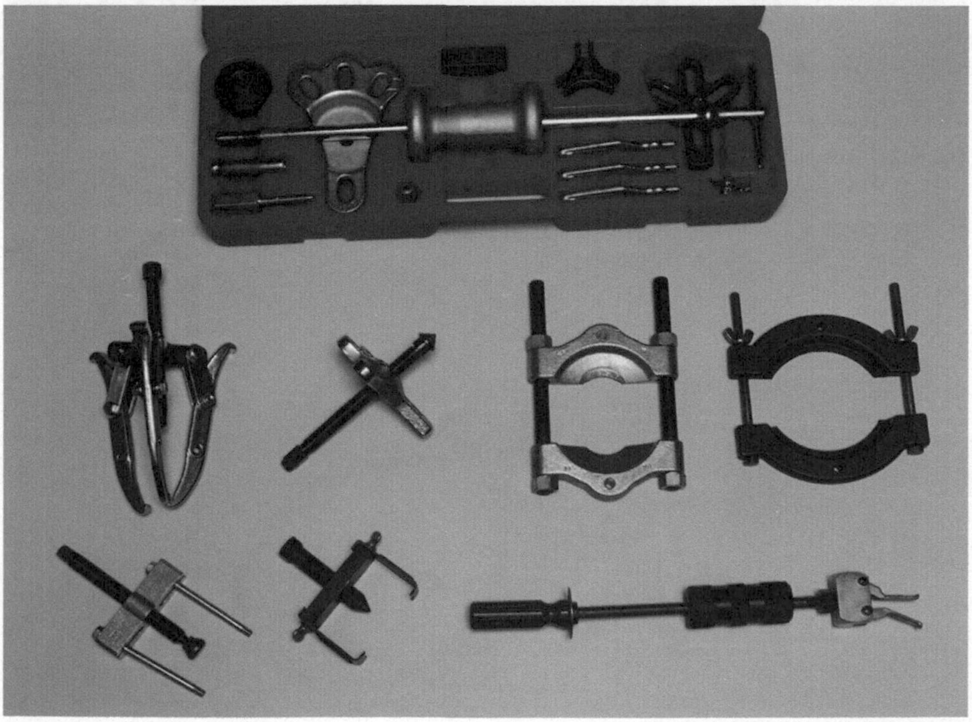

Figure 2-3 Bearing pullers.

Axle pullers are usually slide-hammer-type. These pullers attach to the axle studs to remove the axle.

A **stethoscope** is placed in the technician's ears, and the pickup is placed on or near the suspected noise source to diagnose noise problems.

Axle Pullers

Axle pullers are used to pull rear axle shafts in rear-wheel-drive vehicles. Most rear axle pullers are slide-hammer-type (**Figure 2-4**).

Stethoscope

A **stethoscope** is a mechanical device that amplifies sound and diagnoses the source of noises such as bearing noise. The stethoscope pickup is placed on the suspected noise source, and the ends of the two arms are placed in the technician's ears (**Figure 2-5**). Electronic stethoscopes provide improved sound amplification compared with mechanical stethoscopes and can locate the causes of mechanical noises, air leaks, and wind whistles.

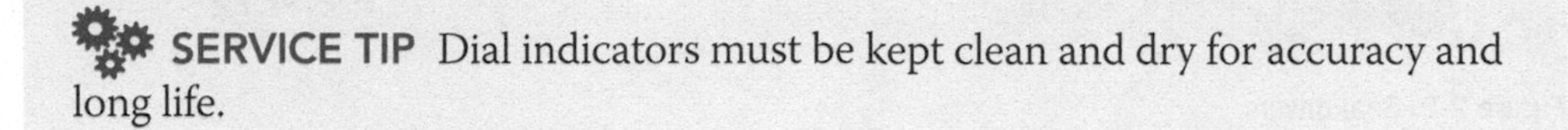

SERVICE TIP Dial indicators must be kept clean and dry for accuracy and long life.

Front Bearing Hub Tool

Front bearing hub tools are designed to remove and install front wheel bearings on front-wheel-drive cars (**Figure 2-6**). These bearing hub tools are usually designed for a specific make of vehicle.

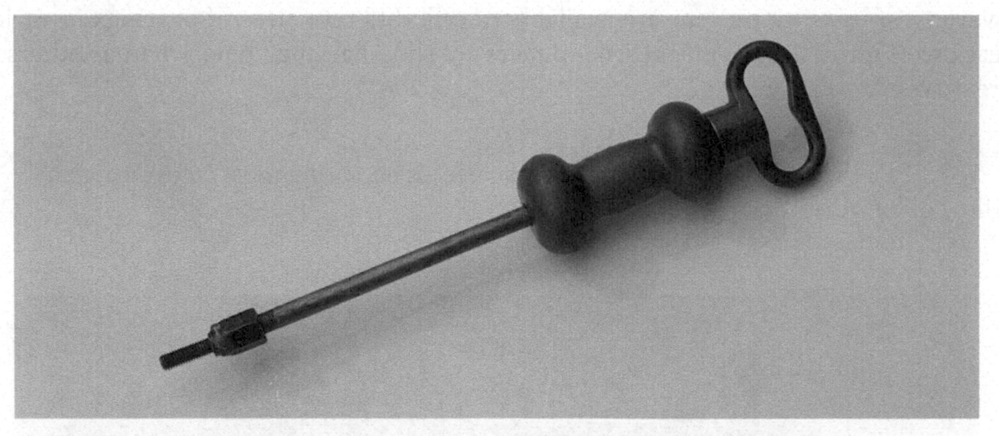

Figure 2-4 Axle puller.

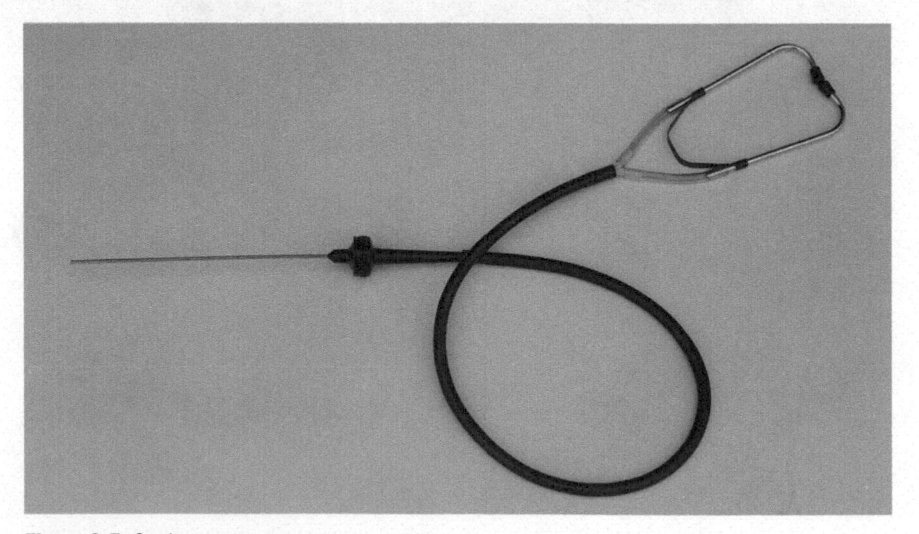

Figure 2-5 Stethoscope.

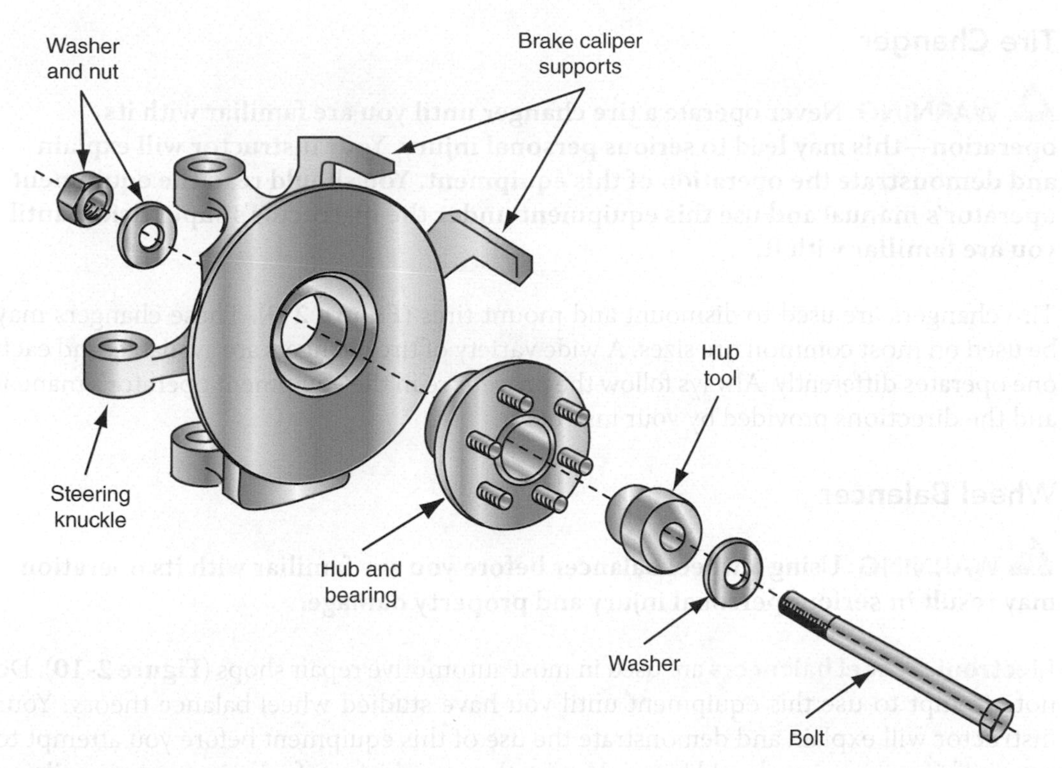

Figure 2-6 Front bearing hub tool.

Dial Indicator

Dial indicators are used for measuring in many different locations. In the suspension and steering area, dial indicators are used for measurements, such as tire runout and ball joint movement (**Figure 2-7**). Dial indicators have many different attaching devices to connect the indicator to the component to be measured.

Tire Tread Depth Gauge

A **tire tread depth gauge** measures tire tread depth. This measurement should be taken at three or four locations around the tire's circumference. The lowest reading is considered to be the tread depth (**Figure 2-8**). Do not place the gauge tip on a wear indicator bar. This gauge is essential when making tire warranty adjustments.

A tire changer is operated pneumatically to dismount and mount tires from the rims.

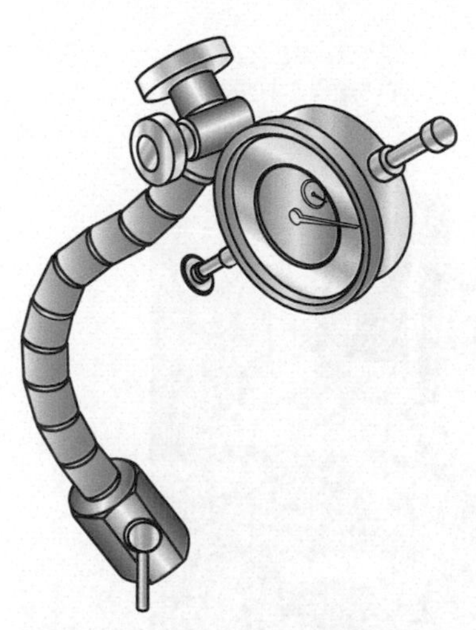

Figure 2-7 Dial indicator designed for ball joint measurement.

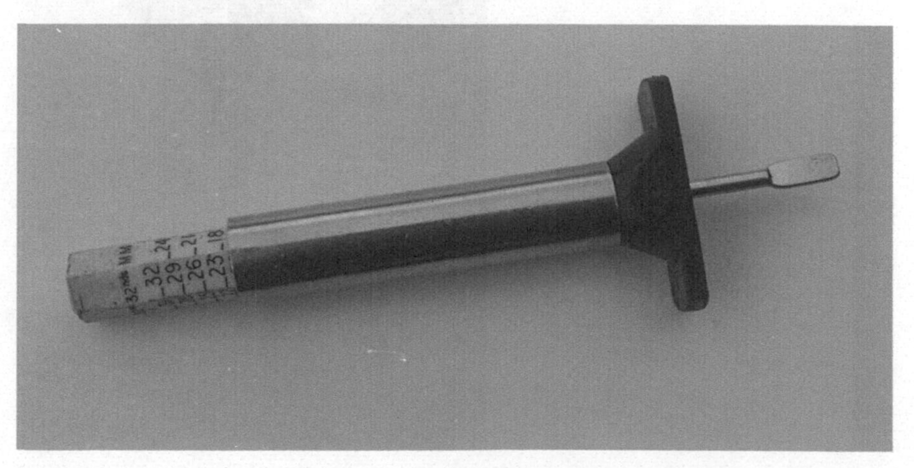

Figure 2-8 Tire tread depth gauge.

Tire Changer

WARNING Never operate a tire changer until you are familiar with its operation—this may lead to serious personal injury. Your instructor will explain and demonstrate the operation of this equipment. You should read the equipment operator's manual and use this equipment under the instructor's supervision until you are familiar with it.

Tire changers are used to dismount and mount tires (**Figure 2-9**). These changers may be used on most common tire sizes. A wide variety of tire changers are available, and each one operates differently. Always follow the procedure in the equipment operator's manual and the directions provided by your instructor.

Wheel Balancer

WARNING Using a wheel balancer before you are familiar with its operation may result in serious personal injury and property damage.

Electronic wheel balancers indicate the required weight location on the rim to obtain proper static and dynamic wheel balance.

Electronic wheel balancers are used in most automotive repair shops (**Figure 2-10**). Do not attempt to use this equipment until you have studied wheel balance theory. Your instructor will explain and demonstrate the use of this equipment before you attempt to use it. This equipment should be used under the supervision of your instructor until you are familiar with it.

Machinist's Rule

A machinist's rule performs many accurate measurements in the shop. Most machinist's rules are graduated in inches and millimeters (**Figure 2-11**).

Air Chisel or Hammer

An air chisel is an electrically operated tool to which various chisels, cutters, and punches may be attached to complete cutting and riveting jobs. The air hammer operates the

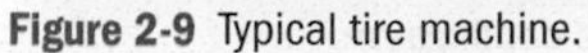
Figure 2-9 Typical tire machine.

Figure 2-10 Typical road force balancer.

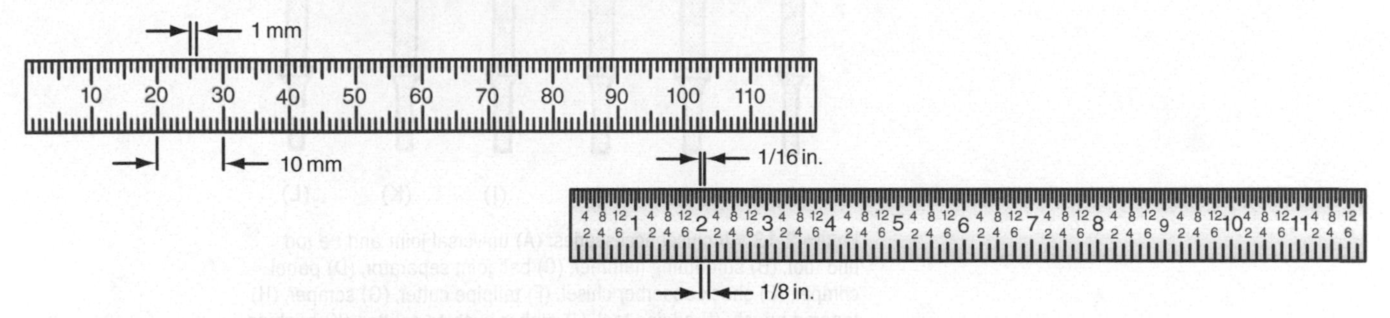

Figure 2-11 Graduations on a typical machinist's rule.

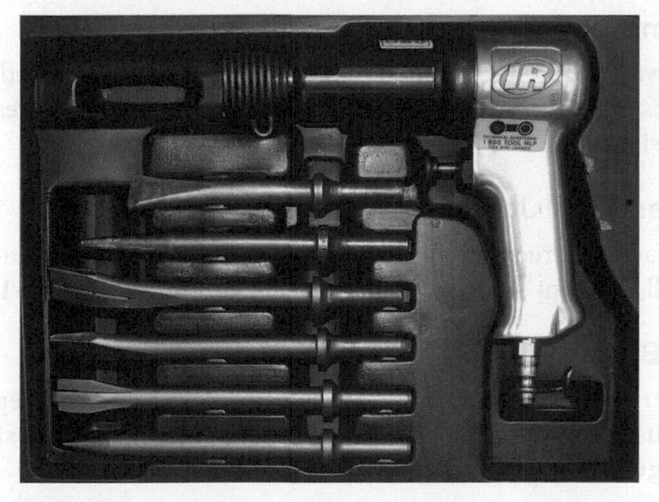

Figure 2-12 Air chisel.

attached tool with a very fast back-and-forth action (**Figure 2-12**). Different tools that may be attached to the air chisel are illustrated in **Figure 2-13**.

WARNING Always be sure the tool being used is properly attached to the air chisel. Improper tool attachment may result in serious personal injury.

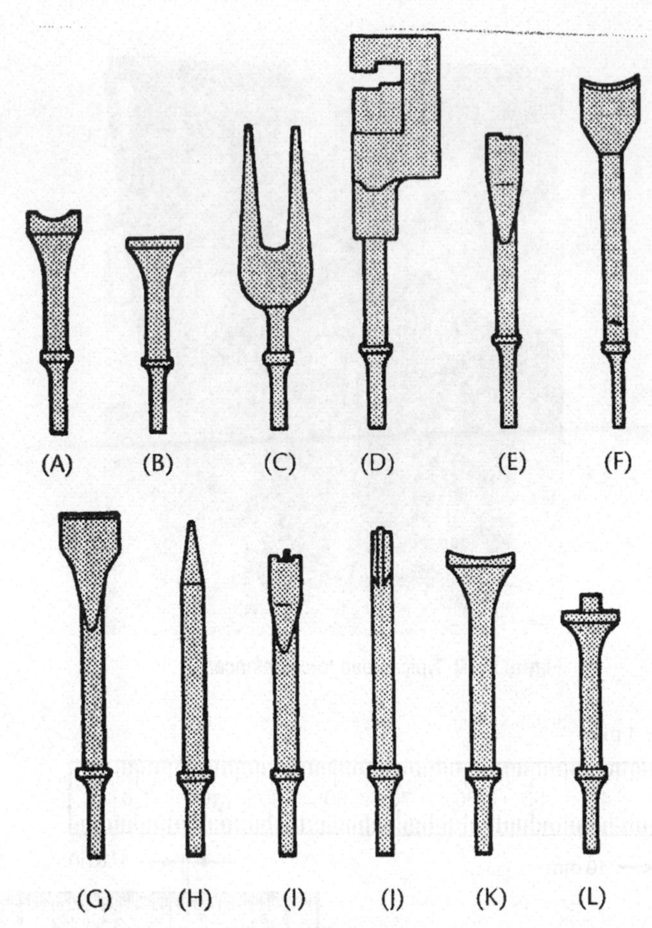

Figure 2-13 Air chisel accessories: (A) universal joint and tie rod end tool, (B) smoothing hammer, (C) ball joint separator, (D) panel crimper, (E) shock absorber chisel, (F) tailpipe cutter, (G) scraper, (H) tapered punch, (I) edging tool, (J) rubber bushing splitter, (K) bushing remover, and (L) bushing installer.

Ball Joint Removal and Installation Tools

Ball joint removal and installation tools remove and install pressed-in ball joints on front suspension systems (**Figure 2-14**). The size of the removal and pressing tool adapter must match the size of the ball joint.

Tie Rod End and Ball Joint Puller

Some car manufacturers recommend a **tie rod end and ball joint puller** to remove tie rod ends and pull ball joint studs from the steering knuckle (**Figure 2-15**).

Control Arm Bushing Tools

A variety of **control arm bushing tools** are available to remove and replace control arm bushings. The bushing removal and installation tool must match the size of the control arm bushing (**Figure 2-16**).

Caution

The vehicle manufacturer's and equipment manufacturer's recommended procedures must be followed for each type of spring compressor tool.

Coil Spring Compressor Tool

WARNING There is a tremendous amount of energy in a compressed coil spring. Never disconnect any suspension component that will suddenly release this tension—this may result in serious personal injury and vehicle or property damage.

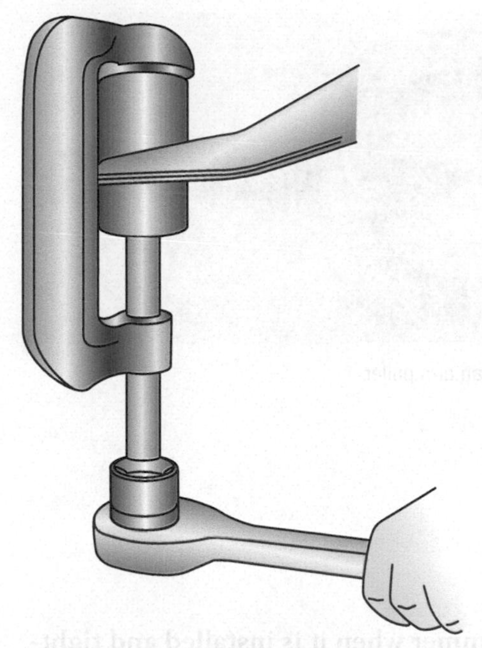

Figure 2-14 Ball joint removal and installation tool.

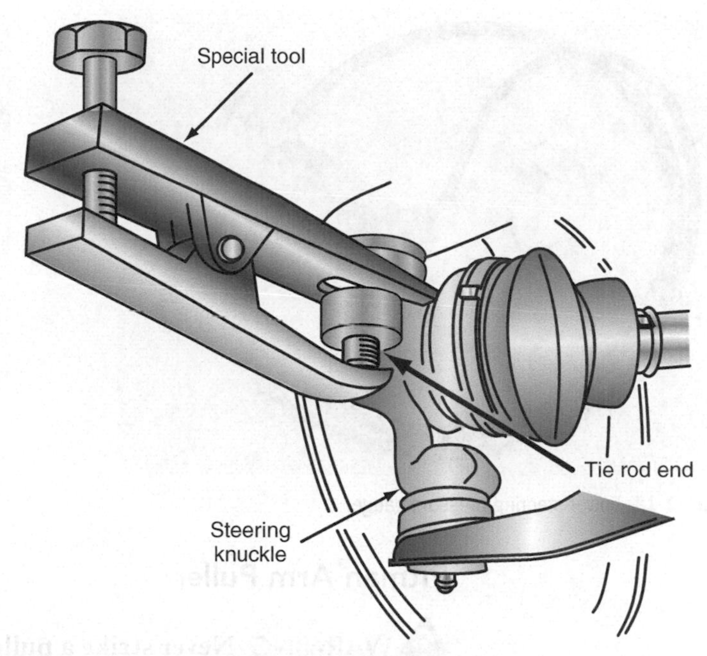

Figure 2-15 Tie rod end and ball joint puller.

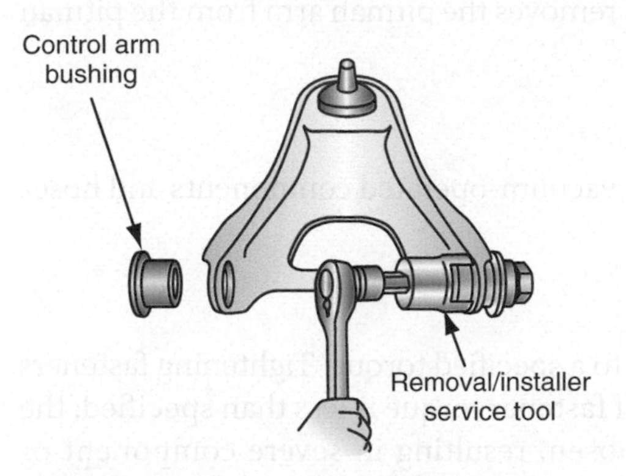

Figure 2-16 Control arm bushing removal and replacement tools.

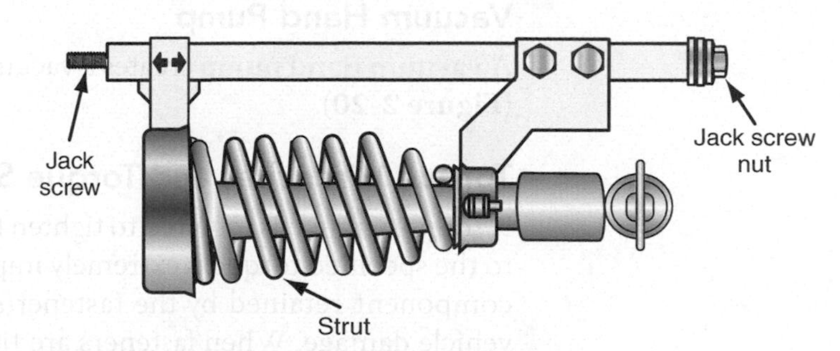

Figure 2-17 MacPherson strut coil spring compressor tool.

Many types of **coil spring compressor tools** are available to the automotive service industry (**Figure 2-17**). These tools compress the coil spring and hold it in the compressed position while removing the strut from the coil spring or performing other suspension work. Each type of front suspension system requires a different type of spring compressor tool. The vehicle manufacturer's and equipment manufacturer's recommended procedures must be followed.

Power Steering Pressure Gauge

WARNING The power steering pump delivers extremely high pressure during the pump pressure test. Always follow the recommended test procedure in the vehicle manufacturer's service manual to avoid personal injury during this test.

A **power steering pressure gauge** is used to test the power steering pump pressure (**Figure 2-18**). Because the power steering pump delivers extremely high pressure during this test, the recommended procedure in the vehicle manufacturer's service manual must be followed.

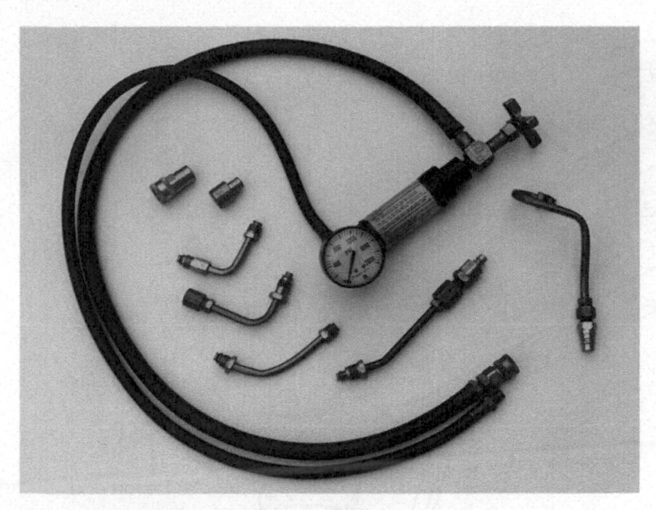

Figure 2-18 Power steering pressure gauge.

Figure 2-19 Pitman arm puller.

Pitman Arm Puller

WARNING Never strike a puller with a hammer when it is installed and tightened, as this may result in personal injury.

A **pitman arm puller** is a heavy-duty puller that removes the pitman arm from the pitman shaft (**Figure 2-19**).

Vacuum Hand Pump

A **vacuum hand pump** creates a vacuum to test vacuum-operated components and hoses (**Figure 2-20**).

Torque Wrenches and Torque Sticks

A torque wrench is required to tighten fasteners to a specified torque. Tightening fasteners to the specified torque is extremely important. If fastener torque is less than specified, the component retained by the fastener(s) may loosen, resulting in severe component or vehicle damage. When fasteners are tightened to a torque above the specified value, the component may become warped, resulting in improper component operation. Torque wrenches can be beam, click, or dial-type (**Figure 2-21**). A beam-type torque wrench bends as torque is applied, and a beam on the wrench points to the torque applied on a scale attached to the wrench handle. On a click-type torque wrench, the handle is rotated to set the specified torque on a scale adjacent to the handle. When the fastener is tightened to the torque setting, the torque wrench provides an audible click. On a dial-type torque wrench, the torque applied to the fastener is indicated on the dial. Digital-type torque wrenches that indicate the torque reading on a digital display are also available.

SERVICE TIP Using a torque stick allows the use of an impact wrench when tightening fasteners, which speeds up the service operation. However, the use of torque sticks may not be as accurate as using a torque wrench. It is a good practice to double-check the accuracy of a torque stick by checking the fastener torque with a torque wrench after it is tightened with a torque stick.

Some tool manufactures provide torque sticks for tightening fasteners to the specified torque with an impact wrench. One end of the torque stick has a ½ in. drive opening, and a socket is attached to the opposite end. A spring steel, heat-treated shaft is connected

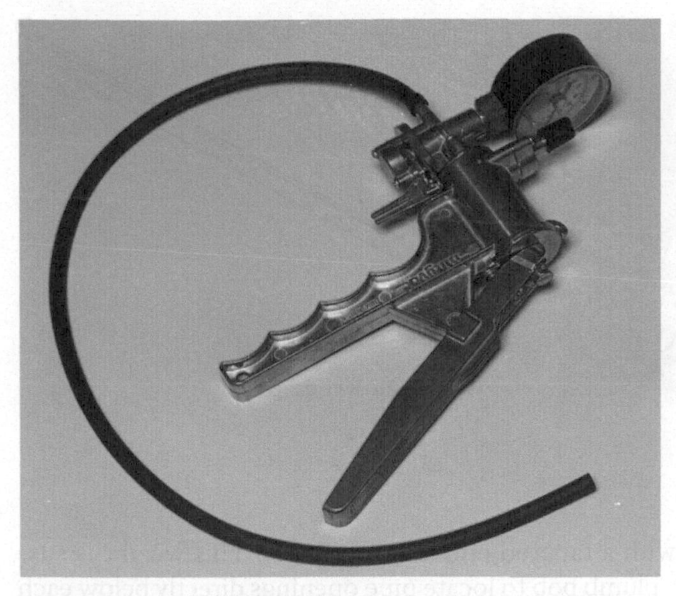

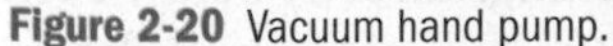

Figure 2-20 Vacuum hand pump.

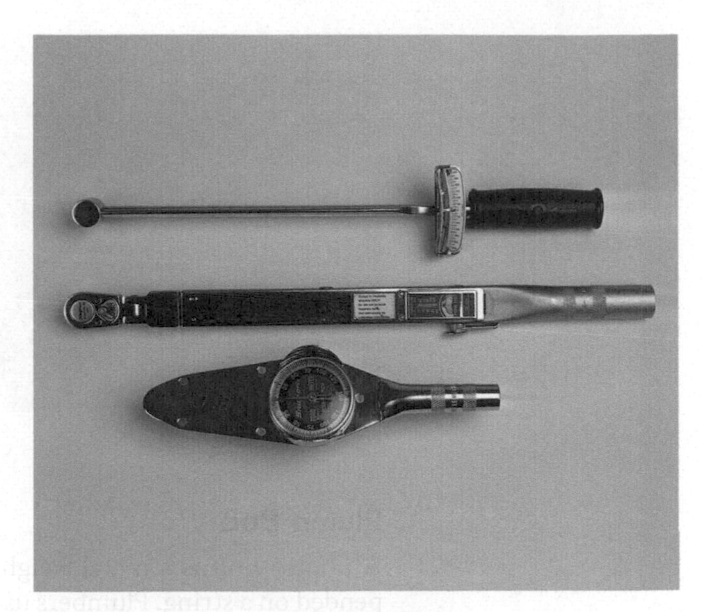

Figure 2-21 Various types of torque wrenches.

between the ½ in. drive opening and the socket end of the torque stick. Torque sticks are calibrated to twist at a specific torque with each blow of an impact wrench, and thus prevent further tightening of the fastener. Torque sticks are available in various USC and metric sizes and are usually sold in color-coded sets. The torque stick color indicates the specified torque at which they twist. For example, a yellow torque stick twists at 65 foot-pounds (ft.-lb). Some torque sticks are a ½ in. extension that fits between the impact wrench drive and a thin-wall impact socket (**Figure 2-22**).

Turning Radius Gauge

Turning radius gauge turntables are placed under the front wheels during a wheel alignment. The top plate in the turning radius gauge rotates on the bottom plate to allow the front wheels to be turned during a wheel alignment. A degree scale and a pointer on the gauge indicate the number of degrees the front wheels are turned (**Figure 2-23**). If the car has four wheel steering (4WS), the turning radius gauges are also placed under the rear wheels during a wheel alignment.

Figure 2-22 Torque stick.

Figure 2-23 Turning radius gauge.

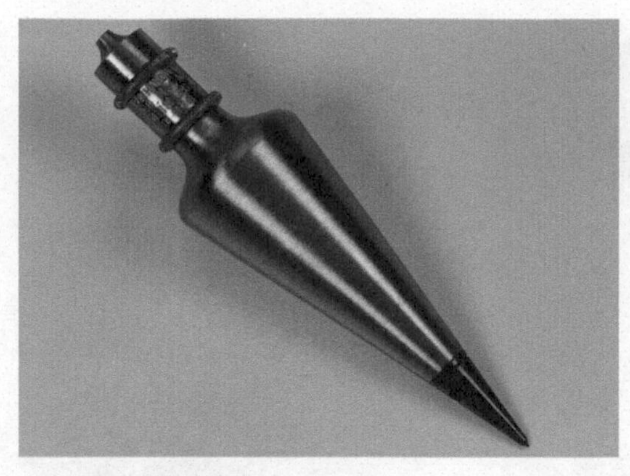
Figure 2-24 Plumb bob.

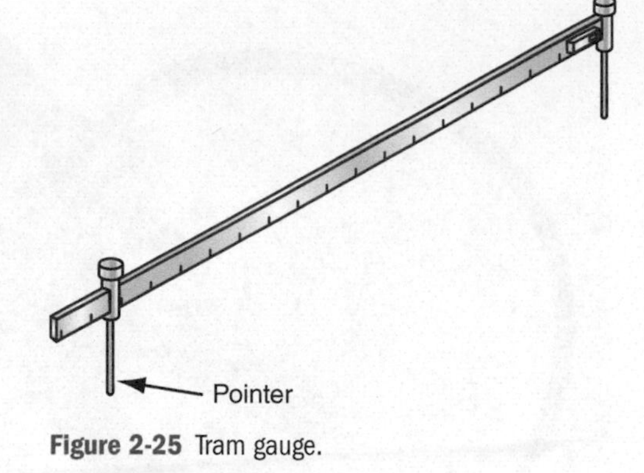

Figure 2-25 Tram gauge.

Plumb Bob

A **plumb bob** is a metal weight with a tapered end (**Figure 2-24**). This weight is suspended on a string. Plumbers use a plumb bob to locate pipe openings directly below each other at the top and bottom of partitions. Some vehicle manufacturers recommend checking vehicle frame measurements with a plumb bob.

Tram Gauge

A **tram gauge** is a long, straight graduated bar with an adjustable pointer at each end (**Figure 2-25**). The tram gauge performs frame and body measurements.

SERVICE TIP Magnetic wheel alignment gauge mounting surfaces must be clean with no metal burrs.

Magnetic Wheel Alignment Gauge

Each **magnetic wheel alignment gauge** contains a strong magnet that holds the gauge securely on the front wheel hubs. The magnetic wheel alignment gauge measures some of the front suspension alignment angles (**Figure 2-26**).

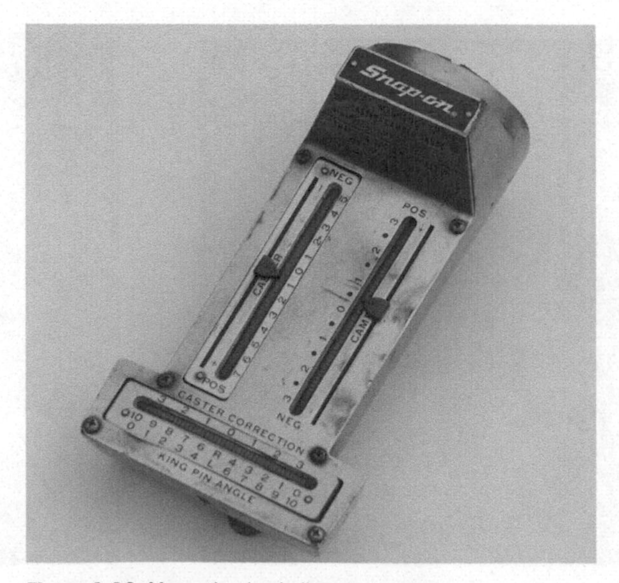

Figure 2-26 Magnetic wheel alignment gauge.

Rim Clamps

When the wheel hub is inaccessible to the magnetic alignment gauge, an adjustable **rim clamp** may be attached to each front wheel. The magnetic gauges may be attached to the rim clamp (**Figure 2-27**). Rim clamps are also used on computer wheel aligners.

Brake Pedal Jack

A **brake pedal jack** must be installed between the front seat and the brake pedal to apply the brakes while checking some front wheel alignment angles (**Figure 2-28**).

Tie Rod Sleeve Adjusting Tool

A **tie rod sleeve adjusting tool** rotates the tie rod sleeves and performs some front wheel adjustments (**Figure 2-29**).

Caution

Do not use anything except a tie rod adjusting tool to adjust the tie rod sleeves. Tools, such as a pipe wrench, will damage the sleeves.

Steering Wheel Locking Tool

A **steering wheel locking tool** locks the steering wheel while performing some front suspension service (**Figure 2-30**).

Toe Gauge

A **toe gauge** is a long, straight, graduated bar that measures front wheel toe (**Figure 2-31**).

Track Gauge

Some **track gauges** use a fiber-optic alignment system to measure front wheel toe and to determine if the rear wheels are tracking directly behind the front wheels. The front and rear fiber-optic gauges may be connected to the wheel hubs or to rim clamps attached to the wheel rims (**Figure 2-32**). A remote light source in the main control is sent through

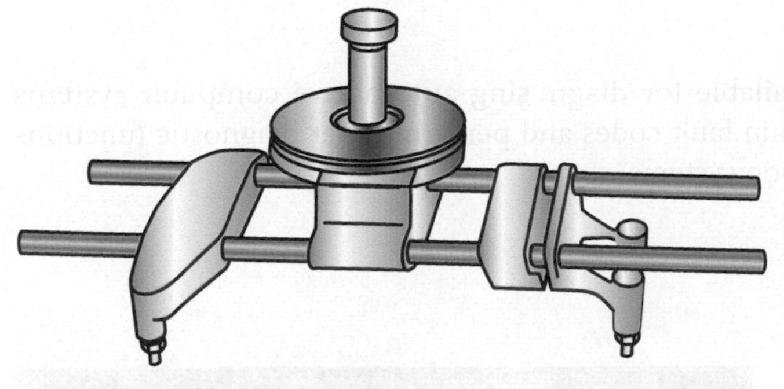

Figure 2-27 Rim clamps.

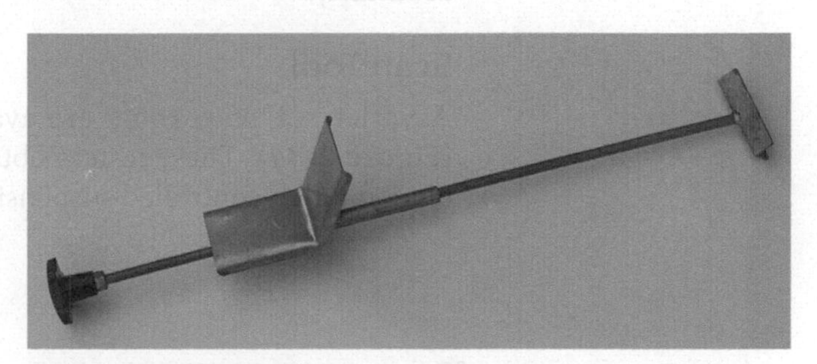

Figure 2-28 Brake pedal jack.

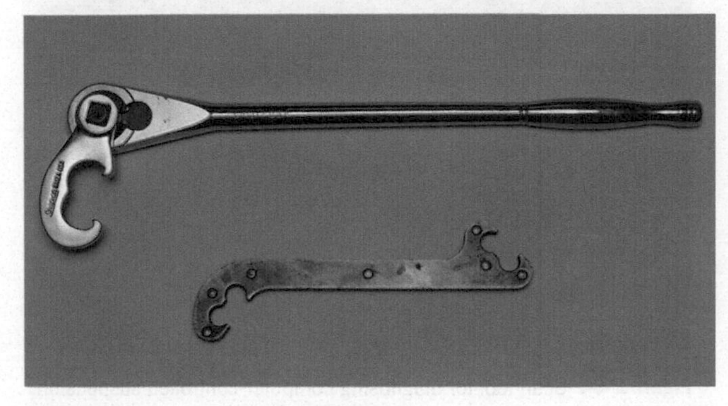

Figure 2-29 Tie rod sleeve adjusting tool.

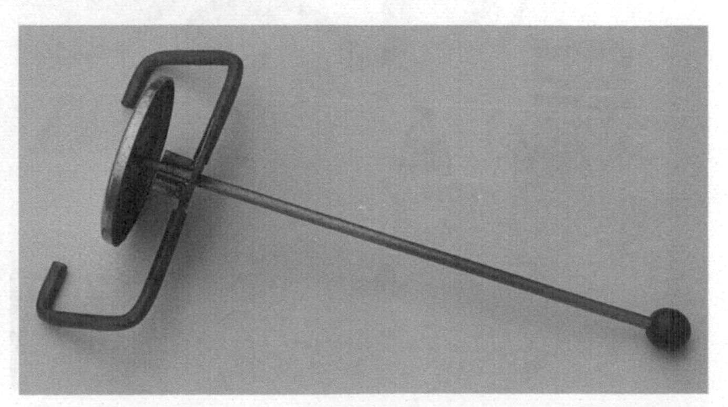

Figure 2-30 Steering wheel locking tool.

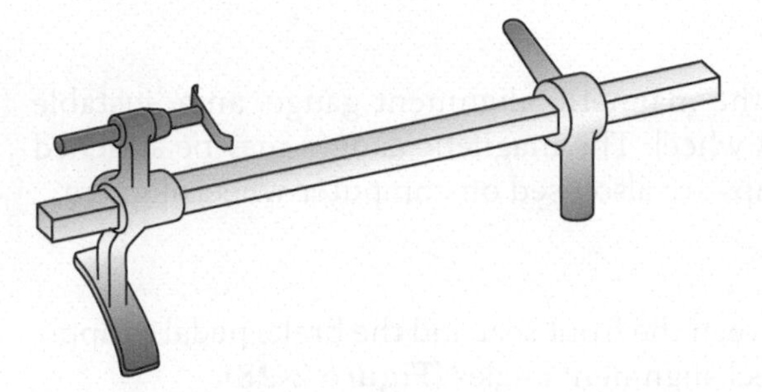

Figure 2-31 Toe gauge.

Figure 2-32 Fiber-optic track and toe gauge.

fiber-optic cables to the wheel gauges. A strong light beam between the front and rear wheel units informs the technician if the rear wheel tracking is correct.

Computer Four Wheel Aligner

Many automotive shops are equipped with a **computer four wheel aligner** (**Figure 2-33**). These wheel aligners perform all front and rear wheel alignment angles quickly and accurately.

Scan Tool

A variety of **scan tools** are available for diagnosing automotive computer systems (**Figure 2-34**). These testers obtain fault codes and perform other diagnostic functions on computer-controlled suspension systems.

Figure 2-33 Computer wheel aligner.

Figure 2-34 Scan tool for diagnosing computer-controlled suspension systems.

Bench Grinder

WARNING Always wear a face shield when using a bench grinder. Failure to observe this precaution may cause personal injury.

WARNING When grinding small components on a grinding wheel, wire brush wheel, or buffing wheel, always hold these components with a pair of vise grips to avoid injury to fingers and hands.

WARNING Grinding and buffing wheels on bench grinders must be mounted on the grinder according to the manufacturer's instructions. Grinding and buffing wheels must be retained with the manufacturer's specified washers and nuts, and the retaining nuts must be tightened to the specified torque. Personal injury may occur if grinding and buffing wheels are not properly attached to the bench grinder.

Bench grinders usually have a grinding wheel and a wire wheel brush driven by an electric motor (**Figure 2-35**). The grinding wheel may be replaced with a grinding disc containing several layers of synthetic material (**Figure 2-36**). A buffing wheel may be used in place of the wire wheel brush. The grinding wheel may be used for various grinding jobs and deburring. A buffing wheel is most commonly used for polishing. Most bench grinders have grinding wheels and wire brush wheels that are 6 to 10 in. (15.24 to 25.4 cm) in diameter. Bench grinders must be securely bolted to the workbench.

Figure 2-35 Bench grinder.

Figure 2-36 Bench grinder accessories.

HYDRAULIC PRESSING AND LIFTING EQUIPMENT

A **hydraulic press** uses a hydraulic cylinder and ram to remove and install precision-fit components from their mounting location.

Hydraulic Press

WARNING When operating a hydraulic press, always be sure that the components being pressed are supported properly on the press bed with steel supports. Improperly supported components may slip, resulting in personal injury.

WARNING When using a hydraulic press, never operate the pump handle if the pressure gauge exceeds the maximum pressure rating of the press. If this maximum pressure is exceeded, some part of the press may suddenly break, causing severe personal injury.

WARNING Be sure the safety cage is in place around the press to prevent personal injury.

When two components have a tight precision fit between them, a hydraulic press is used to either separate these components or press them together. The hydraulic press rests on the shop floor, and an adjustable steel-beam bed is retained to the lower press frame with heavy steel pins. A hydraulic cylinder and ram is mounted on the top part of the press with the ram facing downward toward the press bed (**Figure 2-37**). The component being pressed is placed on the press bed with appropriate steel supports. A hand-operated hydraulic pump is mounted on the side of the press. When the handle is pumped, hydraulic fluid is forced into the cylinder, and the ram is extended against the component on the press bed to complete the pressing operation. A pressure gauge on the press indicates the pressure applied from the hand pump to the cylinder. The press frame is designed for a certain maximum pressure, and this pressure must not be exceeded during hand pump operation.

Floor Jack

WARNING The maximum lifting capacity of the floor jack is usually written on the jack decal. Never lift a vehicle that exceeds the jack lifting capacity. This action may cause the jack to break or collapse, resulting in vehicle damage or personal injury.

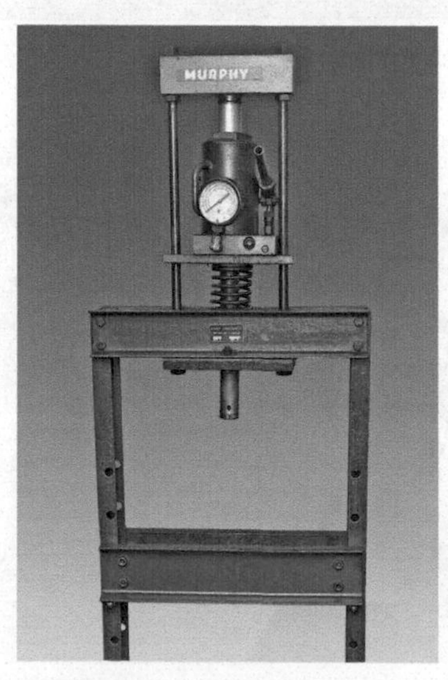

Figure 2-37 Hydraulic press.

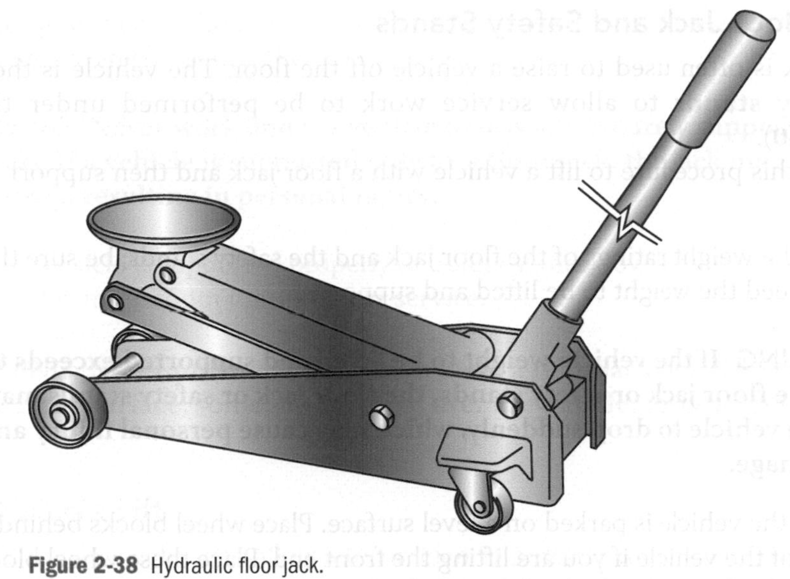

Figure 2-38 Hydraulic floor jack.

A **floor jack** is a portable unit mounted on wheels. The lifting pad on the jack is placed under the chassis of the vehicle, and the jack handle is operated with a pumping action (**Figure 2-38**), forcing fluid into a hydraulic cylinder in the jack. Then this cylinder extends to force the jack lift pad upward and lift the vehicle. Always be sure that the lift pad is positioned securely under one of the car manufacturer's recommended lifting points. To release the hydraulic pressure and lower the vehicle, the handle or release lever must be turned slowly. Do not leave the jack handle where someone can trip over it.

A **floor jack** uses hydraulic pressure supplied to a hydraulic cylinder, ram, and lift pad to lift one end or one corner of a vehicle.

Vehicle Lift (Hoist)

A **vehicle lift** raises a vehicle so the technician can work under it. The lift arms must be placed under the car manufacturer's recommended lifting points prior to raising a vehicle. Twin posts are used on some lifts; other lifts have a single post (**Figure 2-39**). Some lifts have an electric motor, which drives a hydraulic pump to create fluid pressure and force the lift upward. Other lifts use air pressure from the shop air supply to force the lift upward. If shop air pressure is used, it is applied to fluid in the lift cylinder. A control lever or switch is placed near the lift, which supplies shop air pressure to the lift cylinder and turns on the lift pump motor. Always be sure that the safety lock is engaged after the lift is raised. When the safety lock is released, a release lever is operated slowly to lower the vehicle.

A **vehicle lift** may be called a hoist.

Figure 2-39 Vehicle lifts are used to raise a vehicle.

Caution

Do not allow the power steering fluid to become too hot during the pump pressure test. Excessive fluid temperature reduces pump pressure, resulting in false test results.

7. Raise the vehicle to the desired height. Be sure the lift safety mechanism is engaged.

WARNING If heavy components are removed from one end of a vehicle, the vehicle may become unstable on the lift. Therefore, to prevent this action, chain a strong chassis component to the lift on the opposite end of the vehicle before removing heavy components from the other end.

8. Before lowering the lift, remove all tools, toolboxes, and equipment from under the vehicle, and sweep up discarded parts from under the vehicle.
9. Be sure nobody is standing under the vehicle before lowering it.
10. Lower the vehicle slowly until it is resting on the shop floor and be sure the lift is completely lowered.
11. Before driving the vehicle off the lift, be sure there is proper clearance between all parts of the lift and the vehicle chassis.

SUSPENSION AND STEERING SERVICE, DIAGNOSTIC, AND MEASUREMENT TOOLS

A power steering pressure gauge is connected in the power steering system to test power steering pump pressure.

Using a Power Steering Pressure Gauge

The power steering pump pressure test is one of the most important indications of power steering pump condition.

Follow these steps to use the power steering pressure gauge to test power steering pump pressure:

1. Check the power steering belt condition and if it is cracked, damaged, or oil-soaked, replace the belt. If a V-belt is bottomed in the pulley, excessive wear on the sides of a V-belt is indicated. This condition also requires belt replacement.
2. Measure the power steering belt tension with a belt tension gauge. If this tension is less than specified, adjust the power steering belt tension.
3. Be sure the engine is warmed up and the power steering fluid is at normal operating temperature.
4. With the engine shut off, check the fluid level in the power steering pump reservoir. If necessary, add the vehicle manufacturer's specified fluid until the proper level is obtained.
5. Listen to the power steering pump with the engine idling. A growling noise from the pump may indicate aeration of the power steering fluid in the reservoir or air in the power steering system. If air is indicated in the system, turn the steering wheel several times fully in each direction to bleed air from the system before proceeding with the power steering pump pressure test.
6. Shut the engine off and remove the high-pressure hose from the power steering pump. Connect the power steering pressure gauge and valve assembly to the fitting on the end of the high-pressure hose and to the fitting on the pump from which the high-pressure hose was removed (**Figure 2-42**). Be sure the valve on the power steering pressure gauge is fully open.
7. With the engine idling, close the power steering gauge valve for no more than 10 seconds and observe the power steering pump pressure on the gauge. Turn the gauge valve fully open. If the power steering pump pressure is less than the vehicle manufacturer's specified pressure, replace the power steering pump.

Classroom Manual
Chapter 2, page 34

Caution

Figure 2-42 Pressure gauge connections to power steering.

WARNING Wear a face shield and protective gloves during the power steering pressure test. Power steering fluid and components become hot during this test.

WARNING If the gauge valve is closed for more than 10 seconds during the power steering pressure test, excessive power steering pump pressure may rupture power steering hoses. Hot fluid spraying from a ruptured hose may burn anyone near the vehicle.

8. Shut the engine off and remove the power steering pressure gauge. Tighten the high-pressure hose fitting to the specified torque. Start the engine and check for leaks at this fitting. Be sure the power steering reservoir is filled to the proper level.

Using a Dial Indicator

A dial indicator performs precision measurements on some suspension components such as ball joints. A special dial indicator with a stiff, flexible attaching bracket is available for measuring ball joint wear. When measuring ball joint wear, the end of this flexible bracket is clamped to a stable suspension component with a pair of vise grips. The dial indicator must be mounted so there is no movement in the indicator mounting. A dial indicator contains a movable plunger that is positioned against the component to be measured. The pointer on the dial indicator registers the amount of plunger movement. Each rotation of the pointer represents 0.10 in. (2.54 mm) of plunger movement (**Figure 2-43**). The dial indicator plunger must be positioned against the component to be measured so the dial indicator is preloaded approximately 0.25 in. (6.35 mm). When the dial indicator is preloaded properly, about one-half of the plunger should be outside the dial indicator. After the preload procedure is completed, the dial indicator face must be rotated until the 0 position on the scale is aligned with the pointer. Always look squarely at the indicator face. Looking at the dial indicator face from an angle may provide an inaccurate reading.

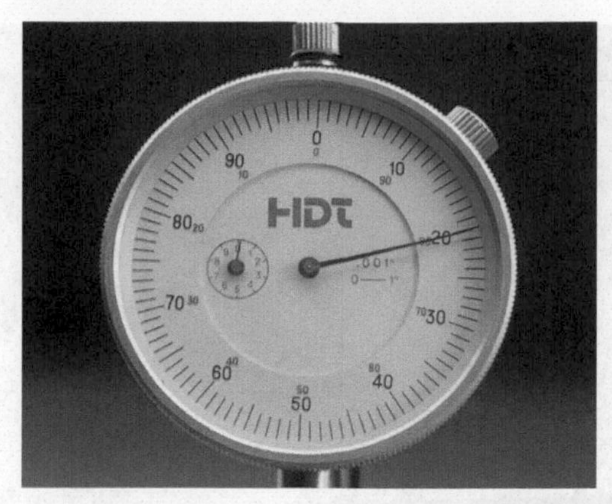

Figure 2-43 The dial indicator pointer rotation indicates the amount of plunger movement.

The following is a typical procedure for measuring ball joint wear on a short-and-long arm suspension system with the coil spring positioned between the lower control arm and the chassis:

1. Lift the front of the vehicle with a floor jack positioned under the manufacturer's specified lift point, and position a safety stand under the lower control arm. Lower the vehicle so the lower control arm and vehicle weight rests on the safety stand. The tire must remain several inches off the floor.
2. Clamp the end of the dial indicator's flexible bracket onto the lower control arm.
3. Position the dial indicator plunger against the lower side of the steering knuckle next to the nut on the ball joint stud. The dial indicator plunger should be positioned at a 90-degree angle with the ball joint stud (**Figure 2-44**).
4. Preload and zero the dial indicator. Be sure the dial indicator is clamped securely to the lower control arm without any movement in the indicator mounting.

SERVICE TIP The procedure for measuring ball joint wear varies depending on the type of suspension system and the type of ball joint mounting. Always consult the vehicle manufacturer's service manual.

Figure 2-44 Dial indicator installed to measure vertical ball joint movement.

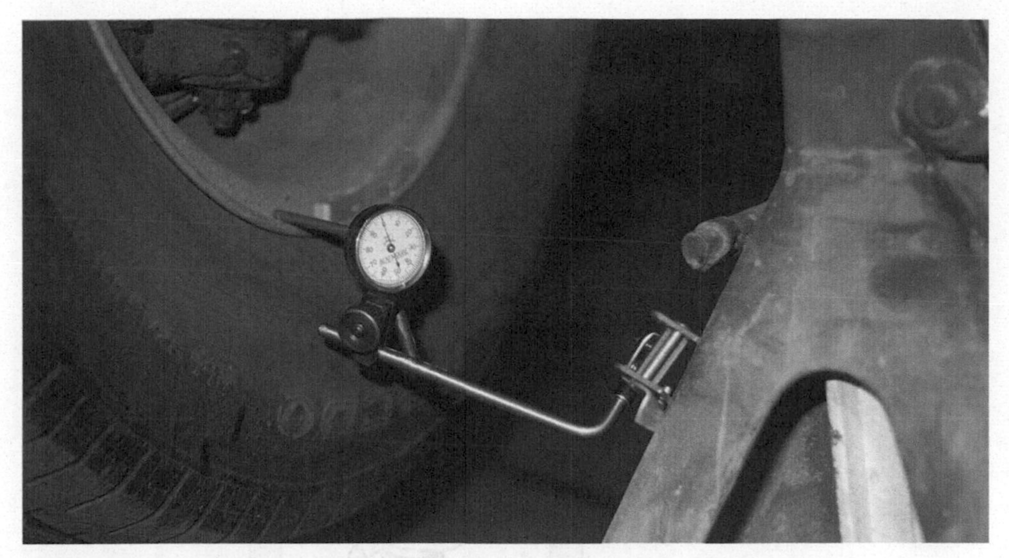

Figure 2-45 Dial indicator installed to measure horizontal joint movement.

5. Position a long steel bar under the tire, and lift upward on the bar while an assistant observes the movement on the dial indicator pointer. If the dial indicator pointer movement exceeds the maximum-specified ball joint vertical movement, the ball joint must be replaced.
6. Be sure the front wheel bearings are adjusted properly. Position the dial indicator plunger against the inner edge of the wheel rim (**Figure 2-45**). Preload and zero the dial indicator. Grasp the top and bottom of the tire, and attempt to tip the tire out at the bottom and in at the top. Release the tire, and repeat this procedure several times while an assistant observes the dial indicator. If the dial indicator pointer movement exceeds the maximum-specified ball joint horizontal movement, the ball joint must be replaced.

Using a Coil Spring Compressing Tool

A coil spring compressing tool is required to compress the coil spring before removing the spring from the strut on a MacPherson strut suspension system.

Follow this procedure for using a coil spring compressing tool to remove a spring from a strut:

1. With the strut-and-coil-spring assembly removed from the vehicle, install the coil spring in the compressing tool according to the tool manufacturer's or vehicle manufacturer's recommended procedure.

WARNING Unless the coil spring is compressed, never loosen the nut on the strut rod that retains the upper strut mount on the strut and coil spring. Loosening this nut will suddenly release the coil spring tension, resulting in personal injury.

2. Adjust the compressing arms on the spring compressing tool so the arms contact the coils farthest away from the center of the spring (**Figure 2-46**).

WARNING Always use the coil spring compressor tool recommended by the vehicle manufacturer. Some compressor tools are designed to work only on specific strut-and-coil-spring assemblies. If the compressor tool does not fit properly on the coil spring, the tool may slip off the spring, suddenly releasing the coil spring tension, which may result in personal injury.

Caution

If the coil spring has an enamel-type coating, tape the spring in the areas where the compressor tool contacts the spring. If the compressor tool chips this coating, the spring may break prematurely.

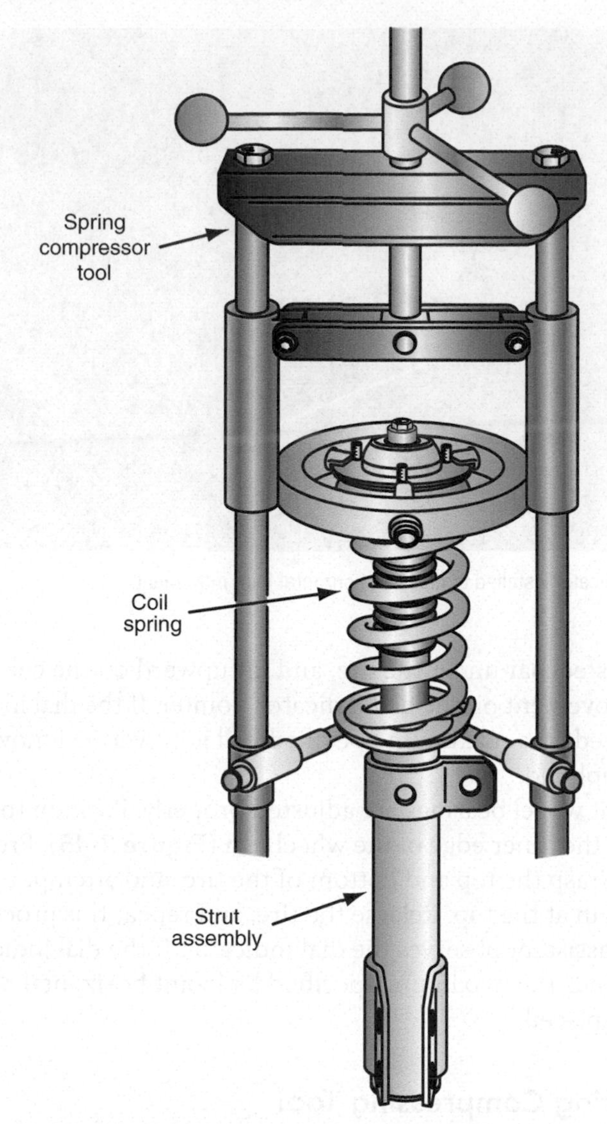

Figure 2-46 Coil spring and strut assembly mounted in a spring compressing tool.

Caution

Never clamp the lower end of the strut in a vise, which may cause internal strut damage.

3. Turn the handle on top of the compressor tool until all the spring tension is removed from the upper strut mount.
4. Loosen and remove the strut rod nut in the center of the upper strut mount (**Figure 2-47**). Be sure all the spring tension is removed from the upper strut mount before loosening this nut.
5. Remove the upper strut mount assembly, mount bearing, and then remove the upper spring seat and insulator.
6. Rotate the handle on the spring compressing tool to release all the tension on the coil spring, and then remove the spring.
7. Remove the dust shield and jounce bumper from the strut rod, and then remove the lower spring insulator (**Figure 2-48**).
8. Replace all worn or defective parts, and be sure the strut rod is fully extended. Install the lower insulator on the lower strut spring seat. Make sure the insulator is properly positioned on the seat.
9. Install the spring bumper on the strut rod.
10. Place the coil spring properly in the spring compressing tool, and rotate the handle on the spring compressing tool to compress the coil spring.

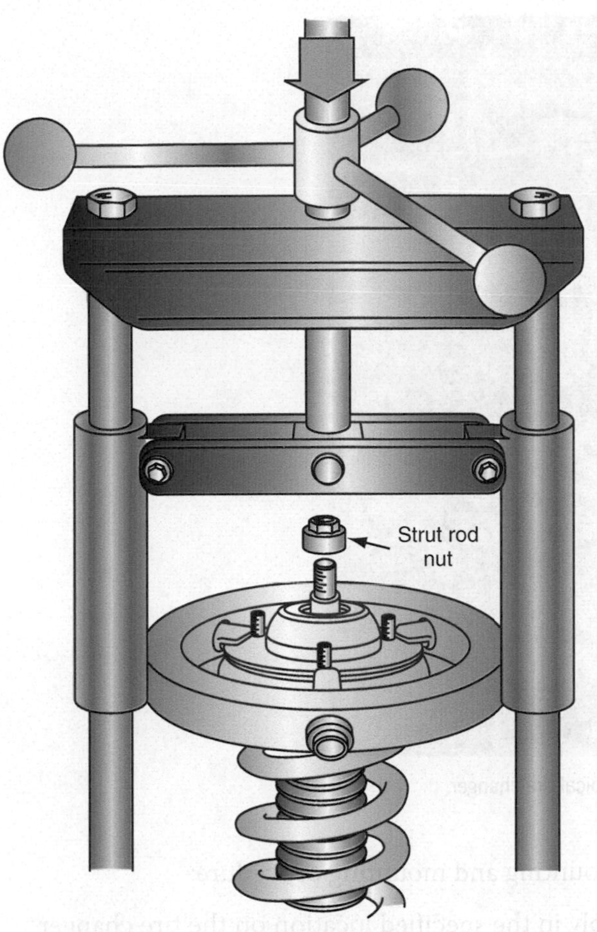

Figure 2-47 After the compressing tool is operated to remove all the spring tension, remove the strut rod nut.

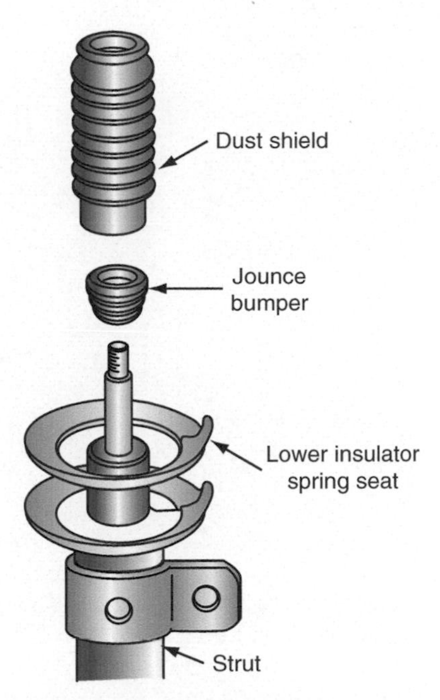

Figure 2-48 Removal of the dust shield, jounce bumper, and lower spring insulator.

SERVICE TIP The tire dismounting and mounting procedure varies depending on the type of tire changer. Always follow the dismounting and mounting procedure recommended by the tire changer manufacturer.

11. Install the strut into the coil spring, and be sure the lower insulator is properly positioned on the coil spring.
12. Install the upper spring insulator and upper strut mount, and make sure these components are properly positioned.
13. Install the strut rod nut and tighten this nut to the specified torque.
14. Rotate the handle on the spring compressing tool to slowly release all the spring tension, and remove the strut and spring assembly from the tool.

Always use the tire changer attachments recommended by the tire changer manufacturer. For example, specific bead removal and installation attachments may be recommended for certain types of wheel rims.

Using a Tire Changer

A tire changer dismounts and mounts tires on wheel rims without damaging the rims. Most tire changers are operated by shop air pressure.

WARNING If the tire changer manufacturer's recommended procedure for dismounting and mounting tires is not followed, personal injury and/or rim damage may occur.

Figure 2-49 Typical tire changer.

The following is a typical tire dismounting and mounting procedure:

1. Position the tire-and-wheel assembly in the specified location on the tire changer (**Figure 2-49**). Be sure the tire-and-wheel assembly is properly positioned and secured.
2. Remove the valve cap and valve core, and be sure the tire is completely deflated.

SERVICE TIP Many shops have a policy of always installing a new valve stem when repairing or replacing a tire, which prevents problems with cracked or defective valve stems.

SERVICE TIP Aligning the painted dot on the tire sidewall ensures proper tire balancing with a minimum amount of balance weight.

3. Position the bead unseating tool to unseat the outer bead, and operate the changer to unseat this bead.
4. Operate the changer to unseat the bead on the inner side of the rim.
5. Position the bead removal tool to remove the outer bead over the rim, and operate the changer to remove this bead over the rim.
6. Position the bead removal tool to remove the inner bead over the rim, and operate the changer to remove this bead over the rim.
7. After the beads are removed over the rim, remove the tire from the changer.
8. Clean dirt and rust from the rim sealing flanges with a wire brush, and apply a coating of rubber compound to the bead area of the tire.

9. Install the tire on top of the rim in the desired position. Many tires have a dot painted on the tire that should be aligned with the valve stem when the tire is mounted on the rim. Therefore, this dot must be on the outer side of the tire prior to mounting.
10. Position the bead installation tool to install the inner bead over the rim, and operate the changer to install this bead over the rim.
11. Position the bead installation tool to install the outer bead over the rim, and operate the changer to install this bead over the rim.
12. Rotate the tire on the rim to align the painted dot with the valve stem, and install the valve core. Be sure the tire is centered properly on the rim.
13. Inflate the tire to the specified tire pressure, install the valve cap, and remove the tire and wheel from the tire changer.

SERVICE TIP The valve core may be left out of the valve stem and the tire partially inflated. This action allows the air pressure to enter the tire more quickly. Seat the beads against the rim flanges. After the beads are seated, allow most of the air to escape from the tire, and then install the valve core.

Using a Scan Tool

A scan tool performs an electronic diagnosis of computer-controlled suspension systems and other electronic systems. Current production vehicle computer diagnostic scan tools have the software preloaded (**Figure 2-50**), depending on the vehicle and system being diagnosed . The scan tool cable is connected to the data link connector (DLC) under the dash. On 1996 and newer vehicles with on-board diagnostic II (OBD II) systems, the scan tool is powered from a terminal in the DLC. On most pre-1996 vehicles, a power cable on the scan tool has to be connected to a 12 V power source such as the cigarette lighter.

Caution

Never connect or disconnect the scan tool data cable with the ignition switch on. This action may damage expensive electronic components on the vehicle or the scan tool.

After the scan tool is connected, the vehicle make, model year, and engine size usually have to be selected on the scan tool display. On vehicles with several different computer systems, the technician must select the computer system to be diagnosed from the list of computer systems on the vehicle. For example, if the technician wants to diagnose the continuously variable road sensing suspension system (CVRSS), then RSS must be selected from the list of computer systems displayed on the scan tool display. After the RSS

Figure 2-50 A vehicle computer diagnostic scan tool.

selection, the scan tool display will ask the technician to select diagnostic trouble codes (DTCs), data, inputs, outputs, or clear codes.

On OBD II vehicles, the DTCs contain five digits. A typical DTC from a CVRSS is C1712. The "C" indicates this DTC is in the chassis group, and the "1" informs the technician this is a manufacturer-specific code. A common DTC dictated by Society of Automotive Engineers (SAE) standards is indicated if the second digit is a "0." The digits 712 in the DTC indicate an open circuit in the left front (LF) damper actuator circuit. On a CVRSS, the damper actuators are solenoids in each shock absorber or strut. Diagnostic trouble codes (DTCs) indicate a defect in a certain electrical area, but they do not indicate a defect in a specific component. For example, the C1712 DTC indicates there is an open circuit in the LF damper actuator solenoid or in the connecting wires between this solenoid and the suspension computer. The technician usually has to use a volt-ohm meter to locate the exact cause of the DTC. DTCs may be identified on the scan tool display as CURRENT or HISTORY. A CURRENT DTC is one that is present at the time of testing. A HISTORY DTC represents an intermittent defect that occurred in the past, which has since disappeared.

If "data" is selected, the scan tool display indicates data related to the CVRSS. These data include readings from the input sensors and readings such as battery voltage or vehicle speed. When "inputs" is selected, the scan tool displays readings from the input sensors in the CVRSS. In the "outputs" parameter, the scan tool displays readings from the outputs in the CVRSS. The output displays may be a voltage reading or solenoids displayed as being "on" or "off." If "clear codes" is selected on the scan tool display, the DTCs are erased from the computer memory.

Depending on the computer-controlled suspension system being diagnosed, the scan tool may be used to perform different functions. For example, on some computer-controlled air suspension systems, the scan tool may be used to command the suspension computer to fill and vent each air spring. This test mode may be used to check the operation of system components. The scan tool may be used to flash program the suspension computer memory. This mode installs new software in the computer as directed by service bulletins from the original equipment manufacturer (OEM). On some computer-controlled suspension systems, such as the system on the 2002 Lincoln Blackwood, the scan tool is used to adjust the suspension ride height.

Using an Electronic Wheel Balancer

The most popular type of wheel balancer in automotive service shops is the electronic balancer. Most electronic wheel balancers have an electric motor that spins the tire and wheel assembly at moderate speed during the balance procedure. On some electronic wheel balancers, the tire-and-wheel assembly is spun by hand for balance purposes. Certain preliminary checks must be performed on the tire-and-wheel assembly before installing this assembly on an electronic wheel balancer.

These preliminary checks include the following:

1. Clean all mud and debris from the tire-and-wheel assembly.
2. Remove all old wheel weights from the rim.

WARNING **Improperly installed wheel weights may fly off the rim when the wheel is spun on the balancer, resulting in personal injury.**

3. Remove objects such as stones from the tire tread.

WARNING **Stones or other objects in the tire tread may fly out when the wheel is spun on the balancer, resulting in personal injury.**

4. Inspect the tire tread and sidewall for defects, such as splits, cuts, chunks out of the tread or sidewall, or bulges indicating ply separation. Replace tires with these defects.
5. Inflate the tire to the specified pressure.

The tire-and-wheel assembly must be installed on the wheel balancer according to the instructions of the wheel balancer manufacturer. After the tire-and-wheel assembly is installed securely on the balancer, slowly rotate the tire by hand and listen for objects such as balls of rubber rolling inside the tire. Objects inside the tire must be removed, because they make wheel balancing difficult or impossible.

WARNING If the tire-and-wheel assembly is not securely installed on the wheel balancer, the wheel may come loose and fly off the balancer during the balance procedure, resulting in personal injury.

Classroom Manual
Chapter 2, page 30

Use a dial indicator to measure vertical and lateral tire and wheel runout. If tire and wheel runout is excessive, wheel balancing may be difficult. Enter the wheel diameter, width, and offset in the wheel balancer. The balancer must have this information to calculate the required amount of wheel weight and the weight position. Lower the safety hood on the wheel balancer, and activate the balancer to spin the wheel. Apply the brake on the balancer to stop the wheel. Install the indicated amount of wheel weight in the proper position as indicated on the wheel balancer display. Be sure the weight(s) are securely attached to the wheel rim.

Using a Computer Four Wheel Aligner

If a vehicle has a frame and a live axle rear suspension system, it is not as likely to require rear wheel alignment as a unibody vehicle with a semi-independent or independent rear suspension system. Because of increased four wheel alignment requirements on unibody cars, most shops that offer wheel alignment service are equipped with a computer four wheel aligner. Before performing a wheel alignment, the technician must complete a preliminary vehicle inspection. The purpose of the preliminary inspection is to locate any defective or worn components that would make wheel alignment inaccurate.

A **computer four wheel aligner** uses a wheel sensor mounted on each wheel to measure all the front and rear wheel alignment angles.

A preliminary wheel alignment inspection includes these checks:

1. Be sure the vehicle has the normal curb weight that it has when the driver is operating the vehicle. Make sure the gas tank is full, and check for excessive mud adhered to the underside of the chassis.
2. Be sure the tires are inflated to the specified tire pressure, and inspect the tires for excessive wear, bulges, cuts, and splits. Be sure the tires are all the same size.
3. Check for the specified front and rear suspension height. Because the suspension height affects many of the wheel alignment angles, this height must be corrected before performing a wheel alignment if it is not within specifications.
4. Check the steering wheel free play. If this free play is more than specified, inspect the steering linkages and tie rod ends for excessive wear. The excessive steering wheel free play must be corrected before a wheel alignment, because this problem affects front wheel toe and steering quality.
5. Be sure the shock absorbers and struts are in satisfactory condition.
6. Check the front and rear wheel bearing adjustments, and correct these adjustments as necessary. Loose wheel bearing adjustments affect some of the wheel alignment angles.
7. Check the ball joints for excessive wear. Worn ball joints affect some of the wheel alignment angles.

8. Inspect the front, rear, upper and lower control arms for damage, and inspect all the control arm bushings for wear. Bent control arms or worn control arm bushings affect wheel alignment angles.
9. Inspect the front and rear stabilizer bars and bushings. Worn stabilizer bushings affect ride quality and vehicle handling.

After the preliminary inspection is completed, and all the necessary components have been replaced or adjusted, drive the vehicle onto the wheel alignment rack.

Then follow this procedure to complete the four wheel alignment with a computer wheel aligner:

1. Follow all instructions provided in the wheel aligner operator's manual.
2. Be sure the front wheels are centered on the wheel aligner turntables.
3. Be sure the rear wheels are positioned properly on the slip plates.
4. Mount the wheel sensor units on each wheel.
5. Select the vehicle make and model year on the wheel aligner screen.
6. On the wheel aligner preliminary inspection screen, check the condition of each item. Most of these items are mentioned in the previous preliminary inspection procedure.
7. Display the ride height screen on the wheel aligner, and be sure the front and rear ride height is within specifications.
8. Display the wheel runout screen of the wheel aligner, and compensate for wheel runout as directed.
9. Display the turning angle screen on the wheel aligner. Apply the brakes with a brake pedal depressor as directed on the screen, and perform the turning angle check.
10. Display the front and rear wheel alignment angles on the screen. Most computer four wheel aligners mark the alignment angles that are not within specifications.
11. Display the adjustment screen on the wheel aligner, and perform the necessary front and rear suspension adjustments to bring all the alignment angles within specifications. **Photo Sequence 2** illustrates a typical procedure for performing a four wheel alignment with a computer wheel aligner.

SERVICE TIP The wheel runout compensation procedure varies, depending on the make and year of the wheel aligner.

PHOTO SEQUENCE 2

Typical Procedure for Performing Four Wheel Alignment with a Computer Wheel Aligner

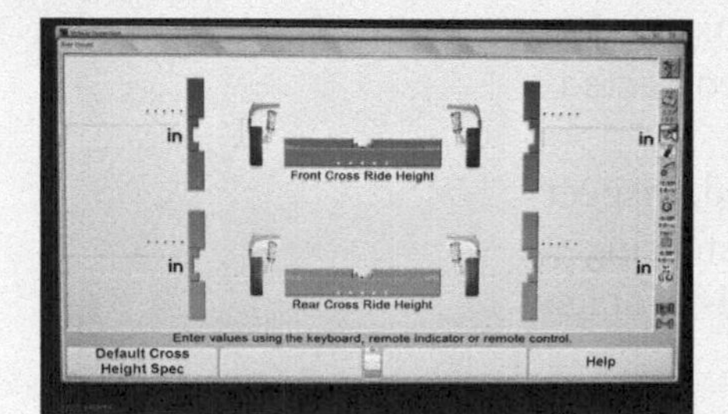

P2-1 Display the ride height screen. Check the tire condition for each tire on the tire condition screen.

P2-2 Position the vehicle on the alignment rack.

P2-3 Make sure the front tires are positioned properly on the turntables.

PHOTO SEQUENCE 2 (CONTINUED)

P2-4 Position the rear wheels on the slip plates.

P2-5 Attach the wheel units.

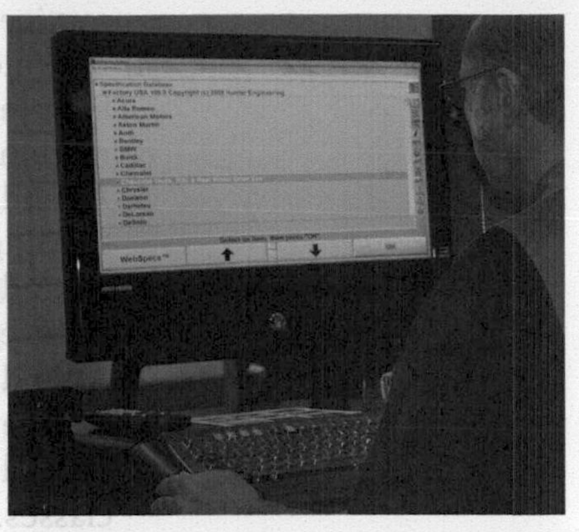

P2-6 Select the vehicle make and model year.

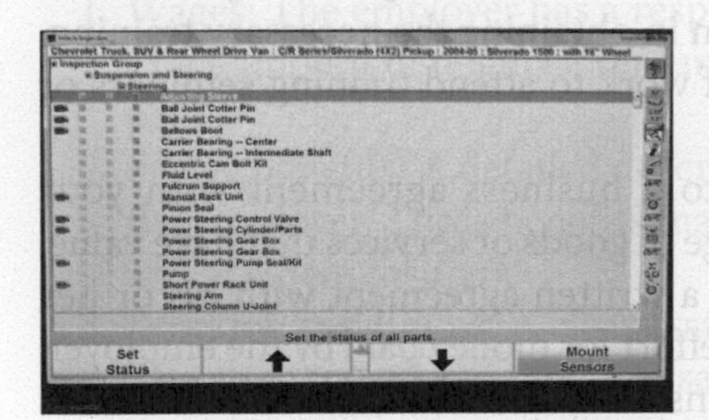

P2-7 Check the items on the screen during the preliminary inspection.

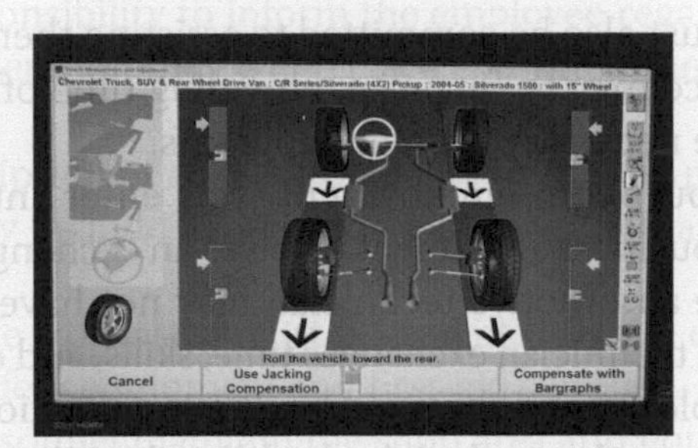

P2-8 Display the wheel runout compensation screen.

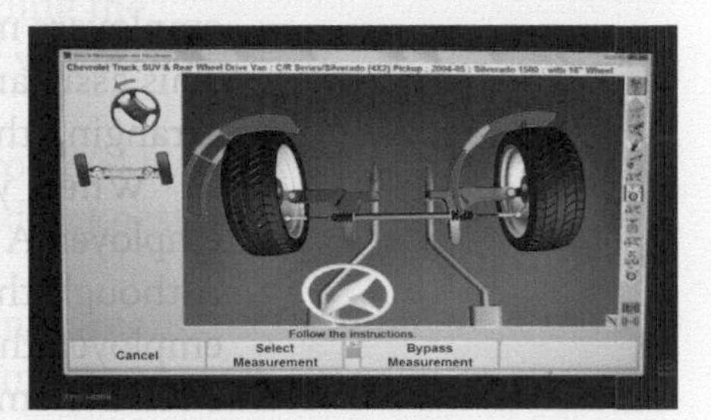

P2-9 Display the turning angle screen and perform the turning angle check.

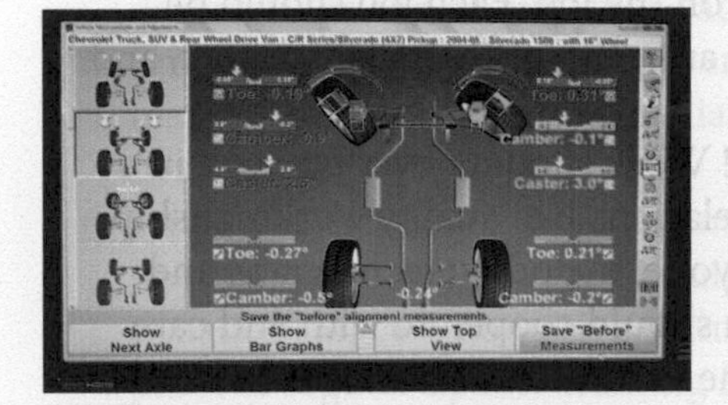

P2-10 Display the front and rear wheel alignment angle screen.

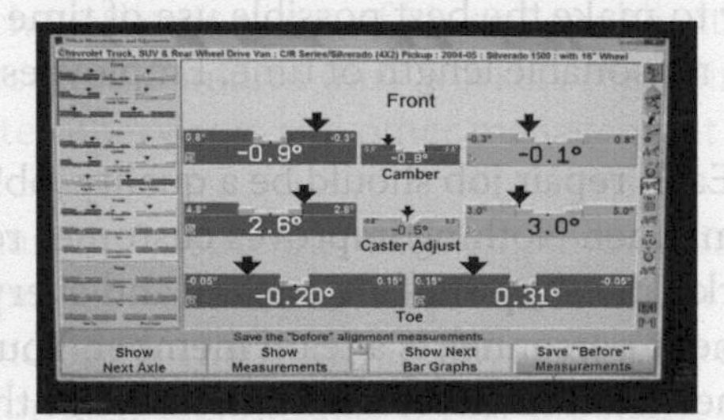

P2-11 Display the adjustment screen.

EMPLOYER AND EMPLOYEE OBLIGATIONS

The ever-increasing electronics content on today's vehicles also requires that technicians are familiar with the latest electronics technology. To be a successful automotive technician, you must be committed to lifelong training. There are many different ways to obtain this training, but it is absolutely essential.

Automotive training may be obtained by these methods:

1. Obtain bulletins, service manuals, and training information from original equipment manufacturers (OEMs), independent parts and component manufacturers, and independent suppliers of service manuals and training books. After the information is obtained, it is essential that you read and study it.

4. Follow the service procedures in the service manual provided by the vehicle manufacturer or an independent manual publisher.
5. When the repair job is completed, always be sure the customer's complaint has been corrected.
6. Do not be too concerned with work speed when you begin working as an automotive technician. Speed comes with experience.

Liable Responsibilities

During many repair jobs you, as a student or technician working on a customer's vehicle, actually have the customer's life and the vehicle safety in your hands. For example, if you are doing a brake job and leave the nuts loose on one wheel, that wheel may fall off the vehicle at high speed. This could result in serious personal injury for the customer and others, as well as extensive vehicle damage. If this type of disaster occurs, the individual who worked on the vehicle and the shop may be involved in a very expensive legal action. As a student or technician working on customer vehicles, you are responsible for the safety of every vehicle that you work on! Even when careless work does not create a safety hazard, it leads to dissatisfied customers, who will often take their business to another shop. Nobody benefits when that happens.

NATIONAL INSTITUTE FOR AUTOMOTIVE SERVICE EXCELLENCE (ASE) CERTIFICATION

The National Institute for Automotive Service Excellence (ASE) has provided voluntary testing and certification of automotive technicians on a national basis for many years. The image of the automotive service industry has been enhanced by the ASE certification program. More than 415,000 technicians now have current certifications and work in a wide variety of automotive service shops.

ASE provides certification in eight areas of automotive repair:

1. Engine repair
2. Automatic transmissions/transaxles
3. Manual drivetrain and axles
4. Suspension and steering
5. Brakes
6. Electrical systems
7. Heating and air conditioning
8. Engine performance

A technician may take the ASE test and become certified in any or all of the eight areas. When a technician passes an ASE test in one of the eight areas, an Automotive Technician's shoulder patch is issued by ASE. Technicians who pass all eight tests receive a Master Technician's shoulder patch (**Figure 2-51**). Retesting at five-year intervals is required to remain certified. The certification test in each of the eight areas contains 40 to 80 multiple-choice questions. The test questions are written by a panel of automotive service experts from various areas of automotive service, including automotive instructors, service managers, automotive manufacturers' representatives, test equipment representatives, and certified technicians. The test questions are pretested and checked for quality by a national sample of technicians. On an ASE certification test, approximately 45 to 50 percent of the questions are Technician A and Technician B format, and the multiple-choice format is used in 40 to 45 percent of the questions. Less than 10 percent of ASE certification questions are an EXCEPT format, in which the technician selects one incorrect answer out of four possible answers. ASE regulations demand that each

Figure 2-51 ASE certification shoulder patches worn by Automotive Technicians and Master Technicians.

technician must have two years of working experience in the automotive service industry prior to taking a certification test or tests. However, relevant formal training may be substituted for one year of working experience. Contact ASE for details regarding this substitution. The contents of the suspension and steering test are listed in **Table 2-1**.

ASE also provides certification tests in automotive specialty areas, such as Parts Specialist; Advanced Engine Performance Specialist; Alternate Fuels, Light Vehicle—Compressed Natural Gas; Machinist, Cylinder Head Specialist; Machinist, Cylinder Block Specialist; and Machinist, Assembly Specialist.

Shops that employ ASE-certified technicians display an official **ASE blue seal of excellence**. This blue seal increases the customer's awareness of both the shop's commitment to quality service and the competency of its certified technicians.

Repair Orders

Repair orders may vary depending on the shop, but they usually have this basic information:

1. Customer's name, address, and phone number(s)
2. Customer's signature
3. Vehicle make, model, year, and color
4. Vehicle identification number (VIN)
5. Vehicle mileage
6. Engine displacement

TABLE 2-1 Suspension and Steering Test Summary

Content area	Questions in test	Percentage of test
A. Steering systems diagnosis and repair	10	25%
1 Steering columns and manual steering gears (3) 2 Power-assisted steering units (4) 3 Steering linkage (3)		
B. Suspension systems diagnosis and repair	13	33%
1 Front suspensions (6) 2 Rear suspensions (5) 3 Miscellaneous service (2)		
C. Wheel alignment diagnosis, adjustment, and repair	12	30%
D. Wheel and tire diagnosis and repair	5	12%
TOTAL	**40**	**100%**

7. Date and time
8. Service writers code number
9. Work order number
10. Labor rate
11. Estimate of repair costs
12. Accurate and concise description of the vehicle problem

In many shops, the repair orders are completed on a computer terminal, and the computer may automatically write the vehicle repair history on the repair order if the vehicle has a previous repair history in the shop computer system.

The repair order informs the technician regarding the problem(s) with the vehicle. In many shops the technician has to enter a starting time on the repair order. This may be done on a computer terminal or by inserting the work order into the time clock. The technician may also have to enter his code number on the work order to indicate who worked on the vehicle. The technician's code number on the work order is also used to pay the technician for the repair job. The technician must diagnose and repair the problem(s) indicated on the repair order. When the technician obtains parts from the parts department to complete the repair, the technician must present the order number. The parts personnel enter the parts and the cost on the repair order. In some shops the technician is required to enter an accurate description of the completed repairs on the work order. For example, the description of the problem on the work order may be "A/C system inoperative." If the technician replaced the A/C compressor fuse to correct the problem, he or she may enter, "Diagnosed A/C electrical defects and found a blown A/C fuse. Replaced the fuse and operated the A/C system. A/C system operation is normal." Some shops require the service writer or shop foreman to sign the work order when the repair job is successfully completed. The work order is routed back to the accountant who calculates all the charges on the work order including the appropriate taxes. Some shops add a miscellaneous charge on the work order. A typical miscellaneous charge is 10 percent of the total charges on the work order. This miscellaneous charge is to cover the cost of small items such as bolts, cotter pins, grease, lubricants, and sealers that are not entered separately on the work order.

In many shops the technicians work on a flat rate basis. In these shops the technician is paid a flat rate for each repair. In dealership the flat rate time is set by the vehicle manufacturer. Independent shops use generic flat rate manuals published by firms such as Mitchell Publications. If the flat rate time is 2 hours for completing a specific vehicle repair, the customer is charged for 2 hours, and the technician is paid for 2 hours even though he or she completed the repair in 1.5 hours. Conversely, if the technician takes 2.5 hours to complete the job, the technician is only paid for 2 hours and the customer is charged for 2 hours. The flat rate time is usually entered on the work order.

SERVICE MANUALS

The service manual is one of the most important tools for today's technician. It provides information concerning component identification, service procedures, specifications, and diagnostic information. In addition, the service manual provides information concerning wiring harness connections and routing, component location, and fluid capacities. Service manuals can be supplied by the vehicle manufacturer or through aftermarket suppliers.

The service manual provides an explanation of the vehicle identification number (VIN). The VIN information is essential when ordering parts. Most service manuals published by vehicle manufacturers now have a standard format (**Figure 2-52**). The service manual usually provides illustrations to guide the technician through the service operation

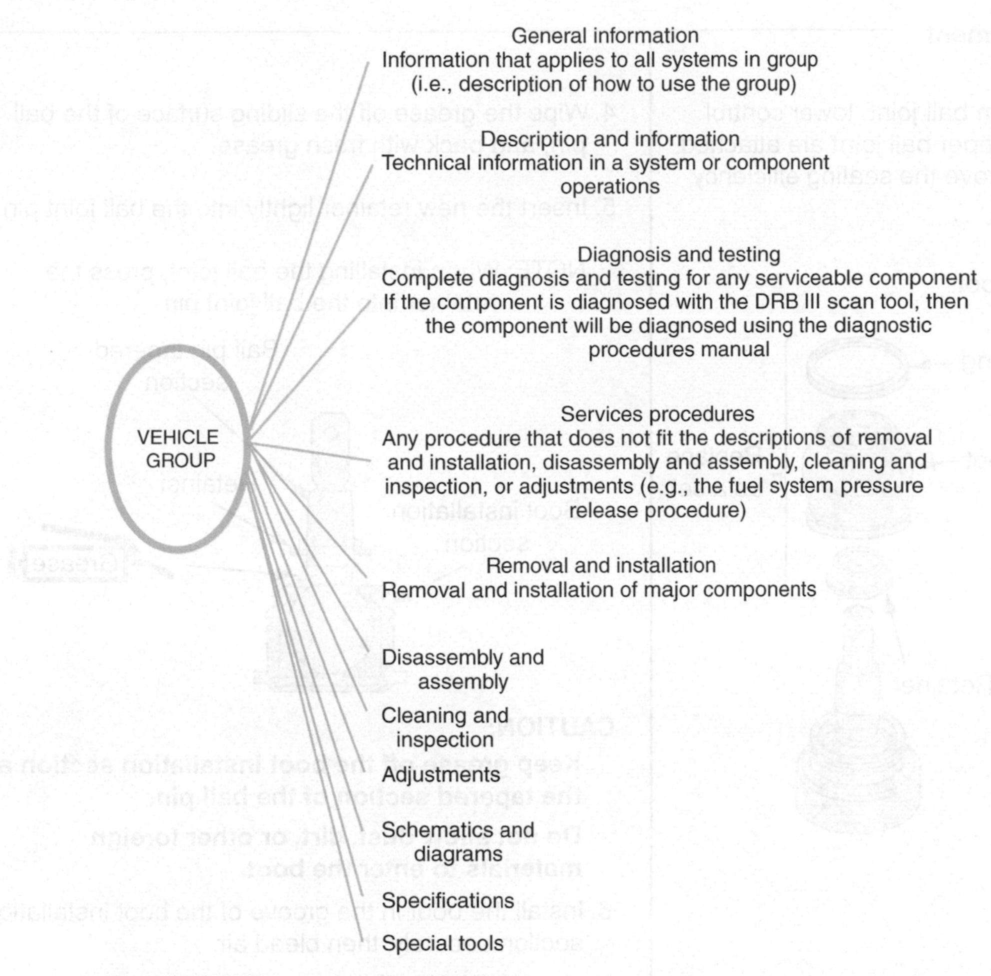
General information
Information that applies to all systems in group
(i.e., description of how to use the group)
Description and information
Technical information in a system or component
operations
Diagnosis and testing
Complete diagnosis and testing for any serviceable component
If the component is diagnosed with the DRB III scan tool, then
the component will be diagnosed using the diagnostic
procedures manual
Services procedures
Any procedure that does not fit the descriptions of removal
and installation, disassembly and assembly, cleaning and
inspection, or adjustments (e.g., the fuel system pressure
release procedure)
VEHICLE
GROUP
Removal and installation
Removal and installation of major components
Disassembly and
assembly
Cleaning and
inspection
Adjustments
Schematics and
diagrams
Specifications
Special tools

Figure 2-52 Uniform service manual layout.

(**Figure 2-53**). Always use the correct manual for the vehicle and system being serviced. Follow each step in the service procedure. Do not skip steps! Measurements such as torque, end play, and clearance specifications are located in or near the service manual text or procedural information. Specification tables are usually provided at the end of the procedural information or component area (**Figure 2-54**).

Because the service manual is divided into a number of main component and system areas, a table of contents is provided at the front of the manual to provide quick access to the desired information. Each component area or system is covered in a section of the service manual (**Figure 2-55**). At the beginning of each section in the service manual, a smaller table of contents guides the technician to the information regarding the specific system or component being serviced. The service manual may be divided into several volumes, because of the extensive amount of information required to service today's vehicles. Diagnostic information in each section of the service manual is usually provided in **diagnostic procedure charts** (**Figure 2-56**). The test results obtained in a specific diagnostic step guide the technician to the next appropriate step.

Service and parts information can also be provided through computer services (**Figure 2-57**). Computerized service information may be provided on computer discs or compact discs (CDs), which are easier to store and access. Computers may also be connected to a central database to obtain service information. Using the mouse, light pen, computer keyboard, or touch-sensitive screen, the technician selects choices from a series of menus on the computer monitor. When the desired information is accessed, it may be printed out for detailed study. Modern automotive service information is presently computer or web-based.

Ball Joint Boot Replacement

NOTE: The upper control arm ball joint, lower control arm ball joint, and knuckle upper ball joint are attached with the boot retainer to improve the sealing efficiency of the boot.

1. Remove the set ring and boot.

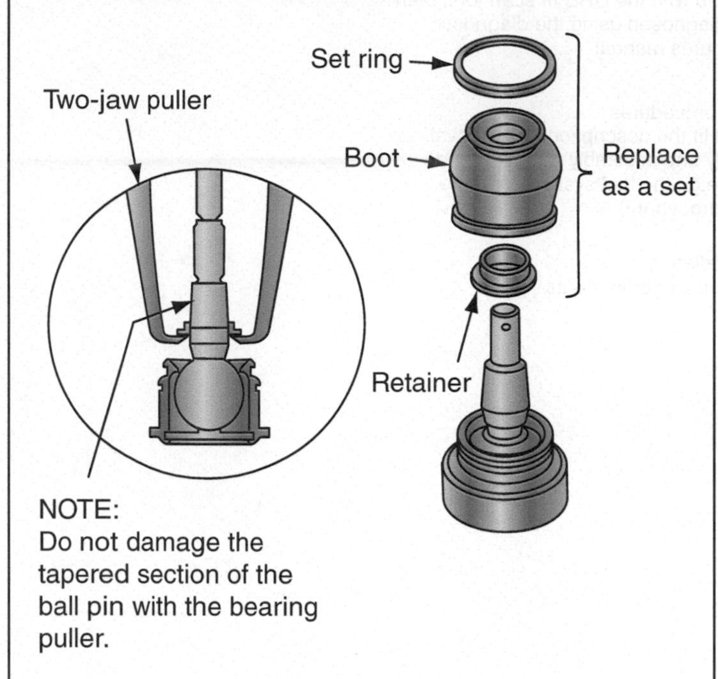

2. Remove the retainer.

NOTE: The knuckle lower ball joint does not have a retainer.

3. Pack the interior of the boot and lip with grease.

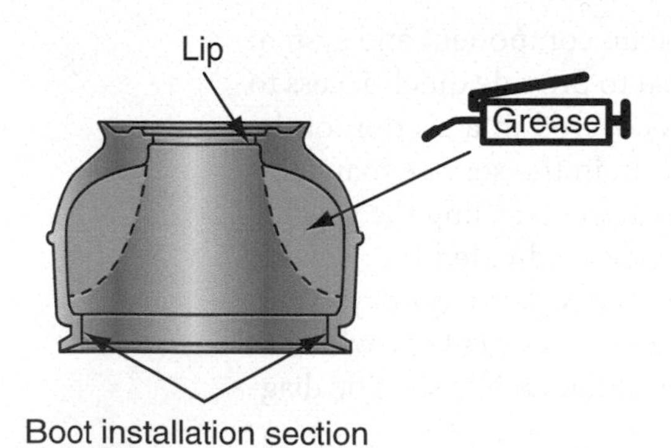

CAUTION: Do not contaminate the boot installation section with grease.

4. Wipe the grease off the sliding surface of the ball pin, and pack with fresh grease.

5. Insert the new retainer lightly into the ball joint pin.

NOTE: When installing the ball joint, press the retainer into the ball joint pin.

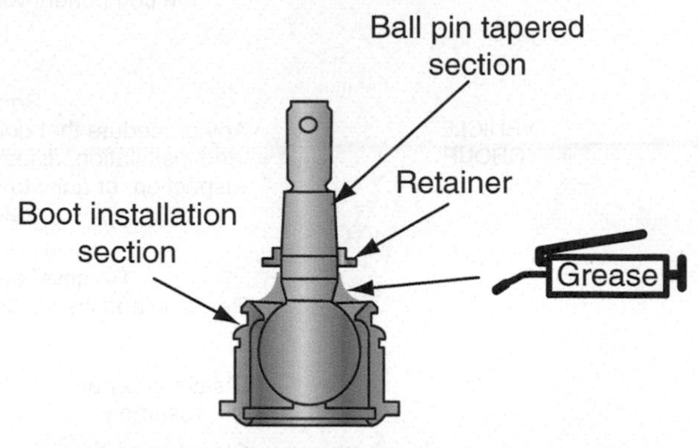

CAUTION:

Keep grease off the boot installation section and the tapered section of the ball pin.

Do not allow dust, dirt, or other foreign materials to enter the boot.

6. Install the boot in the groove of the boot installation section securely, then bleed air.

7. Adjust the special tool with the adjusting bolt until the end of the tool aligns with the groove on the boot.

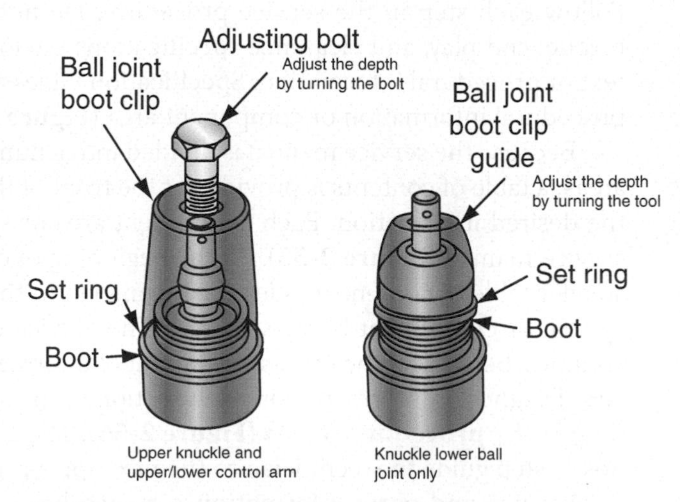

8. Slide the set ring over the tool and into position.

CAUTION: After installing the boot, check the ball pin tapered section and threads for grease contamination and wipe them if necessary.

Figure 2-53 Illustrations in the service manual that guide the technician through service procedures.

<table>
<tr><td rowspan="6">Cold tire inflation pressure</td><td rowspan="2"></td><td rowspan="2">Tire size</td><td colspan="2">Pressure</td></tr>
<tr><td>Front</td><td>Rear</td></tr>
<tr><td rowspan="2">For all roads including full rated loads</td><td>P195/70R14</td><td>220 kPa (2.2 kgf/cm^2, 32 psi)</td><td>240 kPa (2.4 kgf/cm^2, 34 psi)</td></tr>
<tr><td>P205/65R15</td><td>220 kPa (2.2 kgf/cm^2, 32 psi)</td><td>240 kPa (2.4 kgf/cm^2, 34 psi)</td></tr>
<tr><td rowspan="2">Optional inflation for reduced loads (1 to 4 passengers)</td><td>P195/70R14</td><td>180 kPa (1.8 kgf/cm^2, 26 psi)</td><td>180 kPa (1.8 kgf/cm^2, 26 psi)</td></tr>
<tr><td>P205/65R15</td><td>180 kPa (1.8 kgf/cm^2, 26 psi)</td><td>180 kPa (1.8 kgf/cm^2, 26 psi)</td></tr>
<tr><td rowspan="4">Vehicle height</td><td colspan="2" rowspan="2">Tire size</td><td colspan="2">Height</td></tr>
<tr><td>Front</td><td>Rear</td></tr>
<tr><td colspan="2">P195/70R14</td><td>210 mm (8.27 in.)</td><td>270 mm (10.63 in.)</td></tr>
<tr><td colspan="2">P205/65R15</td><td>213 mm (8.39 in.)</td><td>276 mm (10.87 in.)</td></tr>
<tr><td rowspan="9">Front wheel alignment</td><td colspan="2">Toe-in (total)</td><td colspan="2">0° +/- 0.2° (0 +/- 2 mm, 0 +/- 0.08 in.)</td></tr>
<tr><td rowspan="3">Wheel angle</td><td>Tire size</td><td>Inside wheel</td><td>Outside wheel</td></tr>
<tr><td>P195/70R14</td><td>37°20' +/- 2°</td><td>32°15'</td></tr>
<tr><td>P205/65R15</td><td>36°00' +/- 2°</td><td>31°20'</td></tr>
<tr><td colspan="2">Camber</td><td colspan="2">-0°35' +/- 45'</td></tr>
<tr><td colspan="2">Cross camber</td><td colspan="2">45' or less</td></tr>
<tr><td colspan="2">Caster</td><td colspan="2">1°05' +/- 45'</td></tr>
<tr><td colspan="2">Cross caster</td><td colspan="2">45' or less</td></tr>
<tr><td colspan="2">Steering axis inclination</td><td colspan="2">13°00' +/- 45'</td></tr>
<tr><td rowspan="3">Rear wheel alignment</td><td colspan="2">Toe-in (total)</td><td colspan="2">0.4° +/- 0.2° (4 +/- 2 mm, 0.16 +/- 0.08 in.)</td></tr>
<tr><td colspan="2">Camber</td><td colspan="2">-0°15' +/- 45'</td></tr>
<tr><td colspan="2">Cross camber</td><td colspan="2">45' or less</td></tr>
</table>

Figure 2-54 Specification table.

Service procedures may be modified by the vehicle manufacturer at any time. Service bulletins provide up-to-date corrections for the service manuals. If a significant number of corrections are required, a second edition of the manual may be published. When service information is provided on CDs, the CDs are updated frequently to provide the latest information.

CUSTOMER CARE When advising customers regarding when the work will be completed on their car, it is a good idea to estimate a longer time than you anticipate. For example, if you expect it will take 3 hours to repair the vehicle, advise the customer it will be ready in 4 hours. This allows extra time for diagnosing difficult problems or road testing the vehicle. It can be very frustrating for customers when they come to pick up their vehicle at the appointed time, and the vehicle is not ready.

TABLE OF CONTENTS	SECTION NUMBER
GENERAL INFO. AND LUBE	
General Information	0A
Maintenance and Lubrication	0B
HEATING AND AIR COND.	
Heating and Vent. (non-A/C)	1A
Air-Conditioning System	1B
V-5 A/C Compressor Overhaul	1D3
BUMPERS AND FRONT BODY PANELS	
Bumpers (See 10-4)	
Fr. End Body Panels (See 10-5)	
STEERING, SUSPENSION, TIRES, AND WHEELS	
Diagnosis	3
Wheel Alignment	3A
Power Steering Gear & Pump	3B1
Front Suspension	3C
Rear Suspension	3D
Tires and Wheels	3E
Steering Col. On-Vehicle Service	3F
Steering Col. – Std. Unit Repair	3F1
Steering Col. – Tilt, Unit Repair	3F2
DRIVE AXLES	
Drive Axles	4D
BRAKES	
General Info. – Diagnosis and On-Car Service	5
Compact Master Cylinder	5A1
Disc Brake Caliper	5B2
Drum Brake - Anchor Plate	5C2
Power Brake Booster Assembly	5D2
ENGINES	
General Information	6
2.0 Liter I-4 Engine	6A1
3.1 Liter V6 Engine	6A3
Cooling System	6B
Fuel System	6C
Engine Electrical – General	6D
Battery	6D1
Cranking System	6D2
Charging System	6D3
Ignition System	6D4
Engine Wiring	6D5
Driveability & Emissions – Gen.	6E
Driveability & Emissions – TBI	6E2
Driveability & Emissions – PFI	6E3
Exhaust System	6F

TABLE OF CONTENTS	SECTION NUMBER
TRANSAXLE	7A
Auto.Transaxle On-Car Serv.	3T40-HD
Auto. Trans. – Hydraulic Diagnosis	
Auto. Trans. – Unit Repair	3T40
Man. Trans. On-Car Service	7B
5-Sp. 5TM40 Man. Trans. Unit Repair	7B1
5-Sp. Isuzu Man. Trans. Unit Repair	7B2
Clutch	7C
CHASSIS ELECTRICAL, INSTRUMENT PANEL & WIPER/WASHER	
Electrical Diagnosis	8A
Lighting and Horns	8B
Instrument Panel and Console	8C
Windshield Wiper/Washer	8E5
ACCESSORIES	
Audio System	9A
Cruise Control	9B
Engine Block Heater	9C
BODY SERVICE	
General Body Service	10-1
Stationary Glass	10-2
Underbody	10-3
Bumpers	10-4
Body Front End	10-5
Doors	10-6
Rear Quarters	10-7
Body Rear End	10-8
Roof & Convertible Top	10-9
Seats	10-10
Safety Belts	10-11
Body Wiring	10-12
Unibody Collision Repair	11-1
Welded Panel Replacement	11-2
INDEX	
Alphabetical Index	

Figure 2-55 The table of contents directs you to the major systems and component areas in the service manual.

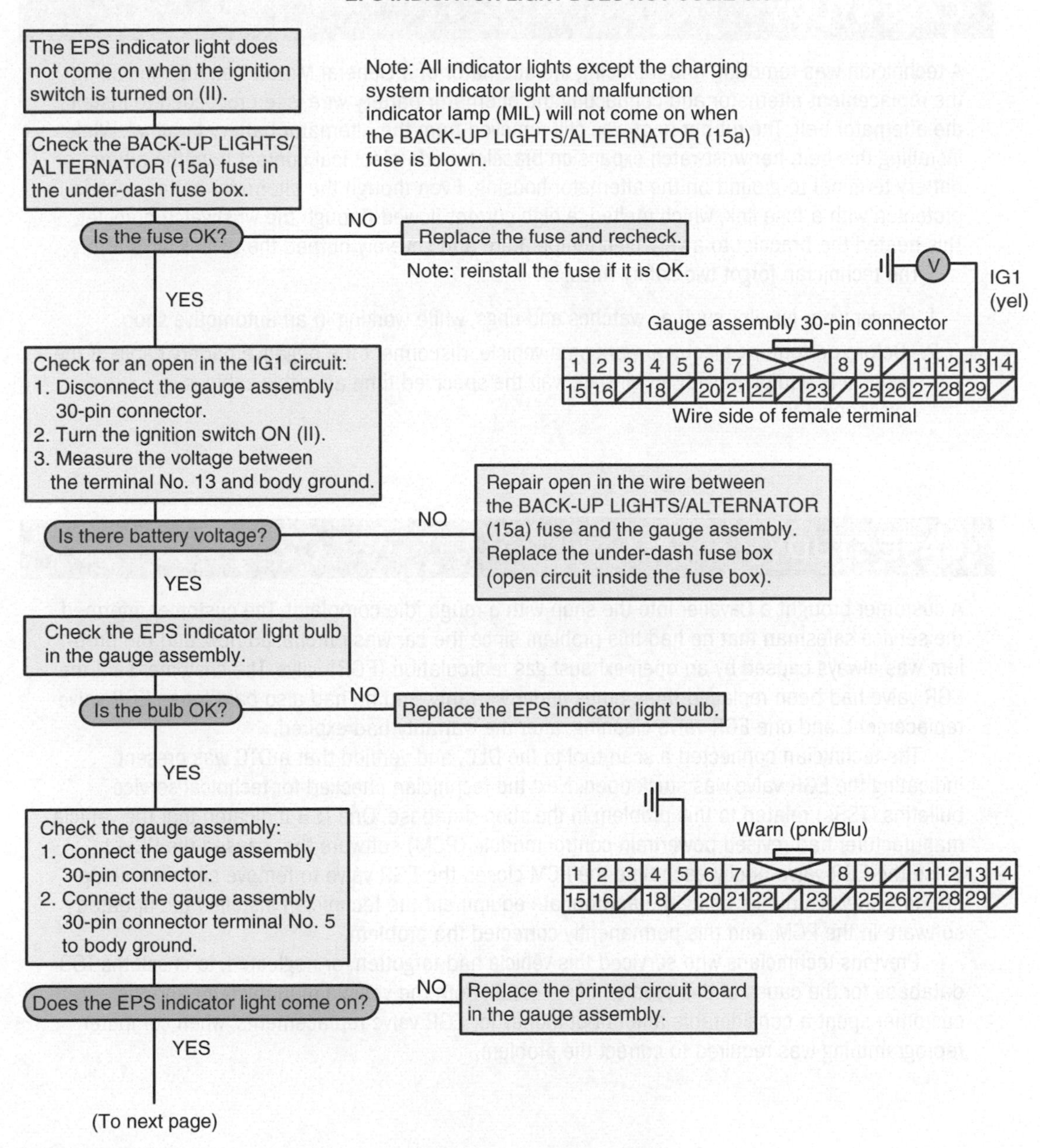

EPS INDICATOR LIGHT DOES NOT COME ON
The EPS indicator light does not come on when the ignition switch is turned on (II).
Note: All indicator lights except the charging system indicator light and malfunction indicator lamp (MIL) will not come on when the BACK-UP LIGHTS/ALTERNATOR (15a) fuse is blown.
Check the BACK-UP LIGHTS/ ALTERNATOR (15a) fuse in the under-dash fuse box.
Is the fuse OK?
NO
Replace the fuse and recheck.
Note: reinstall the fuse if it is OK.
YES
V
IG1 (yel)
Gauge assembly 30-pin connector
1 2 3 4 5 6 7 8 9 11 12 13 14
15 16 18 20 21 22 23 25 26 27 28 29
Wire side of female terminal
Check for an open in the IG1 circuit:
1. Disconnect the gauge assembly 30-pin connector.
2. Turn the ignition switch ON (II).
3. Measure the voltage between the terminal No. 13 and body ground.
Is there battery voltage?
NO
Repair open in the wire between the BACK-UP LIGHTS/ALTERNATOR (15a) fuse and the gauge assembly. Replace the under-dash fuse box (open circuit inside the fuse box).
YES
Check the EPS indicator light bulb in the gauge assembly.
Is the bulb OK?
NO
Replace the EPS indicator light bulb.
YES
Warn (pnk/Blu)
1 2 3 4 5 6 7 8 9 11 12 13 14
15 16 18 20 21 22 23 25 26 27 28 29
Check the gauge assembly:
1. Connect the gauge assembly 30-pin connector.
2. Connect the gauge assembly 30-pin connector terminal No. 5 to body ground.
Does the EPS indicator light come on?
NO
Replace the printed circuit board in the gauge assembly.
YES
(To next page)

Figure 2-56 Typical diagnostic chart.

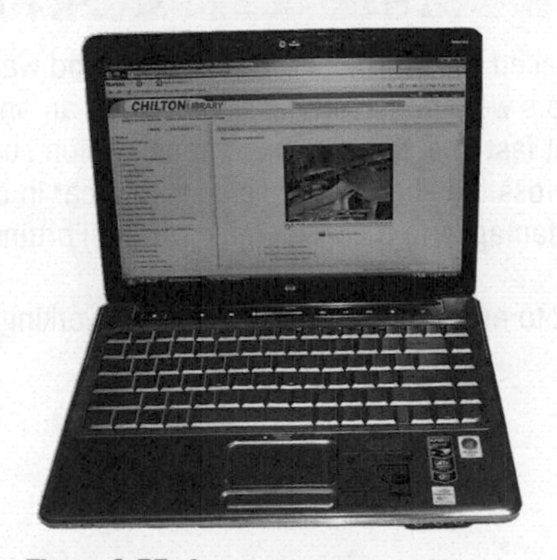

CHILTON

Figure 2-57 Computers are replacing printed service manuals in many shops.

CASE STUDY

A technician was removing and replacing the alternator of a General Motors car. After installing the replacement alternator and connecting the alternator battery wire, she proceeded to install the alternator belt. The rubber boot was still removed from the alternator battery terminal. While installing this belt, her wristwatch expansion bracelet made electrical contact from the alternator battery terminal to ground on the alternator housing. Even though the alternator battery wire is protected with a fuse link, which melted, a high current flowed through the wristwatch bracelet. This heated the bracelet to a very high temperature and severely burned the technician's arm.

The technician forgot two safety rules:

1. Never wear jewelry, such as watches and rings, while working in an automotive shop.
2. Before performing electrical work on a vehicle, disconnect the negative battery cable. If the vehicle is equipped with an air bag, wait the specified time after this cable is disconnected.

CASE STUDY

A customer brought a Cavalier into the shop with a rough idle complaint. The customer informed the service salesman that he had this problem since the car was purchased new, and the problem was always caused by an open exhaust gas recirculation (EGR) valve. The customer said the EGR valve had been replaced three times under warranty, and he had also paid for an EGR valve replacement, and one EGR valve cleaning, after the warranty had expired.

The technician connected a scan tool to the DLC, and verified that a DTC was present indicating the EGR valve was stuck open. Next the technician checked for technical service bulletins (TSBs) related to this problem in the shop database. One TSB indicated that the vehicle manufacturer had revised powertrain control module (PCM) software that caused the PCM to pulse the EGR valve briefly each time the PCM closed the EGR valve to remove carbon buildup on the EGR valve pintle. Using the appropriate equipment the technician installed the updated software in the PCM, and this permanently corrected the problem.

Previous technicians who serviced this vehicle had forgotten, or neglected, to check the TSB database for the cause of this problem. As a result, both the vehicle manufacturer and the customer spent a considerable amount of money for EGR valve replacements, when computer reprogramming was required to correct the problem.

CASE STUDY

A technician had just replaced the engine in a Ford vehicle and was performing final adjustments. In this shop, the cars were parked in the work bays at an angle on both sides of the shop. With the engine running at fast idle, the automatic transmission suddenly slipped into reverse. The car went backward across the shop and collided with a car in one of the electrical repair bays. Both vehicles were damaged to a considerable extent. Fortunately, no personnel were injured.

This technician forgot to apply the parking brake while working on the vehicle!

SUMMARY

- Each unit in the metric system of measurements can be multiplied or divided by 10 to obtain larger or smaller units.
- A stethoscope amplifies sound to help the technician identify the source of abnormal noises.
- A dial indicator is a precision measuring device that measures movement or wear on various components in thousands of an inch.
- A tire changer is a pneumatically operated machine that is used to dismount and mount tires on rims.
- An electronic wheel balancer indicates static and dynamic unbalance in tire and wheel assemblies. The electronic wheel balancer also indicates the mounting location and size of wheel weight(s) required to provide a balanced tire and wheel assembly.
- A spring compressor tool compresses a coil spring so the spring and other suspension components can be safely removed.
- A power steering pressure gauge is used to test power steering pump pressure and determine the condition of the power steering system.
- Turning radius gauges are placed under the front wheels during a wheel alignment.
- A magnetic wheel alignment gauge may be attached to the outer surface of the front wheel hubs to measure certain wheel alignment angles.
- A computer four wheel aligner is used to electronically measure the alignment angles on the front and rear suspensions.
- A scan tool is connected to the DLC under the dash to diagnose various electronic systems.
- A vehicle lift is used to raise vehicles off the floor so under-car service may be performed.

ASE-STYLE REVIEW QUESTIONS

1. While discussing systems of weights and measures:
   Technician A says the international system is called the metric system.
   Technician B says every unit in the metric system can be divided or multiplied by 10.
   Who is correct?
   A. A only C. Both A and B
   B. B only D. Neither A nor B

2. While discussing the metric system:
   Technician A says that 1 inch is equal to 2.54 millimeters.
   Technician B says that 1 meter is equal to 36.37 inches.
   Who is correct?
   A. A only C. Both A and B
   B. B only D. Neither A nor B

3. When diagnosing a vehicle operational problem, the first step is to:
   A. Think of the possible causes of the problem.
   B. Perform diagnostic tests to locate the exact cause of the problem.
   C. Consult TSB information to find the cause of the problem.
   D. Verify the customer complaint.

4. When testing power steering pump pressure, the power steering gauge valve should be closed for a maximum of:
   A. 10 seconds. C. 25 seconds.
   B. 20 seconds. D. 35 seconds.

5. While discussing employer and employee responsibilities:
   Technician A says employers are required to inform their employees about hazardous materials in the shop.
   Technician B says that employers have no obligation to inform their employees about the safe operation of shop equipment.
   Who is correct?
   A. A only C. Both A and B
   B. B only D. Neither A nor B

Task Completed

8. If the answer to both questions in Step 7 is yes, very slowly operate the release lever on the hydraulic floor jack to slowly lower the vehicle until the vehicle weight is completely supported on the safety stands. Be sure all the safety stand legs are contacting the shop floor evenly, and then remove the floor jack. ☐

9. Place the floor jack under the proper lifting point on the rear of the vehicle, and raise the floor jack pad until it contacts the lifting point. ☐

10. Operate the floor jack, and raise the vehicle to the desired height. Place safety stands under the proper support points on the rear of the vehicle chassis or suspension. ☐
    Are the safety stands properly positioned? ☐ Yes ☐ No
    No Instructor check ______________

11. Very slowly operate the release lever on the hydraulic floor jack to slowly lower the vehicle until the rear support points lightly contact the safety stands. Stop lowering the floor jack. ☐

12. Be sure the safety stands contact the proper rear support points on the vehicle, and check to be sure all the safety stand legs contact the shop floor evenly. ☐
    Are the safety stands contacting the proper support points on the vehicle?
    ☐ Yes ☐ No
    Are all the safety stand legs contacting the floor evenly? ☐ Yes ☐ No
    Instructor check ______________

13. If the answer to both questions in Step 12 is yes, very slowly operate the release lever on the hydraulic floor jack to slowly lower the rear of the vehicle until the vehicle weight is completely supported on the safety stands. Be sure all the safety stand legs are contacting the shop floor evenly; then remove the floor jack. ☐

Instructor's Response

____

____

____

Name ______________________________ Date ______________

JOB SHEET 6

FOLLOW THE PROPER PROCEDURE TO HOIST A CAR

Upon completion of this job sheet, you should be able to raise and lower a car on a hoist.

ASE Education Foundation Correlation ______________________________

This job sheet addresses the following **RST** task:

1-4. Identify and use proper procedures for safe lift operation.

We Support ASE | Education Foundation

Tools and Materials
Car
Lift with enough capacity to hoist the car

Caution
Do not raise a four-wheel-drive vehicle with a frame contact lift because this may damage axle joints.

Describe the vehicle being worked on:

Year ______________ Make ______________ Model ______________

VIN ______________ Engine type and size ______________

Procedure **Task Completed**

1. Always be sure the lift is completely lowered before driving the car onto the lift. ☐
2. Do not hit or run over lift arms and adaptors when driving a car onto the lift. ☐
3. Have a coworker guide you when driving a car onto the lift. ☐

WARNING Do not stand in front of a lift with the car coming toward you. This action may result in personal injury.

4. Be sure the lift pads on the lift are contacting the car manufacturer's recommended lifting points shown in the service manual. ☐
   Is the vehicle properly positioned on lift? ☐ Yes ☐ No
   Recommended front lifting points: ☐ Right side ☐ Left side
   Recommended rear lifting points: ☐ Right side ☐ Left side
   Are all four lift pads contacting the recommended lifting points? ☐ Yes ☐ No
   Instructor check ______________
5. Be sure the doors and hood are closed, and be sure there are no people in the car. ☐
6. When a car is lifted a short distance off the floor, stop the lift and check the contact between the lift pads and the car chassis to be sure the lift pads are still on the recommended lifting points. ☐
7. Be sure there is adequate clearance between the top of the vehicle and the shop ceiling or components under the ceiling. ☐

Task Completed

8. When a car is raised on a lift, be sure the safety mechanism is in place to prevent the lift from dropping accidentally.
Is the lift safety mechanism in place? ☐ Yes ☐ No
Instructor check ______________
List one precaution that must be observed when a front-wheel-drive vehicle is raised on a lift, and explain the reason for this precaution.______________

9. Prior to lowering a car on a lift, always make sure there are no tools, objects, or people under the vehicle. ☐

WARNING Do not rock a car on a lift during a service job. This action may cause the car to fall off the lift, resulting in personal injury and vehicle damage.

WARNING When a car is raised on a lift, removal of some heavy components may cause car imbalance on the lift, which may cause the car to fall off the lift, resulting in personal injury and vehicle damage.

10. Be sure the lift is lowered completely and no lift components are contacting the vehicle before backing the vehicle off the lift. ☐

Instructor's Response

____

____

____

Caution

If the proper lifting points are not used, components under the vehicle such as brake lines or body parts may be damaged. Failure to use the recommended lifting points may cause the car to slip off the lift, resulting in severe vehicle damage and personal injury.

Name ______________________________ Date ______________

JOB SHEET 7

DETERMINE THE AVAILABILITY AND PURPOSE OF SUSPENSION AND STEERING TOOLS

ASE Education Foundation Correlation

This job sheet addresses the following **RST** tasks:

1-2. Utilize safe procedures for handling of tools and equipment.

2-1. Identify tools and their usage in automotive applications.

2-3. Demonstrate safe handling and use of appropriate tools.

2-4. Demonstrate proper cleaning, storage, and maintenance of tools and equipment.

We Support ASE Education Foundation

Procedure

Locate the following tools in your shop or tool room, and explain the purpose of each tool.

1. Stethoscope: available? ☐ Yes ☐ No

Location ______________

Purpose ______________

2. Dial indicator for measuring ball joint wear: available? ☐ Yes ☐ No

Location ______________

Purpose ______________

3. Ball joint removal and installing tools: available? ☐ Yes ☐ No

Location ______________

Purpose ______________

4. Coil spring compressing tool: available? ☐ Yes ☐ No

Location ______________

Purpose ______________

5. Power steering pressure gauge: available? ☐ Yes ☐ No

Location ______________

Purpose ______________

6. Vacuum hand pump: available? ☐ Yes ☐ No

Location ______________

Purpose ______________

7. Turning radius gauge: available? ☐ Yes ☐ No

Location ______________

Purpose ______________

8. Brake pedal jack: available? ☐ Yes ☐ No

Location ______________

Purpose ______________

9. Steering wheel locking tool: available? ☐ Yes ☐ No
   Location ______________
   Purpose ______________
10. Scan tool: available? ☐ Yes ☐ No
   Location ______________
   Purpose ______________

Instructor's Response

____

____

____

CHAPTER 3

TIRE-AND-WHEEL SERVICING AND BALANCING

Upon completion and review of this Chapter, you should be able to:

- Diagnose tire thump and vibration problems.
- Diagnose steering pull problems related to tire condition.
- Rotate tires according to the vehicle manufacturer's recommended procedure.
- Remove and replace tire-and-wheel assemblies.
- Dismount, inspect, repair, and remount tires.
- Inspect wheel rims.
- Diagnose and service tire pressure monitoring systems.
- Measure tire-and-wheel radial and lateral runout.
- Diagnose problems caused by excessive radial or lateral tire-and-wheel runout.
- Measure tire tread wear.
- Perform off-car static wheel balance procedures.
- Perform off-car dynamic wheel balance procedures.
- Diagnose tire wear problems caused by tire-and-wheel imbalance.
- Perform on-car balance procedures.
- Perform noise, vibration, and harshness diagnosis.

Terms To Know

Antitheft locking wheel covers
Antitheft wheel nuts
Brake torque test
Downshift test
Dynamic wheel balance
Electronic vibration analyzer (EVA)
Heavy spot
Lateral tire runout
Neutral coast-down test
Neutral run-up test
Ply separation
Radial runout
Slow acceleration test
Standing acceleration test
Static balance
Steering input test
Steering pull
Tire conicity
Tire thump
Tire vibration
Tread wear indicators

 **Special Tools**

Basic technician's tool set
Service manual
Tire repair kit
Tread depth gauge

Proper servicing of tires and wheels is extremely important to maintain vehicle safety and provide normal tire life. Improperly serviced and/or balanced tires and wheels cause wheel vibration and shimmy problems, resulting in excessive tire tread wear, increased wear on suspension and steering components, and decreased vehicle stability and steering control.

TIRE NOISES AND STEERING PROBLEMS

Diagnosis of Tire Noises

Tire thump may be defined as a pounding noise caused by tire-and-wheel rotation.

Tire vibration is a fast shaking of the tire that is transferred to the chassis and passenger compartment.

Uneven tread surfaces may cause tire noises that seem to originate elsewhere in the vehicle. These noises may be confused with differential noise. Differential noise usually varies with acceleration and deceleration, whereas tire noise remains more constant in relation to these forces. Tire noise is most pronounced on smooth asphalt road surfaces at speeds of 15–45 mph (24–72 km/h).

Tire Thump and Vibration

When tire thump and tire vibration are present, check these items:

Radial runout refers to a tire out-of-round condition.

1. Cupped tire treads
2. Excessive tire **radial runout**
3. Manufacturing defects such as heavy spots, weak spots, or tread chunking
4. Incorrect wheel balance

SERVICE TIP Tire noise varies with road surface conditions; differential noise is not affected when various road surfaces are encountered.

Steering Pull

Steering pull is the tendency of the steering to gradually pull to the right or left when the vehicle is driven straight ahead on a reasonably smooth, straight road.

A vehicle should maintain the straight-ahead forward direction on smooth, straight road surfaces without excessive steering wheel correction by the driver. If the steering gradually pulls to one side on a smooth, straight road surface, a tire, steering, or suspension defect is present. Tires of different types, sizes, designs, or inflation pressures on opposite sides of a vehicle cause **steering pull**. Sometimes a tire manufacturing defect occurs in which the belts are wound off-center on the tire. This condition is referred to as **tire conicity**. This condition may also develop thousands of miles after the tire has been installed on the vehicle. A cone-shaped object rolls in the direction of its smaller diameter. Similarly, a tire with conicity tends to lead, or pull, to one side, which causes the vehicle to follow the action of the tire (**Figure 3-1**). In the trade this is often referred to as a radial pull.

Because tire conicity cannot be diagnosed by a visual inspection, it must be diagnosed by switching the two front tires and reversing the front and rear tires (**Figure 3-2**). Incorrect front suspension alignment angles as well as worn or damaged suspension components may also cause steering pull.

Some wheel balancers with force variation capabilities sense and indicate tire conicity defects. These wheel balancers have a roller that presses against the tire tread during the wheel balance procedure, and this roller senses tire conicity.

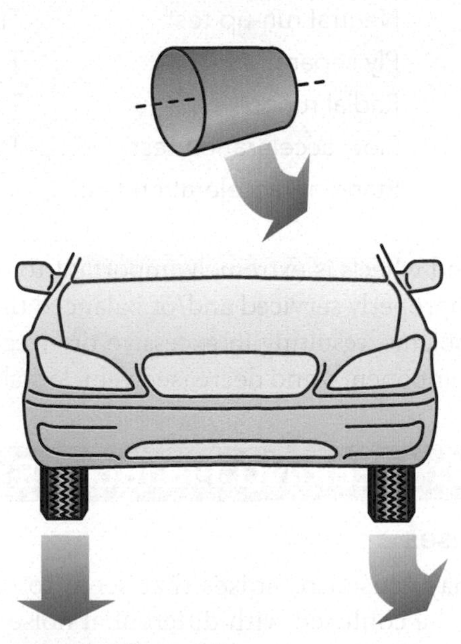

Figure 3-1 Tire conicity.

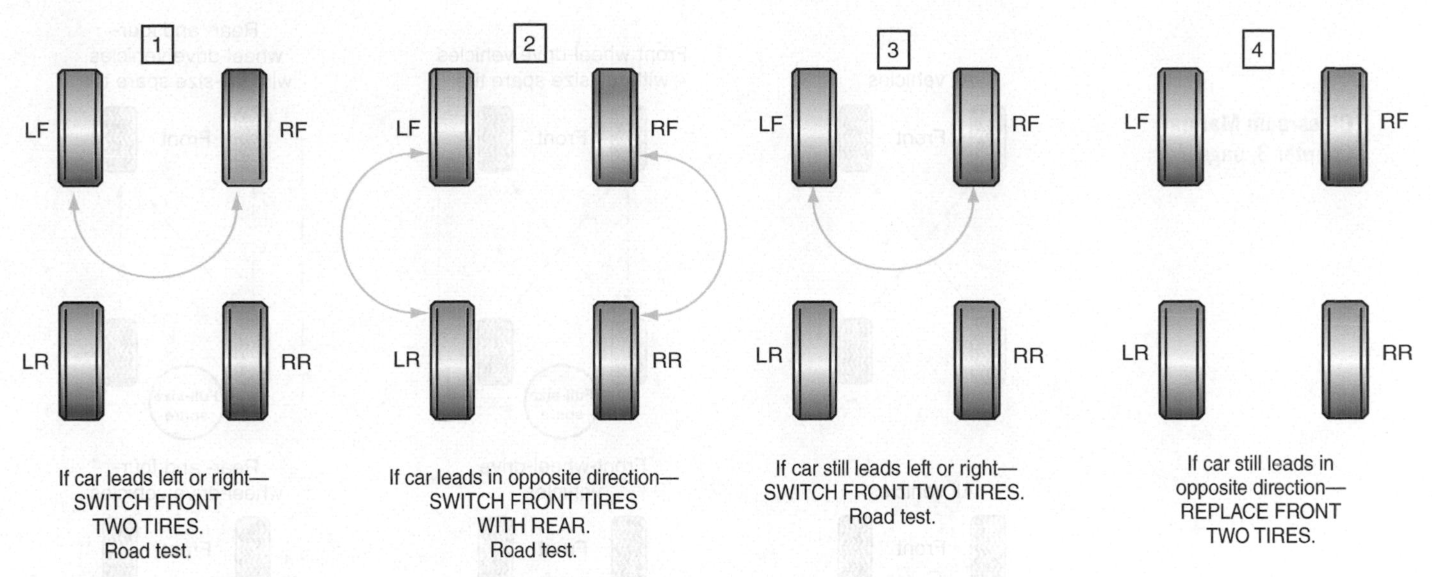

Figure 3-2 Tire conicity diagnosis.

SERVICE TIP Tire conicity is not visible. It can be diagnosed only by changing the tire-and-wheel position.

TIRE ROTATION

Driving habits determine tire life to a large extent. Severe brake applications, high-speed driving, turning at high speeds, rapid acceleration and deceleration, and striking curbs are just a few driving habits that shorten tire life. Most car manufacturers recommend tire rotation at specified intervals to obtain maximum tire life, generally every 6000 mi. (9656 km). The exact tire rotation procedure depends on the model year, the type of tires, and whether the vehicle has a conventional spare or a compact spare (**Figure 3-3**). Tire

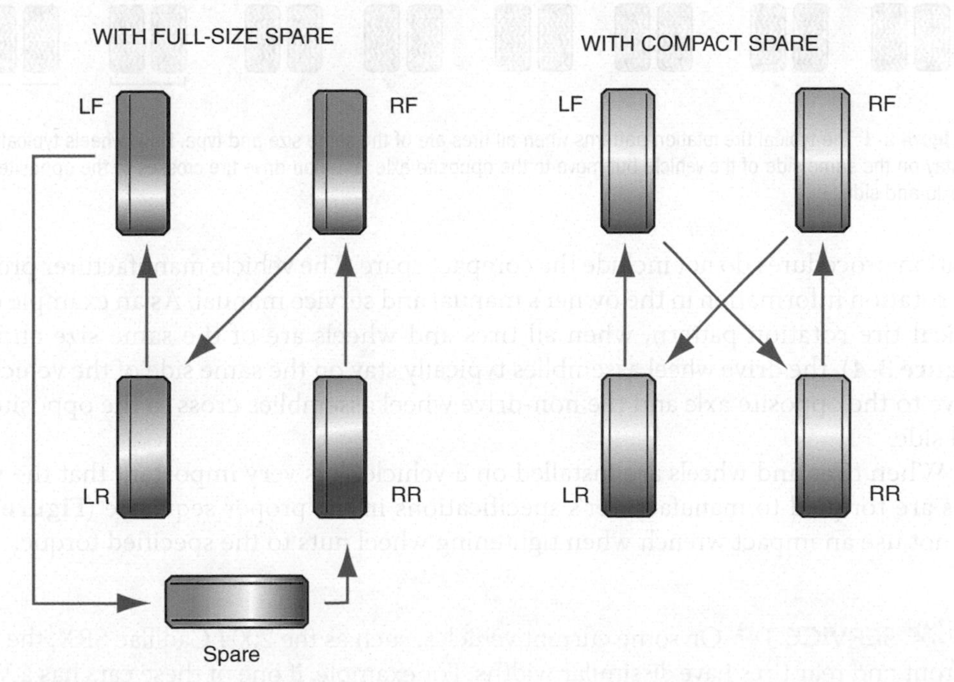

Figure 3-3 Radial tire rotation procedure.

Classroom Manual
Chapter 3, page 74

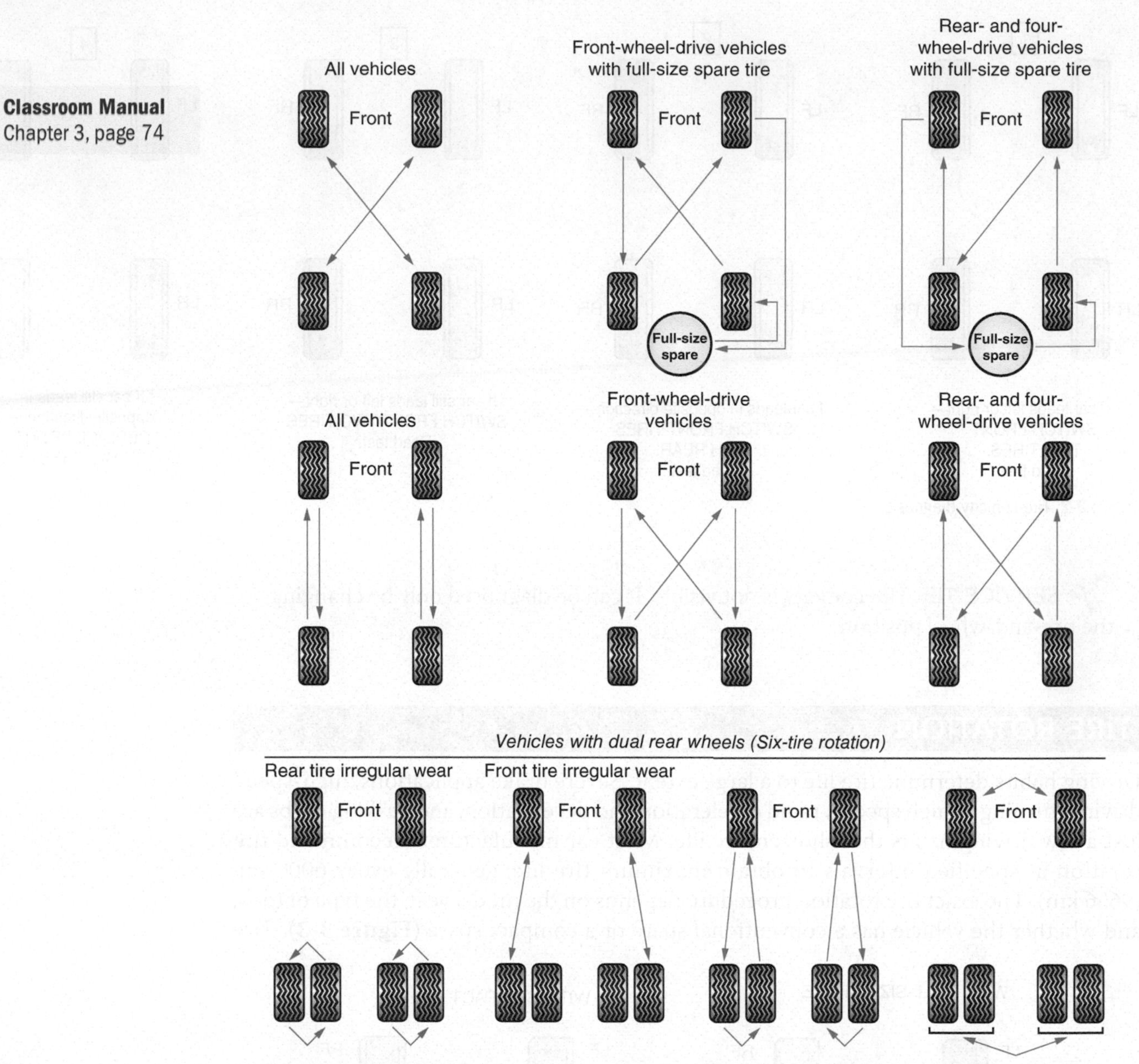

Figure 3-4 The typical tire rotation patterns when all tires are of the same size and type. Drive wheels typically stay on the same side of the vehicle but move to the opposite axle, and non-drive tire crosses to the opposite axle and side.

rotation procedures do not include the compact spare. The vehicle manufacturer provides tire rotation information in the owner's manual and service manual. As an example of the typical tire rotation pattern, when all tires and wheels are of the same size and type (**Figure 3-4**), the drive wheel assemblies typically stay on the same side of the vehicle but move to the opposite axle and the non-drive wheel assemblies cross to the opposite axle and side.

When tires and wheels are installed on a vehicle, it is very important that the wheel nuts are torqued to manufacturer's specifications in the proper sequence (**Figure 3-5**). Do not use an impact wrench when tightening wheel nuts to the specified torque.

SERVICE TIP On some current vehicles, such as the 2004 Cadillac SRX, the front and rear tires have dissimilar widths. For example, if one of these cars has a V8

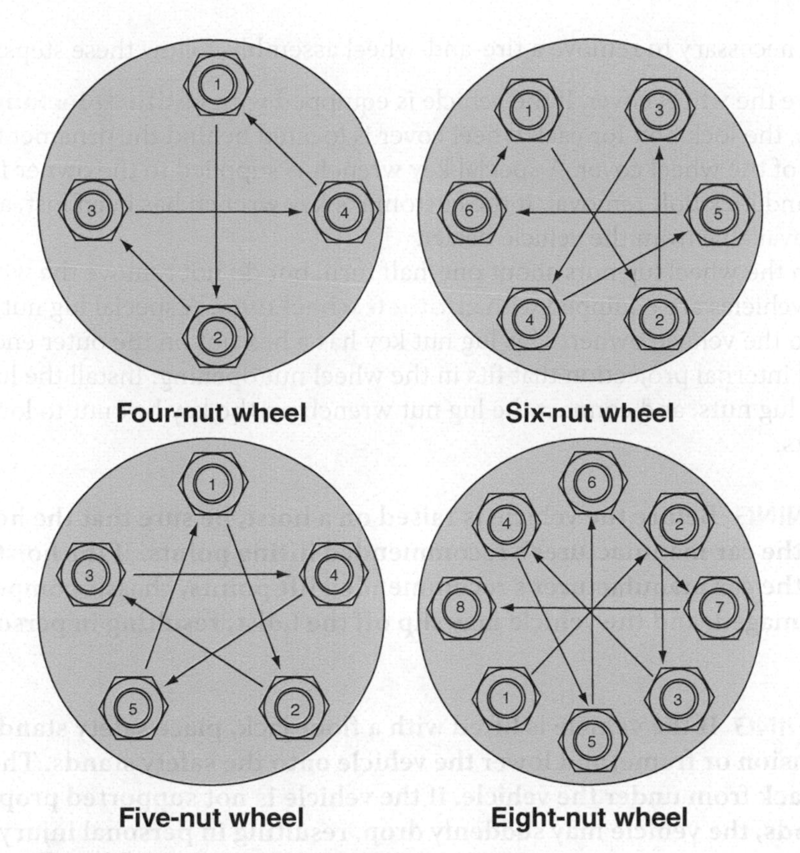

Figure 3-5 Wheel nut tightening sequence.

engine, the vehicle is equipped with 18-in. wheels, and the front tire size is P235/60R18 while the rear tire size is P255/55R18. These tires must not be rotated from front to rear, but they can be rotated from side to side. Directional tires would further complicate rotation options and would require tires to be dismounted and remounted and balanced for installation on the opposite side to insure that the proper directional orientation is maintained.

SERVICE TIP Some wheel covers have fake plastic lug nuts that must be removed to access the lug nuts. Be careful not to break the fake lug nuts.

TIRE-AND-WHEEL SERVICE

Tire-and-Wheel Removal

CUSTOMER CARE During many automotive service operations, including tire-and-wheel service, the technician literally has the customer's life in his or her hands! Always perform tire-and-wheel service carefully and thoroughly. Watch for unsafe tire or wheel conditions, and report these problems to the customer. When you prove to the customer that you are concerned about vehicle safety, you will probably have a steady customer.

When it is necessary to remove a tire-and-wheel assembly, follow these steps:

Antitheft locking wheel covers prevent unauthorized personnel from removing the wheel covers.

Antitheft wheel nuts prevent unauthorized personnel from removing the wheel nuts.

1. Remove the wheel cover. If the vehicle is equipped with **antitheft locking wheel covers**, the lock bolt for each wheel cover is located behind the ornament in the center of the wheel cover. A special key wrench is supplied to the owner for ornament and lock bolt removal. If the customer's key wrench has been lost, a master key is available from the vehicle dealer.
2. Loosen the wheel lug nuts about one-half turn, but do not remove the wheel nuts. Some vehicles are equipped with **antitheft wheel nuts**. A special lug nut key is supplied to the vehicle owner. This lug nut key has a hex nut on the outer end and a special internal projection that fits in the wheel nut opening. Install the lug nut key on the lug nuts, and connect the lug nut wrench on the key hex nut to loosen the lug nuts.

Caution

If heat is used to loosen a rusted wheel, the wheel and/or wheel bearings may be damaged.

WARNING **Before the vehicle is raised on a hoist, be sure that the hoist is lifting on the car manufacturer's recommended lifting points. If the hoist is not lifting on the car manufacturer's recommended lift points, chassis components may be damaged, and the vehicle may slip off the hoist, resulting in personal injury.**

WARNING **If the vehicle is lifted with a floor jack, place safety stands under the suspension or frame, and lower the vehicle onto the safety stands. Then remove the floor jack from under the vehicle. If the vehicle is not supported properly on safety stands, the vehicle may suddenly drop, resulting in personal injury.**

3. Raise the vehicle on a hoist or with a floor jack to a convenient working level.
4. Chalk mark the tire, wheel, and one of the lug nuts so the tire-and-wheel can be reinstalled in the same position.
5. Remove the lug nuts and the tire-and-wheel assembly. If the wheel is rusted and will not come off, hit the inside of the wheel with a large rubber mallet. Do not hit the wheel with a steel hammer, because this action could damage the wheel. Do not heat the wheel.

TIRE-AND-WHEEL SERVICE PRECAUTIONS

There are many different types of tire changing equipment in the automotive service industry. However, specific precautions apply to the use of any tire changing equipment.

These precautions include the following:

1. Before you operate any tire changing equipment, always be absolutely certain that you are familiar with the operation of the equipment.
2. When operating tire changing equipment, always follow the equipment manufacturer's recommended procedure.
3. Always deflate a tire completely before attempting to dismount the tire.
4. Clean the bead seats on the wheel rim before mounting the tire on the wheel rim.
5. Lubricate the outer surface of the tire beads with rubber lubricant before mounting the tire on the wheel rim.
6. When the tire is mounted on the wheel rim, be sure the tire is positioned evenly on the wheel rim.
7. While inflating a tire, do not stand directly over the tire. An air hose extension allows the technician to stand back from the tire during the inflation process.
8. Do not overinflate tires.

9. When mounting tires on cast aluminum alloy wheel rims or cast magnesium alloy wheel rims, always use the tire changing equipment manufacturer's recommended tools and procedures.
10. When mounting or dismounting run-flat tires, be sure the tire changing equipment is compatible with these tires and wheels.

Caution

Never use petroleum-based lubricants on tire beads. This action damages the beads.

Tire Dismounting

Always use a tire changer to dismount tires. Do not use hand tools or tire irons for this purpose. Various types of tire changers are available. When servicing tires it is very important that the tire changing equipment will mount and dismount run-flat tires, low-profile tires, and tires mounted on alloy wheels without damaging the wheel rims or tires. Some tire changers have the features illustrated in **Figure 3-6** and described in the following explanation:

Caution

If hand tools or tire irons are used to dismount tires, tire bead and wheel rim damage may occur.

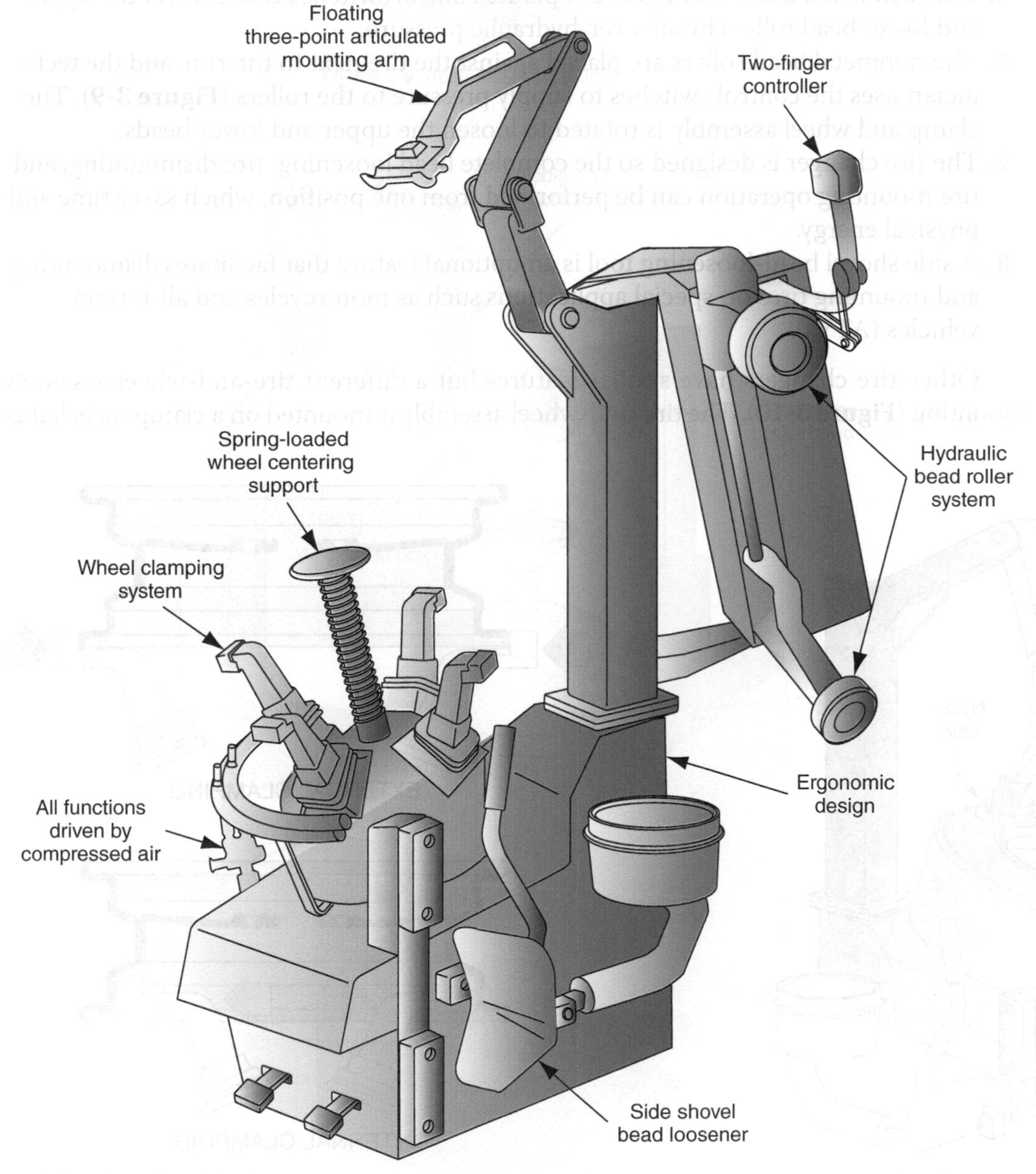

Figure 3-6 Tire changer.

Figure 3-14 Tire dunk tank used for air leak detection.

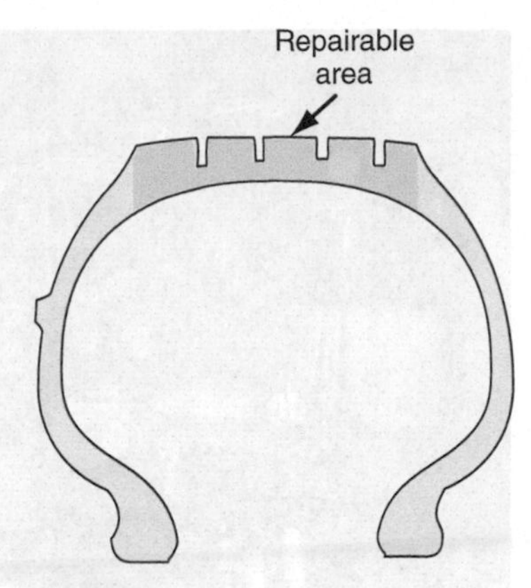

Figure 3-15 Repairable area on bias-ply and belted bias-ply tires.

Because compact spare tires have thin treads, do not attempt to repair these tires.

Inspect the tire; do not repair a tire with any of the following defects, signs of damage, or excessive wear:

1. Tires with the wear indicators showing
2. Tires worn until the fabric or belts are exposed
3. Bulges or blisters
4. **Ply separation**
5. Broken or cracked beads
6. Cuts or cracks anywhere in the tire

When **ply separation** occurs, the tire plies are pulled apart. This condition often appears as a bulge on the tire surface.

Classroom Manual
Chapter 3, page 72

Caution

Radial tire patches should have arrows that must be positioned parallel to the radial plies to provide proper adhesion.

Caution

The use of abrasive cleaners, alkaline-base detergents, or caustic agents on aluminum or magnesium wheel rims may cause discoloration or damage to the protective coating.

Because most vehicles are equipped with tubeless tires, we will discuss this type of tire repair. If the cause of the puncture, such as a nail, is still in the tire, remove it from the tire. Most punctures can be repaired from inside the tire with a service plug or vulcanized patch service kit. The instructions of the tire service kit manufacturer should be followed, but we will discuss three common tire repair procedures.

Plug Installation Procedure

1. Remove the object that caused the puncture, if still present.
2. Insert a tire rasp
3. Select a plug slightly larger than the puncture opening, and insert the plug in the eye of the insertion tool.
4. Wet the plug and the insertion tool with vulcanizing fluid.
5. While holding and stretching the plug, pull the plug into the puncture from the inside of the tire (**Figure 3-16**). The head of the plug should contact the inside of the tire. If the plug pulls through the tire, repeat the procedure.
6. Cut the plug off 1/32 in. from the tread surface. Do not stretch the plug while cutting.

Cold Plug Patch Repair Procedure

A tire plug patch repair is the most common type of repair and the one recommended by both tire and vehicle manufacturers for a puncture injury. Always refer to both vehicle and tire manufacturers' service information for acceptable tire repair procedures.

1. Adjust all tire pressures to specifications.
2. The first step is to find the tire injury that is the source of the air leak. If the injury is not visually apparent, a tire dunk tank or soapy water solution may be needed.

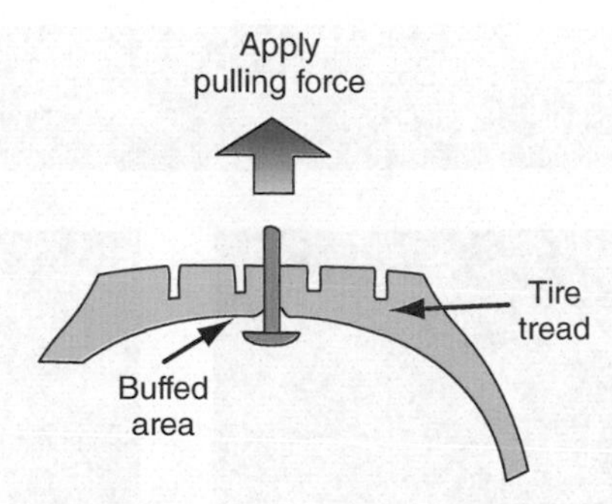

Figure 3-16 Plug installation procedure.

3. Once the injury has been located, mark the exterior of the tire injury with a tire crayon. Dismount the tire from the rim, inspect the condition of the interior of the tire and locate the puncture on the inner lining. Circle and X the puncture with a tire crayon and remove the object that caused the puncture if still present.
4. Prepare the inner tire surface with a buffing tool. Buff the area around the puncture with a tire repair buffing wheel. Surface should be buffed to a smooth finish with a RMA (Rubber Manufacturers Association) 1-2 texture buffing wheel.
5. Once surface has been buffed, clean area with a tire buffer/cleaner such as Camel 12-092.
6. Refer to information on the product or manufacturer's Safety Data Sheet and follow guidelines for handling and disposal.
7. Guide a radial hand rasp/reamer through the puncture to allow the plug patch to fit properly.
8. Apply tire patch cement according to manufacturer's recommendations and then install plug patch. The plug portion of the patch has a removable metal sleeve to help guide it through the puncture injury.
9. Using a corrugated patch stitching tool work out all air bubbles from under patch to ensure proper vulcanization of tire patch to inner lining of tire.
10. Trim the tire plug flush with tire tread.
11. Remount and balance tire-and-wheel assembly.

Photo Sequence 3 shows a typical procedure for inspecting tire-and-wheel assembly for an air leak and repairing a tire with a plug patch.

PHOTO SEQUENCE 3

Typical Procedure for Inspecting Tire-and-Wheel Assembly for an Air Leak and Repair with a Plug Patch

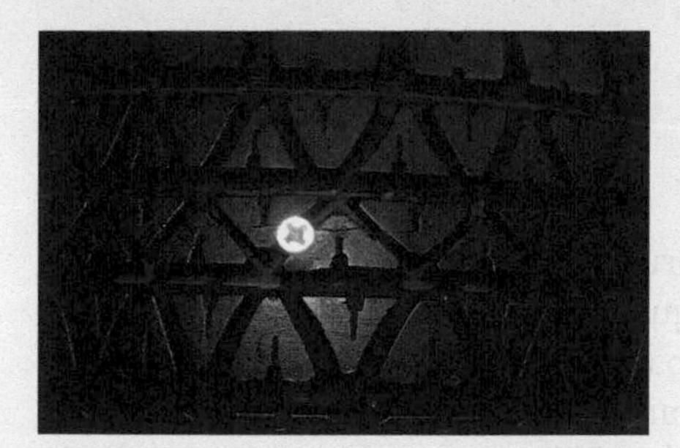

P3-1 The first step is to find the tire injury that is the source of the air leak. If the injury is not visually apparent, a tire dunk tank or soapy water solution may be needed.

P3-2 For a tire air leak that is not visually apparent place the tire in a dunk tank.

P3-3 Dismount the tire from the rim and locate the puncture on the inner lining; circle it with a tire crayon and remove the object that caused the puncture if still present.

PHOTO SEQUENCE 3 (CONTINUED)

P3-4 Prepare the inner tire surface with a buffing tool.

P3-5 Once surface has been buffed, clean area with a tire buffer/cleaner such as Camel 12-092.

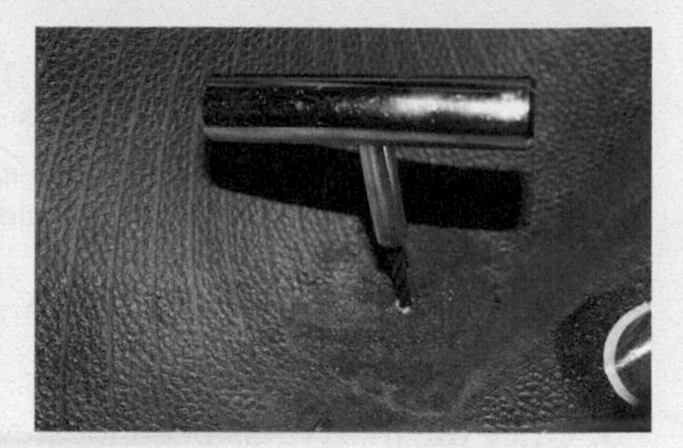

P3-6 Guide a radial hand rasp/reamer through the puncture to allow the plug patch to fit properly.

P3-7 Apply tire patch cement according to manufacturer's recommendations and then install plug patch. The plug portion of the patch has a removable metal sleeve to help guide it through the puncture injury.

P3-8 Using a corrugated patch stitching tool work out all air bubbles from under patch to ensure proper vulcanization of tire patch to inner lining of tire.

P3-9 Trim the tire plug flush with tire tread.

P3-10 Remount and balance tire-and-wheel assembly.

Hot Patch Installation Procedure

1. Buff the area around the puncture with a wire brush or buffing wheel.
2. Apply vulcanizing fluid to the buffed area, if required.
3. Peel the backing from the patch and install the patch so it is centered over the puncture on the inside of the tire. Many hot patches are heated with an electric heating element clamped over the patch. This element should be clamped in place for the amount of time recommended by the equipment or patch manufacturer.
4. After the heating element is removed, allow the patch to cool for a few minutes and be sure the patch is properly bonded to the tire.

WHEEL RIM SERVICE

Steel wheel rims must not be welded, heated, or peened with a ball-peen hammer. These procedures may weaken the rim and create a safety hazard.

Steel rims should be spray cleaned with a water hose. Aluminum or magnesium wheel rims should be cleaned with a mild soap and water solution, and rinsed with clean water. The use of abrasive cleaners, alkaline-base detergents, or caustic agents may damage aluminum or magnesium wheel rims. Clean the rim bead seats on these wheel rims thoroughly with the mild soap and water solution. The rim bead seats on steel wheel rims should be cleaned with a wire brush or coarse steel wool.

Steel wheel rims should be inspected for excessive rust and corrosion, cracks, loose rivets or welds, bent or damaged bead seats, and elongated lug nut holes. Aluminum or magnesium wheel rims should be inspected for damaged bead seats, elongated lug nut holes, cracks, and porosity. If any of these conditions are present on either type of wheel rim, replace the wheel rim.

Many shops always replace the tire valve assembly when a tire is repaired or replaced. This policy helps prevent future problems with tire valve leaks. The inner end of the valve may be cut off with a pair of diagonal pliers, and then the outer end may be pulled from the rim. Coat the new valve with rubber tire lubricant and pull it into the rim opening with a special puller screwed onto the valve threads.

Wheel Rim Leak Repair

A wheel rim leak may be repaired if the leak is not caused by excessive rust on a steel rim, and the rim is in satisfactory condition.

Follow these steps for wheel rim leak repair:

1. Use #80-grit sandpaper to thoroughly clean the area around the leak on the tire side of the rim.
2. Use a shop towel to remove any grit from the leak area.
3. Be sure the wheel rim is at room temperature, and apply a heavy coating of silicone rubber sealer over the leak area.
4. Spread the sealer over the entire sanded area with a putty knife.
5. Allow the sealer to cure for 6 hours before remounting the tire.

Installing an inner tube to correct leaks in a tubeless tire or wheel rim is not an approved procedure.

TIRE REMOUNTING PROCEDURE

1. Be sure the wheel rim bead seats are thoroughly cleaned.
2. Coat the tire beads and the wheel rim bead seats with rubber tire lubricant.
3. Secure the wheel rim on the tire changer with the narrow bead ledge facing upward, and place the tire on top of the wheel rim with the bead on the lower side of the tire in the drop center of the wheel rim.
4. Use the tire changer head or lever under the tire bead to install the tire bead over the wheel rim. Always operate the tire changer with the manufacturer's recommended procedure.
5. Repeat Steps 3 and 4 to install the upper bead over the wheel rim.
6. Rotate the tire on the wheel rim until the crayon mark is aligned with the valve stem. This mark was placed on the tire prior to dismounting.

WARNING When a bead expander is installed around the tire, never exceed 10 psi (69 kPa) pressure in the tire. A higher pressure may cause the expander to break and fly off the tire, causing serious personal injury or property damage.

WARNING When a bead expander is not used, never exceed 40 psi (276 kPa) tire pressure to move the tire beads out tightly against the wheel rim. A higher pressure may blow the tire bead against the rim with excessive force, and this action could burst the rim or tire, resulting in serious personal injury or property damage.

WARNING While inflating a tire, do not stand directly over the tire. In this position, serious injury could occur if the tire or wheel rim flies apart.

Follow the recommended procedure supplied by the manufacturer of the tire changer to inflate the tire. This procedure may involve the use of a bead expander installed around the center of the tire tread to expand the tire beads against the wheel rim. If a bead expander is used, inflate the tire to 10 psi (69 kPa) to move the beads out tightly against the wheel rim. Never exceed this pressure with a bead expander installed on the tire. Always observe the circular marking around the tire bead as the tire is inflated. This mark should be centered around the wheel rim (**Figure 3-17**). Always observe both beads while a tire is inflated. If the circular mark around the tire bead is not centered on the rim, deflate the tire and center it on the wheel rim. If a bead expander is not used, never exceed 40 psi (275 kPa) when moving tire beads out tightly against the rim. When either tire bead will not move out tightly against the wheel rim with 40 psi (276 kPa) tire pressure, deflate the tire and center it on the wheel rim again.

Warning: Some run flat tires, such as the Goodyear Extended Mobility Tire (EMT), may require more than 275 kPa (40 psi) to seat the bead. For tires that require high air pressure, a tire safety cage must be used (**Figure 3-18**). NEVER stand, lean, or reach over the assembly during inflation.

Photo Sequence 4 shows a typical procedure for dismounting and mounting a tire on a wheel assembly.

DIAGNOSING AND SERVICING TIRE PRESSURE MONITORING SYSTEMS

Visual Inspection

Classroom Manual
Chapter 3, page 77

If the TPMS warning light is illuminated and/or a warning message is displayed in the message center, the first step in diagnosing the system is a visual inspection. During this inspection, be sure each tire is inflated to the specified pressure. Make sure all four or five tires are the size specified by the vehicle manufacturer, and that each wheel contains a tire pressure sensor. Inspect the wiring harness connection to the TPMS module for loose or corroded connections and damaged wires.

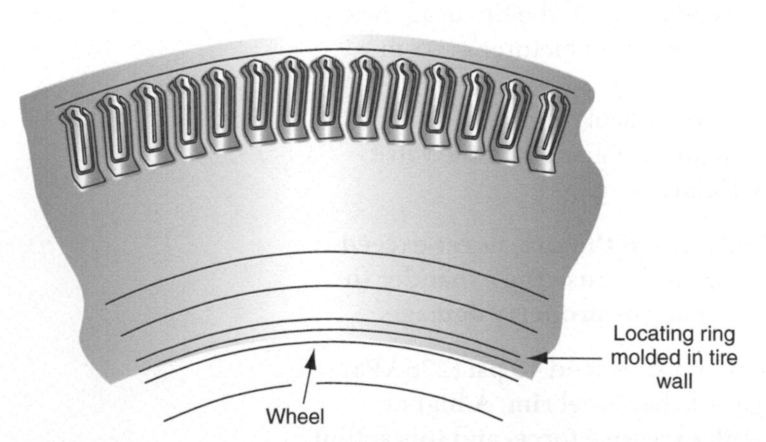

Figure 3-17 The circular ring around the tire bead must be centered on the wheel rim.

Figure 3-18 Tire inflation safety cage.

Scan Tool Diagnosis

The vehicle manufacturer's recommended diagnostic procedure varies considerably depending on the vehicle model year. Most systems are diagnosed using a scan tool with TPMS capabilities. The following is a typical scan tool diagnosis. If the visual inspection does not reveal any defects, use a scan tool to proceed with the diagnostic system check.

Diagnostic System Check

The diagnostic system check provides the following information:

1. Identification of the control modules in the TPMS system.
2. Indication of the ability of the system control modules to communicate through the serial data circuit.
3. Indication of any diagnostic trouble codes (DTCs) stored in the system control modules.

Follow these steps to complete the diagnostic system check.

1. Connect a scan tool to the data link connector (DLC) under the left side of the instrument panel, and turn the scan tool on.

PHOTO SEQUENCE 4
Typical Procedure for Dismounting and Mounting a Tire on a Wheel Assembly

P4-1 Dismounting the tire from the wheel begins with releasing the air, removing the valve stem core, and unseating the tire from its rim. The machine does the unseating. The technician merely guides the operating lever.

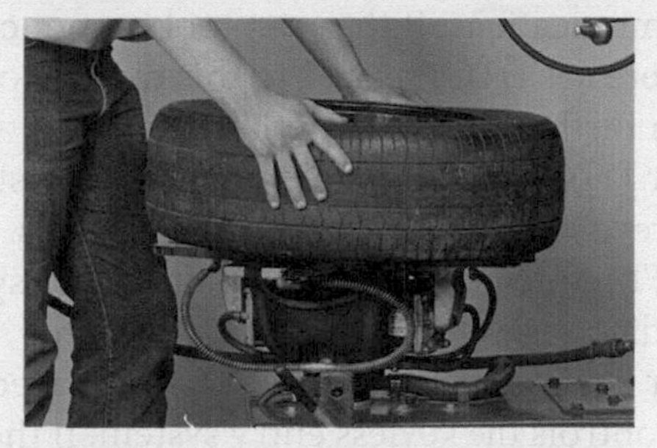

P4-2 Once both sides of the tire are unseated, place the tire-and-wheel onto the machine. Then depress the pedal that clamps the wheel to the tire machine.

P4-3 Lower the machine's arm into position on the tire-and-wheel assembly.

P4-4 Insert the tire iron between the upper bead of the tire and the wheel. Depress the pedal that causes the wheel to rotate. Do the same with the lower bead.

P4-5 After the tire is totally free from the rim, remove the tire.

P4-6 Prepare the wheel for the mounting of the tire by using a wire brush to remove all dirt and rust from the sealing surface. Apply rubber compound to the bead area of the tire.

PHOTO SEQUENCE 4 (CONTINUED)

P4-7 Place the tire onto the wheel and lower the arm into place. As the machine rotates the wheel, the arm will force the tire over the rim. After the tire is completely over the rim, install the air ring over the tire. Activate it to seat the tire against the wheel.

P4-8 Reinstall the valve stem core and inflate the tire to the recommended inflation.

2. Turn the ignition switch on, and do not start the engine.
3. Select antenna module on the scan tool. Check for the DTCs related to the antenna module. If there are DTCs displayed with a "U" prefix, there is a defect in the data link communications. If the scan tool cannot communicate with the antenna module, or there are DTCs related to this module or the data link communications, proceed with further diagnosis of these items.
4. Check for DTCs related to the radio/audio system. The antenna grid in the rear window receives TPMS system sensor signals. This grid shares its connector with the AM/FM antenna grid. If any radio/audio system DTCs are present, proceed with the diagnosis of the radio/audio system.
5. Check for keyless entry system DTCs displayed in the scan tool. The antenna module also controls the keyless entry system. If the scan tool displays DTCs related to the keyless entry system, proceed with the diagnosis of these DTCs.
6. Check for TPMS system DTCs on the scan tool display. If TPMS system DTCs are present, diagnose the cause of these DTCs.

TPMS System Data Display

TPMS data display may be selected on the scan tool display. The data display is very useful when diagnosing the TPMS system and sensors. The following data displays may be selected on the scan tool:

1. Battery voltage—The scan tool displays the amount of battery voltage supplied to the antenna module.
2. LF pressure sensor ID—The scan tool displays an eight-digit number or an asterisk. The eight-digit number is a unique LF sensor identification number, and this display indicates the number has been learned by the antenna module. If an asterisk is displayed, the sensor ID number has not been learned by the antenna module.
3. LF pressure sensor mode—This mode display may indicate stationary, wake, drive, remeasure, learn, low bat, or N/A (not available). If stationary is displayed, the sensor roll switch is open and the sensor has sent a stationary message that only occurs every 60 minutes from the previous stationary transmission. A wake display indicates the sensor has detected an initial roll switch closure, and the sensor is

changing from the stationary mode to the drive mode. A drive display indicates the vehicle speed is above 20 mph (32 km/h), and the sensor roll switch has been closed for a minimum of 10 seconds. In this mode the sensor transmits every 60 seconds. If remeasure is displayed, the sensor has detected a 1.6 psi (11 kPa) pressure change. A learn display indicates the sensor has been activated by a low-frequency voltage signal that occurs during a sensor relearn procedure. A low battery display indicates the internal sensor battery has low voltage and sensor replacement is necessary.

4. LF tire pressure—The scan tool displays the actual tire pressure between 0 and 51 psi (0 to 344 kPa).
5. LF tire pressure sensor status—In this mode the scan tool displays valid or invalid. A valid display indicates the specified LF tire pressure is present. If invalid is displayed, the LF tire pressure is not within specifications and DTC C0750 is currently set in the antenna module memory. This DTC indicates a defective LF sensor or faulty LF sensor circuit.

Caution

Do not remove the valve stem core to relieve the tire pressure. If the valve stem core is inadvertently removed from the valve stem, a new nickel-plated valve core must be installed. Failure to use a nickel-plated valve core will result in corrosion and possible loss of tire pressure.

The same data from each wheel sensor may be displayed on the scan tool. Other DTCs include C0755, indicating a defective LR sensor; C0760, representing a faulty RF sensor; and C0765, indicating a defect in the RR sensor. A current DTC indicates the fault causing the DTC is present at the time of diagnosis. The DTCs vary depending on the vehicle make and model year. A history DTC represents a fault that occurred in the past, but it is not present at the time of diagnosis. The vehicle speed must be above 20 mph (32 km/h) for the antenna module to run the DTC check. A DTC is set in the antenna module memory if any wheel sensor does not enter the drive mode or does not transmit any data for 10 minutes or more. When a DTC is set in the antenna module memory, the following actions are taken:

1. A DTC is stored in memory.
2. The driver information center (DIC) displays a SERVICE TIRE MONITOR message.
3. The DIC indicates row of dashes in place of the suspect tire pressure display.

Caution

The tire pressure sensors are not serviceable. If a sensor is defective, it must be replaced.

A current DTC is cleared from memory when the fault causing the DTC is corrected. A history DTC is cleared after 100 consecutive fault-free ignition cycles. A scan tool may be used to clear DTCs.

Sensor Replacement

Vehicles with a TPMS may require a tire pressure adjustment in relation to atmospheric temperature. For example, if the atmospheric temperature decreases 30°F when a vehicle is parked outside overnight, the tire pressure decreases 3 psi. This pressure decrease may activate the TPMS warning system. Refer to the vehicle manufacturer's specifications for tire inflation pressure at various atmospheric temperatures.

Caution

Tire pressure sealing products may render the TPMS sensor inoperative. If this occurs, remove all of the tire sealing material and replace the sensor.

Proper valve stem and sensor removal procedures must be followed to avoid damage to the sensor: To remove a valve stem and sensor, follow these steps:

1. Before removing the tire-and-wheel assembly from the vehicle, mark the wheel rim in relation to one of the wheel studs.
2. Position the tire-and-wheel assembly with the valve stem at the 6 o'clock position.
3. Remove the sensor retention nut and push the valve stem and sensor into the tire.
4. Place the wheel-and-tire on the turntable of the tire changer so the valve stem is positioned 270° from the mounting/demounting arm on the tire changer (**Figure 3-19**).
5. Loosen both tire beads away from the wheel rim.
6. Place index marks on the tire beside the valve stem opening and the wheel weight positions.
7. Lubricate the outer tire bead, and dismount the outer bead over the wheel rim.
8. Lift up on the outer tire bead, and remove the TPMS sensor and valve stem assembly.

Caution

Each time a TPMS sensor is removed, a new grommet, retention nut, and valve cap must be installed. Replace the valve core if it has been removed or damaged.

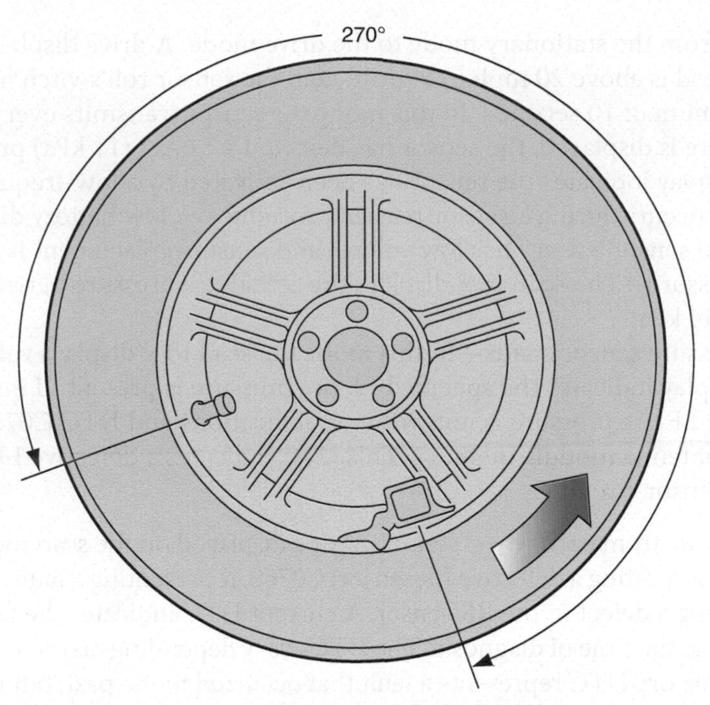

Figure 3-19 Proper tire-and-wheel position on a tire changer.

9. Install the new sensor in the valve stem opening, and install a new grommet and retention nut.
10. Tighten the retention nut to 71 in.-lb (8 Nm).
11. Lubricate and mount the outer tire bead.
12. Inflate the tire to the specified pressure.
13. Remove the tire-and-wheel from the tire changer and install this assembly in the proper position on the wheel studs. Tighten the wheel retaining nuts in the proper sequence to the specified torque.

Sensor Learning Procedure with Magnetic Tool

If a TPMS sensor or component is serviced, a sensor learning procedure must be performed. There are a number of different sensor learning procedures depending on the vehicle make and model year. The sensor learning procedure usually involves the use of a magnetic tool or a scan tool.

Follow these steps to complete the sensor learning procedure with a magnetic tool:

1. Starting with the ignition switch off, cycle the ignition switch on and off three times, and on the third cycle leave the ignition switch in the on position. Do not wait more than 2 seconds between switch cycles.
2. Press and release the brake pedal.
3. Repeat the ignition switch cycling procedure as explained in Step 1. Upon completion of this procedure, the horn should sound once to indicate successful entry to the learn mode.
4. After the horn sounds, a TRAIN LEFT FRONT TIRE message should appear in the instrument panel message center.
5. Place the special magnetic tool on the valve stem of the left front tire (**Figure 3-20**). When the TPMS module recognizes the left front sensor, the horn sounds momentarily.

Special Tool
Magnetic learn tool

WARNING **The special magnetic tool may adversely affect magnetically sensitive devices such as heart pacers, and this action may result in personal injury!**

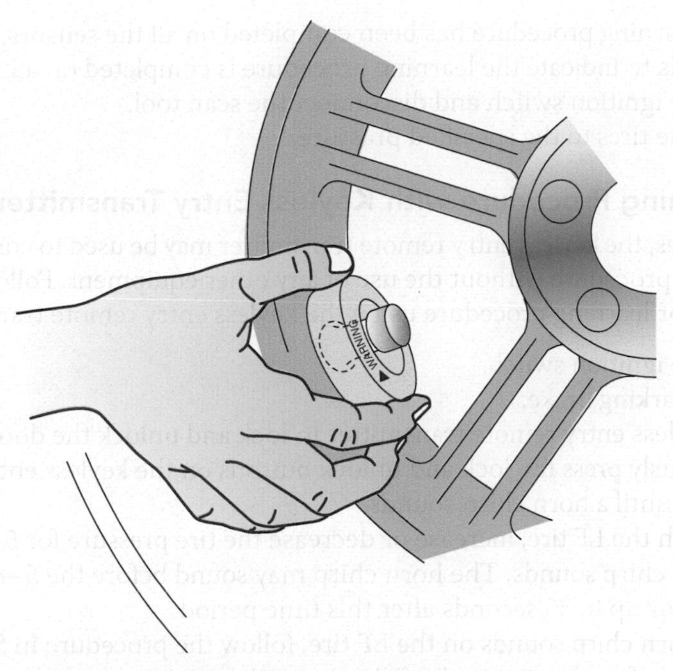

Figure 3-20 Magnetic learning tool for TPMSs.

6. Repeat Step 5 at the right front, right rear, left rear, and spare wheels.
7. If the learn procedure fails on any wheel, the horn sounds twice, and a TIRES NOT LEARNED-REPEAT message appears in the message center. If this action occurs, the learn procedure must be repeated from Step 1.

Sensor Learning Procedure with Scan Tool

Follow these steps to complete the sensor learning procedure with a scan tool:

1. Connect a scan tool to the DLC.
2. Turn on the ignition switch, and do not start the engine.
3. Apply the parking brake.
4. Select Special Functions on the scan tool.
5. Select Sensor Learn Mode on the scan tool, and press the Enter key.
6. Press the on soft key. A horn chirp should sound to indicate the Sensor Learn Mode is enabled.
7. Starting with the LF tire, increase or decrease the tire pressure for 5–8 seconds or until a horn chirp sounds. The horn chirp may occur before the 5- to 8-second time period, or up to 30 seconds after this time period.

WARNING If you are increasing tire pressure during the learning procedure, never inflate a tire above the vehicle manufacturer's maximum specified tire pressure. This action may cause personal injury and tire damage.

8. After the horn chirp sounds, repeat Step 7 on the other 3 or 4 sensors in the following order:
   (a) RF
   (b) RR
   (c) LR
   (d) Spare tire (if applicable)

   If a horn chirp is not heard after 35 seconds for any of the four sensors, turn off the ignition switch and exit the Sensor Learn Mode on the scan tool. Repeat the sensor learning procedure from Step 4.

9. After the learning procedure has been completed on all the sensors, a double horn chirp sounds to indicate the learning procedure is completed on all the sensors.
10. Turn off the ignition switch and disconnect the scan tool.
11. Inflate all the tires to the specified pressure.

Sensor Learning Procedure with Keyless Entry Transmitter

On some vehicles, the keyless entry remote transmitter may be used to complete the wheel sensor learning procedure without the use of any other equipment. Follow these steps to perform a sensor learning procedure using the keyless entry remote transmitter:

1. Turn on the ignition switch.
2. Apply the parking brake.
3. Use the keyless entry remote transmitter to lock and unlock the doors three times.
4. Simultaneously press the lock and unlock buttons on the keyless entry remote transmitter until a horn chirp sounds.
5. Starting with the LF tire, increase or decrease the tire pressure for 5–8 seconds, or until a horn chirp sounds. The horn chirp may sound before the 5–8 seconds is completed, or up to 35 seconds after this time period.
6. After the horn chirp sounds on the LF tire, follow the procedure in Step 5 on the other three or four sensors in the following order:
   (a) RF
   (b) RR
   (c) LR
   (d) Spare (if applicable)

   If a horn chirp does not sound after 35 seconds on any of the tires, turn off the ignition switch and exit the learn mode on the scan tool. Repeat the procedure starting with Step 1.
7. After all the sensors have been learned, a double horn chirp sounds to indicate all the sensors have been learned.
8. Turn off the ignition switch and disconnect the scan tool.
9. Inflate all the tires to the specified tire pressure.

TIRE-AND-WHEEL RUNOUT MEASUREMENT

Ideally, a tire-and-wheel assembly should be perfectly round. However, this condition is rarely achieved. A tire-and-wheel assembly that is out-of-round is said to have radial runout. If the radial runout exceeds manufacturer's specifications, a vibration may occur because the radial runout causes the spindle to move up and down (**Figure 3-21**). A defective tire with a variation in stiffness may also cause this up-and-down spindle action.

Classroom Manual
Chapter 3, page 82

Wheel balancers with force variation capabilities have a roller that is pressed against the tire tread during the wheel balance procedure. This roller senses and indicates stiffness variation in a tire.

A dial indicator gauge may be positioned against the center of the tire tread as the tire is rotated slowly to measure radial runout (**Figure 3-22**). Radial runout of more than 0.060 in. (1.5 mm) will cause vehicle shake. If the radial runout is between 0.045 in. and 0.060 in. (1.1 mm and 1.5 mm), vehicle shake may occur. These are typical radial runout specifications. Always consult the vehicle manufacturer's specifications. Mark the highest point of radial runout on the tire with chalk, and mark the valve stem position on the tire.

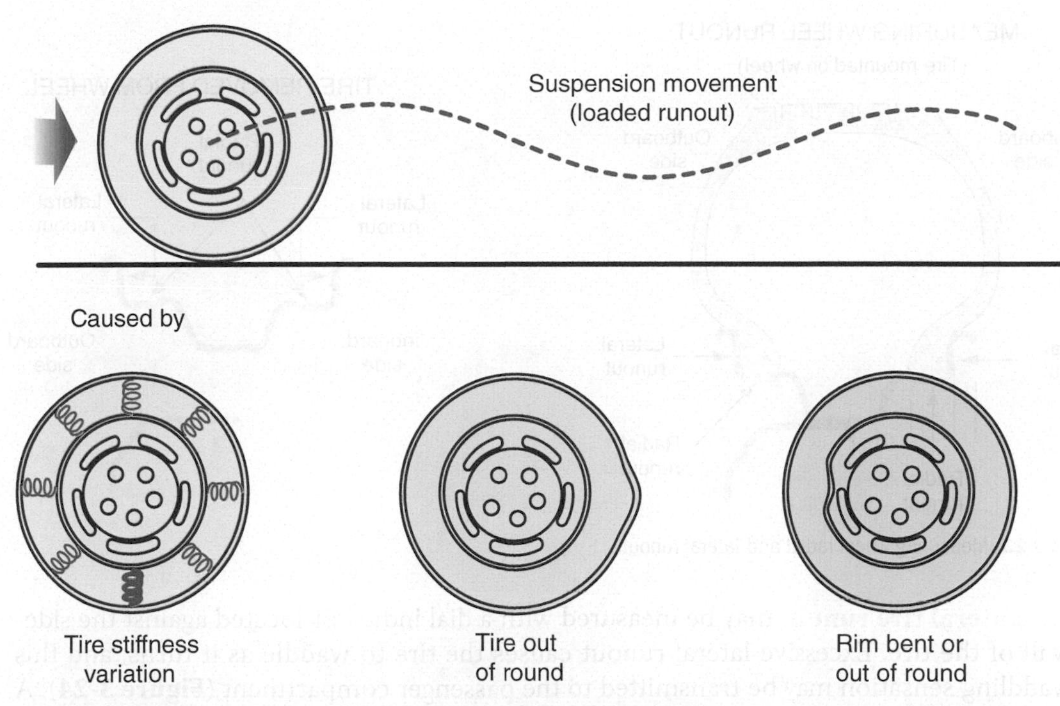

Figure 3-21 Vertical tire-and-wheel vibrations caused by radial tire or wheel runout, or variation in tire stiffness.

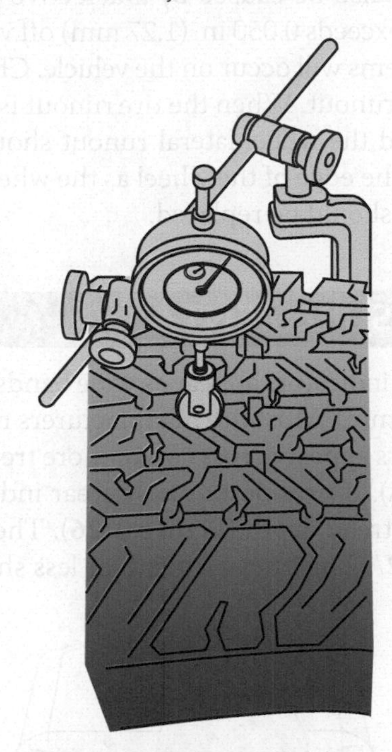

Figure 3-22 Measuring tire radial runout.

If the radial tire runout is excessive, dismount the tire and check the runout of the wheel rim with a dial indicator positioned against the lip of the rim while the rim is rotated (**Figure 3-23**). Use chalk to mark the highest point of radial runout on the wheel rim. Radial wheel runout should not exceed 0.035 in. (0.9 mm), whereas the maximum lateral wheel runout is 0.045 in. (1.1 mm). If the highest point of wheel radial runout coincides with the chalk mark from the highest point of maximum tire radial runout, the tire may be rotated 180° on the wheel to reduce radial runout. Tires or wheels with excessive runout are usually replaced.

Radial tire runout refers to excessive variations in the tread circumference.

Side-to-side tire and chassis movement caused by excessive lateral tire runout may be called tire or chassis waddle.

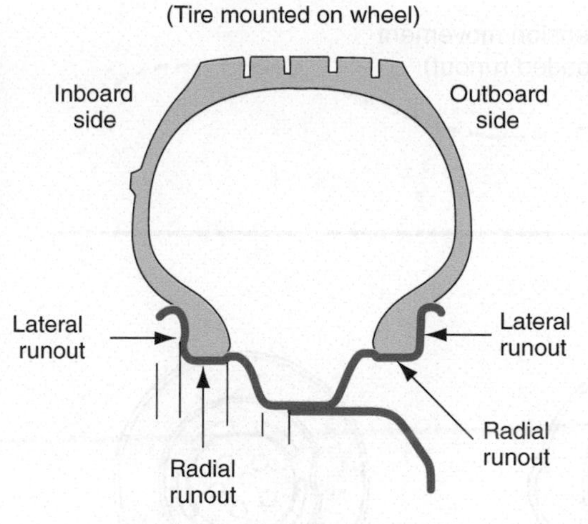

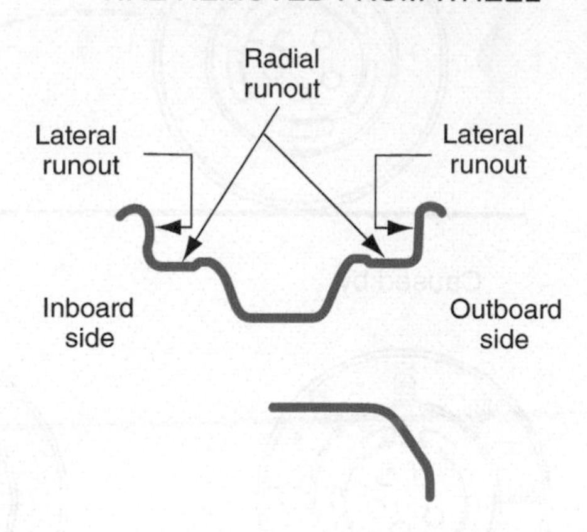

Figure 3-23 Measuring wheel radial and lateral runout.

Lateral tire runout may be measured with a dial indicator located against the sidewall of the tire. Excessive lateral runout causes the tire to waddle as it turns, and this waddling sensation may be transmitted to the passenger compartment (**Figure 3-24**). A chassis waddling action may also be caused by a defective tire in which the belt is not straight. If the lateral runout exceeds 0.050 in. (1.27 mm) off vehicle or 0.060 in. (1.52 mm) on vehicle, wheel shake problems will occur on the vehicle. Chalk mark the tire-and-wheel at the highest point of lateral runout. When the tire runout is excessive, the tire should be removed from the wheel, and the wheel lateral runout should be measured with a dial indicator positioned against the edge of the wheel as the wheel is rotated. Tires or wheels with excessive lateral runout should be replaced.

TREAD WEAR MEASUREMENT

Special Tool
Tire runout gauge

Classroom Manual
Chapter 3, page 73

On most tires, the tread wear indicators appear as wide bands across the tread when tread depth is worn to 2/32 in. (1.6 mm). Most tire manufacturers recommend tire replacement when the tread wear indicators appear across two or more tread grooves at three locations around the tire (**Figure 3-25**). If tires do not have wear indicators, a tread depth gauge may be used to measure the tread depth (**Figure 3-26**). The tread depth gauge reads in 32nds of an inch. Tires with 2/32 in. of tread depth or less should be replaced.

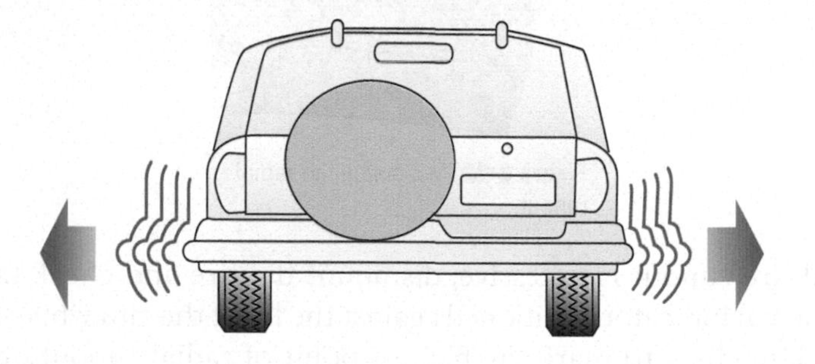

Tire waddle often caused by

- Steel belt not straight within tire
- Excessive lateral runout

Figure 3-24 Chassis waddling action caused by lateral tire or wheel runout, or a defective tire with a belt that is not straight.

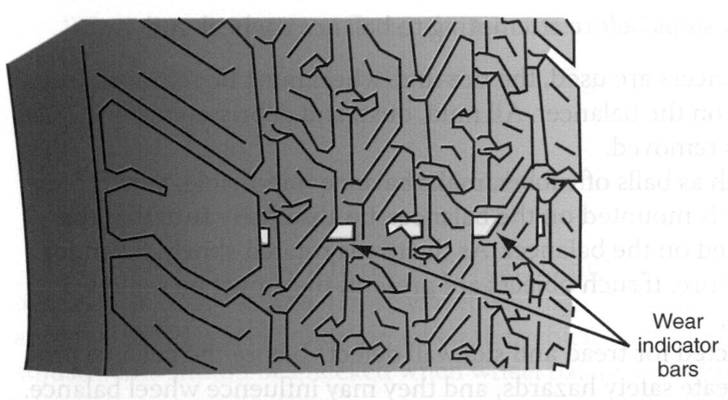

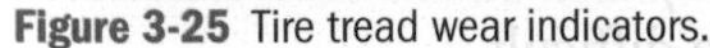
Figure 3-25 Tire tread wear indicators.

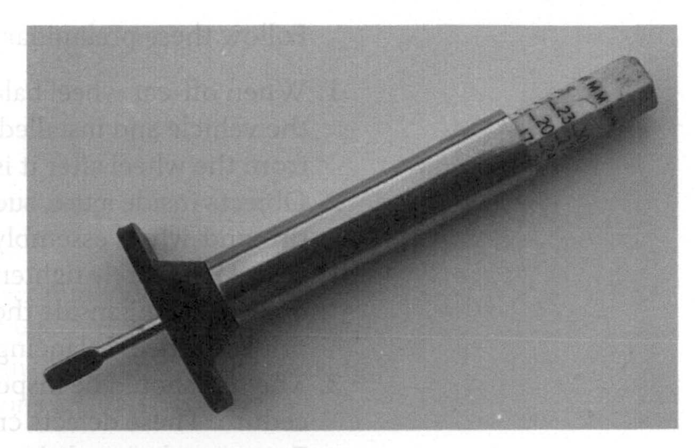
Figure 3-26 Tread depth gauge.

PRELIMINARY WHEEL BALANCING CHECKS

These preliminary checks should be completed before a tire-and-wheel are balanced:

1. Check for objects in the tire tread.
2. Check for objects inside the tire.
3. Inspect the tread and sidewall.
4. Check the inflation pressure.
5. Measure the tire-and-wheel runout.
6. Check the wheel bearing adjustment.
7. Check for mud collected on the inside of the wheel and wash the tire-and-wheel assembly.
8. Inspect the wheel rim for damage and excessive rust.

WARNING On many wheel balancers, the tire-and-wheel are spun at high speed during the dynamic balance procedure. Be sure that all wheel weights are attached securely, and check for other loose objects on the tire-and-wheel, such as stones in the tread. If loose objects are detached from the tire or wheel at high speed, they may cause serious personal injury or property damage.

WARNING On the type of wheel balancer that spins the tire-and-wheel at high speed during the dynamic balance procedure, always attach the tire-and-wheel assembly securely to the balancer. Follow the equipment manufacturer's recommended wheel mounting procedure. If the tire-and-wheel assembly becomes loose on the balancer at high speed, serious personal injury or property damage may result.

WARNING Prior to spinning a tire-and-wheel at high speed on a wheel balancer, always lower the protection shield over the tire. This shield provides protection in case anything flies off the tire or wheel.

All of the items on the preliminary check list influence wheel balance or safety. Therefore, it is extremely important that the preliminary checks be completed. Since a tire-and-wheel assembly is rotated at high speed during the dynamic balance procedure, it is very important that objects such as stones be removed from the treads. Centrifugal force may dislodge objects from the treads and cause serious personal injury. For this reason, it is also extremely important that the old wheel weights be removed from the wheel prior to balancing and the new weights attached securely to the wheel during the balance procedure.

may be extended so it contacts the flange on the side of the wheel rim next to the balancer. This tool provides the required wheel offset measurement that indicates the position of the wheel on the balancer shaft. An electronic wheel balancer performs static balance and dynamic wheel balance calculations simultaneously. The following is a typical wheel balance procedure:

Classroom Manual
Chapter 3, page 92

Static balance refers to the balance of a stationary wheel.

1. Complete all the preliminary balance checks mentioned previously in this chapter.
2. Be sure the tire-and-wheel assembly is mounted on the balancer using the balancer manufacturer's recommended mounting procedure.
3. Use a pair of wheel weight pliers to remove all the old wheel weights from the wheel rim (**Figure 3-29**).
4. Enter the wheel diameter, width, and offset in the balancer computer.
5. Be sure the safety hood is lowered over the tire-and-wheel assembly, and activate the balancer control to spin the wheel-and-tire assembly.
6. Operate the balancer brake to slow and stop the wheel, and observe the balancer display screen to determine the size and location of wheel weight(s) required on the wheel rim.
7. Install the correct wheel weights in the locations indicated on the balancer screen. Use that hammer head on the wheel weight pliers to install the weights.
8. Spin the wheel again on the balancer; the balancer screen should indicate a balanced tire-and-wheel assembly.
9. Remove the wheel-and-tire assembly from the balancer, and install this assembly on the vehicle using the alignment marks placed on the tire wheel and hub stud prior to wheel removal.
10. Tighten the wheel lug nuts to the specified torque in the proper sequence.

SERVICE TIP Some wheel rims such as magnesium rims require the use of stick-on wheel weights.

 **SERVICE TIP** Wheel weights are rated in ounces or grams.

On some pre-electronic wheel balancers, a static balance procedure is performed by allowing the wheel-and-tire assembly to rotate by gravity on the wheel balancer. A

Figure 3-29 Wheel weights are removed and installed with special wheel weight pliers.

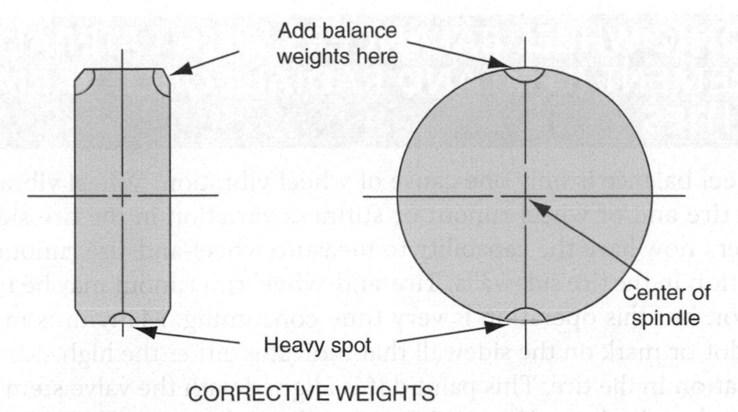

Figure 3-30 Static wheel balance procedure.

heavy spot in the tire rotates the tire-and-wheel assembly until the heavy spot is at the bottom. The necessary wheel balance weights were then added at the top of the wheel 180° from the heavy spot. Equal wheel weights were installed on each side of the wheel (**Figure 3-30**). If a wheel has dynamic unbalance, and the heavy spot is on the outside edge of the tire tread, the correct size of wheel weight is installed on the rim 180° from the heavy spot (**Figure 3-31**). Improper static or dynamic wheel balance causes cupped tire tread wear and bald spots on the tread (**Figure 3-32**). Incorrect static wheel balance results in wheel tramp, and improper dynamic wheel balance causes wheel shimmy. When using an electronic wheel balancer, the technician does not have to determine the wheel weight size or location because the balancer provides this information to the technician.

The Center for Environmental Health (CEH) in California, together with Chrysler Group and three major wheel weight manufacturers, reached an agreement to end the use of lead wheel weights in California by December 31, 2009. It is estimated that lead wheel weights cause 500,000 lb of lead to be placed in the environment each year from wheel weights breaking off vehicle wheels. Lead wheel weights will be replaced with a metal or material that is more environmentally friendly. Be sure you are installing the type of wheel weights required by the legislation in your state.

Photo Sequence 5 illustrates a typical off-car wheel balance procedure.

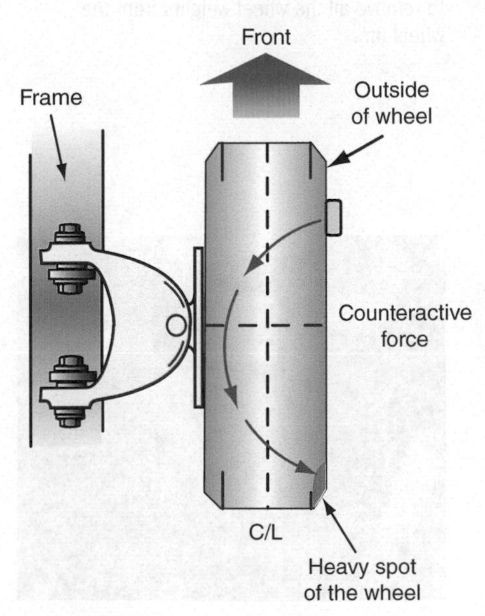

Figure 3-31 Dynamic wheel balance procedure with heavy spot on the outside edge of the tread.

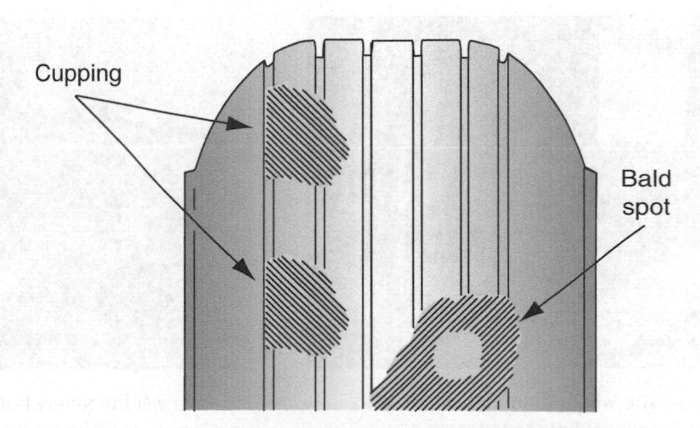

Figure 3-32 Cupped tire tread wear with bald spots on the tread caused by improper static or dynamic wheel balance.

ELECTRONIC WHEEL BALANCERS WITH LATERAL FORCE MEASUREMENT (LFM) AND RADIAL FORCE VARIATION CAPABILITIES

Improper wheel balance is only one cause of wheel vibration. Wheel vibration may also be caused by tire and/or wheel runout or stiffness variation in the tire sidewalls. Some wheel balancers now have the capability to measure wheel-and-tire runout and stiffness or force variation in the tire sidewalls. Tire-and-wheel rim runout may be measured with a dial indicator, but this operation is very time consuming. Many tires marketed today have a paint dot or mark on the sidewall that indicates either the high side or low side of the force variation in the tire. This paint dot is aligned with the valve stem when the tire is mounted on the wheel rim. Using this system, the technician is assuming the paint dot is placed at the high point of radial force, and the valve stem hole is at the low point of runout on the rim. However, this is not always true, and a tire mounted with this tire position in relation to the wheel rim may cause vibration even though it is properly balanced. Other tires may have a slight conicity problem that causes steering pull on the vehicle.

PHOTO SEQUENCE 5

Typical Off-Car Wheel Balancing Procedure

P5-1 Complete all the preliminary wheel balance checks.

P5-2 Follow the wheel balancer manufacturer's recommended procedure to mount the tire-and-wheel assembly securely on the electronic wheel balancer. Some wheel balancers have a centering check in the programming procedure. If the tire-and-wheel are not mounted properly, tire-and-wheel balance will be inaccurate.

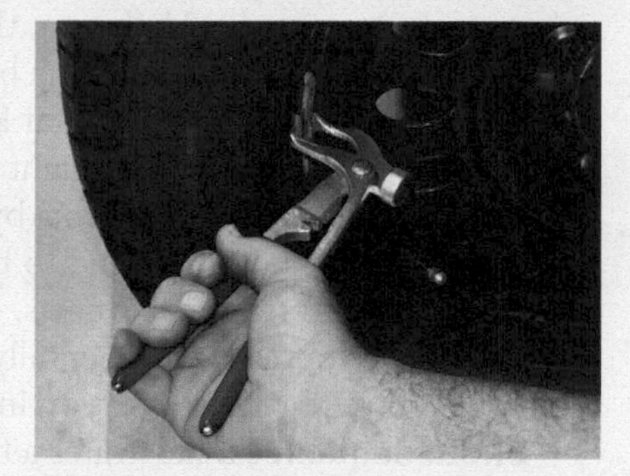

P5-3 Use a pair of wheel weight pliers to remove all the wheel weights from the wheel rim.

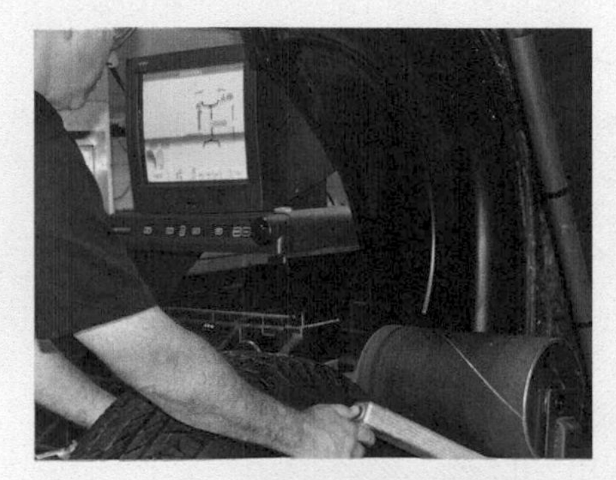

P5-4 Enter the wheel diameter, width, and offset on the wheel balancer screen.

P5-5 Lower the safety hood over the wheel-and-tire on the wheel balancer.

P5-6 Activate the wheel balancer control to spin the tire-and-wheel assembly.

PHOTO SEQUENCE 5 (CONTINUED)

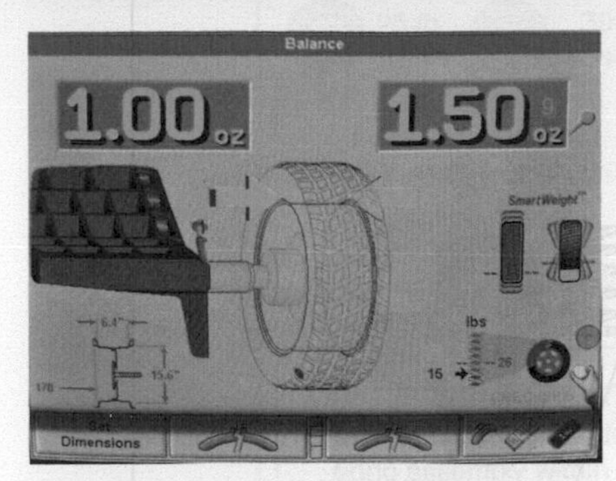

P5-7 Stop the tire-and-wheel assembly. Observe the wheel balancer display screen to determine the correct size and location of the required wheel weights.

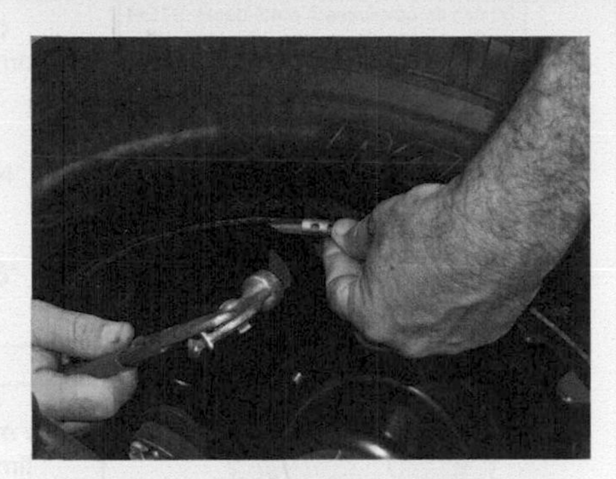

P5-8 Install the correct size wheel weights in the location(s) indicated on the wheel balancer screen.

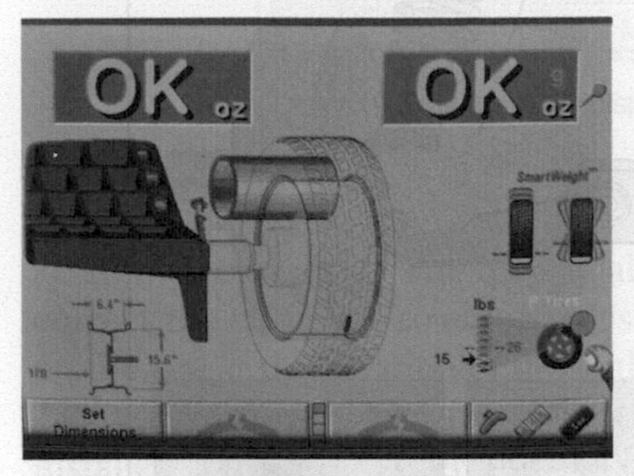

P5-9 Activate the wheel balancer control to spin the wheel again, and observe the wheel balancer display to confirm that it indicates a balanced tire-and-wheel assembly.

P5-10 Stop the wheel, lift the safety hood, and remove the tire-and-wheel assembly from the balancer.

An electronic wheel balancer with radial force variation capabilities has a roller that is forced against the tire tread with considerable force during the balance procedure. This type of balancer provides wheel balance following the usual procedure. The roller forced against the tire allows the balancer to measure tire conicity that results in lateral force and steering pull (**Figure 3-33**). As each tire-and-wheel assembly is balanced, the balancer display indicates a number for each assembly (**Figure 3-34**). The balancer display also indicates the

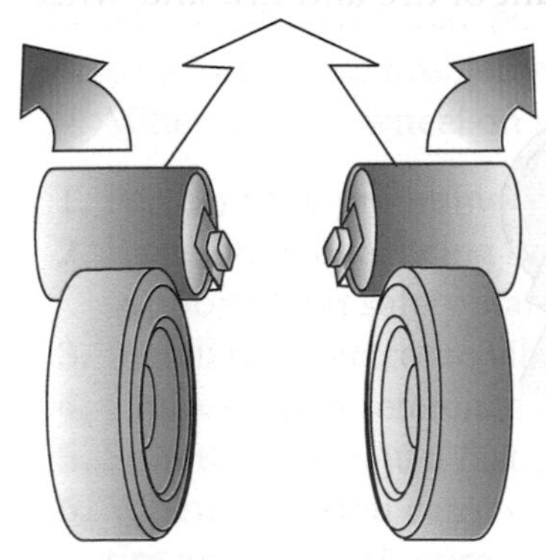

Figure 3-33 Tire conicity sensed by the roller on the balancer causes lateral force and steering pull.

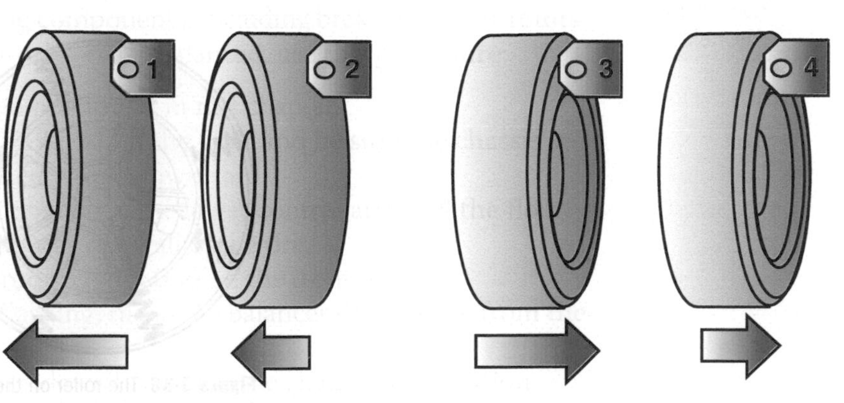

Figure 3-34 The balancer assigns a number to each tire-and-wheel assembly.

TABLE 3-1 TIRE-AND-WHEEL DIAGNOSIS

Problem	Symptoms	Possible Causes
Tire thump	Thumping noise while driving	Cupped tire treads Excessive radial tire or wheel runout Tire defects Improper wheel balance Variation in tire stiffness
Chassis waddling	Lateral rear chassis oscillations while driving	Excessive rear tire or wheel lateral runout
Steering pull	Steering pulls to one side while driving straight ahead	Tire conicity Unmatched tires side-to-side Improper wheel alignment angles
Wheel vibration	A vibration that occurs at certain speeds and may be felt on the steering wheel or throughout the chassis	Improper wheel balance Excessive lateral tire runout Variation in tire stiffness Tire defects
Excessive wear on tire treads	Cupped tire treads Wear on tread center Equal wear on tread edges	Improper wheel balance Overinflation Underinflation

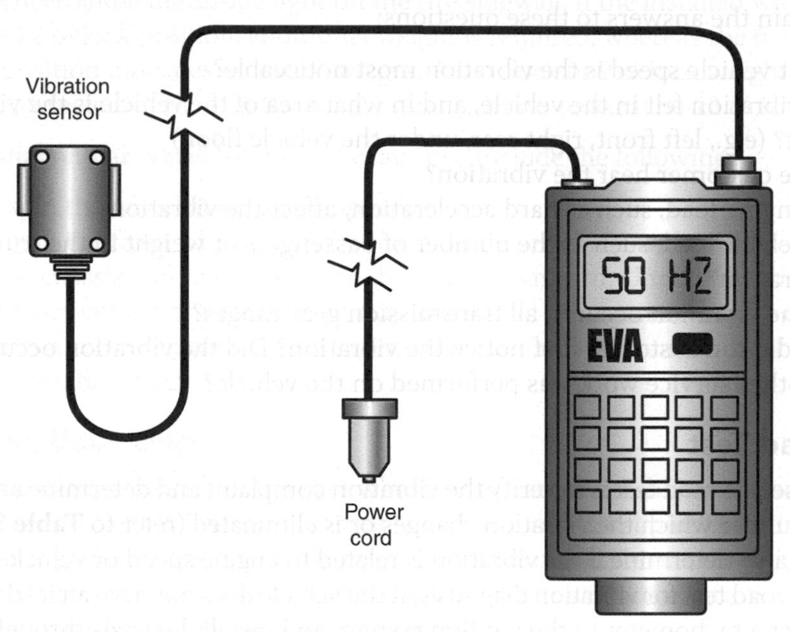

Figure 3-39 Electronic vibration analyzer (EVA).

SPECIAL ROAD TESTS

Slow Acceleration Test

All of the special road tests help the technician to locate the cause of the vibration. A tire-and-wheel inspection should be completed before any road test. Defects discovered during the tire inspection may explain the cause of the vibration that occurs during a road test.

Inspect the tires and wheel for these defects:

1. Check the inflation pressure in each tire.
2. Check for unusual tire wear such as cupping, excessive tread wear, and flat or bald spots. Inspect the tire sidewalls and tread area for bulges and ply separation. These tire defects may cause vibration, slapping noise, and tire growl or howl.
3. Inspect the rims for bends, damage, and excessive corrosion.

Perform the **slow acceleration test** on a smooth level road. During this test, accelerate the vehicle from a stop to legal highway speed. Check for any vibration that matches the customer's description. If any vibration is present, record the frequency, vehicle speed, and engine speed when the vibration occurred.

Neutral Coast-Down Test

Perform the **neutral coast-down test** on a smooth, level road. Accelerate the vehicle to a speed slightly higher than the speed at which the vibration occurs. Shift the transmission into neutral and allow the vehicle to coast down through the speed range at which the vibration occurs. If the vibration still occurs in neutral, the vibration is vehicle-speed sensitive, and the engine, propeller shaft, flexplate, and torque converter have been eliminated as causes of the vibration. When the vibration occurs with the transmission in neutral, the vibration diagnosis should concentrate on the drive axle assemblies and the tire-and-wheel assemblies. Off-car or on-car balance procedures are performed to correct balance problems in brake drums or rotors and tires and wheels.

Downshift Test

To complete the **downshift test**, drive the vehicle on a smooth, level road at the speed when the vibration is present. Note the engine rpm. Decelerate and reduce the vehicle speed until the vibration is no longer present. Manually downshift the transmission into the next lowest gear, and accelerate the vehicle until the engine is running at the previously recorded rpm. If the vibration returns at the same engine rpm, the engine, propeller shaft, flexplate and torque converter (automatic transmission), or clutch disc and pressure plate (manual transmission) are the most likely causes of the vibration.

Neutral Run-Up Test

Perform the **neutral run-up test** when the customer complains about vibration at idle speed or as a follow-up to the downshift test. To perform the neutral run-up test, slowly increase the engine rpm from idle. Note the engine rpm and frequency at which vibration occurs. If vibration occurs, the cause of the vibration is related to the engine, flexplate and torque converter (automatic transmission), or clutch disc and pressure plate (manual transmission).

Brake Torque Test

The **brake torque test** may be used to identify engine vibrations that were not present during the downshift and neutral run-up tests. This test may also be used when diagnosing vibrations that are sensitive to engine load.

To perform the brake torque test, follow this procedure:

1. Firmly apply the parking brake.
2. Block the front wheels.
3. Firmly apply the brake pedal.
4. Shift transmission into drive.
5. Slowly increase the engine speed to 1200 rpm and check for vibrations.
6. Note the engine rpm when vibration occurred.

Caution

Do not increase the engine rpm above 1200, and limit this test to 10 seconds to protect the transmission from overheating.

If engine vibrations occur during the brake torque test, but they did not occur during the downshift and neutral run-up tests, the engine and/or transmission mounts are a prime suspect for the cause of the vibration.

Steering Input Test

The **steering input test** may be performed to diagnose vibrations in suspension components. To perform this test, drive the vehicle through sweeping turns at the speed when the vibration occurs. Drive the vehicle when turning in one direction and then perform some turns in the opposite direction. If the vibration changes during these turns, the vibration is likely caused by defective wheel bearings, tire tread, or front drive axle shafts and joints.

Standing Acceleration Test

The **standing acceleration test** is used to diagnose a vibration that may be called a launch shudder. To perform the standing acceleration test, bring the vehicle to a complete stop. Place the transmission in drive and remove your foot from the brake pedal. Accelerate the vehicle to 30–40 mph (48–64 km/h), and check for vibrations that match the customer's complaint. If a shudder-type vibration occurs during the test, the cause of this vibration could be incorrect curb riding height, defective front drive axle joints, defective engine or transmission mounts, or faulty exhaust hangers or mounts.

Vibration Diagnosis with the Electronic Vibration Analyzer (EVA)

The EVA is helpful in locating the source of vibration problems.

Follow these steps to connect the EVA:

Caution

When installing the sensor connector in the EVA input terminal, do not twist the sensor connector. Do not remove this connector when the EVA is in use. This action may damage the EVA.

1. Be sure the correct software cartridge is securely installed in the bottom of the EVA.
2. Connect the vibration sensor cord to input A or B on the EVA. Position the release button facing downward on the lower side of this connector. Push the connector into the input port until the connector clicks and locks in place.
3. Follow these steps to perform the sensor calibration procedure:
   a. Lay the sensor on a flat stationary surface, and be sure the side of the sensor marked UP is facing upward.
   b. Plug the EVA power cord into a 12 V power supply, such as the cigarette lighter.
   c. Press the "up" arrow key.
   d. Press number 2 on the keypad three times until the word BURNING appears, followed by a request to turn the sensor over.
   e. Turn the sensor over and press any key on the keypad to begin calibration. The calibration procedure takes approximately 20 seconds, and the display returns to the active mode when calibration is completed.
4. Two vibration sensors can be connected to ports A and B on the EVA at the same time, and the A/B button on the tester can be used to switch from one sensor to the other. Place the sensor(s) on any component that is suspected of being a vibration source. Each sensor has a 20-ft. cord, so it may be placed anywhere on the vehicle. Each sensor contains a magnet for attachment to metal surfaces. For attachment to nonmagnetic surfaces, use putty or a hook and loop fastener. Vibrations are usually felt in an up-and-down direction. Because the vibration sensor is directionally sensitive, this sensor must be mounted on the lower part of the suspected component with the marking on the sensor facing upward (**Figure 3-40**). When repeating a vibration test on the same component, always be sure the vibration sensor is mounted in the exact same location.

Hertz or rpm are displayed on the left side of the EVA display. Either of these values may be selected by pressing the "RPM/Hertz" button on the tester. The selected value is displayed at the top of the left column on the display. A bar graph of the suspected source

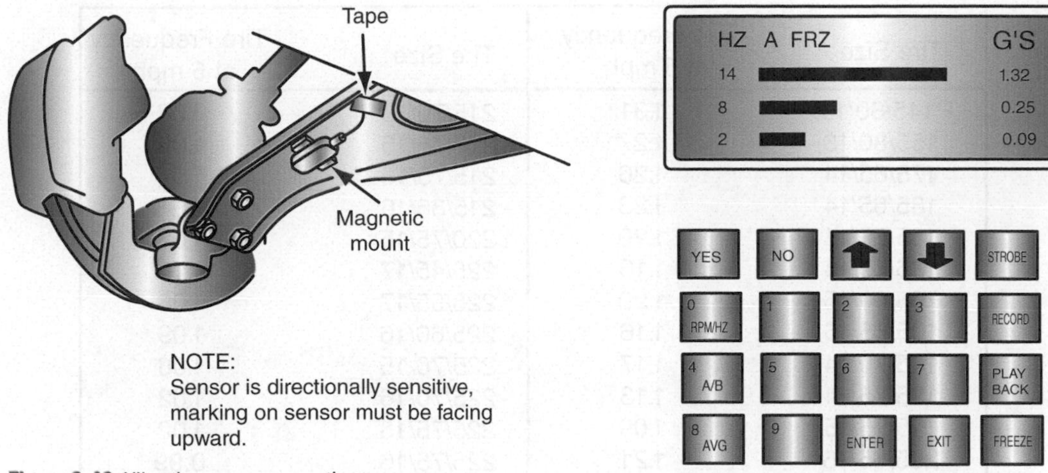

Figure 3-40 Vibration sensor mounting.

Figure 3-41 EVA display and keypad.

of the vibration is shown in the center of the display. When the AUTO mode is selected, the suspected vibration source is shown in the display. If the AUTO mode is not in use, the bar graph is indicated in the center of the display (**Figure 3-41**). The amplitude of each frequency indicated on the left of the screen is displayed on the right side of the screen. These amplitudes are displayed in Gs of acceleration force. Higher amplitudes indicate stronger vibration forces. The EVA displays up to three of the most dominant input frequencies in descending order of amplitude strength.

The technician may press the averaging (AVG) button to select this mode. In the AVG mode, the EVA displays the average of multiple vibration readings taken over a period of time. The EVA is usually operated in the AVG mode to minimize distractions caused by a sudden vibration that is not related to the main vibration source. These distractions could be caused by the vehicle wheels striking road irregularities. When the EVA is operating in the AVG mode and the AUTO Mode, "A" is displayed along the top of the screen. The AUTO mode is designed to be used with the vibration diagnosis tables in the appropriate service manual. If the EVA is operating in the AVG mode and the MANUAL mode, AVG is shown at the top of the screen. When the EVA is operating in the non-AVG mode and the AUTO mode, "I" is indicated at the top of the display. The top row on the display also indicates "A" or "B," representing the vibration sensor input port.

When the FREEZE button is pressed, the EVA displays FRZ on the top line of the display. In this mode the EVA will store data for a short time period when the RECORD button is pressed during various vibration tests. The technician may press the PLAY BACK button to display the recorded data. The technician presses the EXIT button to return the EVA to normal operation.

The EVA has a strobe light trigger wire that may be used with an inductive pickup timing light to balance rotating components such as propeller shafts.

CALCULATING VIBRATION FREQUENCIES OF VARIOUS COMPONENTS

Calculating Tire/Wheel Frequency

The first step in calculating tire/wheel frequency is to divide the vehicle speed by 5-mph increments. There are 12 increments of 5 mph, if the vehicle speed is 60 mph. The next step is to use the chart in **Figure 3-42** to determine how many times a wheel-and-tire rotate per second at 5 mph. As indicated in this chart, if the tire size is 195/70/14, the tires revolve at 1.17 times per second for every 5 mph of vehicle speed. To calculate the

Tire Size	Tire Frequency at 5 mph	Tire Size	Tire Frequency at 5 mph
P145/80/13	1.31	215/70/15	1.08
155/80/13	1.27	215/75/15	1.05
175/65/14	1.26	215/75/16	1.01
185/65/14	1.23	215/85/16	0.95
185/70/14	1.20	220/75/15	1.03
185/75/14	1.16	225/45/17	1.16
195/60/15	1.20	225/55/17	1.08
195/65/15	1.16	225/60/16	1.09
195/70/14	1.17	225/70/15	1.06
195/75/14	1.13	225/70/16	1.02
195/75/15	1.09	225/75/15	1.02
205/55/15	1.21	225/75/16	0.99
205/55/16	1.16	235/70/15	1.04
205/60/14	1.22	235/75/15	1.00
205/60/15	1.17	235/75/16	0.97
205/65/15	1.14	245/70/15	1.02
205/70/14	1.14	245/70/16	0.98
205/70/15	1.10	245/75/16	0.95
205/75/14	1.11	275/40/17	1.13
205/75/15	1.07	295/40/20	0.99
215/50/17	1.14	335/35/17	1.10
215/55/16	1.14	30X9.5/15	0.98
215/65/15	1.11	31X10.5/15	0.95
215/65/16	1.07		

Figure 3-42 Tire/wheel frequencies.

tire/wheel frequency at 60 mph, multiply the number of 5-mph increments, 12, by the number of times the tire-and-wheel revolve in 1 second per 5-mph increment, 1.17. Therefore, we have the calculation $12 \times 1.17 = 14.04$, and this size of tire-and-wheel on a vehicle traveling at 60 mph produces a vibration of approximately 14 Hz. With a vibration sensor attached to the lower control arm, if the EVA displays a vibration of 14 Hz, the tire-and-wheel mounted on the knuckle attached to the control arm have a vibration problem. This vibration problem could also be caused by a rotor or a hub.

With a vibration of 14 Hz, a second-order vibration is 28 Hz, a third-order vibration is 42 Hz, and a fourth-order vibration is 56 Hz. On a front-wheel-drive vehicle, a worn constant velocity drive axle joint may cause a vibration in the fifth order of the tire/wheel vibration, and a defective tripod drive axle joint may result in a vibration of the third order of the tire/wheel vibration. On a rear-wheel-drive-vehicle if the tire/wheel frequency is 14 and the rear axle ratio is 3.42:1, the frequency of the drive shaft is $14 \times 3.42 = 47.88$.

Calculating Engine and Engine Firing Frequency

To calculate the engine frequency, divide the engine rpm by 60 seconds per minute. For example, if the engine rpm is 3000, the engine frequency is $3000 \div 60 = 50$ revolutions per second, or 50 Hz. To determine the engine firing frequency, divide the number of engine cylinders by two, and multiply the answer by the engine frequency. Let us assume we are working on a V6 engine running at 3000 rpm. Divide the number of cylinders by two, and so $6 \div 2 = 3$. Then multiply this answer by the engine frequency of 50 to calculate the firing frequency. Therefore, $3 \times 50 = 150$ is the engine firing frequency. Because a cylinder in a four-cycle engine expels exhaust only once every two revolutions, a single cylinder frequency is one-half the engine frequency. Therefore, in our example if the engine frequency is 50, a single cylinder frequency is 25. This is the frequency of exhaust pulses in a single cylinder and also the frequency of the camshaft that turns at one-half the engine speed.

Calculating Accessory Drive Frequencies

To calculate the frequency of the belt-driven accessories, divide the diameter of the crankshaft pulley by the diameter of the pulley on the belt-driven accessory and multiply the answer by the engine frequency. Let us assume we are still working on the engine in our previous example, which is a V6 engine with a frequency of 50 running at 3000 rpm. With a vibration sensor mounted on the engine the EVA displays a vibration of 160 Hz. The diameter of the crankshaft pulley is 8 in., the diameter of the power steering pulley is 5 in., the diameter of the A/C compressor pulley is 4.5 in., and the diameter of the alternator pulley is 2.5 in.

Therefore, we have the following calculations:

Power steering pump frequency: $8 \div 5 = 1.6 \times 50 = 80$
A/C compressor frequency: $8 \div 4.5 = 1.7 \times 50 = 85$
Alternator frequency: $8 \div 2.5 = 3.2 \times 50 = 160$

When we compare the vibration frequency displayed on the EVA to the belt-driven accessory frequencies, the alternator frequency matches the frequency on the EVA, indicating the alternator is the source of the vibration.

CASE STUDY

A customer complains about severe front end waddling on a 2013 Cadillac CTS equipped with steel-belted radial tires, which have 11/32 in. of tread depth. The technician questions the customer further regarding this problem and learns that the problem occurs only after the car has sat overnight, and it lasts only for approximately six blocks when the customer starts driving the car.

The technician test drove the vehicle but was unable to duplicate the customer complaint. The technician carefully checks all steering and suspension components, wheel bearing adjustment, and tire runout. No problems are found during these checks, and the technician asks the customer to leave the car in the shop overnight. The technician installs the front tires on the rear of the car and rotates the rear tires to the front, and the car is left in the shop overnight. A road test the next morning indicates rear end waddling, which proves the problem must be caused by a tire problem such as a radial cord ply imperfection, since moving the tires' original position was the only thing changed on the car. The technician removed the two rear tires and checked them on a Road Force tire balancer once they cooled down again. The tire that had been moved to the left rear position was found to have excessive road force variation, which was beyond the limits of match mounting. The technician replaced the defective tire and checks the balance of the other three tires. The customer reports later that the problem has been completely corrected.

ASE-STYLE REVIEW QUESTIONS

1. While discussing a tire thumping problem:
Technician A says that this problem may be caused by cupped tire treads.
Technician B says that a heavy spot in the tire may cause this complaint.
Who is correct?
A. A only
B. B only
C. Both A and B
D. Neither A nor B

2. While discussing a vehicle that pulls to one side:
Technician A says that excessive radial runout on the right front tire may cause this problem.
Technician B says that tire conicity may be the cause of this complaint.
Who is correct?
A. A only
B. B only
C. Both A and B
D. Neither A nor B

3. While discussing tire noise:
Technician A says that if it is tire noise the noise will vary with road surface conditions.
Technician B says that if it is tire noise the noise will remain constant when the vehicle is accelerated and decelerated.
Who is correct?
A. A only
B. B only
C. Both A and B
D. Neither A nor B

4. While discussing tire wear:
Technician A says that static imbalance causes feathered tread wear.
Technician B says that dynamic imbalance causes cupped wear and bald spots on the tire tread.
Who is correct?
A. A only
B. B only
C. Both A and B
D. Neither A nor B

5. While discussing on-car wheel balancing:
Technician A says that during an on-car wheel balancing procedure on a rear wheel of a rear-wheel-drive car, the speed indicated on the speedometer should not exceed 65 mph (105 km/h).
Technician B says that the speed indicated on the speedometer on this car must not exceed 35 mph (56 km/h).
Who is correct?
A. A only
B. B only
C. Both A and B
D. Neither A nor B

6. A front tire has excessive wear on both edges of the tire tread. The most likely cause of this problem is:
A. Overinflation.
B. Underinflation.
C. Improper static balance.
D. Improper dynamic balance.

7. When measuring radial tire-and-wheel runout, the maximum runout on most automotive tire-and-wheel assemblies should be:
A. 0.015 in. (0.038 mm).
B. 0.025 in. (0.635 mm).
C. 0.045 in. (1.143 mm).
D. 0.070 in. (1.77 mm).

8. When measuring lateral wheel runout with the tire demounted from the wheel, the maximum runout on most automotive wheels is:
A. 0.020 in. (0.508 mm).
B. 0.030 in. (0.762 mm).
C. 0.045 in. (1.143 mm).
D. 0.055 in. (1.397 mm).

9. All of these statements about improper wheel balance are true EXCEPT:
A. Dynamic imbalance may cause wheel shimmy.
B. Dynamic imbalance may cause steering pull in either direction.
C. Static imbalance causes wheel tramp.
D. Static imbalance causes rapid wear on suspension components.

10. When diagnosing wheel balance problems:
A. Balls of rubber inside the tire have no effect on wheel balance.
B. Loose wheel bearing adjustment may simulate improper static wheel balance.
C. Improper dynamic wheel balance may be caused by a heavy spot in the center of the tire tread.
D. After a tire patch is installed, the tire-and-wheel may be improperly balanced.

ASE CHALLENGE QUESTIONS

1. The owner of a rear-wheel-drive car with aftermarket alloy wheels says he has replaced the wheel bearings three times in the past 2 years. He wants to know why the bearings fail.
Technician A says that excessive radial runout of the wheel may be the cause of the problem.
Technician B says that excessive offset of the wheel may be the cause of the problem.
Who is correct?
A. A only
B. B only
C. Both A and B
D. Neither A nor B

2. A customer returns with recently purchased radial tires saying that the rear of the car feels like "it's riding on Jello™." All of the following could cause this problem EXCEPT:

A. The radial belt of a rear tire is not straight.

B. The wheel is improperly mounted.

C. Excessive lateral wheel runout.

D. Excessive radial tire runout.

3. A customer says the new tires he just purchased vibrate. The installer says he balanced the wheels and tires with a conventional electronic balancer before placing them on the car.

Technician A says that one of the tires may have a conicity problem.

Technician B says that one of the tires may have a force variation problem.

Who is correct?

A. A only

B. B only

C. Both A and B

D. Neither A nor B

4. After a set of radial tires is rotated, the customer returns saying he feels vibration and steering shimmy. To correct this problem, you should:

A. Measure the lateral runout on each tire.

B. Return the tires and wheels to their original positions.

C. Check the wheel bearings.

D. Balance the tires and wheels with an on-car balancer.

5. A customer says that there is a "thumping" vibration in the wheels, and an inspection of the tires shows the two front wheels have flat spots on the tire treads.

Technician A says heavy spots in the tires may have caused this condition.

Technician B says locking the wheels and skidding on pavement caused this condition.

Who is correct?

A. A only

B. B only

C. Both A and B

D. Neither A nor B

Name ______________________ Date ______________

JOB SHEET 8

TIRE INSPECTION AND IDENTIFICATION

Upon completion of this job sheet, you should be able to identify correct tire size and air pressure as listed on the tire information placard/label. In addition, you should be able to identify tire wear patterns and uniform tire grading, including speed rating and load index information.

ASE Education Foundation Task Correlation

This job sheet addresses the following **MLR** task:

D.1. Inspect tire condition; identify tire wear patterns; check for correct tire size, application (load and speed ratings), and air pressure as listed on the tire information placard/label. **(P-1)**

This job sheet addresses the following **AST/MAST** task:

F.1. Inspect tire condition; identify tire wear patterns; check for correct tire size, application (load and speed ratings), and air pressure as listed on the tire information placard/label. **(P-1)**

We Support
ASE | Education Foundation

Tools and Materials

Tire Pressure Gauge

Vehicle to be inspected

Describe the vehicle being worked on:

Year ______________ Make ______________ Model ______________

VIN ______________ Engine type and size ______________

Procedure

1. Locate Tire Information Placard on driver's door or pillar.

Tire size specified on left front drivers door placard: ______________

Tire pressure specified on left front drivers door placard:

Front ______________ Rear ______________ Spare ______________

2. Record the following information:

Tire Info	Left Front	Right Front	Left Rear	Right Rear	Spare
Tire Size					
Brand Name					
Tread Depth					
DOT Build Date					
Tire Wear Rating [100, 150, etc.]					
Traction [AA, A, B, C]]					
Temperature Rating [A, B, C]					
Speed Rating					

Task Completed

3. Record your diagnosis of each tire in the spaces below:

Left front tire:

Defects ______________________________

Wear problems ______________________________

Left rear tire:

Defects ______________________________

Wear problems ______________________________

Right rear tire:

Defects ______________________________

Wear problems ______________________________

Right front tire:

Defects ______________________________

Wear problems ______________________________

Instructor's Check ______________________________

Instructor's Response

____

____

____

Name ______________________ Date ______________

JOB SHEET
9

TIRE ROTATION

Upon completion of this job sheet, you should be able to rotate tires according to manufacturer's recommendation, including vehicles equipped with tire pressure monitoring systems (TPMS).

ASE Education Foundation Task Correlation

This job sheet addresses the following **MLR** task:

D.2. Rotate tires according to manufacturer's recommendation including vehicles equipped with tire pressure monitoring systems (TPMS). **(P-1)**

This job sheet addresses the following **AST/MAST** task:

D.3. Rotate tires according to manufacturer's recommendation including vehicles equipped with tire pressure monitoring systems (TPMS). **(P-1)**

We Support ASE | Education Foundation

Tools and Materials

Tire pressure gauge
TPMS service tool
Diagnostic scan tool
Vehicle to be inspected

Describe the vehicle being worked on:

Year ______________ Make ______________ Model ______________

VIN ______________________ Engine type and size ______________________

Procedure **Task Completed**

1. Locate Tire Information Placard on driver's door or pillar.
   Tire size specified on left front drivers door placard: ______________
   Tire pressure specified on left front drivers door placard:
   Front ______________ Rear ______________ Spare ______________
2. Adjust all tire pressures to specifications ☐
3. Raise the vehicle on a lift and remove the wheels lug nuts. ☐
4. Rotate tires according to vehicle manufacturer's recommendations. ☐
5. Install the wheels in their new positions, and tighten the wheel nuts to the specified torque.
   Specified wheel nut torque ______________
6. Lower the vehicle onto the shop floor. ☐
7. If vehicle is equipped with a tire pressure monitoring system (TPMS), refer to vehicle service information to see if any special procedure is required to relearn new wheel-and-tire position after rotation.
   Vehicle equipped with TPMS: ☐ Yes ☐ No
   If equipped with TPMS, is a TPMS service tool required: ☐ Yes ☐ No
   If equipped with TPMS, is a diagnostic scan tool required: ☐ Yes ☐ No

Task Completed

List special procedure for relearning TPMS if equipped:

____

____

____

8. Road test the vehicle. ☐

Instructor's Response

____

____

____

Name ______________________ Date ______________

JOB SHEET 10

TIRE DISMOUNTING AND MOUNTING

Upon completion of this job sheet, you should be able to demount and mount tires with or without a TPMS.

ASE Education Foundation Task Correlation

This job sheet addresses **MLR** tasks:

D.3. Dismount, inspect, and remount tire on wheel; balance wheel and tire assembly. **(P-1)**

D.4. Dismount, inspect, and remount tire on wheel equipped with TPMS. **(P-1)**

D.8. Demonstrate knowledge of steps required to remove and replace sensors in a TPMS including relearn procedure. **(P-1)**

This job sheet addresses **AST/MAST** tasks:

F.6. Dismount, inspect, and remount tire on wheel; balance wheel-and-tire assembly. **(P-1)**

F.7. Dismount, inspect, and remount tire on wheel equipped with TPMS. **(P-1)**

F.11. Demonstrate knowledge of steps required to remove and replace sensors in a TPMS, including relearn procedure. **(P-1)**

We Support ASE Education Foundation

Tools and Materials

Tire changer

Tire-and-wheel assembly

Procedure **Task Completed**

1. Remove the valve core to release all the air pressure from the tire. Chalk-mark the tire at the valve stem opening in the wheel so the tire may be reinstalled in the same position to maintain proper wheel balance.

   Is all the air pressure released from the tire? ☐ Yes ☐ No

   Is the tire chalk-marked at the valve stem location in the wheel?

   Instructor's Check ______________________

2. Guide the operating lever on the tire changer at 90° to valve stem to avoid damage to tire pressure sensor, if equipped, to unseat both tire beads. Are both tire beads unseated? ☐ Yes ☐ No

3. Place the tire-and-wheel assembly properly on the tire changer. Is the tire-and-wheel assembly positioned properly on the tire changer? ☐ Yes ☐ No

SERVICE TIP The following is a generic tire demounting and mounting procedure.

WARNING **Do not proceed to dismount the tire unless the tire-and-wheel assembly is securely attached to the tire changer. This action may cause personal injury.**

4. Press the pedal on the tire changer that clamps the wheel to the changer. Is the wheel clamped properly to the tire changer? ☐ Yes ☐ No

5. Lower the arm on the tire changer into position on the tire-and-wheel assembly. Is the tire changer arm positioned properly on the tire-and-wheel assembly? ☐ Yes ☐ No

Task Completed

6. Insert the tire iron properly between the upper tire bead and the wheel. Be sure the tire iron is properly positioned. Depress the tire changer pedal that causes the wheel to rotate. This rotation moves the top bead out over the wheel. Is the top tire bead above the wheel rim? ☐ Yes ☐ No

7. Lift the tire up and install the tire iron between the lower bead and the wheel. Be sure the tire iron is properly positioned. Depress the tire changer pedal that causes the wheel to rotate. This rotation moves the lower bead out over the wheel. Is the lower bead above the wheel rim? ☐ Yes ☐ No

SERVICE TIP Different makes of tire changers have various ways of tire bead seating on the wheel. Some tire changers have an air inflation ring that is positioned on top of the wheel-and-tire. Other tire changers have air jets in each clamping jaw that hold the wheel on the tire changer. Air pressure from these jets helps force the tire beads against the wheel sealing surfaces.

8. Remove the tire from the wheel. ☐

9. Remove, inspect, and reinstall tire pressure monitor sensor from the wheel. ☐

10. Use a wire brush to remove rust and debris from the tire bead sealing areas on the wheel rim. Are the tire bead sealing areas on the wheel properly cleaned?
☐ Yes ☐ No

11. Apply rubber lubricant to both tire beads. Are both tire beads lubricated?
☐ Yes ☐ No

12. Place the tire onto the wheel, and lower the tire changer arm into position. Depress the tire changer lever that causes the wheel to rotate. This action causes the lower tire bead over the wheel. Repeat this process to install the upper bead over the wheel. Is the tire properly installed on the wheel? ☐ Yes ☐ No

WARNING **Do not proceed to mount the tire unless the wheel is securely attached to the tire changer. This action may cause personal injury.**

13. Rotate the tire on the wheel so the chalk mark placed on the tire in Step 8 is aligned with the valve stem opening in the wheel. Is the chalk mark on the tire aligned with the valve stem position? ☐ Yes ☐ No

14. Install and tighten the valve core in the valve stem. ☐

SERVICE TIP Some tire changers have an air hose and air pressure gauge designed into the tire changer. On other tire changers, a shop air hose must be used to inflate the tire.

WARNING **When inflating a tire, do not stand or lean over the tire-and-wheel. If the tire bead comes off the rim or the rim cracks, personal injury may occur. When using a shop air hose to inflate the tire, install an extension on the air hose.**

15. Supply air to the air ring or air jets on the tire changer to move the tire beads outward against the wheel. Are the tire beads moved outward against the wheel?
☐ Yes ☐ No

Task Completed

16. Connect an air supply hose to the tire valve stem. When using a shop air hose, connect an extension to the air hose. Inflate the tire to the specified pressure.

    Specified tire pressure ______________________________

    Actual tire pressure ______________________________

Instructor's Response

____

____

____

Name ______________________ Date ______________

JOB SHEET 11

TIRE LEAK DETECTION AND REPAIR

Upon completion of this job sheet, you should be able to repair a tire following vehicle manufacturer approved procedure.

ASE Education Foundation Task Correlation

This job sheet addresses the following **MLR** tasks:

D.5. Inspect tire-and-wheel assembly for air loss; determine necessary action. **(P-1)**

D.6. Repair tire following vehicle manufacturer approved procedure. **(P-1)**

This job sheet addresses the following **AST/MAST** tasks:

F.8. Inspect tire-and-wheel assembly for air loss; determine needed action. **(P-1)**

F.9. Repair tire following vehicle manufacturer approved procedure. **(P-1)**

We Support ASE | Education Foundation

Tools and Materials

Tire pressure gauge

Tire-and-wheel assembly

Procedure **Task Completed**

1. Locate Tire Information Placard on driver's door or pillar.

   Tire size specified on left front driver's door placard: ______________

   Tire pressure specified on left front driver's door placard:

   Front______________ Rear______________ Spare______________

2. Adjust all tire pressures to specifications. ☐
3. The first step is to find the tire injury that is the source of the air leak. If the injury is not visually apparent, a tire dunk tank or soapy water solution may be needed. ☐
4. Once the injury has been located, mark the exterior of the tire injury with a tire crayon. Dismount the tire from the rim, inspect the condition of the interior of the tire and locate the puncture on the inner lining. Circle and X the puncture with a tire crayon, and remove the object that caused the puncture if still present. ☐
5. Prepare the inner tire surface with a buffing tool. Buff the area around the puncture with a tire repair buffing wheel. Surface should be buffed to a smooth finish with a RMA (Rubber Manufacturers Association) 1-2 texture buffing wheel. ☐
6. Once surface has been buffed, clean area with a tire buffer/cleaner such as Camel 12-092. ☐

   Refer to information on the product or manufacturer's Safety Data Sheet and follow guidelines for handling and disposal.

7. Guide a radial hand rasp/reamer through the puncture to allow the plug patch to fit properly. ☐
8. Apply tire patch cement according to manufacturer's recommendations and then install plug patch. The plug portion of the patch has a removable metal sleeve to help guide it through the puncture injury. ☐
9. Using a corrugated patch stitching tool work out all air bubbles from under patch to ensure proper vulcanization of tire patch to inner lining of tire. ☐

Task Completed

10. Trim the tire plug flush with tire tread. ☐

11. Remount and balance tire-and-wheel assembly. ☐

Instructor's Response

____

____

____

Name ______________________ Date ______________

JOB SHEET 12

TIRE-AND-WHEEL RUNOUT MEASUREMENT

Upon completion of this job sheet, you should be able to measure radial and lateral tire-and-wheel runout.

ASE Education Foundation Task Correlation

This job sheet addresses the following **AST/MAST** task:

F.4. Measure wheel, tire, axle flange, and hub runout; determine needed action. **(P-2)**

We Support ASE | Education Foundation

SERVICE TIP The tire-and-wheel may be installed on an off-car wheel balancer to measure tire-and-wheel runout.

Tools and Materials

Dial indicator

Wheel balancer

Tire-and-wheel assembly

Wheel diameter ______________________

Manufacturer and type of tire ______________________

Procedure **Task Completed**

1. Mount the tire-and-wheel assembly properly on the wheel balancer and position a dial indicator against the center of the tire tread as the tire is rotated slowly to measure radial runout.

   Radial runout ______________________

   Specified radial runout ______________________

2. Mark the highest point of radial runout on the tire with chalk, and mark the valve stem position on the tire. ☐

3. If the radial tire runout is excessive, demount the tire, and check the runout of the wheel rim with a dial indicator positioned against the lip of the rim while the rim is rotated.

   Wheel radial runout ______________________

   Specified wheel radial runout ______________________

4. Use chalk to mark the highest point of radial runout on the wheel rim. ☐

5. If the highest point of wheel radial runout coincides with the chalk mark from the highest point of maximum tire radial runout, the tire may be rotated 180° on the wheel to reduce radial runout. Tires or wheels with excessive runout are usually replaced. Does the highest point of tire radial runout coincide with the highest point of wheel radial runout? ☐ Yes ☐ No

   Instructor's Response ______________________

   State the required action to correct excessive radial runout and the reason for this action.

   ____

   ____

Task Completed

6. Position a dial indicator located against the sidewall of the tire to measure lateral runout.

   Tire lateral runout ______________________

   Specified tire lateral runout ______________________

7. Chalk-mark the tire-and-wheel at the highest point of lateral runout. ☐

8. If the tire runout is excessive, the tire should be demounted from the wheel and the wheel lateral runout measured. ☐

9. If the tire runout was excessive, measure the wheel lateral runout with a dial indicator positioned against the edge of the wheel as the wheel is rotated.

   Wheel lateral runout ______________________

   Specified wheel lateral runout ______________________

   State the required action to correct excessive wheel lateral runout and the reason for this action. ______________________

Instructor's Response

____

____

____

Name ______________________ Date ______________

JOB SHEET 13

OFF-CAR WHEEL BALANCING

Upon completion of this job sheet, you should be able to perform off-car wheel balancing procedures.

ASE Education Foundation Task Correlation

This job sheet addresses the following **MLR** task:

D.3. Dismount, inspect, and remount tire on wheel; balance wheel-and-tire assembly. **(P-1)**

This job sheet addresses the following **AST/MAST** task:

F.6. Dismount, inspect, and remount tire on wheel; balance wheel-and-tire assembly. **(P-1)**

We Support ASE Education Foundation

Tools and Materials

Electronic off-car wheel balancer

Wheel weights

Tire-and-wheel

Wheel diameter ______________________

Manufacturer and type of tire ______________________

Describe the vehicle being worked on:

Year ______________ Make ______________ Model ______________

VIN ______________________ Engine type and size ______________________

Procedure **Task Completed**

1. Wash mud and debris from the tire-and-wheel. ☐
2. Mount the tire-and-wheel assembly on the wheel balancer with the wheel balancer manufacturer's recommended mounting components. ☐
3. Tighten the tire-and-wheel retaining nut on the wheel balancer to the specified torque.

   Tire-and-wheel retaining nut on the wheel balancer tightened to the specified torque? ☐ Yes ☐ No

   Instructor's Check ______________________
4. Remove stones and other objects from the tire tread.

   Are stones and other objects removed from the tire tread? ☐ Yes ☐ No

   Instructor's Check ______________________
5. Check for objects inside tire.

   Are there objects inside the tire? ☐ Yes ☐ No

   Instructor's Check ______________________
6. Remove all the old wheel weights from the wheel. Are the old wheel weights removed? ☐ Yes ☐ No

   Instructor's Check ______________________

Task Completed

7. Inspect the tread and sidewall.

   List defective tread and sidewall conditions. ____________________

   List the types of tire tread wear and state the causes of this wear. ____________________

   Based on your inspection of the tire, list the required tire service. ____________________

8. Check tire inflation pressure.

   Tire inflation pressure ____________________

   Specified tire inflation pressure ____________________

9. Measure tire-and-wheel runout.

   Radial tire-and-wheel runout ____________________

   Specified tire-and-wheel radial runout ____________________

   Lateral tire-and-wheel runout ____________________

   Specified lateral tire and lateral wheel runout ____________________

   Recommended tire-and-wheel service. ____________________

10. Lower the safety hood on the wheel balancer.

    Is the safety hood lowered? ☐ Yes ☐ No

11. Enter the required information on wheel balancer keypad, such as wheel diameter, wheel width, and wheel offset or distance. Is the required wheel information entered in the wheel balancer? ☐ Yes ☐ No

    Instructor's Check ______________

12. If the wheel balancer has a mode selector, select the desired balance mode.

    Balance mode selected______________

13. Spin the wheel on the balancer, then apply the balancer brake to stop the wheel. ☐

14. Observe the balancer display indicating the necessary amount of wheel weights and the required location of these weights. Install the required wheel weights in the locations indicated on the balancer. Be sure the wheel weights are securely attached to the rim. Are the wheel weight(s) securely attached to the rim at the locations indicated by the wheel balancer? ☐ Yes ☐ No

    Instructor's Check ______________

15. Spin the tire-and-wheel assembly on the balancer again to be sure this assembly is properly balanced. Does the wheel balancer indicate proper wheel balance? ☐ Yes ☐ No

    If the answer to this question is No, repeat the balance procedure.

Instructor's Response

____

____

____

Name ______________________ Date ______________

JOB SHEET 14

ON-CAR WHEEL BALANCING

Upon completion of this job sheet, you should be able to perform on-car wheel balancing procedures.

ASE Education Foundation Task Correlation

This job sheet addresses the following **MLR** task:

D.3. Dismount, inspect, and remount tire on wheel; balance wheel-and-tire assembly. **(P-1)**

This job sheet addresses the following **AST/MAST** task:

F.6. Dismount, inspect, and remount tire on wheel; balance wheel-and-tire assembly. **(P-1)**

We Support ASE Education Foundation

Tools and Materials

On-car wheel balancer
Car
Wheel weights

Describe the vehicle being worked on:

Year ________ Make ________ Model ________

VIN ______________ Engine type and size ______________

Procedure **Task Completed**

1. Wash mud and debris from the wheel-and-tire. ☐

2. Raise the wheel being balanced 5 in. (12 cm) off the floor; be sure the chassis is securely supported on safety stands. ☐

3. Remove stones and other objects from the tire tread.

Are all objects removed from the tire tread? ☐ Yes ☐ No

Instructor's Check ______________________

4. Inspect the tire tread and sidewall condition.

Tread and sidewall condition ______________________

Types of tire tread wear ______________________

Causes of tire tread wear ______________________

Based on your inspection of the tire tread and sidewall, state the necessary tire service.

____

5. Check the wheel bearing adjustment and adjust to specifications if necessary. Wheel bearing adjustment: ☐ Satisfactory ☐ Unsatisfactory

6. Install the electronic vibration sensor between the lower control arm and the floor. Is the vibration sensor properly installed? ☐ Yes ☐ No

Instructor's Check ______________________

Caution

Do not spin the front wheels on a front-wheel-drive vehicle with the floor jack under the chassis and the suspension dropped downward. Under this condition, severe angles exist in the front drive axle joints, and these joints may be damaged if the wheels are rotated with the balancer. Place the floor jack under the lower control arm to raise the wheel.

Task Completed

7. Chalk-mark a reference mark on the outer sidewall of the tire. ☐
8. Spin the wheel just fast enough to produce vibration on the front bumper. ☐
9. When vibration causes strobe light flashing, move the balancer drum away from the tire and allow the tire to spin freely. ☐
10. Shine the strobe light around the tire sidewall and note the chalk-mark position.

    Chalk-mark clock position ______________________________

11. Note the pointer position on the meter.

    Meter pointer position ______________________________

12. Use the balancer's brake plate to slow and stop the wheel. ☐
13. Rotate the wheel until the chalk mark is in the exact position where it appeared under the strobe light. The heavy spot is now at the bottom of the wheel and the balancing weight should be attached 180° from the heavy spot. Install the amount of weight indicated on the meter.

    Amount of wheel weight installed ______________________________

14. Spin the wheel again. If the wheel balance is satisfactory, the meter pointer will read in the balanced position. If the pointer does not indicate a balanced wheel, shine the strobe light on the tire sidewall. If the installed wheel weight is at the 12 o'clock position, additional weight is required. A 6 o'clock weight position indicates excessive weight. A 3 o'clock or 9 o'clock weight position may be corrected by moving the weight 1 in. (2.5 cm) toward the 12 o'clock position.

    Tire-and-wheel balance: ☐ Satisfactory ☐ Unsatisfactory

    If the wheel balance is unsatisfactory, state the action required to balance tire-and-wheel. ______________________________

    ____

Caution

To obtain proper wheel balance, spin the wheel only to the speed where the vibration occurs. Do not spin the wheel at excessive speeds.

Instructor's Response

____

____

____

Name ______________________ **Date** ______________

JOB SHEET 15

TIRE PULL DIAGNOSIS

Upon completion of this job sheet, you should be able to diagnose tire pull problems and determine needed action.

ASE Education Foundation Task Correlation

The following job sheet addresses the following **AST/MAST** task:

F.5. Diagnose tire pull problems; determine needed action. **(P-1)**

We Support ASE Education Foundation

Tools and Materials

Tire pressure gauge
Tire-and-wheel service tools

Procedure **Task Completed**

1. Locate Tire Information Placard on driver's door or pillar.
   - **a.** Tire size specified on left front driver's door placard: ______________________
   - **b.** Tire pressure specified on left front driver's door placard:
   - **c.** Front ______________ Rear ______________ Spare ______________
2. Verify that tire on the vehicle are the size recommended by the vehicle manufacturer. ☐
3. Adjust all tire pressures to specifications. ☐
4. Test drive the vehicle and verify if the pulling complaint is present. ☐
5. Once the suspension and steering system have been inspected and the alignment angles have been checked, and no causation for vehicle pull condition has been determined, a radial tire pull could be the root problem. ☐
6. Start by cross rotating the two front tires from side to side. ☐
7. If the vehicle now pulls in the opposite direction, the defective tire is one of the front tires. Go on to Step 9. ☐
8. If the vehicle still pulls in the same direction, the problem is either with the rear tires or a non-tire related issue. ☐
   - **a.** Cross rotating the two rear tires from side to side.
     - If the vehicle now pulls in the opposite direction, the defective tire is one of the rear tires.
       - On the side of the car that is in the direction of the pull, rotate the rear tire to the front tire position.
         - If the pull goes away or diminishes, the tire that was moved to the front is defective.
         - If the pull does not change, then the other rear tire is defective.
     - If the vehicle still pulls in the same direction, the problem is not tire related. Inspect vehicle alignment and steering and suspension system for worn or faulty parts.

Task Completed

9. On the side of the car that is in the direction of the pull, rotate the front tire to the rear tire position. ☐
   a. If the pull goes away or diminishes, the tire that was moved to the rear is defective.
   b. If the pull does not change, then the other front tire is defective.

Instructor's Response

____

____

____

Name ______________________________ Date ______________

JOB SHEET 16

DIAGNOSE TIRE-AND-WHEEL VIBRATION, STEERING PULL, AND CHASSIS WADDLE

Upon completion of this job sheet, you should be able to diagnose tire-and-wheel vibration problems, steering pull, and chassis waddle.

ASE Education Foundation Task Correlation

This job sheet addresses the following **AST/MAST** task:

F.2. Diagnose wheel/tire vibration, shimmy, and noise; determine necessary action. **(P-2)**

We Support ASE Education Foundation

Tools and Materials

Late-model vehicle

Describe the vehicle being worked on:

Year ______________ Make ______________ Model ______________

VIN ______________ Engine type and size ______________

Procedure

1. Drive the vehicle at 10 mph (16 km/h), 20 mph (32 km/h), 30 mph (48 km/h), 40 mph (64 km/h), 50 mph (80 km/h), and 60 mph (96.5 km/h). Maintain each vehicle speed long enough to diagnose tire, steering, and chassis problems. Do not exceed the legal speed limit. Did the vehicle indicate any of the following problems at these various speeds?

____

____

2. Chassis waddle. Speed at which waddle occurred: ______________

Did the waddle occur at the front or the rear of the vehicle? ______________

List the possible causes of waddle and the necessary diagnostic steps and service procedures to correct the problem.

____

____

____

3. Tire thump and vibration. Speed at which tire thump or vibration occurred: ________

At which corner of the vehicle did the thump and vibration occur? ______________

List the possible causes of tire thump and vibration and the necessary diagnostic steps and service procedures to correct the problem.

____

____

____

4. Wheel shimmy. Speed at which wheel shimmy occurred: ____________________

   List the possible cause of wheel shimmy, and the necessary diagnostic steps and service procedures to correct the problem.

   ____

   ____

   ____

5. Steering pull. Speed at which steering pull occurred: ____________________

   Direction of steering pull: ____________________

   List the possible causes of steering pull, and the necessary diagnostic steps and service procedures to correct the problem.

   ____

   ____

   ____

Instructor's Response

____

____

____

Name ______________________ Date ______________

JOB SHEET 17

INSPECT, DIAGNOSE, AND CALIBRATE TIRE PRESSURE MONITORING SYSTEMS

Upon completion of this job, you should be able to inspect, diagnose, and calibrate TPMSs.

ASE Education Foundation Task Correlation

This job sheet addresses the following **MLR** task:

D.7. Identify indirect and direct TPMSs; calibrate system; verify operation of instrument panel lamps. **(P-2)**

This job sheet addresses the following **AST/MAST** task:

F.10. Identify indirect and direct TPMSs; calibrate system; verify operation of instrument panel lamps. **(P-1)**

F.11. Demonstrate knowledge of steps required to remove and replace sensors in a tire pressure monitoring system (TPMS) including relearn procedure. **(P-1)**

We Support ASE Education Foundation

Tools and Materials

Late-model vehicle

Scan tool

Describe the vehicle being worked on:

Year ____________ Make ____________ Model ____________

VIN ____________________ Engine type and size ____________________

Procedure **Task Completed**

1. Connect a scan tool to the data link connector (DLC) under the left side of the instrument panel, and turn the scan tool on. ☐
2. Turn the ignition switch on, and do not start the engine. ☐
3. Select antenna module on the scan tool. Check for the DTCs related to the antenna module. If there are DTCs displayed with a "U" prefix, there is a defect in the data link communications. If the scan tool cannot communicate with the antenna module, or there are DTCs related to this module or the data link communications, proceed with further diagnosis of these items. ☐
4. Check for DTCs related to the radio/audio system. The antenna grid in the rear window receives TPMS sensor signals. This grid shares its connector with the AM/FM antenna grid. If any radio/audio system DTCs are present, proceed with the diagnosis of the radio/audio system. ☐
5. Check for keyless entry system DTCs displayed in the scan tool. The antenna module also controls the keyless entry system. If the scan tool displays DTCs related to the keyless entry system, proceed with the diagnosis of these DTCs. ☐
6. Check for TPMS DTCs on the scan tool display. If TPMS DTCs are present, diagnose the cause of these DTCs.

   List and interpret the DTCs displayed on the scan tool: ______________________

   ____

   ____

TPMS Learning Procedure **Task Completed**

1. Connect a scan tool to the DLC. ☐
2. Turn on the ignition switch, and do not start the engine. ☐
3. Apply the parking brake. ☐
4. Select Special Functions on the scan tool. ☐
5. Select Sensor Learn Mode on the scan tool, and press the Enter key. ☐
6. Press the on soft key. A horn chirp should sound to indicate the Sensor Learn Mode is enabled. ☐
7. Starting with the LF tire, increase or decrease the tire pressure for 5–8 seconds or until a horn chirp sounds. The horn chirp may occur before the 5- to 8-second time period, or up to 30 seconds after this time period. ☐

WARNING **If you are increasing tire pressure during the learning procedure, never inflate a tire above the vehicle manufacturer's maximum specified tire pressure. This action may cause personal injury and tire damage.**

8. After the horn chirp sounds, repeat Step 7 on the other three or four sensors in the following order: ☐
   (a) RF
   (b) RR
   (c) LR
   (d) Spare tire (if applicable)

   If a horn chirp is not heard after 35 seconds for any of the four sensors, turn off the ignition switch and exit the Sensor Learn Mode on the scan tool. Repeat the sensor learning procedure from Step 4.
9. After the learning procedure has been completed on all the sensors, a double horn chirp sounds to indicate the learning procedure is completed on all the sensors. ☐
10. Turn off the ignition switch and disconnect the scan tool. ☐
11. Inflate all the tires to the specified pressure.

    Learning procedure completed? ☐ Yes ☐ No

    If the answer is no, state the reason for the test not being completed

Instructor's Response

___

___

___

CHAPTER 4
WHEEL BEARING AND SEAL SERVICE

Upon completion and review of this chapter, you should be able to:

- Diagnose bearing defects.
- Clean and repack wheel bearings.
- Reassemble and adjust wheel bearings.
- Remove and replace wheel bearing seals.
- Diagnose wheel bearing problems.
- Diagnose problems in wheel bearing hub units.
- Remove and replace front drive axles.
- Remove and replace wheel bearing hub units.
- Remove and replace rear-axle bearings on rear-wheel-drive cars.

Basic Tools

Basic technician's tool set

Service manual

Inch-pound torque wrench

Foot-pound torque wrench

Fine-toothed round and flat files

Wheel bearing grease

Differential lubricant

Terms To Know

Bearing brinelling
Bearing fatigue spalling
Bearing frettage
Bearing smears
Circlip
"C" locks
Stethoscope

Technicians must accurately diagnose wheel bearing problems to avoid repeat bearing failures, and thus provide customer satisfaction. Accurate wheel bearing service procedures are essential to maintain vehicle safety! Improper wheel bearing service may cause brake problems, steering complaints, and premature bearing failure. Improper wheel bearing service may even cause a wheel to fly off a vehicle, resulting in personal injury and vehicle damage. A knowledge of drive axle removal is necessary to service the front wheel bearings on front-wheel-drive vehicles, because the drive axle must be removed from the front hub to service the wheel bearings. Similarly, technicians must also understand rear-axle bearing service procedures on rear-wheel-drive vehicles.

DIAGNOSIS OF BEARING DEFECTS

Bearings are designed to provide long life, but there are many causes of premature bearing failure. If a bearing fails, the technician must decide if the bearing failure was caused by normal wear or if the bearing failed prematurely. For example, if a front wheel bearing fails on a car that is two years old with an original odometer reading of 30,000 mi. (48,280 km), experience tells us the bearing failure is premature, because front wheel bearings normally last for a much longer mileage period, often in excess of 100,000 mi. (160,934 km). Always listen to the customer's complaints, and obtain as much information as possible from the customer. Ask the customer specific questions about abnormal or unusual vehicle noises and operation. If a bearing fails prematurely, there must be some cause for the failure.

Bearing galling refers to metal smears on the ends of the rollers.

Bearing abrasive step wear is a fine circular wear pattern on the ends of the rollers.

Bearing etching appears as a loss of material on the bearing rollers and races. Bearing surfaces are gray or grayish black.

Bearing indentations are surface depressions on the rollers and races.

Bearing brinelling may be caused by the continuous vibration when transporting new vehicles by rail or truck from the factory to the dealership.

Classroom Manual
Chapter 4, page 104

The causes of premature bearing failure are:

1. Lack of lubrication
2. Improper type of lubrication
3. Incorrect endplay adjustment (where applicable)
4. Misalignment of related components, such as shafts or housings
5. Excessive bearing load
6. Improper installation or service procedures
7. Excessive heat
8. Dirt or contamination

When a bearing fails prematurely, the technician must correct the cause of this failure to prevent the new bearing from failing. Refer to **Table 4-1** for additional information on noise and vibration diagnosis. The types of bearing failures and the necessary corrective service procedures are provided in **Figure 4-1** and **Figure 4-2**. **Bearing fatigue spalling** appears as flaking of surface metal on bearing rollers and races. **Bearing brinelling** shows up as straight-line indentations on the races and rollers. **Bearing smears** appear as metal loss in a circular, blotched pattern around the bearing races and rollers. **Bearing frettage** shows up as a fine, corrosive wear pattern around the bearing races and rollers. This wear pattern is circular on the races.

The first indication of bearing failure is usually a howling noise while the bearing is rotating. The howling noise will likely vary depending on the bearing load. A front wheel bearing usually provides a more noticeable howl when the vehicle is turning a corner, because this places additional thrust load on the bearing. A defective rear-axle bearing usually provides a howling noise that is more noticeable at lower speeds. The howling noise is more noticeable when driving on a narrow street with buildings on each side, because the noise vibrates off the nearby buildings. A rear-axle bearing noise is present during acceleration and deceleration, because the vehicle weight places a load on the bearing regardless of the operating condition. The rear-axle bearing noise may be somewhat more noticeable during deceleration because there is less engine noise at that time.

TABLE 4-1 DIAGNOSTIC TABLE

Problem	Symptoms	Possible causes
Noise in drive train, front-wheel-drive car.	Clicking noise while turning a corner at low speeds.	Worn outer front drive axle joint.
Vibration in front drive train, front-wheel-drive car.	Vibration while decelerating at 35 to 45 mph (56 to 72 km/h).	Worn inner drive axle joint.
Noise in front suspension.	Growling noise in front suspension, most noticeable when turning a corner or driving at low speeds.	Worn wheel bearings.
Lack of directional stability.	Steering wanders to either side when driving straight ahead.	Excessive front wheel bearing endplay, improper wheel bearing adjustment.
Noise in drive train, rear-wheel-drive car.	Constant growling or clicking noise in rear axle, most noticeable when driving at low speeds or down a narrow street.	Worn rear-axle bearing.

TAPERED ROLLER BEARING DIAGNOSIS

Consider the following factors when diagnosing bearing condition:

1. General condition of all parts during disassembly and inspection.
2. Classify the failure with the aid of the illustrations.
3. Determine the cause.
4. Make all repairs following recommended procedures.

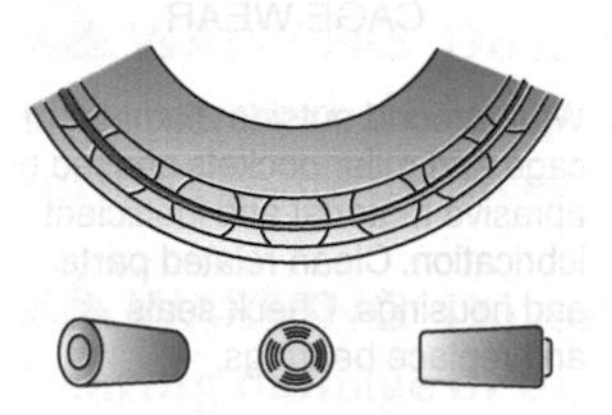

ABRASIVE STEP WEAR

Pattern on roller ends caused by fine abrasives. Clean all parts and housings, check seals and bearings, and replace if leaking, rough, or noisy.

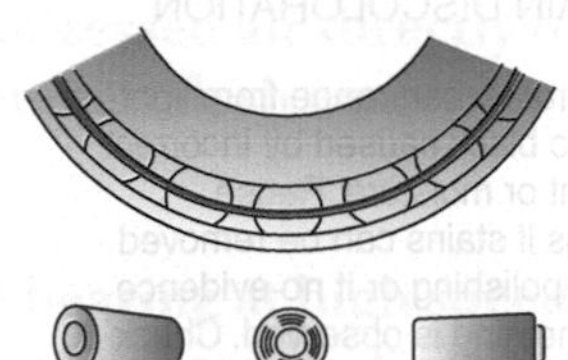

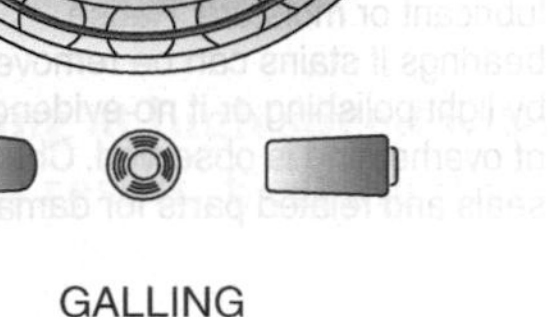

GALLING

Metal smears on roller ends due to overheating, lubricant failure, or overload. Replace bearing, check seals, and check for proper lubrication.

BENT CAGE

Cage damaged due to improper handling or tool usage. Replace bearing.

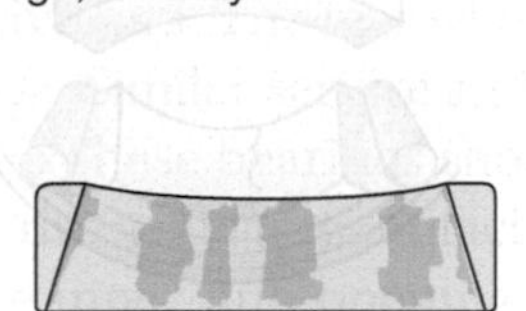

ABRASIVE ROLLER WEAR

Pattern on races and rollers caused by fine abrasives. Clean all parts and housings, check seals and bearings, and replace if leaking, rough, or noisy.

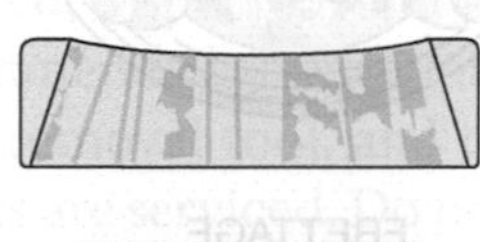

ETCHING

Bearing surfaces appear gray or grayish black in color with related etching away of material usually at roller spacing. Replace bearings, check seals, and check for proper lubrication.

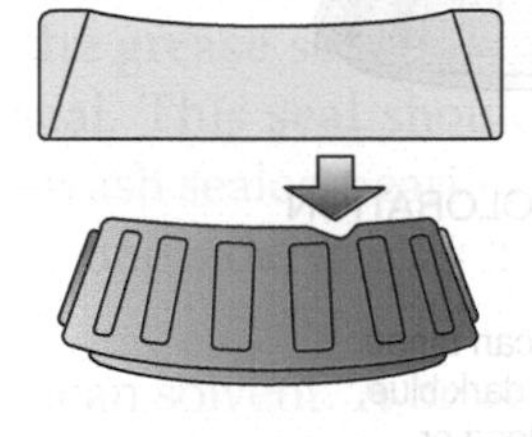

BENT CAGE

Cage damaged due to improper handling or tool usage. Replace bearing.

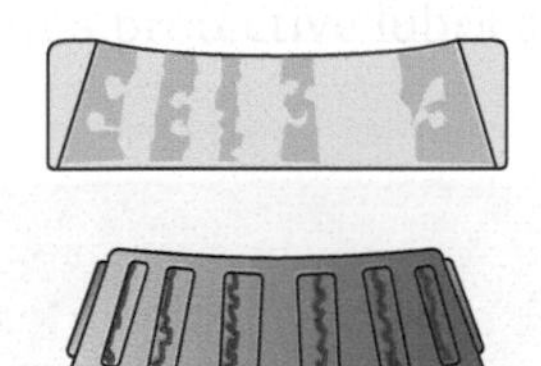

INDENTATIONS

Surface depressions on race and rollers caused by hard particles of foreign material. Clean all parts and housings. Check seals, and replace bearings if rough or noisy.

BRINELLING

Surface indentations in raceway caused by rollers either under impact loading or vibration while the bearing is not rotating. Replace bearing if rough or noisy.

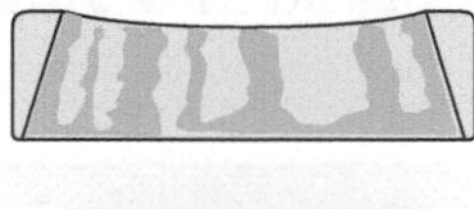

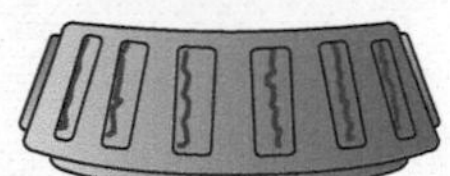

MISALIGNMENT

Outer race misalignment due to foreign object. Clean related parts and replace bearing. Make sure races are properly sealed.

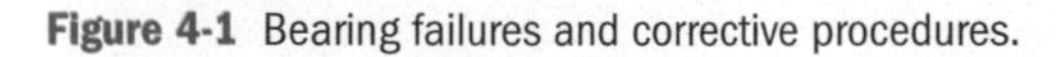

Figure 4-1 Bearing failures and corrective procedures.

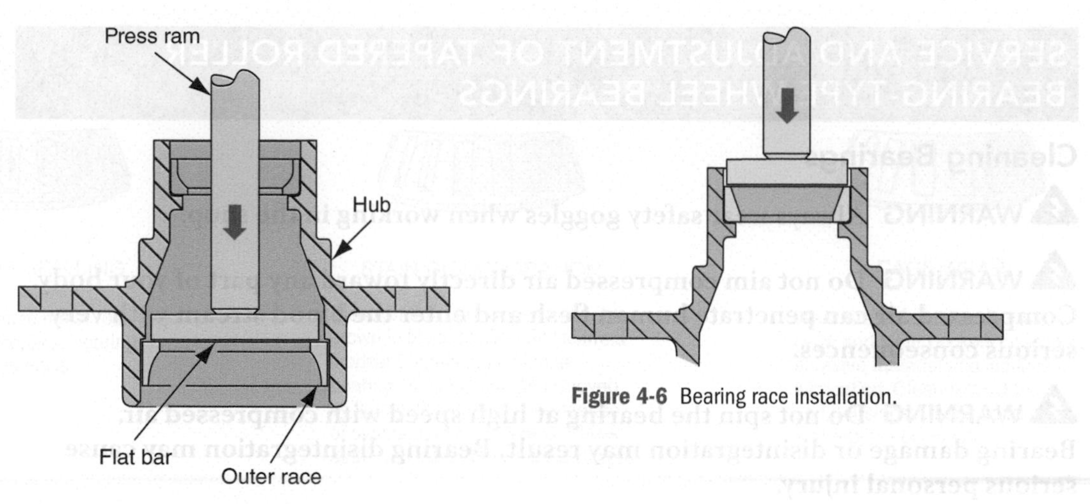

Figure 4-5 Bearing race removal.

Figure 4-6 Bearing race installation.

identify the bearings, or lay them in order, so you reinstall them in their original location. Do not clean bearings or races with paper towels. If you are using a shop towel for this purpose, be sure it is lint-free. Lint from shop towels or paper towels may contaminate the bearing. Bearing races and the inner part of the wheel hub should be thoroughly cleaned with solvent and dried with compressed air. Inspect the seal mounting area in the hub for metal burrs. Remove any burrs with a fine round file.

Small dirt particles left behind the outer bearing race cause race misalignment and premature bearing failure.

Bearing races must be replaced if any of the defects described in Figures 4-1 and 4-2 are found. The proper bearing race driving tool must be used to remove the bearing races (**Figure 4-5**). If a driver is not available for the bearing races, a long brass punch and hammer may be used to drive the races from the hub. When a hammer and punch are used for this purpose, be careful not to damage the hub inner surface with the punch.

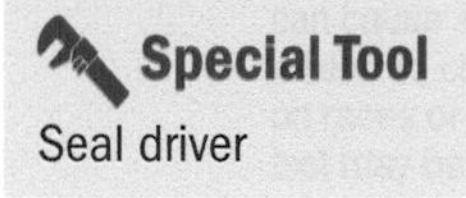

Special Tool

Seal driver

The new bearing races should be installed in the hub with the correct bearing race driving tool (**Figure 4-6**). When bearings and races are replaced, be sure they are the same as the original bearings. The part numbers should be the same on the old bearings and the replacement bearings.

Inspect the bearing and seal mounting surfaces on the spindle. Small metal burrs may be removed from the spindle with a fine-toothed file. If the spindle is severely scored in the bearing or seal mounting areas, spindle replacement is necessary.

Bearing Lubrication and Assembly

After the bearings and races have been cleaned and inspected, the bearings should be packed with grease. Always use the vehicle manufacturer's specified wheel bearing grease. Vehicle manufacturers usually recommend a lithium-based wheel bearing grease. Place a lump of grease in the palm of one hand and grasp the bearing in the other hand. Force the widest edge of the bearing into the lump of grease, and squeeze the grease into the bearing. Continue this process until grease is forced into the bearing around the entire bearing circumference. Place a coating of grease around the outside of the rollers, and apply a light coating of grease to the races. A bearing packing tool may be used to force grease into the bearings rather than using the hand method. Bearing packers may be hand operated or pressure operated (**Figure 4-7**).

Caution

Cleanliness is very important during wheel bearing service. Always maintain cleanliness of hands, tools, work area, and all related bearing components. One small piece of dirt in a bearing will cause bearing failure.

Place some grease in the wheel hub cavity and position the inner bearing in the hub (**Figure 4-8**). Check the fit of the new bearing seal on the spindle and in the hub. The seal lip must fit snugly on the spindle, and the seal case must fit properly in the hub opening.

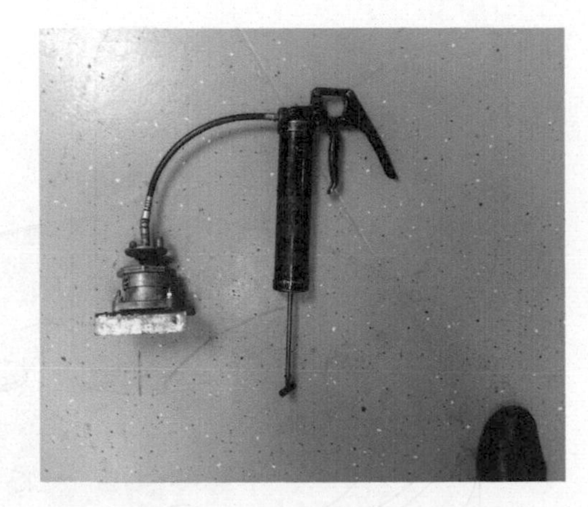

Figure 4-7 Mechanical wheel bearing packer.

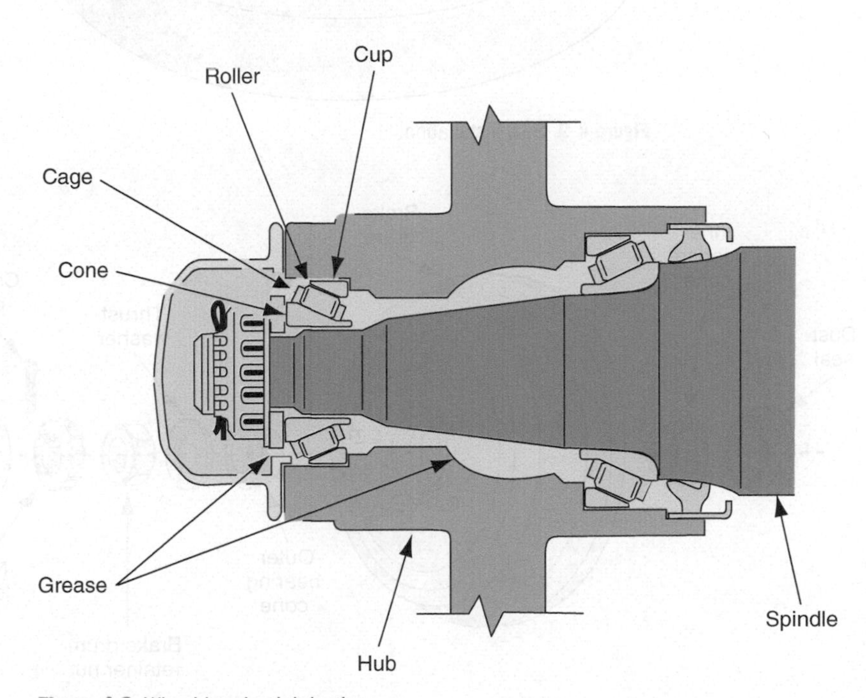

Figure 4-8 Wheel bearing lubrication.

> **Caution**
> Always keep grease containers covered when not in use. Uncovered grease containers are easily contaminated with dirt and moisture.

The part number on the old seal and the replacement seal should be the same. Be sure the seal is installed in the proper direction with the garter spring and higher part of the lip toward the lubricant in the hub. The new inner bearing seal must be installed in the hub with a suitable seal driver (**Figure 4-9**). Place a light coating of wheel bearing grease on the spindle and slide the hub assembly onto the spindle. Install the outer wheel bearing and be sure there is adequate lubrication on the bearing and race. Be sure the washer and nut are clean and install these components on the spindle (**Figure 4-10**). Tighten the nut until it is finger tight.

Photo Sequence 6 shows a typical procedure for adjusting rear wheel bearings on a front-wheel-drive car.

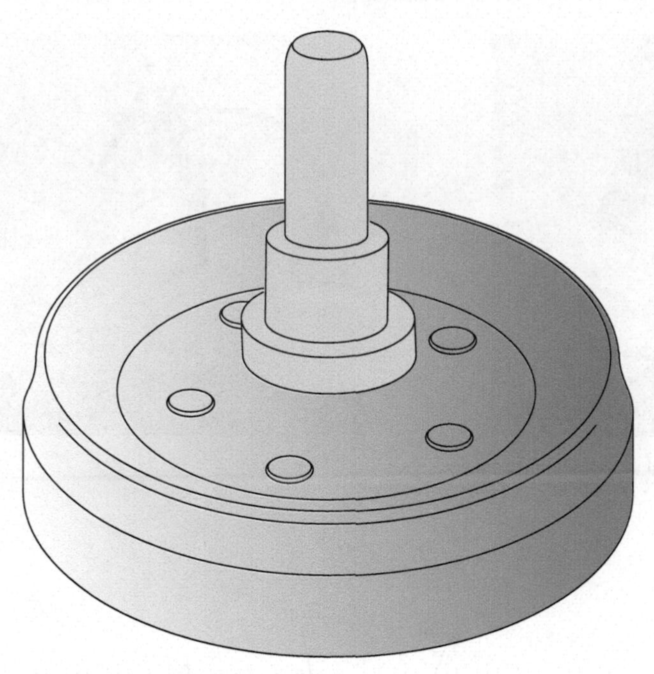

Figure 4-9 Seal installation.

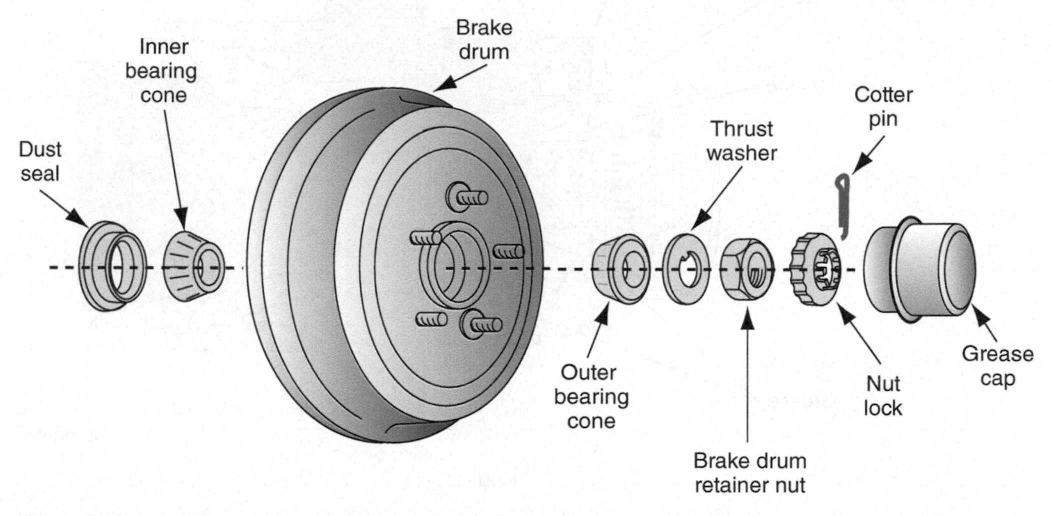

Figure 4-10 Installation of wheel bearings and related components.

SERVICE TIP When a lip seal is installed, the garter spring should always face toward the flow of lubricant.

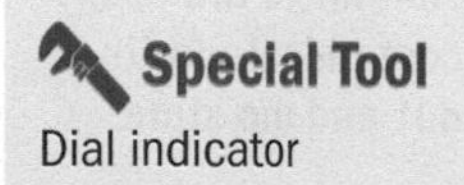

Special Tool
Dial indicator

Classroom Manual
Chapter 4, page 112

Wheel Bearing Adjustment with Two Separate Tapered Roller Bearings in the Wheel Hub

Wheel bearing endplay is the amount of horizontal wheel bearing hub movement. If a bearing has a preload condition, a slight tension is placed on the bearing. Loose front wheel bearing adjustment results in lateral front wheel movement and reduced directional stability. If the wheel bearing adjusting nut is tightened excessively, the bearings may overheat, resulting in premature bearing failure. The bearing adjustment procedure may vary depending on the make of vehicle. Always follow the procedure in the vehicle manufacturer's service manual.

PHOTO SEQUENCE 6

Typical Procedure for Adjusting Rear Wheel Bearings on a Front-Wheel-Drive Car

P6-1 Always make sure the car is positioned safely on a lift before working on the vehicle.

P6-2 Remove the dust cap from the wheel hub.

P6-3 Remove the cotter pin and nut retainer from the bearing adjusting nut.

P6-4 Tighten the bearing adjusting nut to 17–25 ft.-lb.

P6-5 Loosen the bearing adjusting nut one-half turn.

P6-6 Tighten the bearing adjusting nut to 10–15 in.-lb.

P6-7 Position the adjusting nut retainer over the adjusting nut so the slots are aligned with the holes in the nut and spindle.

P6-8 Install a new cotter pin and bend the ends around the retainer flange.

P6-9 Install the dust cap and be sure the hub rotates freely.

CUSTOMER CARE Never sell a customer automotive service that is not required on his or her car. Selling preventive maintenance, however, is a sound business practice and may save a customer some future problems. An example of preventive maintenance is selling a cooling system flush when the cooling system is not leaking but the manufacturer's recommended service interval has elapsed. If customers find out they were sold some unnecessary service, they will probably never return to the shop. They will likely tell their friends about their experience, and that kind of advertising the shop can do without.

A typical bearing adjustment procedure follows:

1. With the hub and bearings assembled on the spindle, tighten the adjusting nut to 17–25 ft.-lb. (23–34 Nm) while the hub is rotated in the forward direction.
2. Loosen the adjusting nut one half turn and retighten it to 10–15 in.-lb (1.0–1.7 Nm). This specification varies depending on the make of vehicle. Always use the manufacturer's specifications.
3. Position the adjusting nut retainer over the nut so the retainer slots are aligned with the cotter pin hole in the spindle.
4. Install a new cotter pin, and bend the ends around the retainer flange.
5. Install the grease cap, and make sure the hub and the drum rotate freely.

After a wheel bearing adjustment is performed, the hub must have an endplay of 0.001 in.–0.005 in. (0.0254 mm–0.127 mm). A dial indicator may be mounted with the stem positioned against the outer edge of the hub to perform this measurement. When the hub is pulled outward and pushed inward, the specified endplay must be displayed on the dial indicator. If the endplay is not correct, the wheel bearing adjustment must be repeated.

WHEEL HUB UNIT DIAGNOSIS AND SERVICE

When wheel bearings and hubs are an integral assembly or unitized hub assembly (**Figure 4-11**), the bearing endplay should be measured with a dial indicator stem mounted against the hub. If the endplay exceeds 0.005 in. (0.127 mm) as the hub is moved in and out, the hub and bearing assembly should be replaced. This specification is typical, but the vehicle manufacturer's specifications must be used. Hub and bearing replacement is also necessary if the bearing is rough or noisy. Integral-type bearing and hub assemblies

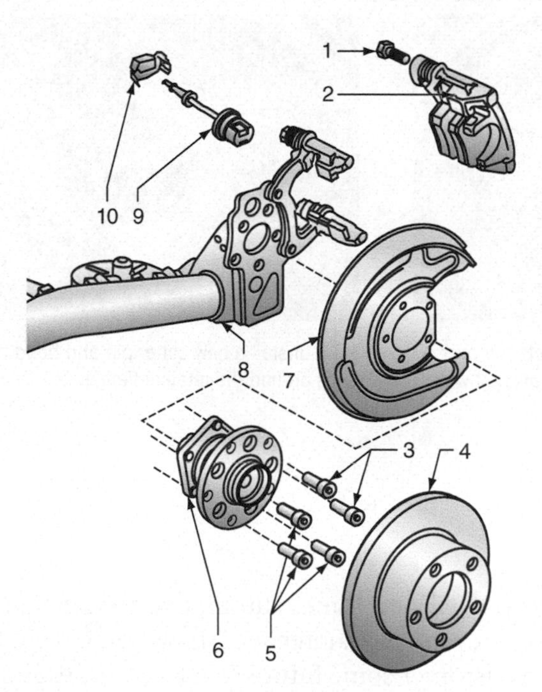

1. Bolt
2. Caliper
3. Socker head bolt
4. Rotor
5. Socket head bolt
6. Wheel Bearing/Hub Assembly
7. Backing plate
8. Axle beam
9. ABS speed sensor
10. Bracket

Figure 4-11 A typical non-serviceable integral hub bearing assembly.

are used on the front and rear wheels on some front-wheel-drive cars. **Photo Sequence 7** shows a typical procedure for measuring front wheel hub endplay.

When removable front wheel bearings are mounted in the steering knuckle, the wheel bearings may be checked with the vehicle raised on the hoist and a dial indicator positioned against the outer wheel rim lip as shown in **Figure 4-12**.

When the wheel is moved in and out, the maximum bearing movement on the dial indicator should be as follows:

0.020 in. (0.508 mm) for 13-in. (33-cm) wheels
0.023 in. (0.584 mm) for 14-in. (35.5-cm) wheels
0.025 in. (0.635 mm) for 15-in. (38-cm) wheels

If the bearing movement is excessive, check the hub nut torque before replacing the bearing. When this torque is correct and bearing movement is excessive, the bearing should be replaced.

When a wheel is removed to service the wheel bearings, proper balance must be maintained between the wheel and tire and the hub. Therefore, the tire, wheel, hub stud, and brake rotor should be chalk-marked prior to removal.

Service of the integral hub assembly is relatively straightforward but does require the removal of the brake caliper and rotor assembly if the vehicle is equipped with disc brakes.

First remove the wheel assembly.

Remove the disc brake caliper mounting bolts and secure with a piece of mechanics wire to avoid damage to the brake hose.

Remove the brake rotor.

If a wheel speed sensor for the antilock brake system (ABS) is integral in the front wheel hubs, disconnect the wheel speed sensor connectors. If the wheel speed sensor is not integrated into the hub it will be necessary to remove it from the steering knuckle to avoid damage.

If this is the drive axle hub it will be necessary to remove the drive axle retaining nut.

Remove the hub-to-knuckle bolts, and place the transaxle in park.

Use the proper puller to remove the wheel-bearing hub from the drive axle.

Install the new hub and bearing assembly on the drive axle splines, and install a new drive axle nut. Tighten the drive axle nut to pull the hub onto the splines. Do not tighten this nut to the specified torque at this time.

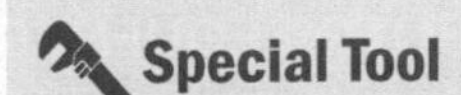
Special Tool
Front wheel bearing hub pulling and installing tools

Classroom Manual
Chapter 4, page 110

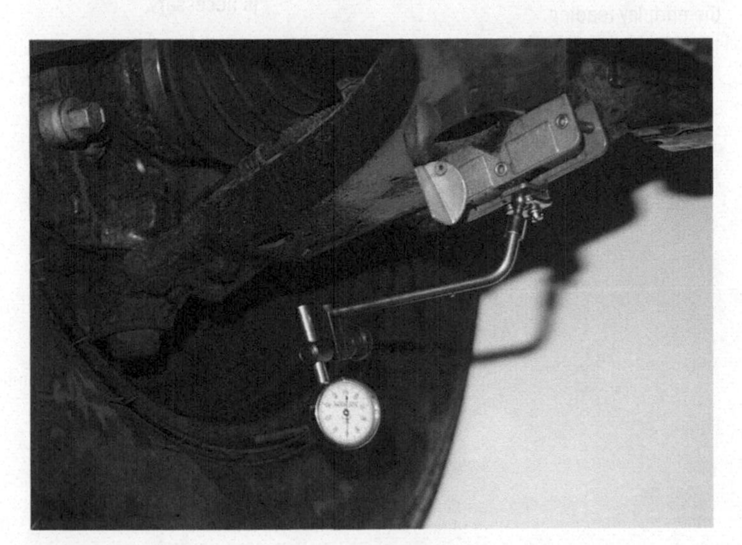

Figure 4-12 Wheel bearing diagnosis on vehicle.

PHOTO SEQUENCE 7

Typical Procedure for Measuring Front Wheel Hub Endplay—Integral, Sealed Wheel Bearing Hub Assemblies

P7-1 Be sure the vehicle is properly positioned on a lift before the wheel bearing hub endplay measurement is performed. The vehicle should be properly positioned on a lift with the lift raised to a comfortable working height for performing this measurement.

P7-2 Remove the wheel cover and dust cap.

P7-3 Attach a magnetic dial indicator base securely to the inside of the fender at the lower edge of the wheel opening. Position the dial indicator stem against the vertical wheel surface as close as possible to the top wheel stud, and preload the dial indicator stem.

P7-4 Zero the dial indicator pointer.

P7-5 Grasp the top of the tire with both hands. Push and pull on the top of the tire without rotating the tire, and note the dial indicator readings with the tire pushed inward and the tire pulled outward. The difference between the two readings is the wheel hub endplay. Repeat this procedure twice to verify the endplay reading.

P7-6 Maximum wheel bearing endplay should be 0.005 in. (0.127 mm). If the endplay measurement is not correct, wheel bearing hub replacement is necessary.

P7-7 Remove the dial indicator and install the dust cap and wheel cover.

Place the transaxle in neutral and install the wheel bearing hub bolts. Tighten these bolts to the specified torque.

Install the wheel sensor and connectors, rotors, and calipers. Tighten the caliper mounting bolts to the specified torque. Place a large punch between the rotor fins and the caliper, and tighten the drive axle nut to the specified torque.

Install the wheels in their original positions, and tighten the wheel nuts to the specified torque.

Lower the vehicle onto the shop floor, and tighten the new drive axle nut to the specified torque. Install a new cotter pin if used instead of a new locking nut.

Road test the car and listen for abnormal wheel bearing noises.

Front Drive Axle Diagnosis

On many front-wheel-drive vehicles, the front drive axles must be removed before the wheel hub unit or steering knuckle and bearing can be detached. Therefore, we will discuss front drive axle diagnosis and removal. Because drive axle noises may be confused with front wheel bearing noise, a brief discussion of drive axle noises and problems may be helpful. A defective inner drive axle joint usually causes a vibration when the vehicle is decelerating at 35–45 miles per hour (mph), or 56–72 kilometers per hour (km/h). A worn inner drive axle joint may also cause vibration during acceleration. When an outer drive axle joint is worn, a clicking noise is heard during a hard turn below 20 mph (32 km/h). To determine which drive axle has the defective joint, lift the vehicle on a hoist, and allow the front wheels to drop down. This action will position the axle joints at a different angle than when the car is driven on the road. Lift the lower control arms one at a time with a floor jack, and place the transmission in drive to simulate the driving conditions that provided the vibration or noise. If the vibration or noise occurs with one lower control arm lifted, that side has the defective drive axle joint.

DRIVE AXLE REMOVAL

Many drive axles have a **circlip** on the inner joint extension that holds the inner joint into the differential side gear. Drive axle systems vary depending on the vehicle. Follow the drive axle removal procedure in the vehicle manufacturer's service manual.

A **circlip** is a split, circular clip mounted in a groove on the inner drive axle joint extension to retain the drive axle in the transaxle on some front-wheel-drive vehicles.

A general front drive axle removal procedure follows:

1. Loosen the front wheel nuts and hub nuts.
2. Lift the vehicle on a hoist and be sure the hoist safety mechanism is in place. Then remove the front wheels and tires.
3. Remove the brake calipers and rotors. Connect a piece of wire from the calipers to a suspension or chassis component. Do not allow the calipers to hang on the end of the brake line.
4. Install protective drive axle boots if these are supplied by the car manufacturer.
5. Remove the ball joint to steering knuckle clamp bolt (**Figure 4-13**).
6. Pry the ball joint stud from the steering knuckle.
7. Pull the inner axle from the transaxle (**Figure 4-14**); do not allow the axle to drop down at a severe angle.
8. Remove the hub nut and washer, and separate the outer axle joint from the wheel hub. Some outer axle joint splines are slightly spiraled. On this type of outer axle joint, a special puller is required to separate the axle joint from the wheel (**Figure 4-15**).
9. Remove the drive axle from the chassis.

Technician's stethoscope

Reverse the drive axle removal procedure for front drive axle installation.

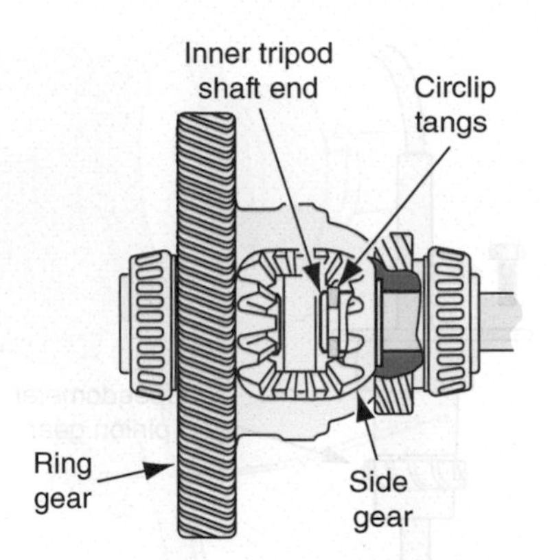

Figure 4-18 Compressible circlips in early model Chrysler transaxles.

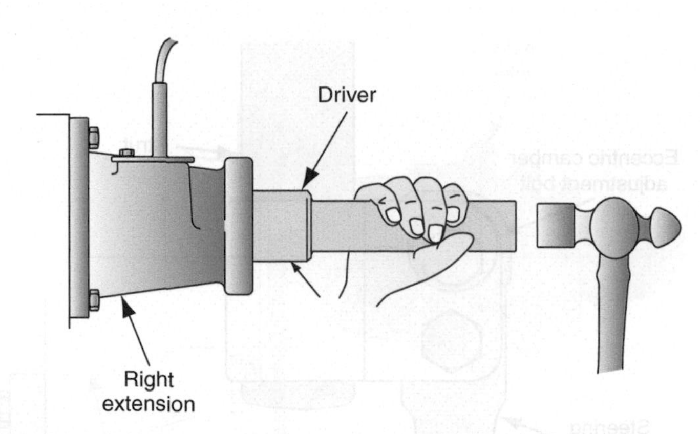

Figure 4-19 Removal of left inner drive axle joint on Ford automatic transaxles.

FRONT WHEEL BEARING HUB UNIT REMOVAL AND REPLACEMENT

The front wheel bearing removal and replacement procedure varies depending on the vehicle and the type of front wheel bearing. Always follow the front wheel bearing removal and replacement procedure in the manufacturer's service manual. The following procedure applies to front wheel bearing units that are pressed into the steering knuckle.

When front wheel bearing replacement is necessary, the steering knuckle must be removed and the wheel hub must be pressed from the bearing with a special tool (**Figure 4-20**).

A special puller is used to remove and replace the wheel bearing in the knuckle (**Figure 4-21** and **Figure 4-22**).

The wheel hub must be pulled into the wheel bearing with a special tool (**Figure 4-23**). The proper driving tool is used to install the seal behind the bearing in the knuckle (**Figure 4-24**).

Caution

Never use an impact wrench to tighten a hub nut. This action may cause wheel bearing damage.

When two separate roller bearings are mounted in the steering knuckle, the bearing races must be driven from the knuckle with a hammer and a punch. These bearings must be lubricated with wheel bearing grease prior to installation, as described earlier in this

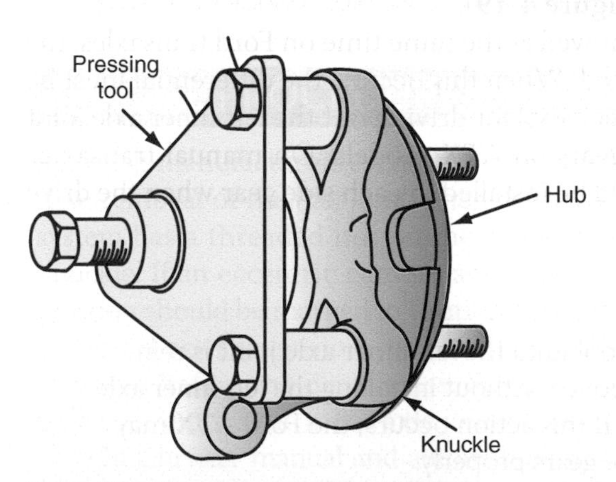

Figure 4-20 Wheel hub removal from bearing.

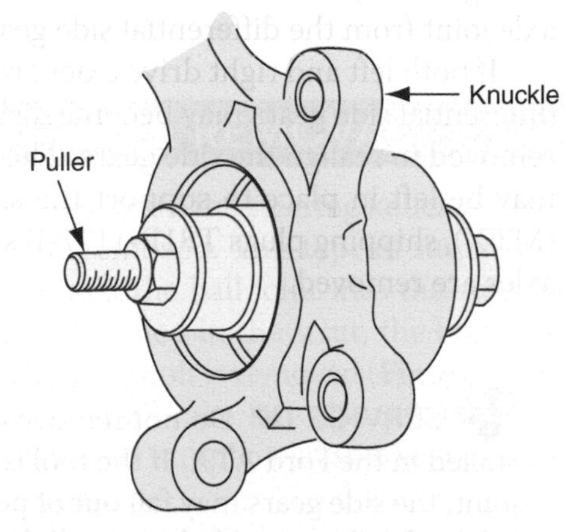

Figure 4-21 Wheel bearing removal from knuckle.

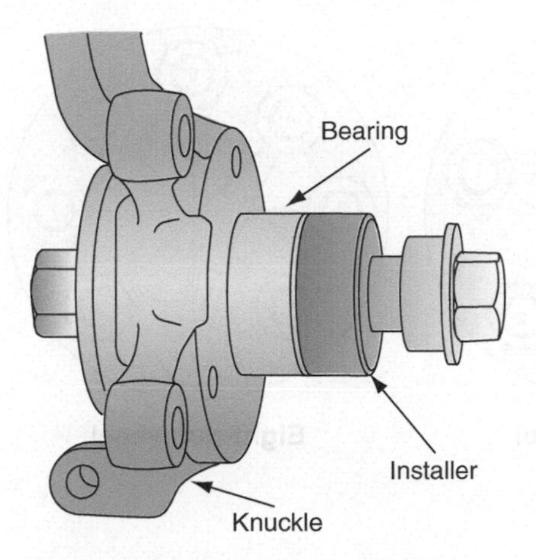

Figure 4-22 Wheel bearing installation in knuckle.

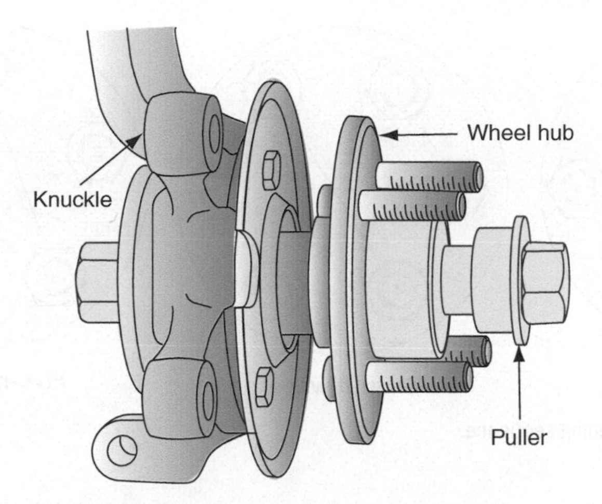

Figure 4-23 Wheel hub installation in wheel bearing.

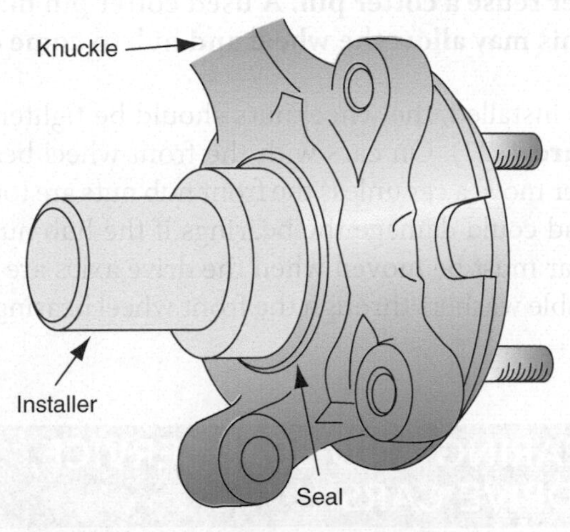

Figure 4-24 Seal installation in steering knuckle.

chapter. When the wheel bearings are removed, all wheel bearing seals must be replaced. A staked-type hub nut must be replaced if it is removed.

On these front-wheel-drive cars, the hub nut torque applies the correct adjustment on the front wheel bearings. Therefore, this torque is extremely important. With the brakes applied, the hub nut should be tightened to the specified torque (**Figure 4-25**). When the hub nut is torqued to specifications, the nut lock and cotter pin should be installed (**Figure 4-26**).

Classroom Manual
Chapter 4, page 111

Figure 4-25 Hub nut torquing.

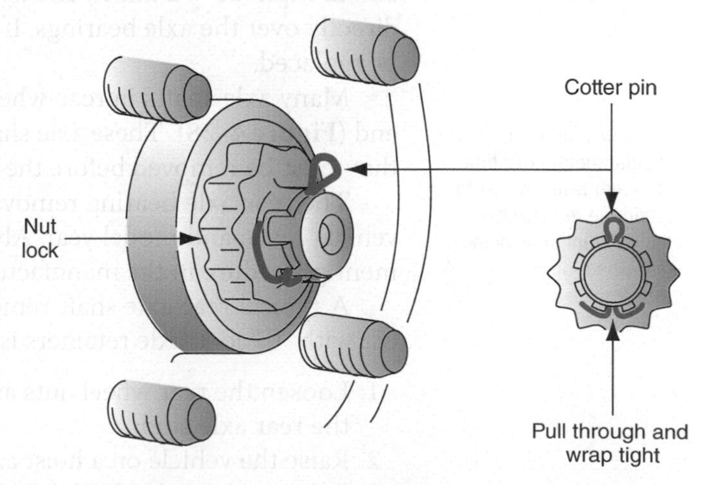

Figure 4-26 Nut lock and cotter pin installation.

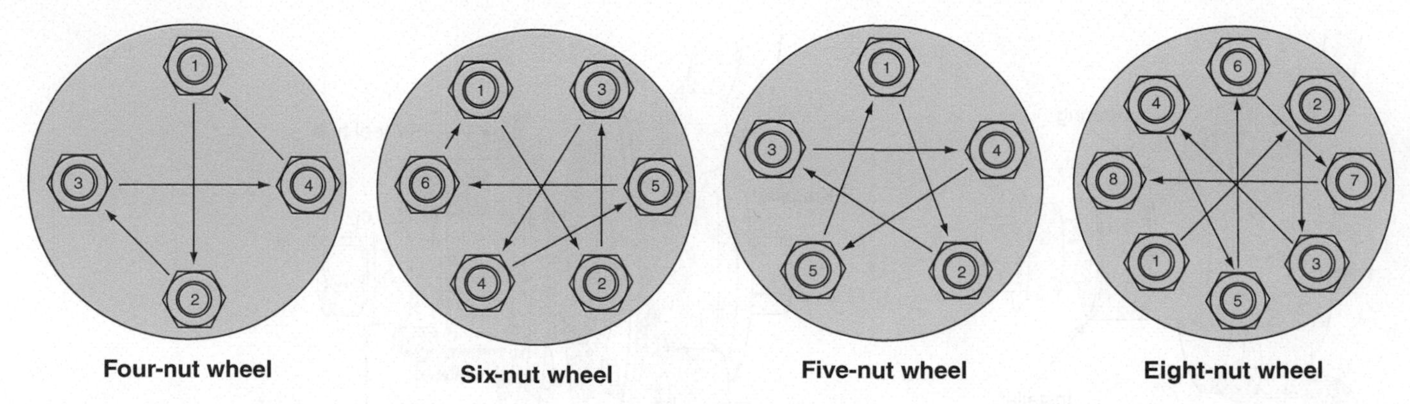

Figure 4-27 Wheel nut tightening sequence.

WARNING Never reuse a cotter pin. A used cotter pin may break, allowing the hub nut to loosen. This may allow the wheel and hub to come off.

After the wheel is installed, the wheel nuts should be tightened in sequence to the specified torque (**Figure 4-27**). On cars with the front wheel bearings mounted in the steering knuckles, never move a car unless the front hub nuts are torqued to specifications. Lack of bearing preload could damage the bearings if the hub nuts are not tightened to specifications. If the car must be moved when the drive axles are removed, place a large bolt and nut with suitable washers through the front wheel bearing and tighten the nut to specifications.

REAR-AXLE BEARING AND SEAL SERVICE, REAR-WHEEL-DRIVE CARS

Special Tools

Seal puller

Bearing puller

WARNING Use extreme caution when diagnosing problems with a vehicle raised on a hoist and the engine running with the transmission in drive. Keep away from rotating wheels, drive shafts, or drive axles.

Rear-axle bearing noise may be diagnosed with the vehicle raised on a hoist. Be sure the hoist safety mechanism is engaged after the vehicle is raised on the hoist. With the engine running and the transmission in drive, operate the vehicle at a moderate speed of 35–45 mph (56–72 km/h) and listen with a **stethoscope** placed on the rear axle housing directly over the axle bearings. If grinding or clicking noises are heard, the bearing must be replaced.

"C" locks are split, circular metal rings that fit in rear axle grooves to retain the rear axles in the differential on some rear-wheel-drive vehicles.

Many axle shafts in rear-wheel-drive cars have a roller bearing and seal at the outer end (**Figure 4-28**). These axle shafts are often retained in the differential with **"C" locks** that must be removed before the axles.

The rear-axle bearing removal and replacement procedure varies depending on the vehicle make and model year. Always follow the rear-axle bearing removal and replacement procedure in the manufacturer's service manual.

A typical rear axle shaft removal and replacement procedure on a rear-wheel-drive car with "C" lock axle retainers is as follows:

1. Loosen the rear wheel nuts and chalk-mark the rear wheel position in relation to the rear axle studs.
2. Raise the vehicle on a hoist and make sure the hoist safety mechanism is in place.
3. Remove the rear wheels and brake drums, or calipers and rotors.

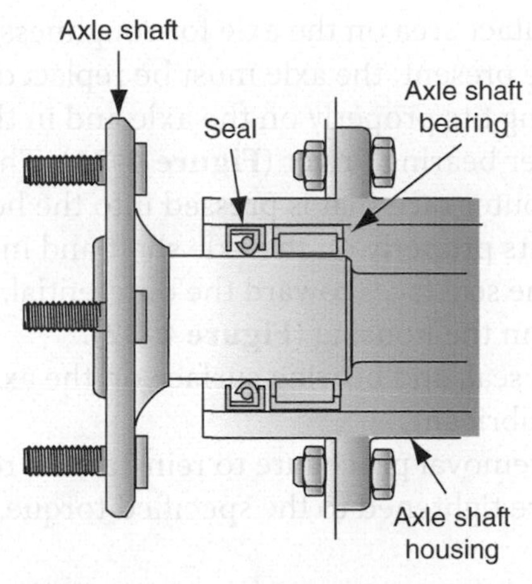

Figure 4-28 Rear axle roller bearing and seal, rear-wheel-drive car.

4. Place a drain pan under the differential and remove the differential cover. Discard the old lubricant.
5. Remove the differential lock bolt, pin, pinion gears, and shaft (**Figure 4-29**).
6. Push the axle shaft inward and remove the axle "C" lock.
7. Pull the axle from the differential housing.

Reverse the axle removal procedure to reinstall the axle. Always use a new differential cover gasket, and fill the differential to the bottom of the filler plug opening with the manufacturer's recommended lubricant. Be sure all fasteners, including the wheel nuts, are tightened to the specified torque.

A typical axle bearing and seal removal procedure follows:

1. Remove the axle seal with a seal puller.
2. Use the proper bearing puller to remove the axle bearing (**Figure 4-30**).
3. Clean the axle housing seal and bearing mounting area with solvent and a brush. Next clean this area with compressed air.
4. Check the seal and bearing mounting area in the housing for metal burrs and scratches. Remove any burrs or irregularities with a fine-toothed round file.
5. Wash the axle shaft with solvent and blow it dry with compressed air.

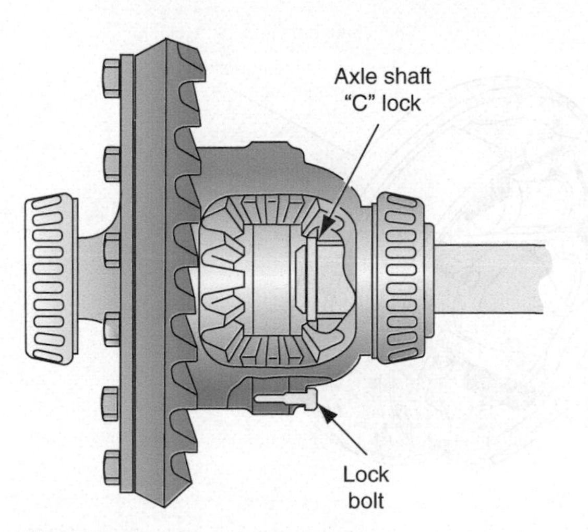

Figure 4-29 Rear axle "C" lock, lock bolt, and pinion gears.

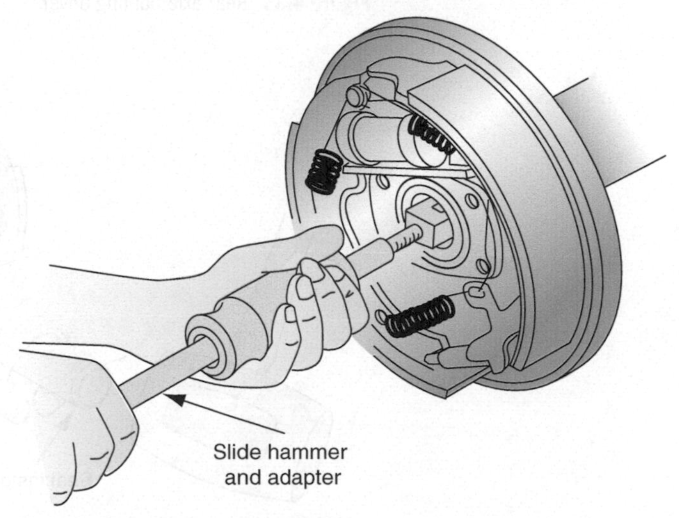

Figure 4-30 Rear-axle bearing puller.

6. Check the bearing contact area on the axle for roughness, pits, and scratches. If any of these conditions are present, the axle must be replaced.
7. Be sure the new bearing fits properly on the axle and in the housing. Install the new bearing with the proper bearing driver (**Figure 4-31**). The bearing driver must apply pressure to the outer race that is pressed into the housing.
8. Be sure the new seal fits properly on the axle shaft and in the housing. Make sure the garter spring on the seal faces toward the differential. Use the proper seal driver to install the new seal in the housing (**Figure 4-32**).
9. Lubricate the bearing, seal, and bearing surface on the axle with the manufacturer's specified differential lubricant.
10. Reverse the rear axle removal procedure to reinstall the rear axle.
11. Be sure all fasteners are tightened to the specified torque.

Classroom Manual
Chapter 4, page 114

WARNING **Never use an acetylene torch to heat axle bearings or adaptor rings during the removal and replacement procedure. The heat may cause fatigue in the steel axle and the axle may break suddenly, causing the rear wheel to fall off. This action will likely result in severe vehicle damage and personal injury.**

Some rear axles have a sealed bearing that is pressed onto the axle shaft and held in place with a retainer ring. These rear axles usually do not have "C" locks in the differential. A retainer plate is mounted on the axle between the bearing and the outer end of the axle.

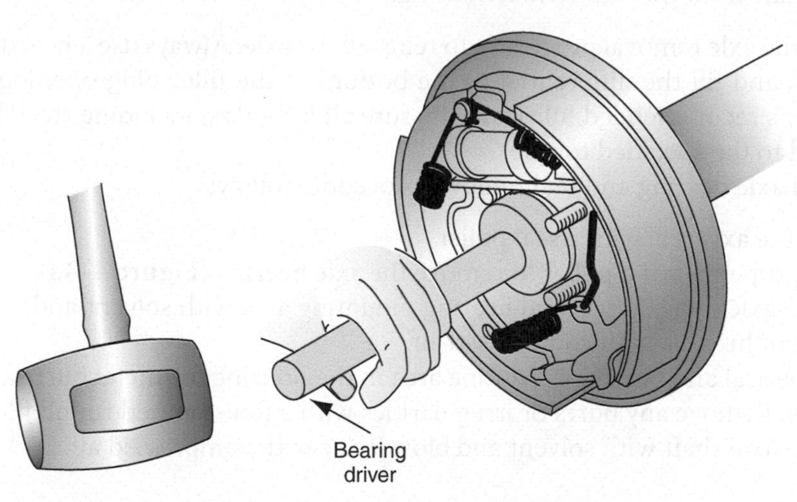

Figure 4-31 Rear-axle bearing driver.

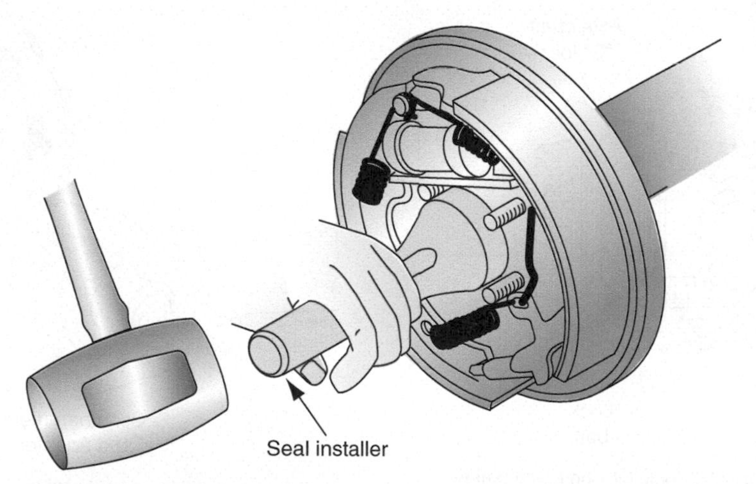

Figure 4-32 Installing rear axle seal.

This plate is bolted to the outer end of the differential housing. After the axle retainer plate bolts are removed, a slide-hammer-type puller is attached to the axle studs to remove this type of axle. When this type of axle bearing is removed, the adaptor ring must be split with a hammer and a chisel while the axle is held in a vise. Do not heat the adaptor ring or the bearing with an acetylene torch during the removal or installation process. After the adaptor ring is removed, the bearing must be pressed from the axle shaft, and the bearing must not be reused. A new bearing and the adaptor ring must be pressed onto the axle shaft. The bearing removal and replacement procedure is shown in **Figure 4-33**.

Classroom Manual
Chapter 4, page 114

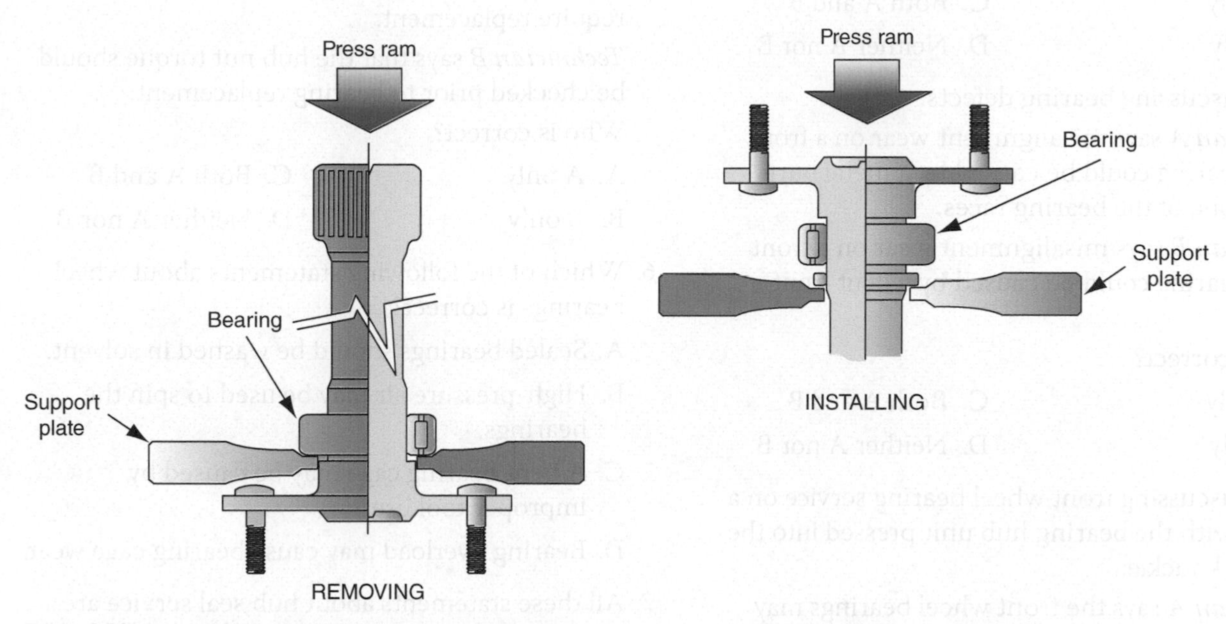

Figure 4-33 Axle bearing and adaptor ring removal and replacement.

CASE STUDY

A customer complains of a bearing noise in the right front wheel of his Lincoln Town Car. He says the right front outer wheel bearing has been replaced twice in the last year, and this is the third failure. The technician asks the customer about the mileage intervals between bearing replacements, and the customer indicates that the wheel bearing has lasted about 8000 mi. (12,800 km) each time it has been replaced. The technician finds out from the customer that no other work was done on the car each time the bearing was replaced.

When the technician removes the right front wheel and hub, the outer bearing rollers and races are badly scored. After cleaning both bearings, races, and hub, the technician closely examines the outer bearing race. It shows an uneven wear pattern, which indicates misalignment. The technician removes the outer bearing race and finds a small metal burr behind the bearing race. This burr caused race misalignment and excessive wear on the race and rollers. The technician removes the burr with a fine-toothed file. The inner bearing and race have indentation wear because metal particles from the outer bearing contaminated the lubricant in the hub. The technician removes the inner bearing race and cleans the hub and spindle thoroughly. He replaces both bearings and the inner seal and repacks the bearings and hub with grease. After reinstalling the hub, the technician carefully adjusts the bearings to the manufacturer's specifications and tightens the wheel nuts to the specified torque. A road test indicates that the bearing noise has been eliminated.

One small metal burr caused this customer a considerable amount of unnecessary expense. This experience proves that a technician's diagnostic capability is extremely important!

ASE-STYLE REVIEW QUESTIONS

1. While discussing defective bearings:
   Technician A says brinelling appears as indentations across the bearing races.
   Technician B says brinelling occurs while the bearing is rotating.
   Who is correct?
   A. A only
   B. B only
   C. Both A and B
   D. Neither A nor B

2. While discussing bearing defects:
   Technician A says misalignment wear on a front wheel bearing could be caused by a metal burr behind one of the bearing races.
   Technician B says misalignment wear on a front wheel bearing could be caused by a bent front spindle.
   Who is correct?
   A. A only
   B. B only
   C. Both A and B
   D. Neither A nor B

3. While discussing front wheel bearing service on a vehicle with the bearing hub unit pressed into the steering knuckle:
   Technician A says the front wheel bearings may be damaged if this front-wheel-drive vehicle is moved without the hub nuts torqued to specifications.
   Technician B says the hub nut torque supplies the correct wheel bearing adjustment.
   Who is correct?
   A. A only
   B. B only
   C. Both A and B
   D. Neither A nor B

4. While discussing front wheel bearing service on a front-wheel-drive car with two separate tapered roller bearings mounted in the front steering knuckles:
   Technician A says the hub nut torque supplies the correct bearing adjustment.
   Technician B says that the brake should be applied while the front hub nuts are torqued.
   Who is correct?
   A. A only
   B. B only
   C. Both A and B
   D. Neither A nor B

5. A front-wheel-drive vehicle has 14-in. (35.5-cm) rims and front wheel bearings mounted in the steering knuckles. The vehicle is lifted on a hoist, and a dial indicator is positioned against the outer rim lip. The total inward and outward rim movement is 0.035 in. (0.889 mm).
   Technician A says that the wheel bearing may require replacement.
   Technician B says that the hub nut torque should be checked prior to bearing replacement.
   Who is correct?
   A. A only
   B. B only
   C. Both A and B
   D. Neither A nor B

6. Which of the following statements about wheel bearings is correct?
   A. Sealed bearings should be washed in solvent.
   B. High-pressure air may be used to spin the bearings.
   C. A bent bearing cage may be caused by improper tool use.
   D. Bearing overload may cause bearing cage wear.

7. All these statements about hub seal service are true EXCEPT:
   A. The garter spring must face toward the flow of lubricant.
   B. A ball peen hammer should be used to install the seal.
   C. Seal contact area on the spindle must be clean and free from metal burrs.
   D. The outer edge of the seal case should be coated with sealant.

8. Which of the following statements about servicing press-in rear-axle bearings on a rear-wheel-drive car is correct?
   A. These bearings may be reused after they are removed from the axle shaft.
   B. The bearing adaptor ring should be removed by splitting it with a hammer and a chisel.
   C. A cutting torch may be used to cut the bearing off the axle shaft.
   D. An acetylene torch may be used to heat the new adaptor ring prior to installation.

9. A front-wheel-drive vehicle has two serviceable tapered roller bearings in each rear wheel hub. Which of the following statements about adjusting these wheel bearings is correct?
   A. The adjusting nut should be tightened to 17–25 ft.-lb (23–24 Nm), backed off one-half turn, and then tightened to 10–15 in.-lb (1.0–1.7 Nm).
   B. The adjusting nut should be tightened to 40 ft.-lb (54 Nm), backed off one turn, and then tightened to 10 ft.-lb (13.5 Nm).
   C. The adjusting nut should be tightened to 50 ft.-lb (67.5 Nm), backed off three-quarter turn, and then tightened to 10–15 in.-lb (1.0–1.7 Nm).
   D. The wheel and hub should not be rotated while adjusting the wheel bearings.
10. A unitized front wheel bearing hub that is bolted to the steering knuckle has 0.010 in. (0.254 mm) of hub endplay. The proper repair procedure is to:
   A. Repack and readjust the wheel bearings in the hub.
   B. Tighten the hub nut to the specified torque.
   C. Inspect the drive axle and hub splines, and replace the worn components.
   D. Replace the wheel bearing hub assembly.

ASE CHALLENGE QUESTIONS

1. The customer says her front-wheel-drive car makes "a moaning noise" in a turn. Which of the following could cause this problem?
   A. Outer front wheel bearing.
   B. Rear-axle bearing.
   C. Differential gear noise.
   D. Transaxle output shaft bearing.
2. The customer complains of a "whining noise" in the back of her rear-wheel-drive car when driving between 30 and 40 mph.
   Technician A says a good way to diagnose this problem is on a hoist with a stethoscope.
   Technician B says a good way to diagnose this problem is on the road with a microphone.
   Who is correct?
   A. A only
   B. B only
   C. Both A and B
   D. Neither A nor B
3. Upon inspecting a noisy front wheel bearing from a four-wheel-drive sport utility vehicle, brinelling damage to the outer bearing race was noticed.
   Technician A says the damage was probably caused by tightening the bearing nut with an impact wrench.
   Technician B says the damage was probably caused by driving the vehicle through deep, muddy water.
   Who is correct?
   A. A only
   B. B only
   C. Both A and B
   D. Neither A nor B
4. While discussing the cause of overheating damage of a front wheel bearing:
   Technician A says overheating may be caused by insufficient or incorrect bearing lubricant.
   Technician B says overheating may be caused by overtightening the wheel bearing nut on installation.
   Who is correct?
   A. A only
   B. B only
   C. Both A and B
   D. Neither A nor B
5. The customer says her front-wheel-drive car has a noise and vibration when decelerating and is most noticeable under 50 mph. Which would be the most probable cause of this problem?
   A. Defective outer front wheel bearing.
   B. Defective inner front wheel bearing.
   C. Defective outer drive axle joint.
   D. Defective inner drive axle joint.

Task Completed

7. If a wheel speed sensor for the antilock brake system (ABS) is integral in the front wheel hubs, disconnect the wheel speed sensor connectors. ☐

8. Remove the hub-to-knuckle bolts, and place the transaxle in park. ☐

9. Use the proper puller to remove the wheel bearing hub from the drive axle. ☐

10. Install the new hub and bearing assembly on the drive axle splines, and install a new drive axle nut. Tighten the drive axle nut to pull the hub onto the splines. Do not tighten this nut to the specified torque at this time. ☐

11. Place the transaxle in neutral and install the wheel bearing hub bolts. Tighten these bolts to the specified torque.

    Specified wheel bearing hub bolt torque ______________________

    Actual wheel bearing hub bolt torque ______________________

12. Install the wheel sensor connectors, rotors, and calipers. Tighten the caliper mounting bolts to the specified torque. Place a large punch between the rotor fins and the caliper, and tighten the drive axle nut to the specified torque.

    Specified caliper bolt torque ______________________

    Actual caliper bolt torque ______________________

    Specified drive axle nut torque ______________________

    Actual drive axle nut torque ______________________

13. Install the wheels in their original positions, and tighten the wheel nuts to the specified torque.

    Specified wheel nut torque ______________________

    Actual wheel nut torque ______________________

14. Lower the vehicle onto the shop floor, and tighten the drive axle nut to the specified torque. Install the lock nut and a new cotter pin.

    Specified drive axle nut torque ______________________

    Actual drive axle nut torque ______________________

    Are the lock nut and cotter pin properly installed and tightened?

    Instructor's Check ______________________

15. Road test the car and listen for abnormal wheel bearing noises.

    Wheel bearing noise: ☐ Satisfactory ☐ Unsatisfactory

Instructor's Response

____

____

____

Name ______________________ **Date** ____________

JOB SHEET 19

DIAGNOSE WHEEL BEARINGS

Upon completion of this job sheet, you should be able to diagnose wheel bearings.

ASE Education Foundation Task Correlation

This job sheet addresses the following **AST/MAST** task:

D.2. Remove, inspect, service, and/or replace front and rear wheel bearings. **(P-1)**

We Support ASE Education Foundation

Tools and Materials

A vehicle with wheel bearing noise or bearing noise from another source

Describe the Vehicle Being Worked on:

Year ____________ Make ____________ Model ____________

VIN ____________ Engine type and size ____________

Procedure — **Task Completed**

1. Road test the vehicle, listen for any abnormal wheel noises, and check for steering looseness and wander that may be caused by loose wheel bearings. ☐
2. Wheel bearing noise: ☐ Satisfactory ☐ Unsatisfactory
3. Is the bearing noise coming from the front or rear of the vehicle? ☐ Front ☐ Rear
4. Is this noise more noticeable when turning a corner at low speeds? ☐ Yes ☐ No
5. Is this noise more noticeable during acceleration? ☐ Yes ☐ No
6. Is this noise more noticeable during deceleration? ☐ Yes ☐ No
7. Is this noise more noticeable when driving at a steady speed? ☐ Yes ☐ No

   If the answer to this question is yes, state the speed when the bearing noise is most noticeable. ____________
8. Is this noise most noticeable when driving down a narrow street at a steady low speed? ☐ Yes ☐ No
9. State the exact cause of the bearing noise, and explain the reason for your diagnosis.

   ____

   ____

   ____

Instructor's Response

____

____

____

7. Repack the wheel bearings and hub with the car manufacturer's specified wheel bearing grease.

   Are the wheel bearings and hub repacked with the manufacturer's recommended grease? ☐ Yes ☐ No

   Instructor's Check ______________________________

8. Install a new inner hub seal with the proper seal driver. Lubricate the seal lips with a light coating of the car manufacturer's recommended wheel bearing grease.

   Is the new inner hub seal properly installed? ☐ Yes ☐ No

   Instructor's Check ______________________________

9. Inspect, clean, and lubricate the spindles and seal contact area.

   Spindle and seal contact condition: ☐ Satisfactory ☐ Unsatisfactory

   If unsatisfactory, state necessary repairs. ______________________________

   Is the spindle and seal contact area properly cleaned and lubricated? ☐ Yes ☐ No

   Instructor's Check ______________________________

10. Install the wheel hubs, wheel bearings, washer, and retaining nut. Adjust the wheel bearing retaining nut to the specified initial torque.

    Specified initial wheel bearing torque ______________________________

    Actual initial wheel bearing torque ______________________________

11. Back off the wheel bearing retaining nut the specified amount.

    Specified portion of a turn to back off the wheel bearing retaining nut ______________

    Actual portion of a turn the wheel bearing retaining nut is backed off ______________

12. Tighten the wheel bearing retaining nut to the final specified torque.

    Specified final wheel bearing retaining nut torque ______________________________

    Actual final wheel bearing retaining nut torque ______________________________

13. Install a new cotter pin through the wheel bearing retaining nut and spindle opening.

    Is the new cotter pin properly installed? ☐ Yes ☐ No

    Instructor's Check ______________________________

14. Install the wheel bearing dust covers. Install the wheels in their original position, and tighten the wheel nuts to the specified torque.

    Specified wheel nut torque ______________________________

    Actual wheel nut torque ______________________________

15. Road test the car and listen for abnormal wheel bearing noises.

    Wheel bearing noise: ☐ Satisfactory ☐ Unsatisfactory

Instructor's Response

____

____

____

CHAPTER 5

SHOCK ABSORBER AND STRUT DIAGNOSIS AND SERVICE

Upon completion and review of this chapter, you should be able to:

- Perform a visual shock absorber inspection.
- Perform a shock absorber bounce test, and determine shock absorber condition.
- Determine shock absorber condition from a manual shock absorber test.
- Remove and replace shock absorbers.
- Diagnose shock absorber and strut noise complaints.
- Remove and replace front and rear struts.
- Remove struts from coil springs.
- Install coil springs on struts.
- Follow the vehicle manufacturer's recommended strut disposal procedure.
- Perform off-car strut cartridge replacement procedures.
- Perform on-car strut cartridge replacement procedures.
- Diagnose electrically controlled shock absorbers.

Basic Tools

Basic technician's tool set

Service manual

Hydraulic floor jack

Jack stands

Center punch

Terms To Know

Bounce test	Spring compressing tools	Strut cartridge
Eccentric camber bolt	Spring insulators	Strut chatter
Manual test	Struts	Strut tower

Shock absorbers and struts must be in good condition to provide satisfactory ride quality and maintain vehicle safety. Worn-out shock absorbers and struts may cause excessive chassis oscillations and harsh ride quality, resulting in passenger discomfort. On a severely rough road surface, the chassis oscillations caused by worn-out shock absorbers and struts may contribute to a loss of steering control, resulting in a vehicle collision. Therefore, technicians must be familiar with shock absorber and strut diagnosis and service procedures.

SHOCK ABSORBER VISUAL INSPECTION

Bolt Mounting

Shock absorbers should be inspected for loose mounting bolts and worn mounting bushings. If these components are loose, rattling noise is evident, and replacement of the bushings and bolts is necessary.

Bushing Condition

In some shock absorbers, the bushing is permanently mounted in the shock, and the complete unit must be replaced if the bushing is worn. When the mounting bushings are worn, the shock absorber will not provide proper spring control.

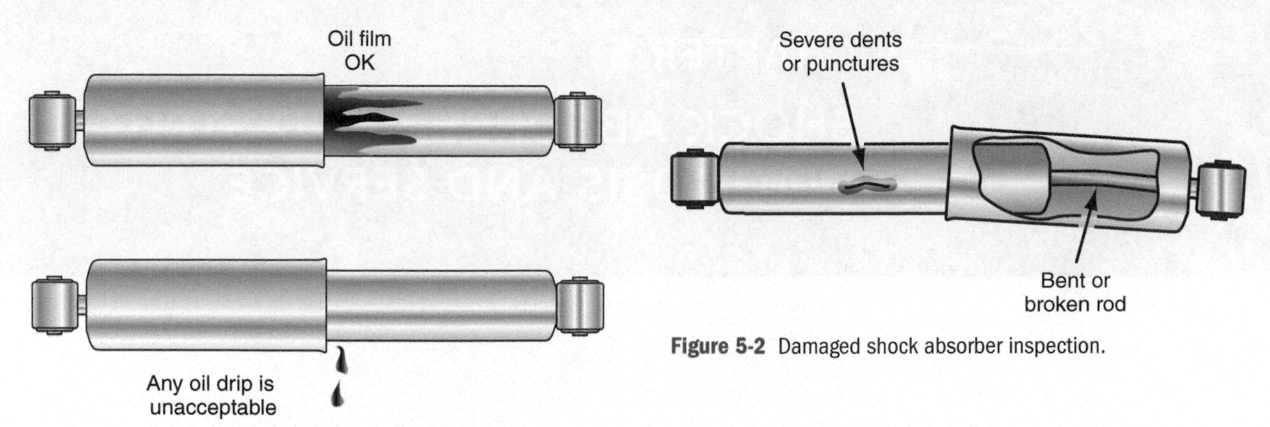

Figure 5-2 Damaged shock absorber inspection.

Figure 5-1 Shock absorber and strut oil leak diagnosis.

SERVICE TIP During a visual shock absorber inspection, check the rebound, or strikeout, bumpers on the control arms or chassis. If the rebound bumpers are severely worn, the shock absorbers may be worn out.

Oil Leakage

Struts are similar internally to shock absorbers, but struts also support the steering knuckle. In many applications, the coil spring is mounted on the strut.

Shock absorbers and **struts** should be inspected for oil leakage. A slight oil film on the lower oil chamber is acceptable. Any indication of oil dripping is not acceptable, and unit replacement is necessary (**Figure 5-1**).

Shock absorbers and struts should be inspected visually for physical damage, such as a bent condition and severe dents or punctures. When any of these conditions are present, unit replacement is required (**Figure 5-2**).

SHOCK ABSORBER TESTING

A shock absorber may be called a shock.

CUSTOMER CARE Always be willing to spend a few minutes explaining problems, including safety concerns, regarding the customer's vehicle. When customers understand why certain repairs are necessary, they feel better about spending the money. For example, if you explain that worn-out shock absorbers cause excessive chassis oscillations and may result in loss of steering control on irregular road surfaces, the customer is more receptive to spending the money for shock absorber replacement.

Although many technicians still rely on the bounce test described below to test the condition of the hydraulic shock absorber or strut, it is not the most reliable method of diagnosis because it does not simulate all the road force conditions that today's velocity-sensitive shock absorbers are exposed to under all driving conditions. The test may uncover a noise from a shock absorber or a strut that has completely failed but does not ensure that the suspension system is still capable of providing vehicle stability and designed damping control.

All modern hydraulic shock absorbers and struts are velocity sensitive and react to the piston travel speed. There are various areas that can wear internally on a shock absorber. The top seal can wear creating an external leak and the piston and cylinder wall can wear causing an internal leak (**Figure 5-3**). However, a more common point of wear or failure is the damping valves in the piston assembly. The best way to test a shock absorber's performance is to go on a test drive and watch for:

- Brake dive on sudden stops
- Acceleration squat

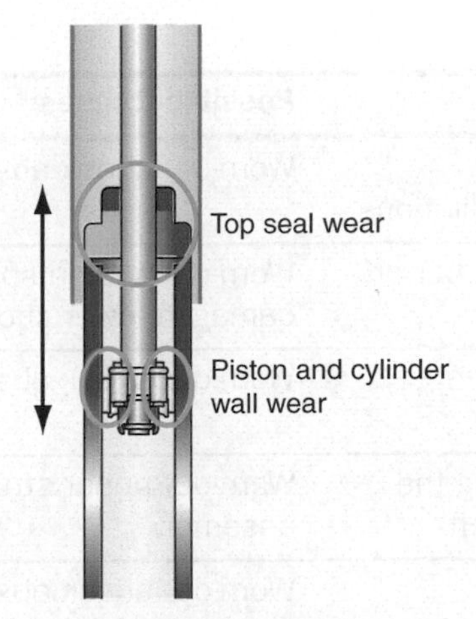

Figure 5-3 Shock absorbers can develop wear at the top seal, creating an external leak, or at the piston seal or bore, creating an internal leak.

- Body roll and swerve
- Ride comfort under all driving conditions

SERVICE TIP A road test should be completed to test for shock absorber noise. If the gas leaks out of a shock absorber, oil will leak past the piston and the shock valve may knock (slap) against the oil. This knocking noise is heard when driving the vehicle over small bumps. This noise cannot be duplicated during a bounce test.

When the **bounce test** is performed, the bumper is pushed downward with considerable weight applied on each corner of the vehicle. The bumper is released after this action, and one free upward bounce should stop the vertical chassis movement if the shock absorber or strut provides proper spring control. Shock absorber replacement is required if more than one free upward bounce occurs. The shock absorber bounce test is illustrated in **Photo Sequence 8**. Refer to **Table 5-1** for additional diagnostic information.

SHOCK ABSORBER MANUAL TEST

A **manual test** may be performed on shock absorbers. When this test is performed, disconnect the lower end of the shock, and move the shock up and down as rapidly as possible. A satisfactory shock absorber should offer a strong, steady resistance to movement on the entire compression and rebound strokes. The amount of resistance may be different on the compression stroke compared with the rebound stroke. If a loss of resistance is experienced during either stroke, shock replacement is essential.

Some defective shock absorbers or struts may have internal clunking, clicking, and squawking noises, or binding conditions. When these shock absorber noises or conditions are experienced, shock absorber or strut replacement is necessary.

SERVICE TIP On twin I-beam front suspension systems, front wheel camber may be noticeably different after a bounce test.

TABLE 5-1 SHOCK ABSORBER AND STRUT DIAGNOSIS

Problem	Symptoms	Possible Causes
Harsh ride quality	Excessive chassis bouncing Excessive vertical wheel oscillations	Worn-out shock absorbers or struts
Shock absorber or strut oil leaks	Oil dripping from shock absorber or strut	Worn-out seal in shock absorber or strut, damaged lower shock absorber or strut housing
Rattling noise	Rattling noise in chassis when driving	Worn-out shock absorber or strut mountings
Chattering noise	Chattering noise when turning the front wheels to the right or left	Worn-out upper strut mount and bearing assembly
Worn, damaged rebound bumpers	Chassis bottoming out	Worn-out shock absorbers or struts, low ride height

PHOTO SEQUENCE 8
Rear Shock Absorber Visual Inspection and Bounce Test

P8-1 Raise the vehicle on a lift.

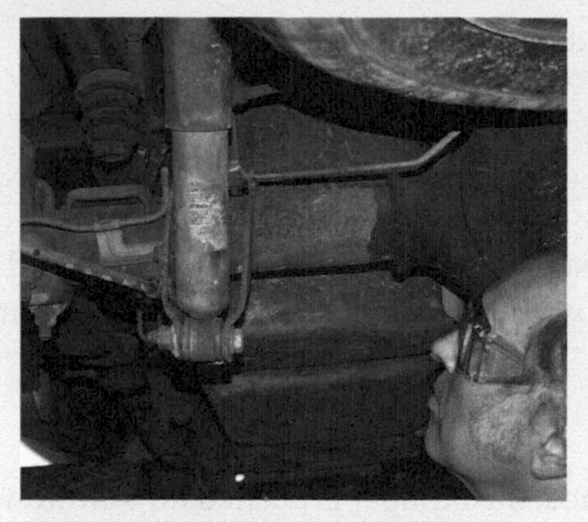

P8-2 Inspect the rear shock absorbers for oil leaks and damage such as dents and punctures.

P8-3 Grasp the lower shock absorber tube and attempt to move the shock absorber vertically and horizontally to check for looseness and wear on the shock absorber bushing and mounting bolt.

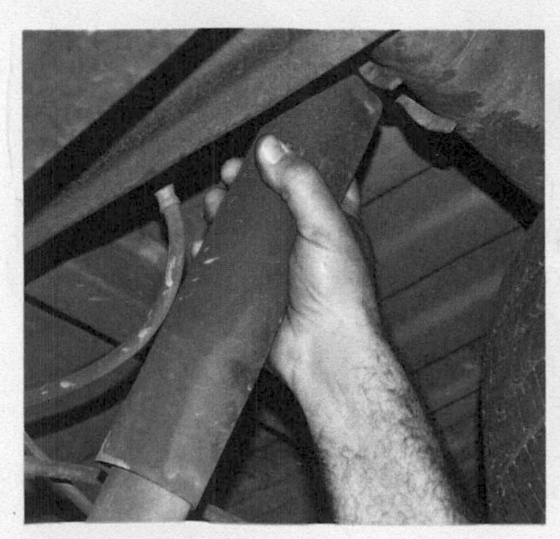

P8-4 Grasp the upper shock absorber cover and attempt to move it vertically and horizontally to check for looseness and wear on the shock absorber mounting bushing and mounting bolt.

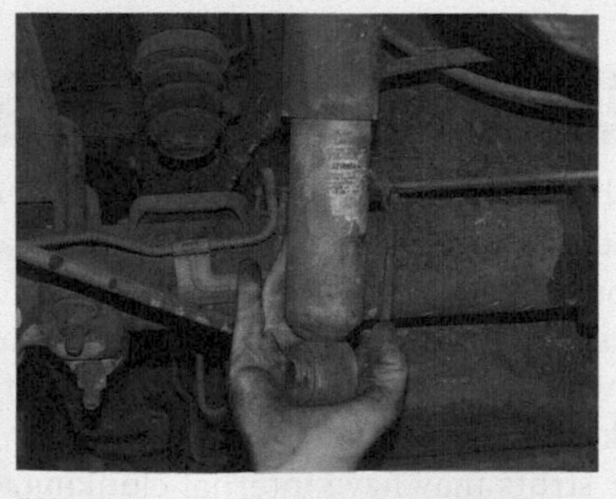

P8-5 Disconnect the lower shock absorber mounting and grasp the lower end of the shock absorber. Pull the shock absorber downward on the extension stroke and push the lower end of the shock absorber upward on the compression stroke. The shock absorber should offer resistance to movement in relation to the shock absorber ratio.

P8-6 Install the lower shock absorber mounting and tighten the mounting bolt to the specified torque.

PHOTO SEQUENCE 8 (CONTINUED)

P8-7 Lower the vehicle onto the shop floor, and move the vehicle off the lift.

P8-8 Using your knee press down on the rear bumper with considerable weight, and suddenly remove your knee from the bumper. Count the number of free chassis oscillations before the chassis stops bouncing upward and downward.

AIR SHOCK ABSORBER DIAGNOSIS AND REPLACEMENT

Air shock absorbers are similar to conventional shocks except they have air pressure applied to them to control chassis curb height. Some air shock absorbers must be pressurized with the shop air hose. This type of unit contains a valve for inflation purposes. Other air shock absorbers are inflated by an onboard compressor with interconnecting plastic lines between the compressor and the shocks. Shock absorber lines must be inspected for breaks, cracks, and sharp bends. If any of these defects are present, the line must be replaced. The shock absorber lines must be secured to the chassis, and they must not rub against other components.

Caution
Gas-filled shock absorbers will extend when disconnected.

When air shock absorbers slowly lose their air pressure and reduce the curb riding height, shock replacement is required. Before removing an air shock, relieve the air pressure in the shock.

Upward wheel movement is referred to as wheel jounce, and downward wheel movement is called rebound.

SHOCK ABSORBER REPLACEMENT

WARNING Never apply heat to the lower shock absorber or strut chamber with an acetylene torch. Excessive heat may cause a shock absorber or strut explosion, which could result in personal injury.

When shock absorber replacement is necessary, follow this procedure:

1. Before replacing rear shock absorbers, lift the vehicle on a hoist and support the rear axle on safety stands so the shock absorbers are not fully extended.
2. When a front shock absorber must be changed, lift the front end on the vehicle with a floor jack, then place safety stands under the lower control arms. Lower the vehicle onto the safety stands and remove the floor jack.
3. Disconnect the upper shock mounting nut and grommet.
4. Remove the lower shock mounting nut or bolts, and remove the shock absorber.
5. Reverse Steps 1 through 4 to install the new shock absorber and grommets.
6. With the full vehicle weight supported on the suspension, tighten the shock mounting nuts to the specified torque.

Always follow the vehicle manufacturer's recommended procedure in the service manual for removal of the strut-and-spring assembly.

Classroom Manual
Chapter 5, page 129

A typical procedure for strut-and-spring assembly removal follows:

1. Raise the vehicle on a hoist or floor jack. If a floor jack is used to raise the vehicle, lower the vehicle onto safety stands placed under the chassis so the lower control arms and front wheels drop downward. Remove the floor jack from under the vehicle.
2. Remove the brake line and antilock brake system (ABS) wheel speed sensor wire from clamps on the strut (**Figure 5-8**). In some cases, the clamps may also have to be removed from the strut.
3. Remove the strut-to-steering knuckle retaining bolts, and remove the strut from the knuckle (**Figure 5-9**).
4. Remove the upper strut mounting bolts on top of the strut tower, and remove the strut-and-spring assembly (**Figure 5-10**).

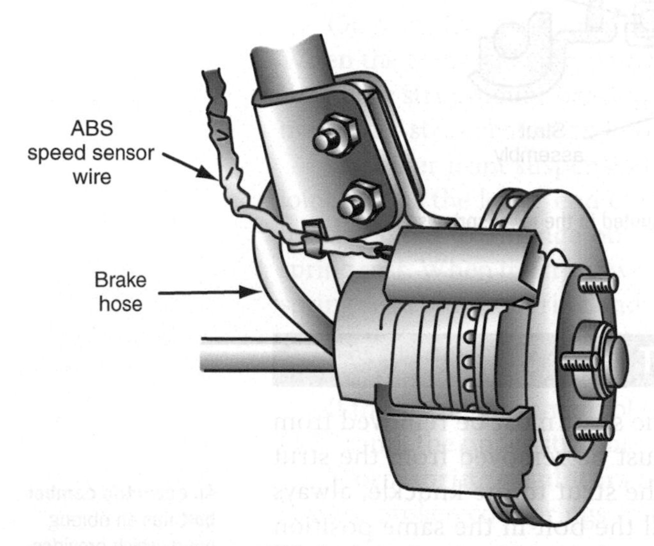

Figure 5-8 Brake line and ABS wheel speed sensor wire removed from the strut.

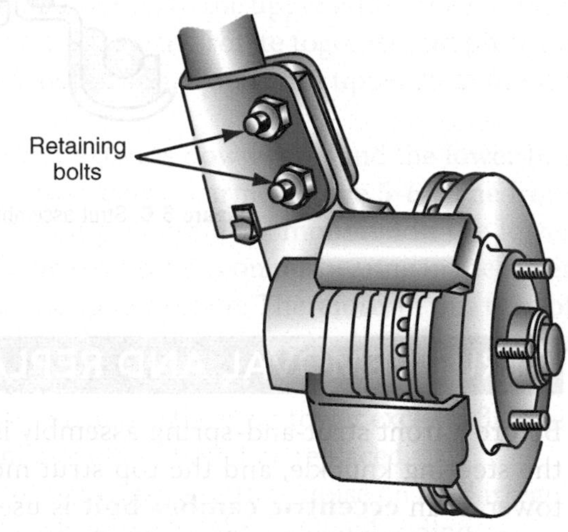

Figure 5-9 Removing strut-to-steering knuckle retaining bolts.

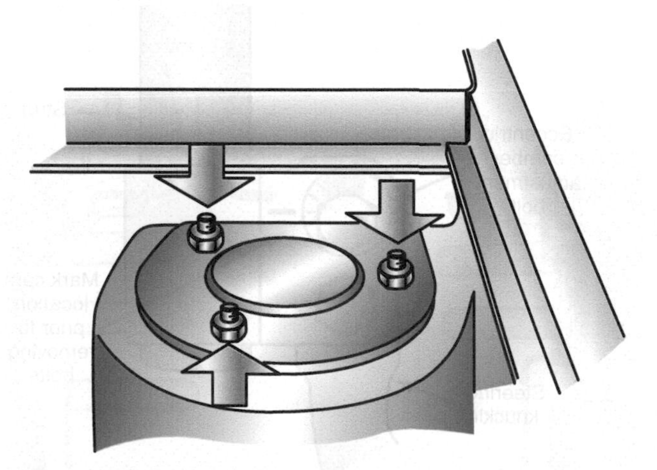

Figure 5-10 Removing upper strut mounting bolts on top of strut tower.

REMOVAL OF STRUT FROM COIL SPRING

Classroom Manual
Chapter 5, page 130

WARNING Always use a coil spring compressing tool according to the tool or vehicle manufacturer's recommended service procedure. Be sure the tool is properly installed on the spring. If a coil spring slips off the tool when the spring is compressed, severe personal injury or property damage may occur.

WARNING Never loosen the upper strut mount retaining nut on the end of the strut rod unless the spring is compressed enough to remove all spring tension from the upper strut mount. If this nut is loosened with spring tension on the upper mount, this mount becomes a very dangerous projectile that may cause serious personal injury or property damage.

Special Tool
Coil spring compressing tool

The coil spring must be compressed with a special tool before the strut can be removed. All the tension must be removed from the upper spring seat before the upper strut piston rod nut is loosened. Many different **spring compressing tools** are available, and they must always be used according to the manufacturer's recommended procedure. *If the coil spring has an enamel-type coating, tape the spring where the compressing tool contacts the spring.* The spring may break prematurely if this coating is chipped.

Spring compressing tools are used to compress a coil spring and relieve the spring tension on the upper strut mount to allow spring removal from the strut.

A typical procedure for removing a strut from a coil spring follows:

Caution
Never clamp the lower shock absorber or strut chamber in a vise with excessive force. This action may distort the lower chamber and affect piston movement in the shock absorber or strut.

1. Install the coil spring and strut assembly in the spring compressing tool according to the tool or vehicle manufacturer's recommended procedure.
2. Adjust the compressing arms on the spring compressing tool so the arms contact the coils farthest away from the center of the spring (**Figure 5-11**).
3. Turn the handle on top of the compressing tool until all the spring tension is removed from the upper strut mount (Figure 5-11).
4. Loosen and remove the nut on the upper strut rod (**Figure 5-12**). Be sure all the spring tension is removed from the upper strut mount before loosening this nut.

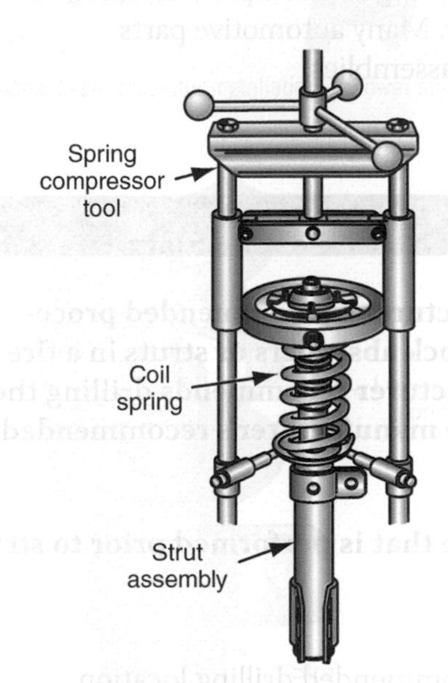

Figure 5-11 Coil spring and strut assembly mounted in a spring compressing tool.

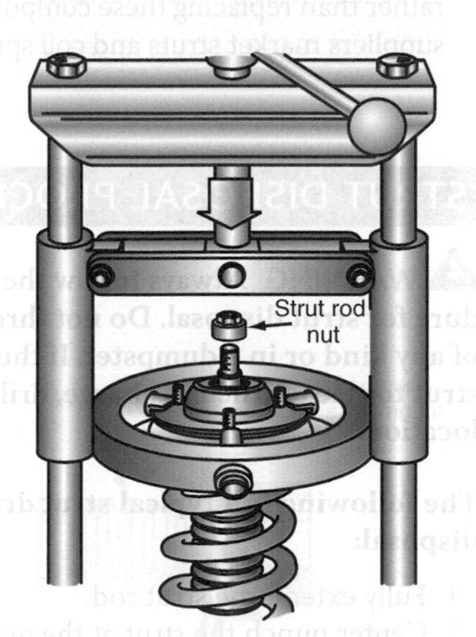

Figure 5-12 After the compressing tool is operated to remove all the spring tension, remove the strut rod nut.

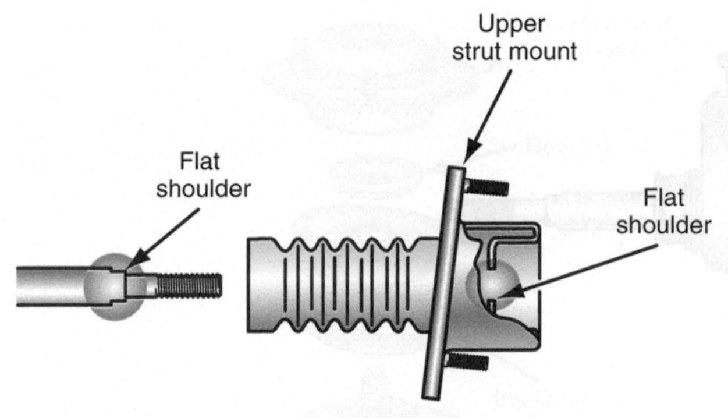

Figure 5-20 Upper strut mount properly positioned on strut piston rod.

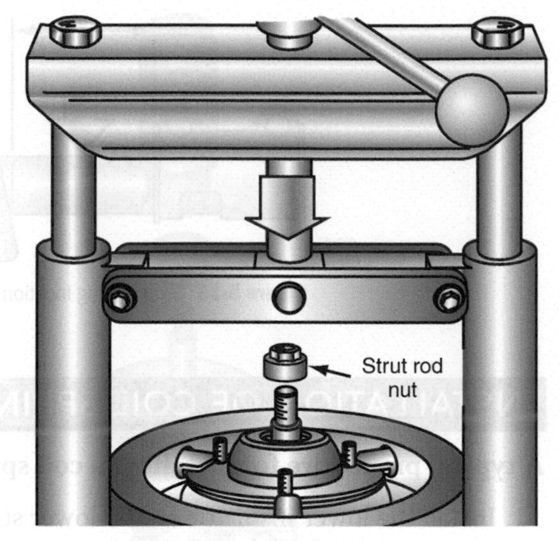

Figure 5-21 Installing strut rod nut.

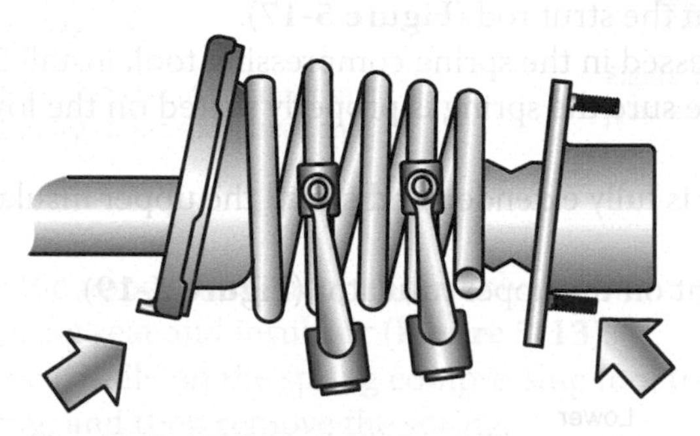

Figure 5-22 Aligning lowest bolt on upper strut mount with tab on lower spring seat.

6. Be sure the spring, upper insulator, and upper strut mount are properly positioned and seated on the coil spring and strut piston rod (**Figure 5-20**).
7. Install the strut rod nut, and tighten this nut to the specified torque (**Figure 5-21**).
8. Rotate the upper strut mount until the lowest bolt in this mount is aligned with the tab on the lower spring seat (**Figure 5-22**).
9. Gradually loosen the handle on the compressing tool until all the spring tension is released from the tool, and remove the strut-and-spring assembly from the tool.

INSTALLATION OF STRUT-AND-SPRING ASSEMBLY IN VEHICLE

A typical installation procedure for a strut-and-spring assembly follows:

1. Install the strut-and-spring assembly with the upper strut mounting bolts extending through the bolt holes in the strut tower. Tighten the nuts on the upper strut mounting bolts to the specified torque (**Figure 5-23**).
2. Install the lower end of the strut in the steering knuckle to the proper depth, and tighten the strut-to-knuckle retaining bolts to the specified torque (**Figure 5-24**). If one of the strut-to-knuckle bolts is an eccentric camber bolt, be sure the eccentric is aligned with the mark placed on the strut during the removal procedure.
3. Install the brake hose in the clamp on the strut. Place the ABS wheel speed sensor wire in the strut clamp if the vehicle is equipped with ABS (**Figure 5-25**).

Photo Sequence 9 shows a typical procedure for removing and replacing a MacPherson strut.

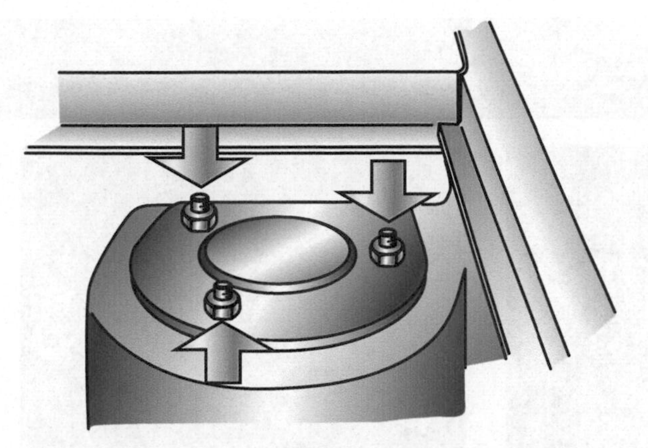

Figure 5-23 Nuts installed on upper strut mount bolts.

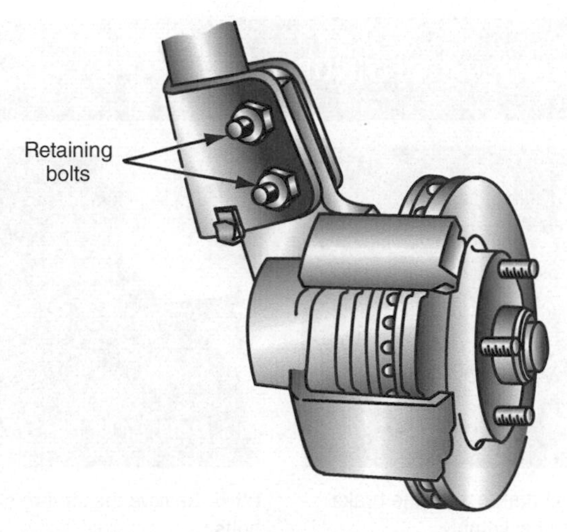

Figure 5-24 Lower end of strut installed in steering knuckle.

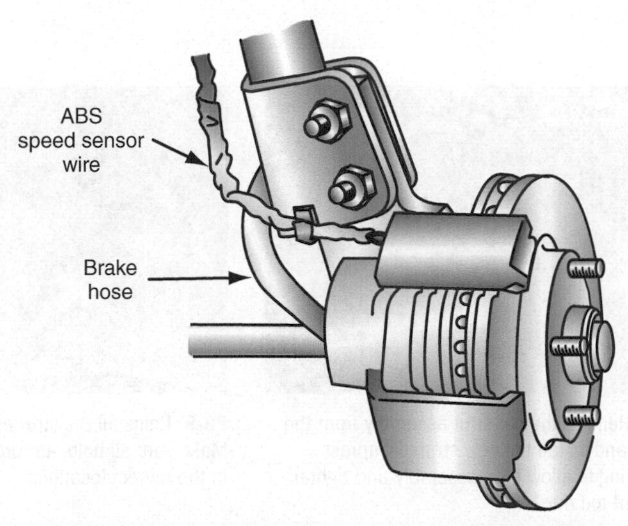

Figure 5-25 Brake hose and ABS wire installed in strut clamps.

PHOTO SEQUENCE 9
Typical Procedure for Removing and Replacing a MacPherson Strut

P9-1 The top of the strut assembly is mounted directly to the chassis of the car. Prior to loosening the strut-to-chassis bolts, scribe alignment marks on the strut bolts and the chassis.

P9-2 With the top strut bolts or nuts removed, raise the car to a working height. It is important that the car be supported on its frame and not on its suspension components.

P9-3 Remove the wheel assembly. The strut is accessible from the wheel well after the wheel is removed.

PHOTO SEQUENCE 9 (CONTINUED)

P9-4 Remove the bolt that fastens the brake line or hose to the strut assembly.

P9-5 Remove the strut-to-steering knuckle bolts.

P9-6 Support the steering knuckle with wire and remove the strut assembly from the car.

P9-7 Install the strut assembly into the proper type spring compressor. Then compress the spring until all spring tension is removed from the upper strut mount. Loosen and remove the strut rod nut.

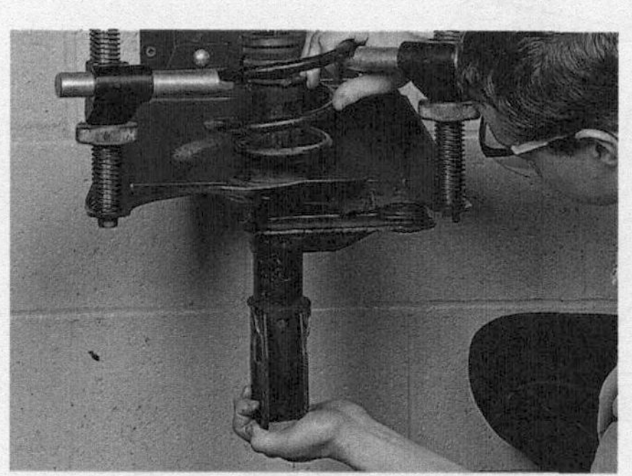

P9-8 Remove the old strut assembly from the spring and install the new strut. Compress the spring to allow for reassembly and tighten the strut rod nut.

P9-9 Reinstall the strut assembly into the car. Make sure all bolts are properly tightened and in the correct locations.

REAR STRUT REPLACEMENT

The rear strut replacement procedure varies depending on the type of rear suspension. Always follow the vehicle manufacturer's recommended procedure outlined in the appropriate service manual.

A typical rear strut replacement procedure follows:

1. Lift the vehicle with a floor jack and lower the vehicle onto safety stands placed under the chassis to support the vehicle weight.
2. Place the floor jack under the lower control arm and operate the jack to support some of the spring tension.
3. Remove the tire-and-wheel assembly.
4. Remove the strut-to-spindle bolts.
5. Pull upward on the strut to remove the strut from the spindle. If necessary, lower the floor jack slightly to remove the strut from the spindle.
6. Disconnect the upper strut mount from the chassis, and remove the strut.

7. When the new strut or mount is installed, reverse Steps 1 through 6. Tighten all the bolts to the specified torque.
8. Check vehicle alignment.

The rear coil springs may be removed from the rear struts using the same basic procedure for spring removal from the front struts.

INSTALLING STRUT CARTRIDGE, OFF-CAR

CUSTOMER CARE Check the cost of the strut cartridges versus the price of new struts. Give customers the best value for their repair dollar!

Many struts are a sealed unit, and thus rebuilding is impossible. However, some manufacturers supply a replacement cartridge that may be installed in the strut housing after the strut has been removed from the vehicle. Always follow the **strut cartridge** manufacturer's recommended installation procedure.

A **strut cartridge** contains the inner working part of the strut, which may be installed in the outer housing of the old strut.

Pipe cutter

The following is a typical off-car strut cartridge installation procedure:

1. Install a bolt and two nuts in the upper strut-to-knuckle mounting bolt hole. Place a nut on the inside and outside of the strut flange.
2. Clamp this bolt in a vise to hold the strut.
3. Locate the line groove near the top of the strut body, and use a pipe cutter installed in this groove to cut the top of the strut body.
4. After the cutting procedure, remove the strut piston assembly from the strut (**Figure 5-26**).
5. Remove the strut from the vise and dump the oil from the strut.
6. Place the special tool supplied by the vehicle manufacturer or cartridge manufacturer on top of the strut body. Strike the tool with a plastic hammer until the tool shoulder contacts the top of the strut body. This action removes burrs from the strut body and places a slight flare on the body.
7. Remove the tool from the strut body.
8. Install the required amount of oil in the strut, place the new cartridge in the strut body, and turn the cartridge until it settles into indentations in the bottom of the strut body.
9. Place the new nut over the cartridge.
10. Using a special tool supplied by the vehicle or cartridge manufacturer, tighten the nut to the specified torque.
11. Move the strut piston rod in and out several times to check for proper strut operation.

INSTALLING STRUT CARTRIDGE, ON-CAR

WARNING **If a vehicle is hoisted or lifted in any way during an on-car strut cartridge replacement, the coil spring may fly off the strut, causing vehicle damage and personal injury.**

On some vehicles, the front strut cartridge may be removed and replaced with the strut installed in the vehicle. Always consult the vehicle manufacturer's service manual for the proper strut service procedure.

6. Thread the special tool onto the strut piston rod, and reextend this rod (**Figure 5-33**). Use the special tool supplied by the vehicle manufacturer to remove the closure nut on top of the strut (**Figure 5-34**).
7. Grasp the top of the strut piston rod and remove the strut valve mechanism (**Figure 5-35**).
8. Remove the oil from the strut tube with a hand-operated suction pump.
9. Install the new strut cartridge in the strut, and tighten the closure nut to the specified torque (**Figure 5-36**). Reverse Steps 1 through 6 to complete the strut cartridge replacement. Place a light coating of the vehicle manufacturer's recommended engine oil on the upper strut mount bushing prior to bushing installation.

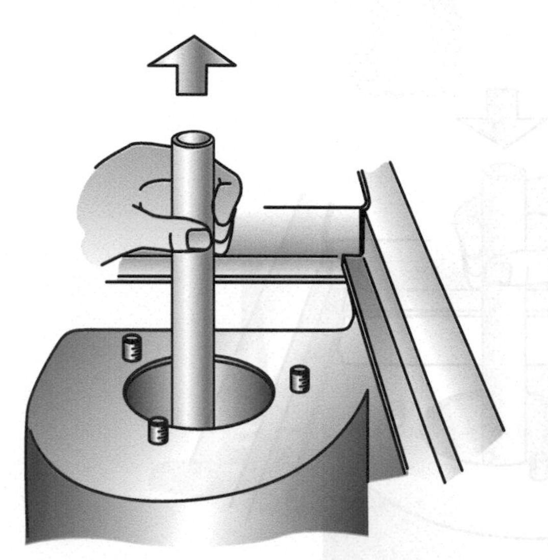

Figure 5-33 Reextending strut piston rod.

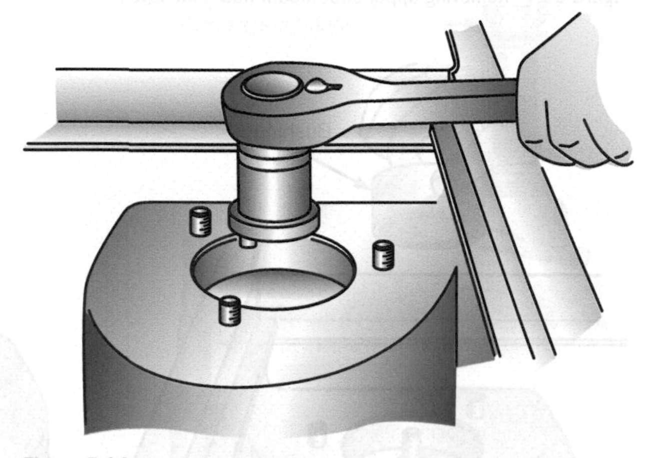

Figure 5-34 Removing strut closure nut.

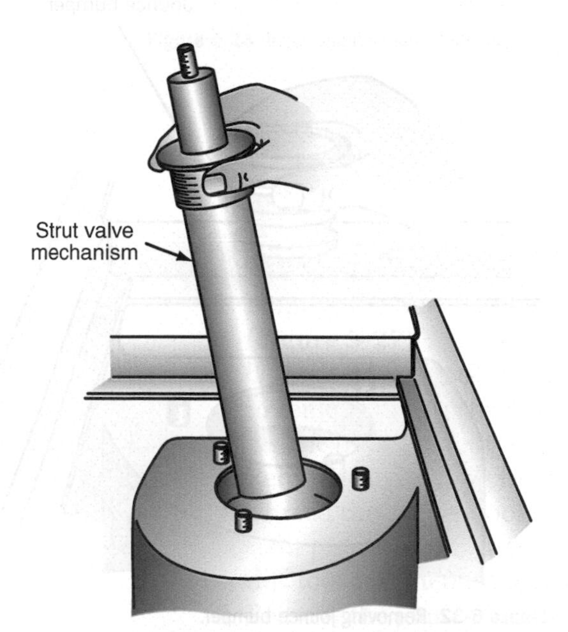

Figure 5-35 Removing strut valve mechanism.

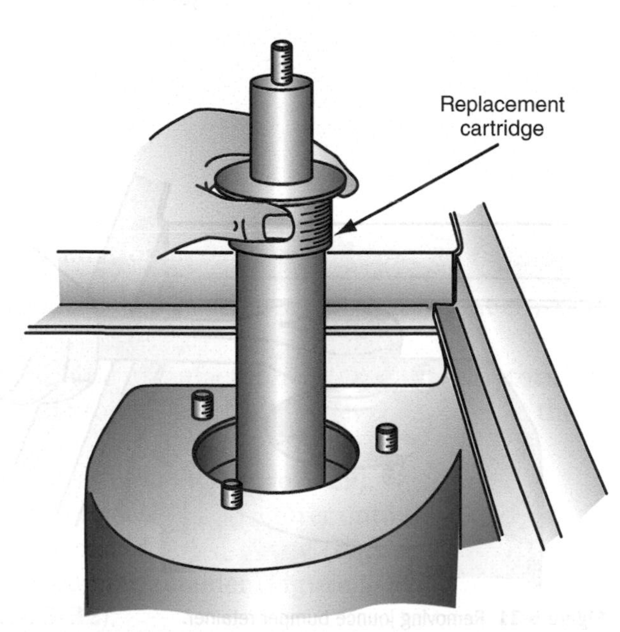

Figure 5-36 Installing new strut cartridge.

DIAGNOSIS OF ELECTRONICALLY CONTROLLED SHOCK ABSORBERS

The actuators on electronically controlled shock absorbers can be removed by pushing inward simultaneously on the two actuator retaining tabs and lifting the actuator off the top of the strut (**Figure 5-37**).

An actuator may be called a solenoid.

WARNING If the strut piston rod nut and actuator mounting bracket are removed, do not move or raise the vehicle. This action releases the coil spring tension and may result in personal injury and vehicle damage.

Follow these steps to diagnose the electronic actuator:

1. With the actuator removed from the strut and the actuator wiring harness connected, turn the ignition switch on. Move the ride control switch to the auto position and wait five seconds.
2. Move the ride control switch to the firm position, and wait 5 seconds. If the actuator control tube on the bottom of the actuator rotated, the actuator is operating. If the actuator control tube did not rotate, proceed with the actuator tests.
3. With the ride control switch in the firm position, place matching Hs beside the control tube and the actuator.
4. With the ride control switch in the auto position, place matching Ss beside the control tube and the actuator.
5. Turn the ignition switch off and disconnect the actuator electrical connector. The actuator control tube may be rotated with a small screwdriver.
6. Connect a pair of ohmmeter leads to the position sense wire and the signal return wire. The position sense wire is white or white with a colored tracer, and the signal return wire is yellow with a black tracer. With the actuator in the S position, the position sense should be closed and the ohmmeter should indicate less than 10 ohms.
7. Rotate the actuator to the H position. Under this condition, the position sense switch should be open and the ohmmeter should indicate over 1000 ohms. If the ohmmeter readings are not within specifications, replace the actuator.
8. Connect the ohmmeter leads to the position sense wire and soft power terminal in the actuator electrical connector. The position sense wire is white or white with a colored tracer, and the soft power wire is tan with a red tracer. If the ohmmeter indicates over 1000 ohms, the actuator is satisfactory. If the ohmmeter reading is below 10 ohms, replace the actuator.

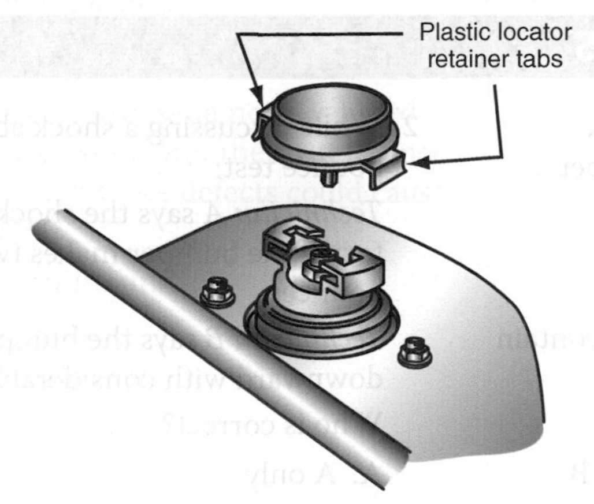

Figure 5-37 Electronically controlled strut actuator.

ASE CHALLENGE QUESTIONS

1. A customer says his MacPherson strut front suspension is making a chattering noise when the steering wheel is turned hard to the left. He says he also feels a "kind of vibration" in the steering wheel when the chatter is heard. To correct this problem, you should:
   A. Replace the front struts.
   B. Check the lower strut spring seating.
   C. Check the stabilizer links.
   D. Check the upper bearing and strut mounting.

2. A faulty shock absorber or strut may cause:
   A. Steering pull.
   B. Suspension bottoming.
   C. Lateral chassis oscillations.
   D. Excessive steering free play.

3. The front rebound bumpers on a coil spring suspension system are badly worn and damaged.
   Technician A says the shock absorbers may be worn out.
   Technician B says the coil springs may be weak.
   Who is correct?
   A. A only C. Both A and B
   B. B only D. Neither A nor B

4. A manual test of a shock absorber shows a stronger resistance to rebound than compression.
   Technician A says the shock is defective.
   Technician B says the shock has a 50/50 ratio.
   Who is correct?
   A. A only C. Both A and B
   B. B only D. Neither A nor B

5. *Technician A* says oil leaking from a strut or shock requires shock or strut replacement.
   Technician B says a slight film of oil on a strut or shock is OK if it performs satisfactorily.
   Who is correct?
   A. A only C. Both A and B
   B. B only D. Neither A nor B

Name ______________________________ Date ______________

JOB SHEET 21

SHOCK ABSORBER DIAGNOSIS AND REPLACEMENT

Upon completion of this job sheet, you should be able to inspect, remove, and/or replace shock absorbers as well as inspect mounts and bushings. and determine needed action.

ASE Education Foundation Task Correlation

This job sheet addresses the following **MLR** task:

B.20. Inspect, remove, and/or replace shock absorbers; inspect mounts and bushings. **(P-1)**

This job sheet addresses the following **AST/MAST** task:

D.1. Inspect, remove, and/or replace shock absorbers; inspect mounts and bushings. **(P-1)**

We Support ASE Education Foundation

Tools and Materials

Hand tool

Underhoist high stand support

Describe the vehicle being worked on:

Year ______________ Make ______________ Model ______________

VIN ______________ Engine type and size ______________

Procedure **Task Completed**

1. Test drive vehicle over rough bumpy road and determine ride quality.

   a. Note findings: ______________________________

2. Raise vehicle on a lift. ☐

3. Inspect the shock assemblies for oil leaks and physical damage such as dents and punctures.

   a. Note findings: ______________________________

4. Grasp shock absorber near lower mounting bushing and attempt to move the shock absorber both vertically and horizontally. A prybar may be used, but care must be taken to not damage shock absorber tube. Check for looseness and wear of the bushings and mounting hardware.

   a. Note findings: ______________________________

5. Grasp shock absorber near upper mounting bushing and attempt to move the shock absorber both vertically and horizontally. A prybar may be used, but care must be taken not to damage shock absorber tube. Check for looseness and wear of the bushings and mounting hardware.

   a. Note findings: ______________________________

6. Follow manufacturer's recommended service procedures for removing shock absorber assembly. This generally requires the use of an underhoist high stand for supporting suspension system to allow safe removal of shock absorber. ☐

7. Disconnect the lower shock absorber mounting bolt(s). ☐

8. If the road test was inconclusive as to the integrity of the shock absorber, pull the shock absorber housing downward on the extension stroke and then push the lower end of the shock absorber upward on the compression stroke. The shock absorber should offer resistance to movement in relation to the shock absorber ratio. ☐

Task Completed

4. Punch mark the cam bolt in relation to the strut, remove the strut-to-steering knuckle retaining bolts, and remove the strut from the knuckle.

Is the cam bolt marked in relation to the strut? ☐ Yes ☐ No

Is the strut removed from the knuckle? ☐ Yes ☐ No

Instructor check ____________________

State the reason for marking the cam bolt in relation to the strut.

____

____

5. Remove the upper strut mounting bolts on top of the strut tower; remove the strut-and-spring assembly. ☐

WARNING Always use a coil spring compressing tool according to the tool or vehicle manufacturer's recommended service procedure. Be sure the tool is properly installed on the spring. If a coil spring slips off the tool when the spring is compressed, severe personal injury or property damage may occur.

WARNING Never loosen the upper strut mount retaining nut on the end of the strut rod unless the spring is compressed enough to remove all spring tension from the upper strut mount. If this nut is loosened with spring tension on the upper mount, this mount becomes a very dangerous projectile that may cause serious personal injury or property damage.

Caution

Never clamp the lower shock absorber or strut chamber in a vise with excessive force. This action may distort the lower chamber and affect piston movement in the shock absorber or strut.

6. Install the spring compressing tool on the coil spring according to the tool or vehicle manufacturer's recommended procedure.

Is the compressing tool properly installed on the strut-and-spring assembly?
☐ Yes ☐ No

Instructor check ____________________

7. Turn the nut on top of the compressing tool until all the spring tension is removed from the upper strut mount.

Is all the spring tension removed from the upper strut mount? ☐ Yes ☐ No

Instructor check ____________________

8. Install a bolt and two nuts in the upper strut-to-knuckle mounting bolt holes. Install a nut on each side of the strut flange. Clamp this bolt securely in a vise to hold the strut-and-spring assembly and the compressing tool. ☐

Caution

If the coil spring has an enamel-type coating and the compressing tool contacts the coil spring, tape the spring where the compressing tool contacts the spring.

9. Use the bar on the spring compressing tool to hold the strut-and-spring assembly from turning, and loosen the nut on the upper strut mount. Be sure all the spring tension is removed from the upper strut mount before loosening this nut.

Is all the spring tension removed from the upper strut mount? ☐ Yes ☐ No

Is the upper strut mount retaining nut loosened? ☐ Yes ☐ No

Instructor check ____________________

Task Completed

10. Remove the nut, upper strut mount, upper insulator, coil spring, spring bumper, and lower insulator. ☐

11. Inspect the strut, upper strut mount, coil spring, spring insulators, and spring bumper. On the basis of this inspection, list the necessary strut-and-spring service, and give the reasons for your diagnosis.

____

____

____

Instructor's Response

____

____

____

Task Completed

6. Install the upper strut mount on the upper insulator. ☐

7. Be sure the spring, upper insulator, and upper strut mount are properly positioned and seated on the coil spring and strut piston rod.

   Are the spring, upper insulator, and upper strut mount properly positioned?
   ☐ Yes ☐ No

   Instructor check ______________________________

8. Use the compressing tool bar to hold the strut and spring from turning, then tighten the strut piston rod nut to the specified torque.

   Specified strut piston rod nut torque ______________________________

   Actual strut piston rod nut torque ______________________________

9. Rotate the upper strut mount until the lowest bolt in this mount is aligned with the tab on the lower spring seat.

   Is the upper strut mount properly aligned? ☐ Yes ☐ No

   Instructor check ______________________________

10. Gradually loosen the nut on the compressing tool until all the spring tension is released from the tool, and remove the tool from the spring. ☐

11. Install the strut-and-spring assembly with the upper strut mounting bolts extending through the bolt holes in the strut tower. Tighten the nuts on the upper strut mounting bolts to the specified torque.

    Specified upper strut mount nut torque ______________________________

    Actual upper strut mount nut torque ______________________________

12. Install the lower end of the strut in the steering knuckle to the proper depth. Align the punch marks on the cam bolt and strut that were placed on these components during disassembly, and tighten the strut-to-knuckle retaining bolts to the specified torque.

    Is the cam bolt properly positioned? ☐ Yes ☐ No

    Specified strut-to-knuckle bolt torque ______________________________

    Actual strut-to-knuckle bolt torque ______________________________

    Instructor check ______________________________

13. Install the brake hose in the clamp on the strut. Place the ABS wheel speed sensor wire in the strut clamp if the vehicle is equipped with ABS.

    Is the brake hose properly installed and tightened? ☐ Yes ☐ No

    Is the ABS wheel speed sensor properly installed and tightened? ☐ Yes ☐ No

    Instructor check ______________________________

14. Raise the vehicle with a floor jack, remove the safety stands, and lower the vehicle onto the shop floor. ☐

Instructor's Response

____

____

____

Name ______________________________ **Date** ______________

JOB SHEET 24

INSTALL STRUT CARTRIDGE, OFF-CAR

Upon completion of this job sheet, you should be able to remove and replace a strut cartridge with the strut removed from the car.

ASE Education Foundation Task Correlation

This job sheet addresses the following **MLR** tasks:

B.13. Inspect suspension system coil springs and spring insulators (silencers). **(P-1)**

B.16. Inspect, remove, and/or replace strut cartridge or assembly; inspect mounts and bushings. **(P-2)**

B.17. Inspect front strut bearing and mount. **(P-1)**

This job sheet addresses the following **AST/MAST** task:

C.10. Inspect, remove, and/or replace strut cartridge or assembly, strut coil spring, insulators (silencers), and upper strut bearing mount. **(P-3)**

We Support ASE Education Foundation

Tools and Materials

Front strut removed from car

Pipe cutter

Torque wrench

Describe the vehicle being worked on:

Year ______________ Make ______________ Model ______________

VIN ______________ Engine type and size ______________

Procedure **Task Completed**

1. List the strut conditions and suspension operating conditions that indicate a strut should be serviced or replaced.

   ____

   ____

   ____

   ____

2. Install a bolt and two nuts in the upper strut-to-knuckle mounting bolt hole. Place a nut on the inside and outside of the strut flange. ☐

3. Clamp this bolt in a vise to hold the strut. ☐

4. Locate the line groove near the top of the strut body, and use a pipe cutter installed in this groove to cut the top of the strut body. ☐

5. After the cutting procedure, remove the strut piston assembly from the strut.

   Is the strut piston assembly removed from the strut? ☐ Yes ☐ No

   Instructor check ______________________________

6. Remove the strut from the vise and dump the oil from the strut. ☐

Task Completed

7. Place the special tool supplied by the vehicle manufacturer or cartridge manufacturer on top of the strut body and strike the tool with a plastic hammer until the tool shoulder contacts the top of the strut body. This action removes burrs from the strut body and places a slight flare on the body.

   Is the strut body properly flared? ☐ Yes ☐ No

   Instructor check ____

8. Remove the tool from the strut body. ☐

9. Place the new cartridge in the strut body and turn the cartridge until it settles into indentations in the bottom of the strut body.

   Is the new strut cartridge properly seated in strut body indentations? ☐ Yes ☐ No

   Instructor check ____

10. Place the new nut over the cartridge. ☐

11. Using a special tool supplied by the vehicle or cartridge manufacturer, tighten the nut to the specified torque.

    Specified strut nut torque ______________________________________

    Actual strut nut torque ____

12. Move the strut piston rod in and out several times to check for proper strut operation.

    Strut operation: ☐ Satisfactory ☐ Unsatisfactory

Instructor's Response

____

____

____

CHAPTER 6
FRONT SUSPENSION SYSTEM SERVICE

Upon completion and review of this chapter, you should be able to:

- Measure riding height.
- Diagnose and correct riding height problems.
- Adjust torsion bars.
- Diagnose front suspension noise and body sway.
- Remove and replace ball joints, and check ball joint condition.
- Remove and replace steering knuckles, and check knuckle condition.
- Remove and replace lower control arms, and check control arm and bushing condition.
- Remove and replace coil springs, and check spring and insulator condition.
- Remove and replace upper control arms, and check control arm and bushing condition.
- Remove and replace control arm bushings.
- Inspect and replace rebound bumpers.
- Diagnose, remove, and replace stabilizer bars.
- Diagnose, remove, and replace strut rods.
- Diagnose, remove, and replace leaf springs.
- Replace torsion bars, and check torsion bar condition.

Basic Tools

Basic technician's tool set
Service manual
Floor jack
Safety stands
Pry bar
3/8 in. electric drill and drill bits

Terms To Know

Ball joint vertical movement
Ball joint wear indicator
Camber angles
Caster angles
Curb riding height
Directional stability
Rebound bumpers
Stabilizer bar
Steering effort
Strut rod

Proper front suspension system service is extremely important to provide adequate vehicle safety and to maintain ride comfort and normal tire life. If worn or loose front suspension system components are ignored, steering control may be adversely affected, which may result in loss of steering control and an expensive collision. Defective front suspension components, such as worn-out shock absorbers or broken springs, may cause rough riding that results in driver and passenger discomfort. Other worn front suspension components, such as worn ball joints and control arm bushings, cause improper alignment angles that cause excessive front tire wear. Therefore, technicians must be familiar with front suspension service. Diagnostic **Table 6-1** contains symptoms and causes of common front suspension issues.

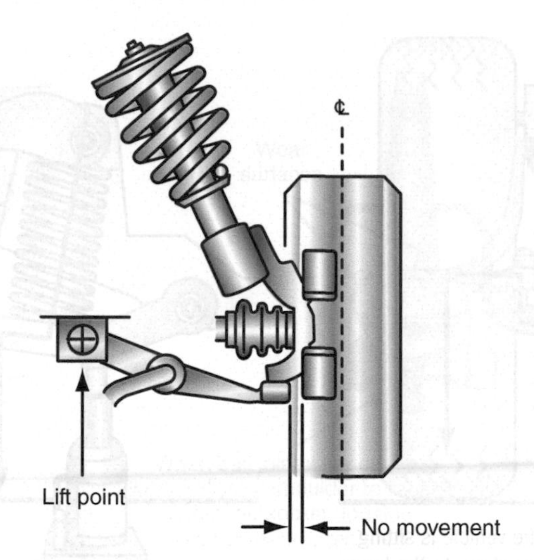

Figure 6-12 Ball joint wear measurement on MacPherson strut front suspension.

lifted off the floor (**Figure 6-12**). Since the spring load is carried by the upper and lower spring seats when the tire is lifted off the floor, it is not necessary to unload this type of ball joint. **Photo Sequence 10** shows a typical procedure for measuring the lower ball joint radial movement on a MacPherson strut front suspension. In this suspension system the lower ball joint is the non-load carrier, but the upper strut mount is the primary load-carrying pivot point. It is important to check the upper strut mount for proper pivoting and excessive play in the assembly. This is an often-overlooked part of the suspension inspection. A binding upper strut mount can lead to a memory steer condition, while a mount with excessive movement can cause a rattle over bumps.

PHOTO SEQUENCE 10

Typical Procedure for Measuring the Lower Ball Joint Horizontal Movement on a MacPherson Strut Front Suspension

P10-1 Raise the front suspension with a floor jack and place safety stands under the chassis at the vehicle manufacturer's recommended lifting points.

P10-2 Grasp the front tire at the top and bottom and rock the tire inward and outward while a coworker visually checks for movement in the front wheel bearing. If there is movement in the front wheel bearing, adjust or replace the bearing.

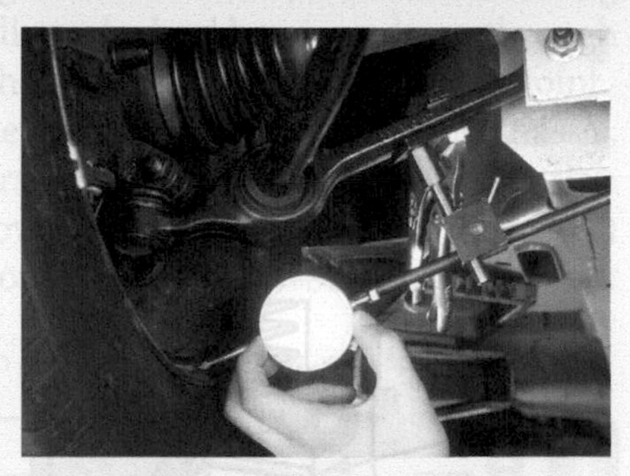

P10-3 Position a dial indicator against the inner edge of the rim at the bottom. Preload and zero the dial indicator.

PHOTO SEQUENCE 10 (CONTINUED)

P10-4 Grasp the bottom of the tire and pull outward.

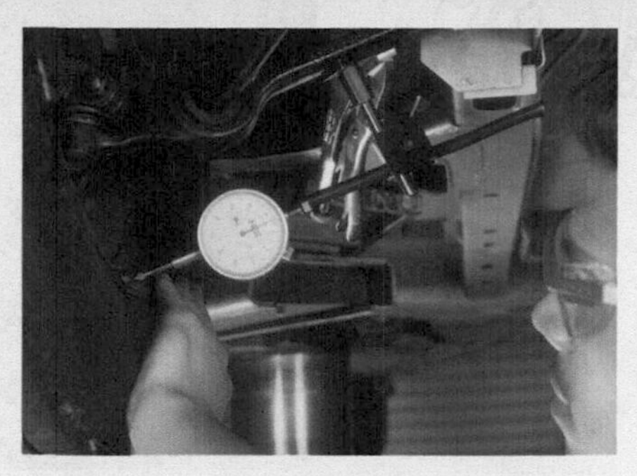

P10-5 With the tire held outward, read the dial indicator.

P10-6 Push the bottom of the tire inward and be sure the dial indicator reading is zero. Adjust the dial indicator as required.

P10-7 Grasp the bottom of the tire and pull outward.

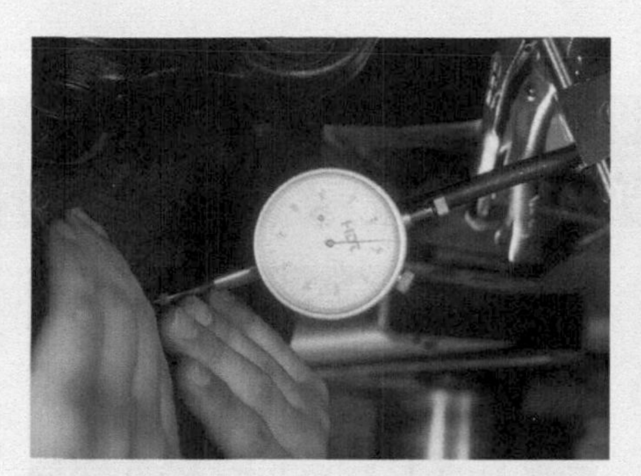

P10-8 With the tire held in this position, read the dial indicator.

P10-9 If the dial indicator reading is more than specified, replace the lower ball joint.

Ball Joint Vertical Measurement

The vehicle manufacturer may provide ball joint vertical and horizontal tolerances. A dial indicator is one of the most accurate ball joint measuring devices (**Figure 6-13**). Always install the dial indicator at the vehicle manufacturer's recommended location for ball joint measurement. When measuring the **ball joint vertical movement** in a compression-loaded ball joint, attach the dial indicator to the lower control arm and position the dial indicator stem on the lower side of the steering knuckle beside the ball joint stud (**Figure 6-14**). Depress the dial indicator stem approximately 0.250 in. (6.35 mm) and zero the dial indicator. Place a pry bar under the tire and pry straight upward while observing the vertical ball joint movement on the dial indicator. If this movement is more than specified, ball joint replacement is necessary. **Photo Sequence 11** illustrates the vertical ball joint measurement procedure.

On a tension-loaded ball joint, clean the top end of the lower ball joint stud, and position the dial indicator stem against the top end of this stud. Depress the dial indicator plunger approximately 0.250 in. (6.35 mm) and zero the dial indicator. Lift upward with a pry bar under the tire and observe the dial indicator reading. If the vertical ball joint movement exceeds the manufacturer's specifications, ball joint replacement is required.

The **ball joint vertical movement** refers to up-and-down movement in a ball joint.

Special Tool

Dial indicator for ball joint measurement

Camber angles are an imaginary line through the centerline of the tire and wheel in relation to the true vertical center-line of the tire.

Caster angles are an imaginary line through the upper and lower ball joint centers in relation to the true vertical tire centerline viewed from the side.

Ball Joint Horizontal Measurement

Worn ball joints cause improper **camber angles** and **caster angles**, which result in reduced directional stability and tire tread wear. Connect the dial indicator to the lower

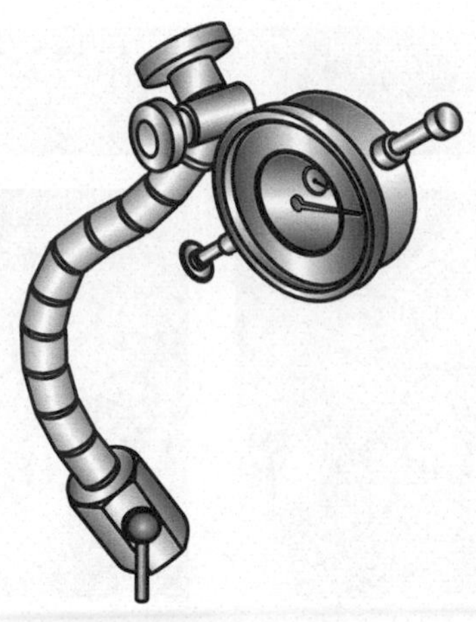

Figure 6-13 Dial indicator designed for ball joint measurement.

Figure 6-14 Dial indicator installed to measure vertical ball joint movement on a compression-loaded ball joint.

PHOTO SEQUENCE 11

Vertical Ball Joint Measurement

P11-1 Place a floor jack under a front, lower control arm, and raise the lower control arm until the front tire is approximately 4 in. (10 cm) off the floor.

P11-2 Lower the lower control arm until it is securely supported on a safety stand, and remove the floor jack.

P11-3 Attach the magnetic base of a dial indicator for measuring ball joints to the safety stand.

P11-4 Position the dial indicator stem against the steering knuckle beside the ball joint stud.

P11-5 Preload the dial indicator stem approximately 0.250 in. (6.35 mm), and zero the dial indicator scale.

P11-6 Position a pry bar under the tire and lift on the tire-and-wheel assembly while a coworker observes the dial indicator. Record the ball joint vertical movement indicated on the dial indicator.

PHOTO SEQUENCE 11 (CONTINUED)

P11-7 Repeat Step 6 several times to confirm an accurate dial indicator reading.

P11-8 Compare the ball joint vertical movement indicated on the dial indicator to the vehicle manufacturer's specifications for ball joint wear. If the ball joint vertical movement exceeds specifications, ball joint replacement is necessary.

control arm of the ball joint being checked and position the dial indicator stem against the edge of the wheel rim (**Figure 6-15**).

Be sure the front wheel bearings are adjusted properly prior to the ball joint horizontal measurement. While an assistant grasps the top and bottom of the raised tire and attempts to move the tire and wheel horizontally in and out, observe the reading on the dial indicator (**Figure 6-16**).

Diagnosis of Ride Harshness

When diagnosing a ride harshness condition, check these measurements and components:

1. Riding height—insufficient riding height causes ride harshness
2. Excessive vehicle load

Figure 6-15 Dial indicator installed to measure radial ball joint movement.

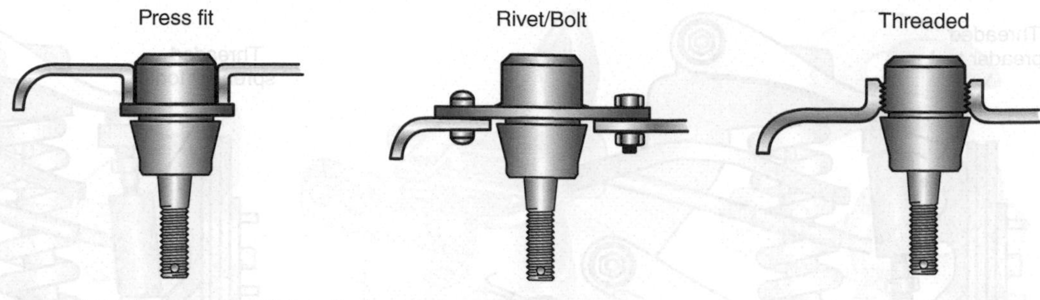
Press fit
Rivet/Bolt
Threaded

Figure 6-26 Methods of ball joint attachment.

The following are typical steps that apply to all three methods of ball joint attachment:

Special Tool

Ball joint pressing tools

1. Remove the wheel cover and loosen the wheel nuts.
2. Lift the vehicle with a floor jack and place safety stands under the chassis so the front suspension is allowed to drop downward. Lower the vehicle onto the safety stands and remove the floor jack.
3. Remove the wheel and place a floor jack under the outer end of the lower control arm. Operate the floor jack and raise the lower control arm until the ball joints are unloaded. Remove other components, such as the brake caliper, rotor, and drive axle, as required to gain access to the ball joints.
4. Remove the cotter pin in the ball joint or joints that requires replacement, and loosen, but do not remove the ball joint stud nuts.
5. Loosen the ball joint stud tapers in the steering knuckle. A threaded expansion tool is available for this purpose.
6. Remove the ball joint nut and lift the knuckle off the ball joint stud. Block or tie up the knuckle and hub assembly to access the ball joint.
7. If the ball joint is riveted to the control arm, drill and punch out the rivets, and bolt the new ball joint to the control arm (**Figure 6-27**).
8. If the ball joint is pressed into the lower control arm, remove the ball joint dust boot, and use a pressing tool to remove and replace the ball joint (**Figure 6-28** and **Figure 6-29**).

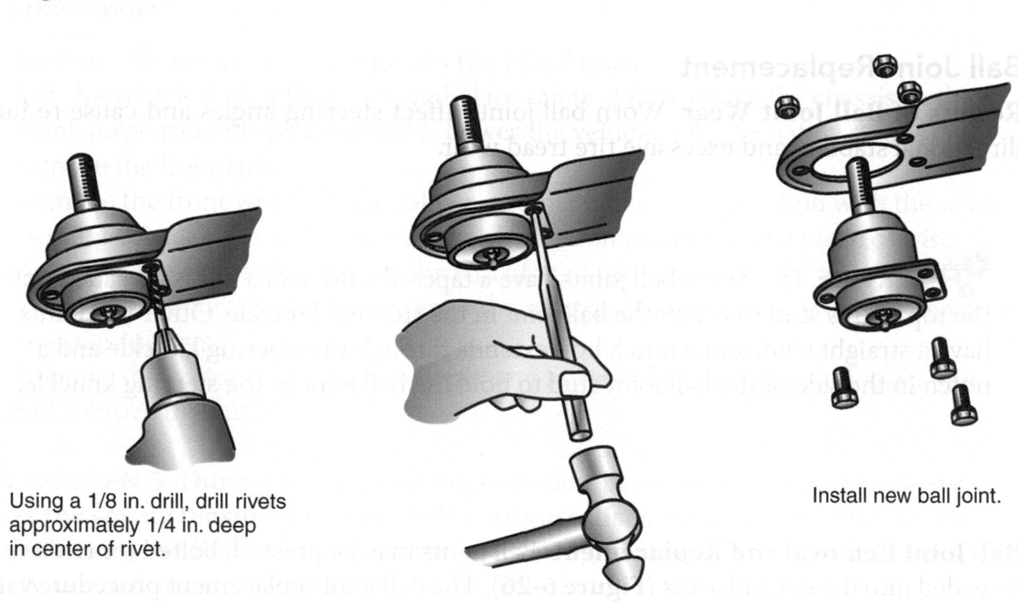
Using a 1/8 in. drill, drill rivets approximately 1/4 in. deep in center of rivet.
Using a 1/2 in. drill, drill just deep enough to remove rivet head.
Remove rivets with punch.
Install new ball joint.

Figure 6-27 Replacing the riveted ball joint.

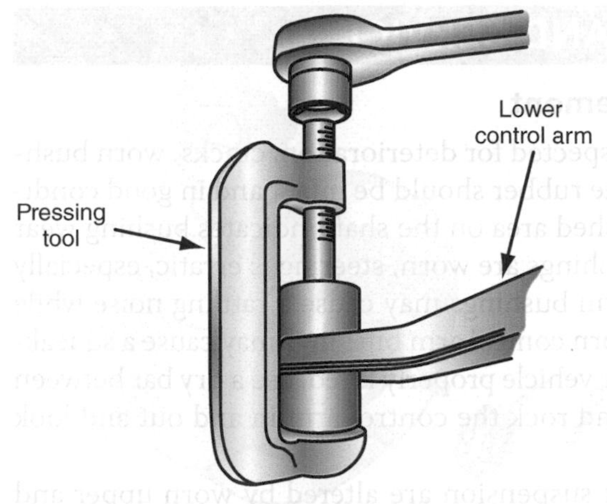

Figure 6-28 Removing the pressed-in ball joint from the lower control arm.

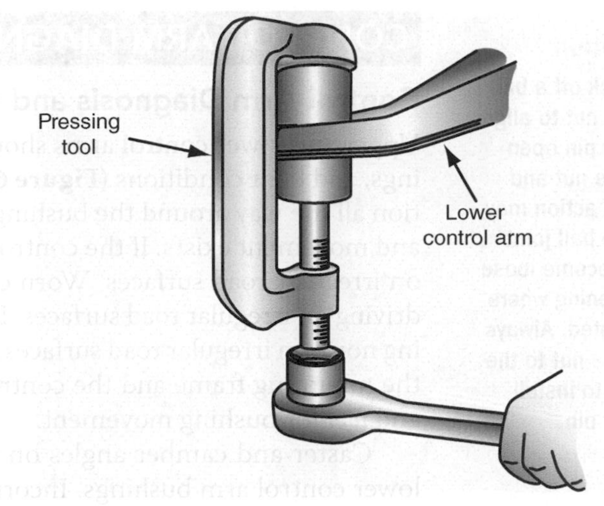

Figure 6-29 Installing the pressed-in ball joint in the lower control arm.

9. If the ball joint housing is threaded into the control arm, use the proper size socket to remove and install the ball joint. The replacement ball joint must be torqued to the manufacturer's specifications. If a minimum of 125 ft.-lb of torque cannot be obtained, the control arm threads are damaged and control arm replacement is necessary.
10. If the ball joint is bolted to the lower control arm, install the new ball joint and tighten the bolt and nuts to the specified torque (**Figure 6-30**).
11. Clean and inspect the ball joint stud tapered opening in the steering knuckle. If this opening is out-of-round or damaged, the knuckle must be replaced.
12. Check the fit of the ball joint stud in the steering knuckle opening. This stud should fit snugly in the opening and only the threads on the stud should extend through the knuckle. If the ball joint stud fits loosely in the knuckle tapered opening, either this opening is worn or the wrong ball joint has been supplied.
13. Install the ball joint stud in the steering knuckle opening, making sure the stud is straight and centered. Install the stud nut and tighten this nut to the specified torque. Install a new cotter pin through the stud and nut. Do not loosen the nut to align the nut and stud openings.
14. Reassemble the components that were removed in Step 3. Make sure the wheel nuts are tightened to the specified torque.
15. After ball joint replacement, the front suspension alignment should be checked.

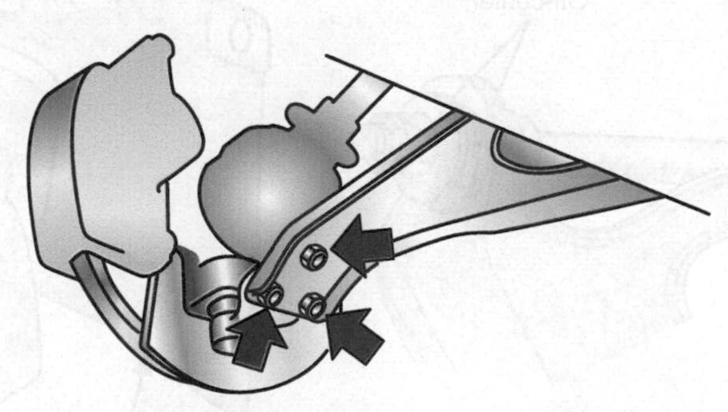

Figure 6-30 Installing the ball joint retaining bolts and nuts in the lower control arm.

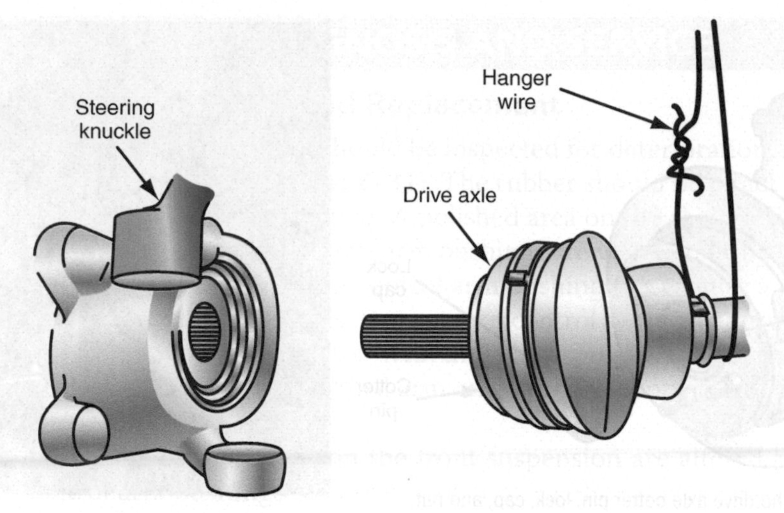

Figure 6-38 Securing the drive axle with a piece of wire.

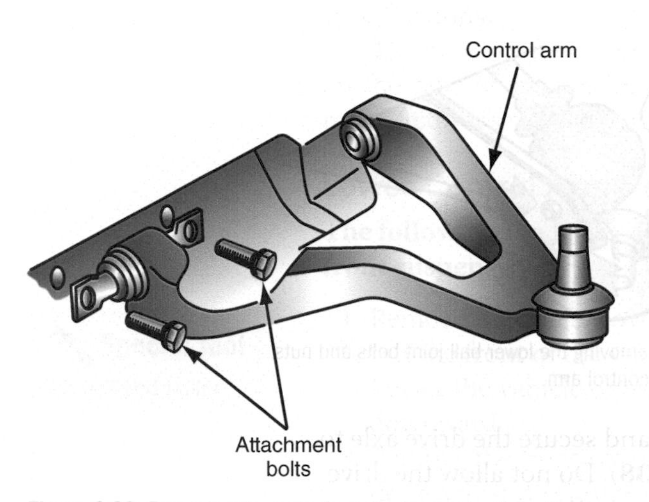

Figure 6-39 Removing the bolts from the front side of the lower control arm.

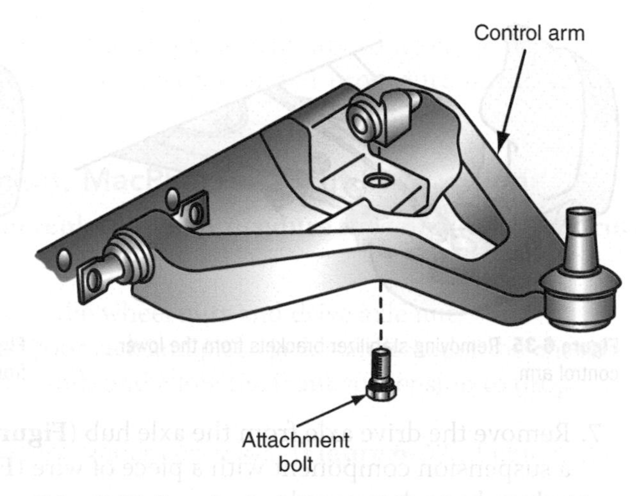

Figure 6-40 Removing the bolt and nut from the rear side of the lower control arm.

Lower Control Arm and Spring Replacement, Short-and-Long Arm Suspension

WARNING During control arm and spring replacement, the coil spring tension is supported by a compressing tool or floor jack. Always follow the vehicle manufacturer's recommended control arm and spring replacement procedures very carefully. Serious personal injury or property damage may occur if the spring tension is released suddenly.

Broken coil springs may cause a rattling noise while driving on irregular road surfaces. Weak or broken coil springs also reduce riding height. A rattling noise while driving on road irregularities may also be caused by worn or broken spring insulators. Since weak or broken coil springs affect front suspension alignment angles, this problem may cause reduced directional stability, excessive tire wear, and harsh riding.

Follow this procedure when a lower control arm and/or spring is replaced on a vehicle with a short-and-long arm suspension system with the coil springs positioned between the lower control arm and the frame:

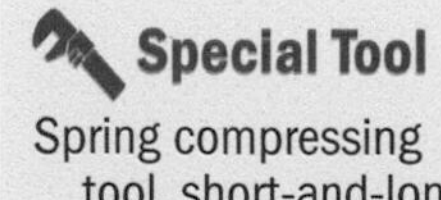

Special Tool

Spring compressing tool, short-and-long arm suspension

1. Lift the vehicle on a hoist until the tires are a short distance off the floor, and allow the front suspension to drop downward. An alternate method is to lift the vehicle with a floor jack, and then support the chassis securely on safety stands so the front suspension drops downward.

2. Disconnect the lower end of the shock absorber. On some applications, the shock absorber must be removed.
3. Disconnect the stabilizer bar from the lower control arm.
4. Install a spring compressor and turn the spring compressor bolt until the spring is compressed (**Figure 6-41**). Make sure all the spring tension is supported by the compressing tool.
5. Place a floor jack under the lower control arm, and raise the jack until the control arm is raised and the rebound bumper is not making contact.
6. Remove the lower ball joint cotter pin and nut, and use a threaded expansion tool to loosen the lower ball joint stud.
7. Lower the floor jack very slowly to lower the control arm and coil spring.
8. Disconnect the lower control arm inner mounting bolts, and remove the lower control arm.
9. Rotate the compressing tool bolt to release the spring tension, and remove the spring from the control arm.
10. Inspect the lower control arm for a bent condition or cracks. If either of these conditions is present, replace the control arm. Visually inspect all control arm bushings. Loose or worn bushings must be replaced. Visually inspect upper and lower spring insulators for cracks and wear, and inspect the spring seat areas in the chassis and lower control arm. Worn or cracked spring insulators must be replaced.
11. Reverse Steps 1 through 9 to install the lower control arm. Be sure the coil spring and insulators are properly seated in the lower control arm and in the upper spring seat.

NOTE: See Chapter 5 for strut and spring service.

Upper Control Arm Removal and Replacement, Short-and-Long Arm Suspension System

Proceed as follows when replacing an upper control arm on a short-and-long arm suspension system with the front coil springs located between the lower control arm and the frame:

1. Remove the wheel cover and loosen the wheel nuts.
2. Lift the vehicle with a floor jack and install safety stands under the chassis. Lower the vehicle onto the safety stands so the front suspension drops downward.
3. Remove the front wheel and tire.

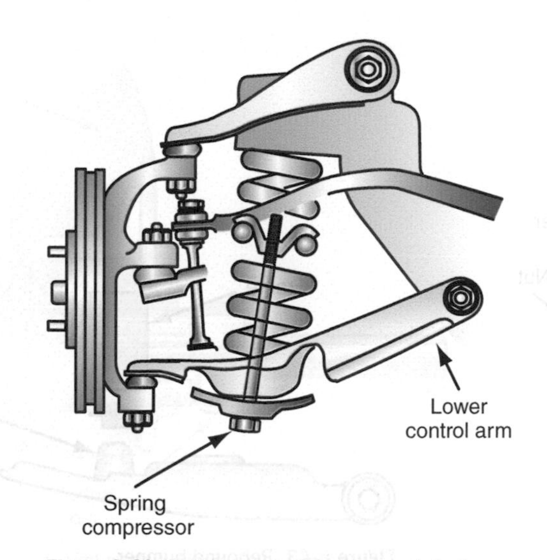

Figure 6-41 Spring compressing tool installed on a spring in a short-and-long arm suspension system.

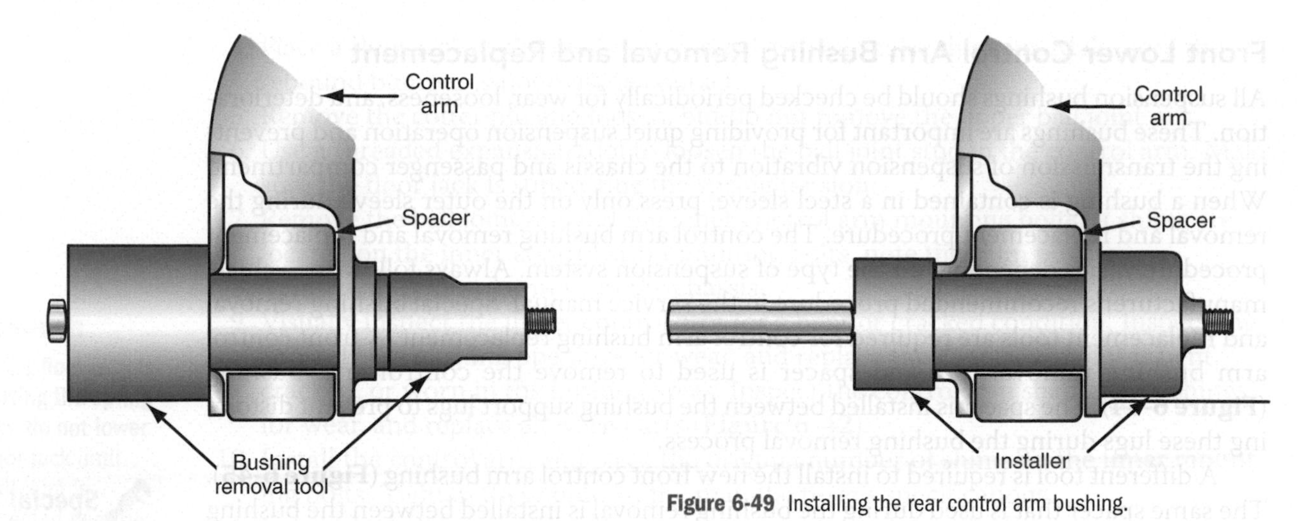

Figure 6-48 Removing the rear control arm bushing.

Figure 6-49 Installing the rear control arm bushing.

The rear control arm bushing is replaced with the same procedure as the front control arm bushing. Some of the same special tools are used for rear control arm bushing replacement (**Figure 6-48** and **Figure 6-49**).

Stabilizer Bar Diagnosis and Replacement

A Stabilizer bar may be referred to as a sway bar. It reduces body sway when one wheel strikes a road irregularity.

Worn **stabilizer bar** mounting bushings, grommets, or mounting bolts cause a rattling noise as the vehicle is driven on irregular road surfaces. A weak stabilizer bar or worn bushings and grommets cause harsh riding and excessive body sway while driving on irregular road surfaces. Worn or very dry stabilizer bar bushings may cause a squeaking noise on irregular road surfaces. All stabilizer bar components should be visually inspected for wear. Stabilizer bar removal and replacement procedures vary depending on the vehicle. Always follow the vehicle manufacturer's recommended procedure in the service manual.

Classroom Manual
Chapter 6, page 161

The following is a typical stabilizer bar removal and replacement procedure:

1. Lift the vehicle on a hoist and allow both sides of the front suspension to drop downward as the vehicle chassis is supported on the hoist.
2. Remove the mounting bolts at the outer ends of the stabilizer bar and remove the bushings, grommets, brackets, or spacers (**Figure 6-50**).
3. Remove the mounting bolts in the center area of the stabilizer bar.

Retainer
Nut
Bushings
Retainer
Tube
Bolt

Figure 6-50 Stabilizer bar components.

4. Remove the stabilizer bar from the chassis.
5. Visually inspect all stabilizer bar components, such as bushings, bolts, and spacer sleeves. Replace the stabilizer bar, grommets, bushings, brackets, or spacers as required. Split bushings may be removed over the stabilizer bar. Bushings that are not split must be pulled from the bar.
6. Reverse Steps 2 through 4 to install the stabilizer bar. Make sure all stabilizer bar components are installed in the original position, and tighten all fasteners to the specified torque.

Some vehicle manufacturers specify that stabilizer bars must have equal distances between the outer bar ends and the lower control arms. Always refer to the manufacturer's recommended measurement procedure. If this measurement is required, adjust the nut on the outer stabilizer bar mounting bolt until equal distances are obtained between the outer bar ends and the lower control arms. Worn grommets cause these distances to be unequal.

Strut Rod Diagnosis and Replacement

Some front suspension systems have a **strut rod** connected from the lower control arm to the frame. Rubber grommets isolate the rod from the chassis components. Worn strut rod grommets may allow the lower control arm to move rearward or forward. This movement changes the caster angle, which may affect steering quality and cause the vehicle to pull to one side. A bent strut rod also causes steering pull. Worn strut rod grommets or loose mounting bolts cause a rattling noise while driving on irregular road surfaces. Inspect the strut rod grommets visually for wear and deterioration. With the vehicle lifted on a hoist, grasp the strut rod firmly and apply vertical and horizontal force to check the rod and grommets for movement. Worn grommets must be replaced. If a strut rod is bent, replace the rod.

A **strut rod** prevents fore-and-aft lower control arm movement.

Classroom Manual
Chapter 6, page 154

Follow these steps for strut rod replacement:

1. Lift the vehicle on a hoist.
2. Remove the strut rod nut from the front end of the rod.
3. Remove the strut rod bolts from the lower control arm.
4. Pull the strut rod rearward to remove the rod (**Figure 6-51**).
5. Remove the bushings from the opening in the chassis.
6. Visually inspect the strut rod, bushings, washers, and retaining bolts. Replace all worn parts. Reverse Steps 1 through 5 to reinstall the strut rod. Tighten the strut rod nut and bolts to the specified torque.

Since strut rod and bushing conditions affect front suspension alignment, check front suspension alignment after strut rod service.

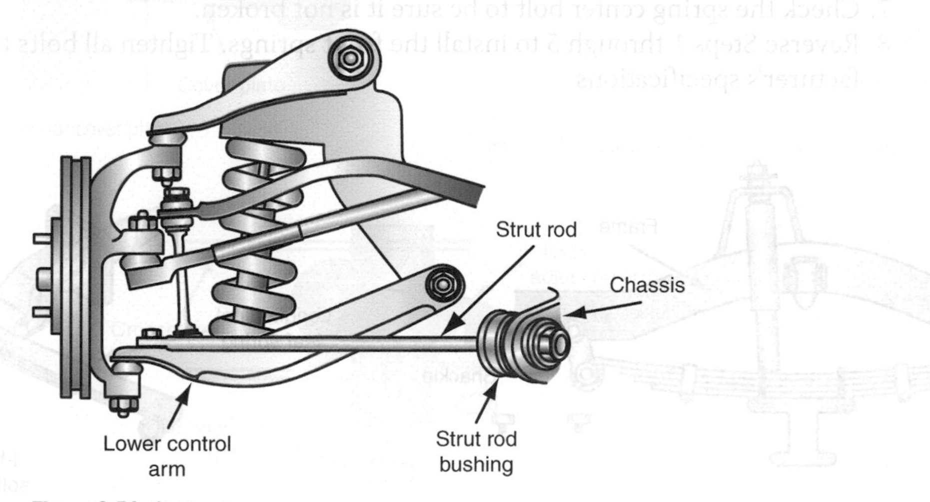

Figure 6-51 Strut rod.

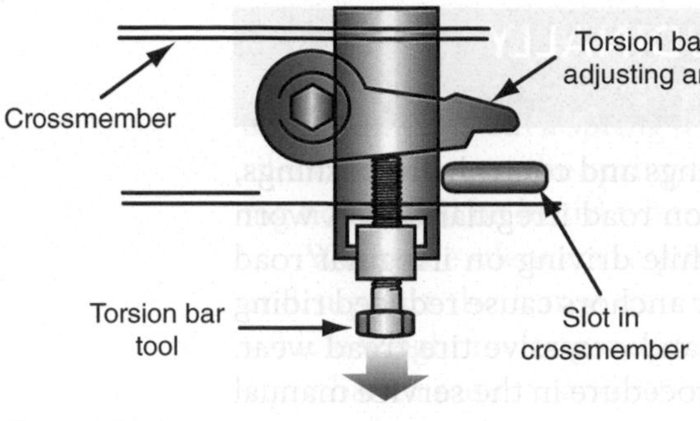

Figure 6-56 Loosening the torsion bar tool.

Figure 6-57 Alignment marks placed on the torsion bar and adjuster.

6. Loosen the torsion bar tool to remove all the tension from the torsion bar (**Figure 6-56**).
7. Use the end of a screwdriver to place matching alignment marks on the torsion bar and the adjuster so these components may be reassembled in the same position. Remove the torsion bar insulator. Pull the torsion bar to the rear to remove this bar from the lower control arm.
8. Position the torsion bar in the lower control arm.
9. Install the torsion bar adjuster on the torsion bar with the alignment marks properly aligned (**Figure 6-57**).
10. Install the torsion bar tool and adapters. Tighten the torsion bar tool until a new adjustment bolt can be installed. Tighten this bolt until the measurement obtained in Step 3 is obtained between the head of the bolt and the casting the bolt is threaded into.
11. Install the torsion bar cover plate, and tighten the plate retaining bolts to the specified torque.
12. Measure the suspension riding height as explained previously in this chapter. If the suspension riding height is not within specifications, adjust the torsion bar adjusting bolt to obtain the specified riding height.

CASE STUDY

A customer complained about steering pull to the left while braking on a 2009 Chrysler 300. The technician questioned the customer about other symptoms, but the customer stated that the steering did not pull while driving, and the car had no other problems. Further questioning of the customer revealed that extensive brake work had been done in an attempt to correct the problem. The front brake pads had been replaced and the front brake rotors had been turned.

While performing a road test, the technician discovered the car had a definite pull to the left while braking, but the steering was normal while driving. During the road test, no other problems were evident. The technician removed the front wheels and brake rotors. A careful examination of the brake linings and rotor surfaces indicated these components were in excellent condition. The technician checked the rotor surfaces to be sure they were both in the same condition. The pistons in both front calipers moved freely. An inspection of the brake lines and hoses did not reveal any visible problems, and a pressure check at each front brake caliper indicated equal pressures at both front wheels during a brake application.

Next, the technician made a visual inspection of the steering and suspension components and discovered a worn, loose left front tension rod bushing. This tension rod was replaced, and the strut rod nut tightened to the specified torque. A road test of the car indicated no steering pull during brake application.

ASE-STYLE REVIEW QUESTIONS

1. While discussing riding height:
   Technician A says worn control arm bushings reduce riding height.
   Technician B says incorrect riding height affects most other front suspension angles.
   Who is correct?
   A. Technician A
   B. Technician B
   C. Both A and B
   D. Neither A nor B

2. The shoulder on the ball joint grease fitting is inside the ball joint cover:
   Technician A says the ball joint should be replaced.
   Technician B says a longer grease fitting should be installed.
   Who is correct?
   A. Technician A
   B. Technician B
   C. Both A and B
   D. Neither A nor B

3. While discussing ball joint unloading on a short-and-long arm suspension system with the coil springs between the lower control arm and the chassis:
   Technician A says a steel spacer should be installed under the upper control arm.
   Technician B says a floor jack should be placed under the lower control arm to unload the ball joints.
   Who is correct?
   A. Technician A
   B. Technician B
   C. Both A and B
   D. Neither A nor B

4. While discussing ball joint radial measurement:
   Technician A says the dial indicator should be positioned against the top of the ball joint stud.
   Technician B says the front wheel bearing play does not affect the ball joint radial measurement if measured at the wheel rim.
   Who is correct?
   A. Technician A
   B. Technician B
   C. Both A and B
   D. Neither A nor B

5. While discussing ball joint installation:
   Technician A says the ball joint nut may be backed off to install the cotter pin.
   Technician B says only the ball joint threads should extend above the opening in the steering knuckle.
   Who is correct?
   A. Technician A
   B. Technician B
   C. Both A and B
   D. Neither A nor B

6. When installing a new threaded ball joint in a lower control arm, the technician can only torque the ball joint to 90 ft.-lb. The necessary repair for this problem is to:
   A. Weld the ball joint into the control arm.
   B. Place Loctite on the ball joint threads.
   C. Replace the lower control arm.
   D. Install a larger-diameter ball joint.

7. All of these defects could result in worn-out rebound bumpers EXCEPT:
   A. Sagged springs.
   B. Worn-out shock absorbers.
   C. Continual driving on rough roads.
   D. Riding height more than specified.

8. When removing and replacing control arm bushings:
   A. If the bushing is contained in a steel sleeve, press on the rubber bushing.
   B. A spacer should be installed between the control arm support lugs during bushing removal and installation.
   C. The same tool is used for bushing removal and replacement.
   D. Bushing flaring is necessary to expand the bushing.

Task Completed

6. Place a pry bar under the front tire and lift straight upward on the pry bar while a coworker observes the dial indicator. ☐

7. Compare the reading on the dial indicator with the vehicle manufacturer's specifications. If the vertical ball joint movement exceeds specifications, replace the ball joint.

   Specified ball joint vertical movement ____________________

   Actual ball joint vertical movement ____________________

   Necessary ball joint service ____________________

8. Be sure the front wheel bearings are properly adjusted. ☐

9. Attach the dial indicator to the lower control arm, and position the dial indicator against the inner edge of the wheel rim. Preload the dial indicator 0.250 in. (6.35 mm) and place the dial in the zero position.

   Is the dial indicator preloaded and placed in the zero position? ☐ Yes ☐ No

   Instructor check ____________________

10. Grasp the tire at the top and bottom, and try to rock the tire inward and outward while a coworker observes the dial indicator. ☐

11. Compare the reading on the dial indicator to the vehicle manufacturer's specifications. If the horizontal ball joint movement exceeds specifications, replace the ball joint.

    Specified ball joint horizontal movement ____________________

    Actual ball joint horizontal movement ____________________

12. Repeat the measurements in Steps 3 through 10 on the opposite side of the front suspension. ☐

13. Based on your ball joint measurements, state all the necessary ball joint service and explain the reasons for your diagnosis.

    ____

    ____

    ____

Instructor's Response

____

____

____

Name ______________________ Date ______________

JOB SHEET 26

IN A SUSPENSION SYSTEM DETERMINE IF THE UPPER OR LOWER SUSPENSION PIVOT POINT IS A PRIMARY LOAD CARRIER OR A FOLLOWER JOINT, AND MEASURE JOINT FOR VERTICAL AND HORIZONTAL MOVEMENT

Upon completion of this job sheet, you should be able to measure ball joint or pivot point wear and determine the necessary joint service.

ASE Education Foundation Task Correlation

This job sheet addresses the following **MLR** task:

B.12. Inspect upper and lower ball joints (with or without wear indicators). **(P-1)**

This job sheet addresses the following **AST/MAST** task:

C.5. Inspect, remove, and/or replace upper and/or lower ball joints (with or without wear indicators). **(P-2)**

We Support ASE Education Foundation

Tools and Materials

Ball joint dial indicator
Floor jack
Safety stands

Describe the vehicle being worked on:

Year ______________ Make ______________ Model ______________

VIN ______________ Engine type and size ______________

Procedure **Task Completed**

1. Determine the load-carrying ball joint or pivot point.

Upper ______________

Lower ______________

Both ______________

Position the floor jack correctly for testing.

Describe: ______________

2. Install safety stands and remove the floor jack. ☐

3. Locate and record the specification for axial play:

Load-carrying: ______________ Follower: ______________

4. Locate and record the specification for lateral play:

Load-carrying: ______________ Follower: ______________

5. Attach a dial indicator for lower ball joint measurement to the lower control arm, and position the dial indicator stem against the lower end of the steering knuckle next to the ball joint retaining nut if the suspension has a lower compression-loaded ball joint. When the suspension has a tension-loaded ball joint, place the indicator stem against the top of the ball joint stud. ☐

6. Preload the dial indicator stem 0.250 in. (6.35 mm), and zero the dial indicator.

Is the dial indicator preloaded and placed in the zero position? ☐ Yes ☐ No

Instructor check ______________

Task Completed

7. Place a pry bar under the front tire and lift straight upward on the pry bar while a coworker observes the dial indicator. ☐

8. Grasp the tire at the top and bottom, and try to rock the tire inward and outward while a coworker observes the dial indicator. ☐

9. Repeat the measurements in Steps 4 through 9 for all four pivot points of the front suspension. Record your results of the inspect both load-carrying and follower joints in the following table:

Ball Joint	Left	Pass/Fail	Right	Pass/Fail
Load-Carrying Axial Play				
Load-Carrying Lateral Play				
Follower Axial Play				
Follower Lateral Play				

10. Compare the reading on the dial indicator with the vehicle manufacturer's specifications. If the vertical ball joint movement exceeds specifications, replace the ball joint. ☐

11. Based on your ball joint measurements, state all the necessary ball joint service and explain the reasons for your diagnosis.

____

____

____

Instructor's Response

____

____

____

Name ______________________ Date ______________

JOB SHEET 27

BALL JOINT REPLACEMENT

Upon completion of this job sheet, you should be able to remove and replace ball joints.

ASE Education Foundation Task Correlation

This job sheet addresses the following **MLR** task:

B.12. Inspect upper and lower ball joints (with or without wear indicators). **(P-1)**

This job sheet addresses the following **AST/MAST** task:

C.5. Inspect, remove, and/or replace upper and/or lower ball joints (with or without wear indicators). **(P-2)**

We Support ASE Education Foundation

Tools and Materials

Floor jack
Ball joint loosening tool
Safety stands
Torque wrench

Describe the vehicle being worked on:

Year ______________ Make ______________ Model ______________

VIN ______________ Engine type and size ______________

Procedure **Task Completed**

1. Remove the wheel cover and loosen the wheel nuts. ☐
2. Lift the vehicle with a floor jack and place safety stands under the chassis so the front suspension is allowed to drop downward. Lower the vehicle onto the safety stands and remove the floor jack. ☐
3. Remove the wheel and place a floor jack under the outer end of the lower control arm. Operate the floor jack and raise the lower control arm until the ball joints are unloaded.

   Are the ball joints properly unloaded? ☐ Yes ☐ No

   Instructor check ______________
4. Remove other components, such as the brake caliper, rotor, and drive axle, as required to gain access to the ball joints. ☐
5. Remove the cotter pin in the ball joint or joints that require replacement, and loosen, but do not remove, the ball joint stud nuts.

   Is the ball joint stud nut loosened? ☐ Yes ☐ No

   Instructor check ______________
6. Loosen the ball joint stud tapers in the steering knuckle. A threaded expansion tool is available for this purpose.

   Are the ball joint stud tapers loosened? ☐ Yes ☐ No

   Instructor check ______________
7. Remove the ball joint nut and lift the knuckle off the ball joint stud. Block or tie up the knuckle and hub assembly to access the ball joint. ☐

Task Completed

7. If an eccentric cam is used on one of the strut-to-knuckle bolts, mark the cam and bolt position in relation to the strut and remove the strut-to-knuckle bolts.

   Is the eccentric camber marked in relation to the strut? ☐ Yes ☐ No

   Instructor check ______________________

8. Remove the knuckle from the strut and lift the knuckle out of the chassis. ☐

9. Pry the dust deflector from the steering knuckle with a large flat-blade screwdriver. ☐

10. Use a puller to remove the ball joint from the steering knuckle. Check the ball joint and tie rod end openings in the knuckle for wear and out-of-round. Replace the knuckle if these openings are worn or out-of-round.

    Condition of ball joint opening in the knuckle:

    ☐ Satisfactory ☐ Unsatisfactory

    Condition of tie rod end opening in the knuckle:

    ☐ Satisfactory ☐ Unsatisfactory

    State the necessary steering knuckle and related repairs, and explain the reason for your diagnosis.

    ____

    ____

11. Use the proper driving tool to reinstall the dust deflector in the steering knuckle. ☐

Instructor's Response

____

____

____

Name ______________________________ Date ______________

JOB SHEET 29

CONTROL ARM AND SPRING REPLACEMENT ON SLA SUSPENSION SYSTEM

Upon completion of this job sheet, you should be able to inspect, remove, and/or replace coil spring, sway bar links, and control arms and bushings from the front of an SLA suspension system.

ASE Education Foundation Task Correlation

This job sheet addresses the following **MLR** tasks:

B.9. Inspect upper and lower control arms, bushings, and shafts. **(P-1)**

B.10. Inspect and replace rebound bumpers. **(P-1)**

B.13. Inspect suspension system coil springs and spring insulators (silencers). **(P-1)**

B.15. Inspect and/or replace front/rear stabilizer bar (sway bar) bushings, brackets, and links. **(P-1)**

This job sheet addresses the following **AST/MAST** tasks:

C.3. Inspect, remove, and/or replace upper and lower control arms, bushings, shafts, and rebound bumpers. **(P-3)**

C.7. Inspect, remove, and/or replace short-and-long arm suspension system coil springs and spring insulators. **(P-3)**

C.9. Inspect, remove, and/or replace front/rear stabilizer bar (sway bar) bushings, brackets, and links. **(P-3)**

We Support ASE Education Foundation

Tools and Materials

Hand tool

In-vehicle coil spring compression tool

Ball joint/bushing press

Torque wrench

Describe the vehicle being worked on:

Year ____________ Make ____________ Model ____________

VIN ____________ Engine type and size ____________

Procedure **Task Completed**

1. Raise vehicle and remove wheel assembly. ☐
2. Disconnect shock absorber. ☐
3. Disconnect the stabilizer bar from the lower control arm if equipped. ☐

WARNING During control arm and spring replacement, the coil spring tension must be supported by a spring compressing tool or floor jack and jack stand. Serious personal injury or property damage may occur if the spring tension is released suddenly. Always follow vehicle manufacture's recommended service procedures.

4. Following manufacturer's service procedures install a spring compressor and turn spring compressor bolts until spring tension is released from control arm. Make sure the compressing tool supports all spring tension. ☐

Task Completed

5. Place a screw jack or other jacking device under the lower control arm. ☐
6. Remove the lower ball joint cotter pin and nut and separate the ball joint stud from steering knuckle. ☐
7. Lower the floor jack very slowly to lower the control arm and coil spring. ☐
8. Disconnect the lower control arm inner mounting bolts and remove the lower control arm. (First mark the location of spring index and insulator.) ☐
9. Remove the coil spring and compressing tool. ☐
10. Remove and inspect upper and lower control arms, bushings, shafts, and rebound bumpers. ☐
11. If control arm bushings, shafts, and rebound bumpers require replacement follow manufacturer's service procedures. ☐
12. Reverse Steps 1 through 10 to install the lower control arm. Be sure the coil spring and insulators are properly seated in the lower control arm and in the upper spring seat. ☐
13. With vehicle weight supported by jack stands on lower control arms tighten and torque lower and upper control arm to frame mounting bolts.
    a. Torque specifications: ______________________________

Instructor's Response

____

____

____

Name ______________________________ Date ____________

JOB SHEET
30

VEHICLE RIDE HEIGHT MEASUREMENT

Upon completion of this job sheet, you should be able to inspect vehicle ride height and determine necessary action.

ASE Education Foundation Task Correlation

This job sheet addresses the following **MLR** task:

C.1. Perform prealignment inspection; measure vehicle ride height. **(P-1)**

This job sheet addresses the following **AST/MAST** tasks:

C.1. Diagnose short-and-long arm suspension system noises, body sway, and uneven ride height concerns; determine needed action. **(P-1)**

E.2. Perform prealignment inspection; measure vehicle ride height; determine needed action. **(P-1)**

We Support
ASE Education Foundation

Tools and Materials

Hand tool

Tape measure

Describe the vehicle being worked on:

Year ________________ Make ________________ Model ________________

VIN ________________ Engine type and size ________________

Procedure **Task Completed**

1. Adjust tire pressure to conform to setting specified by the manufacturer's.
   a. Tire pressure specifications: Front ______________ Rear ______________
2. Be sure the rear cargo compartment is empty of non-manufacturer equipped items. ☐
3. Check fuel level. If fuel tank is not full, add 60 lb. (27 kg) for every 10 gallons of gasoline, to simulate a full tank. ☐
4. Measure ride height on a flat and level surface. ☐
5. To measure proceed as follows.
   a. Place hand on bumper and bounce (jounce) the vehicle a few times, both front and rear.
   b. Push vehicle rear bumper down approximately 1.5 in. and release very slowly.
   c. Measure both left and right vehicle ride height at rear according to manufacturer location. Record results in table below.
   d. Repeat the jouncing procedure.
   e. Lift vehicle rear bumper down approximately 1.5 in. the release very slowly.
   f. Measure ride height at rear for a second time. Record results in table on the next page.
   g. The true height is the average of the two readings taken in Steps c and f above.
   h. Repeat Steps a through g on the front of the vehicle.

Name ______________________ Date ______________

JOB SHEET 32

STRUT ROD AND BUSHING REPLACEMENT

Upon completion of this job sheet, you should be able to inspect, remove, and/or replace strut rods and bushings.

ASE Education Foundation Task Correlation

This job sheet addresses the following AST/MAST task:

C.4. Inspect, remove, and/or replace strut rods and bushings. **(P-3)**

We Support ASE Education Foundation

Tools and Materials

Hand tool

Torque wrench

Describe the vehicle being worked on:

Year ________ Make ______________ Model ______________

VIN ______________ Engine type and size ______________

Procedure **Task Completed**

1. Lift the vehicle chassis on a hoist and allow the front suspension to drop downward. ☐
2. Loosen and remove the strut rod nut from the end of the strut rod. ☐
3. Loosen and remove the strut rod bolt(s) from the lower control arm. ☐
4. Remove the strut rod and bushings from the subframe mount. ☐
5. Perform a visual inspection of the strut rod, bushings. ☐
6. Replace worn or faulty components. A ball joint/bushing press may be required. ☐
7. Reverse Steps 1 through 5 to install the strut rod and bushing assembly. ☐
8. Torque strut rod mounting bolts to lower control arm to manufacturer's specification.

   Torque specification: ______________
9. Torque strut rod nut to manufacturer's specification.

   Torque specification: ______________
10. Lower vehicle. ☐
11. If the strut rod is adjustable perform an alignment to set caster angle to manufacturer specifications.

    Is an alignment required: ☐ Yes ☐ No

Instructor's Response

____

____

____

Name ______________________ Date ______________

JOB SHEET 33

STABILIZER BAR, LINKS, AND BUSHING REPLACEMENT

Upon completion of this job sheet, you should be able to inspect, remove, and/or replace front/rear stabilizer bar (sway bar) bushings, brackets, and links.

ASE Education Foundation Task Correlation

This job sheet addresses the following **MLR** task:

B.15. Inspect and/or replace front/rear stabilizer bar (sway bar) bushings, brackets, and links. **(P-1)**

This job sheet addresses the following **AST/MAST** tasks:

C.1. Diagnose short-and-long arm suspension system noises, body sway, and uneven ride height concerns; determine needed action. **(P-1)**

C.2. Diagnose strut suspension system noises, body sway, and uneven ride height concerns; determine needed action. **(P-1)**

C.9. Inspect, remove, and/or replace front/rear stabilizer bar (sway bar) bushings, brackets, and links. **(P-3)**

We Support ASE | Education Foundation

Tools and Materials

Hand tool

Torque wrench

Describe the vehicle being worked on:

Year ______________ Make ______________ Model ______________

VIN ______________ Engine type and size ______________

Procedure **Task Completed**

1. Test drive the vehicle over rough bumpy road and note any unusual noised or suspension performance issues.

   Were any suspension noises noted? ______________________

   Was excessive body sway noted? ______________________

2. Lift the vehicle chassis on a hoist and allow the front suspension to drop downward. ☐
3. Remove the link(s) at the outer ends of the stabilizer bar. This may include bushings, grommets, brackets, or spacers. ☐
4. Remove the mounting bolts and/or nuts in the center area of the stabilizer bar that retain the insulator bushings. ☐
5. Remove the stabilizer bar from the chassis. ☐
6. Visually inspect all stabilizer bar components. ☐
7. Measure stabilizer bar diameter. This information is required when ordering replacement bushings.

   Stabilizer bar diameter: ______________________

Task Completed

8. Replace the stabilizer bar, insulator bushings, and links as required. ☐

9. Reinstall stabilizer bar, insulator bushings, and links. ☐

10. Torque stabilizer bar bushing mount bolts to manufacturer's specification.

Torque specification: ______________________

11. Torque stabilizer link to manufacturer's specification.

Torque specification: ______________________

12. Lower vehicle. ☐

Instructor's Response

____

____

____

CHAPTER 7
REAR SUSPENSION SERVICE

Upon completion and review of this chapter, you should be able to:

- Diagnose rear suspension noises.
- Diagnose rear suspension sway and lateral movement.
- Measure and correct rear suspension riding height.
- Remove and replace rear coil springs.
- Inspect rear springs, insulators, and seats.
- Inspect strut or shock absorber bushings and upper mount, and replace strut cartridge.
- Remove, inspect, and replace lower control arms.
- Inspect, remove, and replace rear ball joints.
- Inspect, remove, and replace suspension adjustment links.
- Diagnose, remove, and replace rear leaf springs.
- Diagnose, remove, and replace stabilizer bars.
- Diagnose, remove, and replace track bars.
- Inspect, remove, and replace tie rods.

Basic Tools

Basic technician's tool set
Service manual
Machinist's rule
Floor jack
Safety stands
Transmission jack
Foot-pound torque wrench
Pry bar

Terms To Know

Body sway
Lateral chassis movement
Spring silencers
Suspension adjustment link
Wear indicators

Rear suspension system diagnosis and service is extremely important to maintain vehicle safety, ride quality, and tire life. Many rear suspension components may create a safety hazard if they are not serviced properly. For example, if wheel nuts or spindle nuts are not tightened properly, a wheel assembly may come off the vehicle, with disastrous results. Improper rear suspension riding height, worn struts or shock absorbers, and sagged or broken springs will result in harsh ride quality. Rear wheel toe and camber must be adjusted to specifications to provide normal tire tread life.

Diagnosis of Ride Harshness

When diagnosing a ride harshness condition, check these measurements and components:

1. Riding height—insufficient riding height causes ride harshness
2. Excessive vehicle load
3. Worn shock absorbers or struts
4. Broken or weak springs
5. Worn suspension bushings, such as strut mounts, control arm bushings, and shock absorber mounting bushings
6. Wheel alignment—excessive positive caster on front wheels causes ride harshness

Noise Diagnosis

A squeaking noise in the rear suspension may be caused by a suspension bushing or a defective strut or shock absorber.

If a rattling noise occurs in the rear suspension, check these components:

1. Worn or missing suspension bushings, such as control arm bushings, track bar bushings, stabilizer bar bushings, trailing arm bushings, and strut rod bushings
2. Worn strut or shock absorber bushings or mounts
3. Defective struts or shock absorbers
4. Broken springs or worn spring insulators
5. Broken exhaust system hangers or improperly positioned exhaust system components

Special tools can be used to locate noises at any location on the vehicle. A Chassis Ear tool has six clamp-on sensors that are miniature microphones (**Figure 7-1**). These six sensors are connected to a switchbox that is also connected to a headset worn by the technician. The technician can rotate the switch on the switchbox to connect any sensor to the headset. The sensors can be mounted on various suspected noise sources. When the control on the switchbox is rotated, the technician can determine which sensor is connected to the noise source.

An electronic stethoscope can be used to locate the source of a noise. This tool has one pickup that is placed on or near a suspected noise source. The pickup is connected to an amplifier box, and the headset is also connected to this box (**Figure 7-2**). When the pickup is placed on the source of a noise, the noise is amplified in the headset.

Body sway is leaning of the chassis to one side.

Classroom Manual
Chapter 7, page 185

Sway and Lateral Movement Diagnosis

Excessive **body sway**, or roll, on road irregularities may be caused by a weak stabilizer bar or loose stabilizer bar bushings. If lateral movement is experienced on the rear of the chassis, the track bar or track bar bushings may be defective. Lateral movement is sideways movement.

SERVICE TIP Proper riding height must be maintained to provide normal steering quality and tire wear.

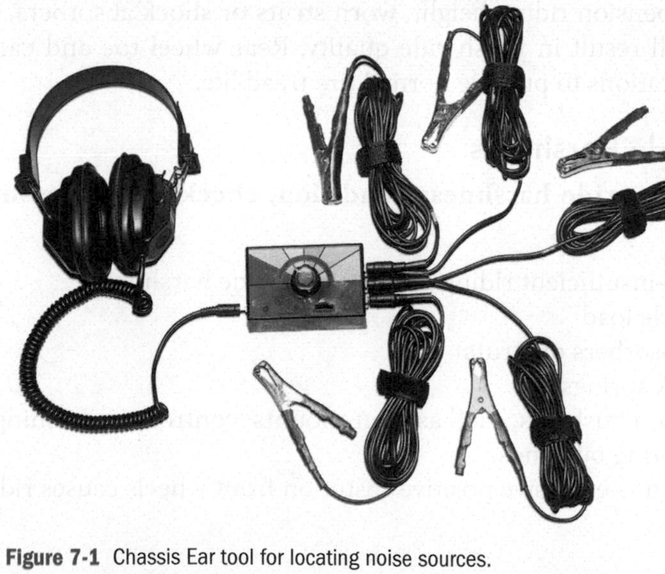

Figure 7-1 Chassis Ear tool for locating noise sources.

Figure 7-2 Electronic stethoscope.

Rear

Figure 7-3 Curb riding height measurement, rear suspension.

Riding Height Measurement

Regular inspection and proper maintenance of suspension systems are extremely important to maintain vehicle safety. The riding height is determined mainly by spring condition. Other suspension components, such as control arm bushings, affect riding height if they are worn. *Since incorrect riding height affects most of the other suspension angles, this measurement is critical.* Reduced rear suspension height increases steering effort and causes rapid steering wheel return after turning a corner. Harsh riding occurs when the riding height is less than specified. The curb riding height must be measured at the vehicle manufacturer's specified location, which varies depending on the type of vehicle and suspension system. On some vehicles, the riding height on the rear suspension is measured from the floor to the center of the strut rod mounting bolt when the vehicle is on a level floor or an alignment rack (**Figure 7-3**). **Photo Sequence 12** shows a typical procedure for measuring front and rear curb riding height.

PHOTO SEQUENCE 12

Typical Procedure for Measuring Front and Rear Riding Height

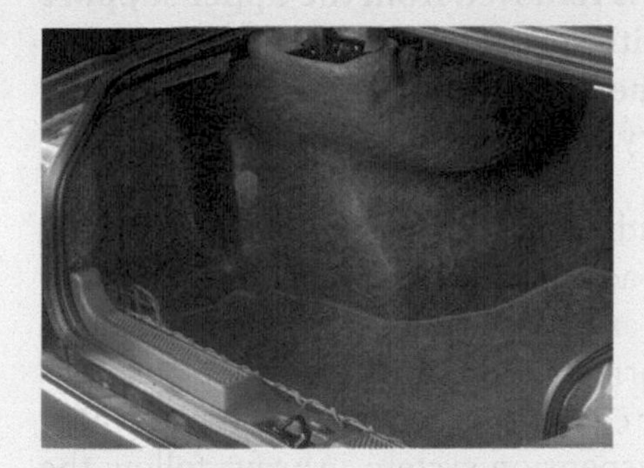

P12-1 Check the trunk for extra weight.

P12-2 Check the tires for normal inflation pressure.

P12-3 Park the car on a level shop floor or an alignment rack.

PHOTO SEQUENCE 12 (CONTINUED)

P12-4 Find the vehicle manufacturer's specified riding height measurement locations in the service manual.

P12-5 Measure and record the right front riding height.

P12-6 Measure and record the left front riding height.

P12-7 Measure and record the right rear riding height.

P12-8 Measure and record the left rear riding height.

P12-9 Compare the measurement results to the specified riding height in the service information.

Rear Strut, Coil Spring, and Upper Mount Diagnosis and Service

WARNING **When the rear coil springs are mounted on the struts, never loosen the upper strut nut until the entire spring tension is removed from the upper support with a spring compressor. If this nut is removed with spring tension on the upper support, the spring tension turns the spring and the upper support into a dangerous projectile, which may cause serious personal injury and/or property damage.**

Weak springs cause harsh riding and reduced curb riding height. Broken springs or spring insulators cause a rattling noise while driving on road irregularities. Worn-out struts or shock absorbers cause excessive chassis oscillations and harsh riding. (Refer to Chapter 5 for strut and shock absorber service.) Loose or worn strut or shock absorber bushings cause a rattling noise on road irregularities. The rear coil spring removal and replacement procedure varies depending on the type of rear suspension system. Always follow the vehicle manufacturer's recommended procedure in the service manual.

SERVICE TIP If the plastic coating on a coil spring is chipped, the spring may break prematurely. The spring may be taped in the compressing tool contact areas to prevent chipping.

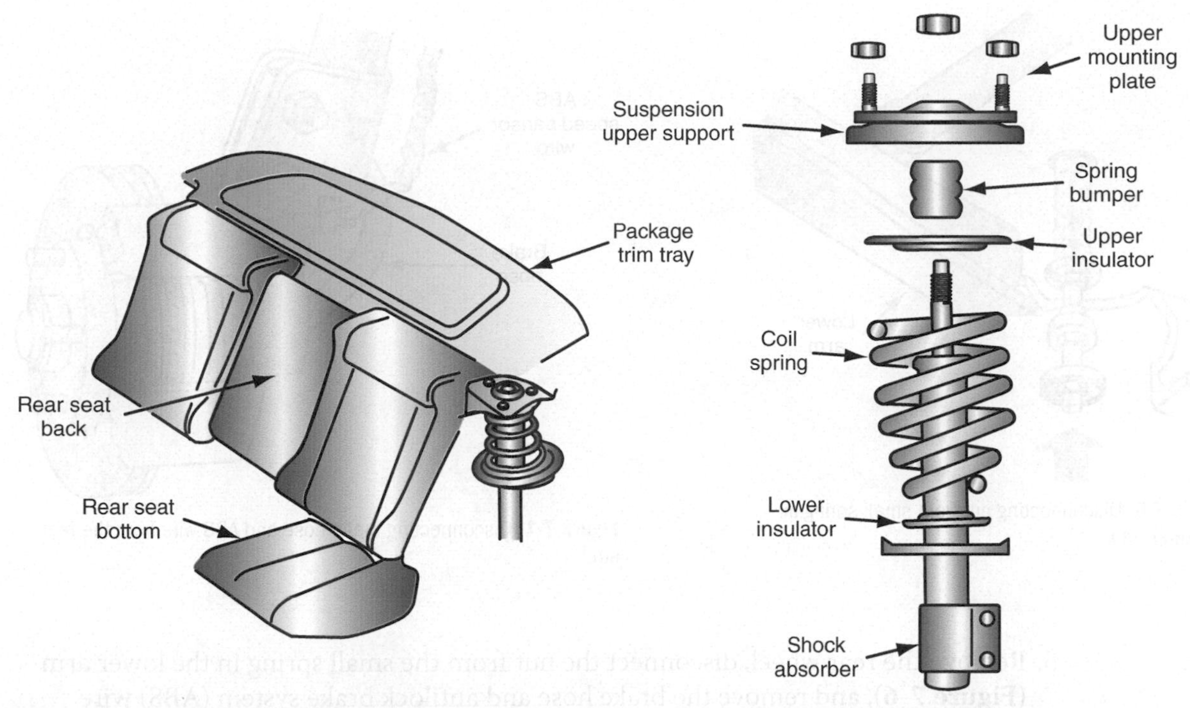

Figure 7-4 Rear suspension with rear seat and package trim tray.

The following is a typical rear strut and spring removal and replacement procedure on a MacPherson strut independent rear suspension system with the coil springs mounted on the struts:

1. Remove the rear seat and the package trim tray (**Figure 7-4**).
2. Remove the wheel cover, and loosen the wheel nuts.
3. Lift the vehicle with a floor jack, and lower the chassis onto safety stands so the rear suspension is allowed to drop downward.
4. Place a floor jack lift pad under the rear spindle on the side where the strut and spring removal is taking place. Raise the floor jack to support some of the suspension system weight (**Figure 7-5**).

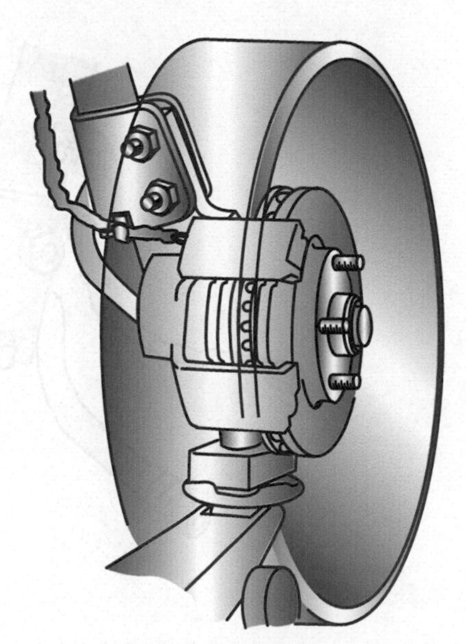

Figure 7-5 Floor jack supporting some of the rear suspension weight.

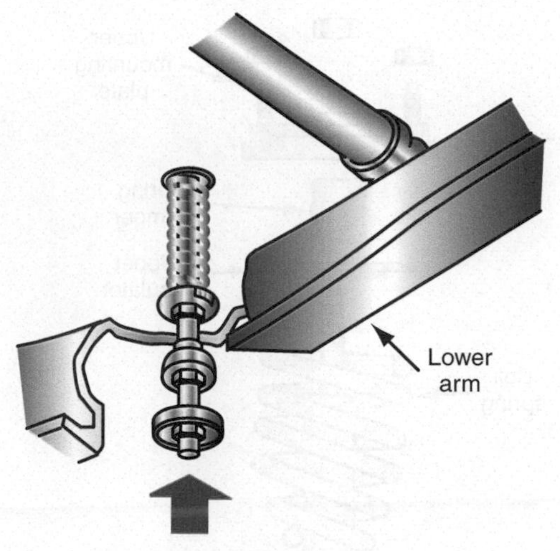

Figure 7-6 Disconnecting nut from small spring in the lower arm.

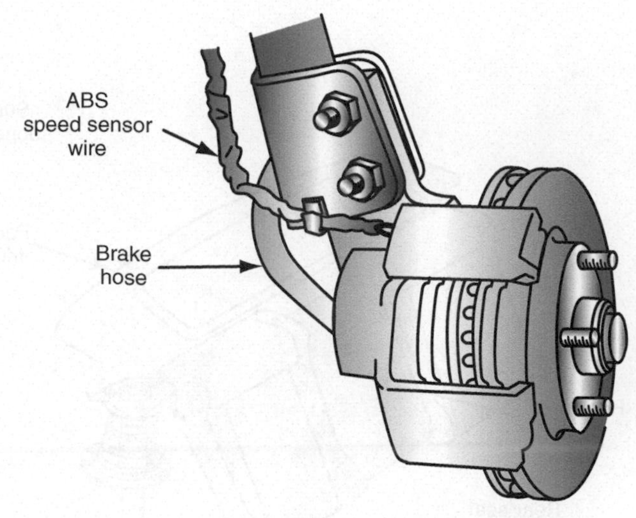

Figure 7-7 Disconnecting brake hose and ABS wire from the rear strut.

5. Remove the rear wheel, disconnect the nut from the small spring in the lower arm (**Figure 7-6**), and remove the brake hose and antilock brake system (ABS) wire from the strut (**Figure 7-7**).
6. Remove the stabilizer bar link from the strut (**Figure 7-8**), and loosen the strut-to-spindle mounting bolts (**Figure 7-9**).
7. Remove the three upper support nuts under the package tray trim (**Figure 7-10**), and lower the floor jack to remove the strut from the knuckle (**Figure 7-11**). Remove the strut from the chassis.
8. Following the vehicle or equipment manufacturer's recommended procedure, install a spring compressing tool on the coil spring, and tighten the tool until the entire spring tension is removed from the upper support (**Figure 7-12**).
9. Operate the compressing tool to remove the entire spring tension from the upper strut mount and then remove the strut rod nut (**Figure 7-13**).
10. Remove the strut rod nut, upper support, and upper insulator.

Special Tool

Coil spring compressor

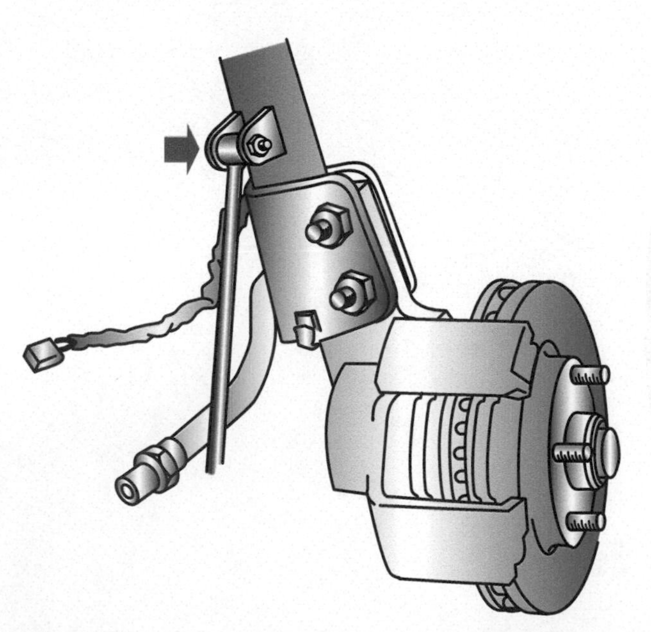

Figure 7-8 Removing stabilizer bar from the rear strut.

Figure 7-9 Loosening strut-to-rear spindle bolts.

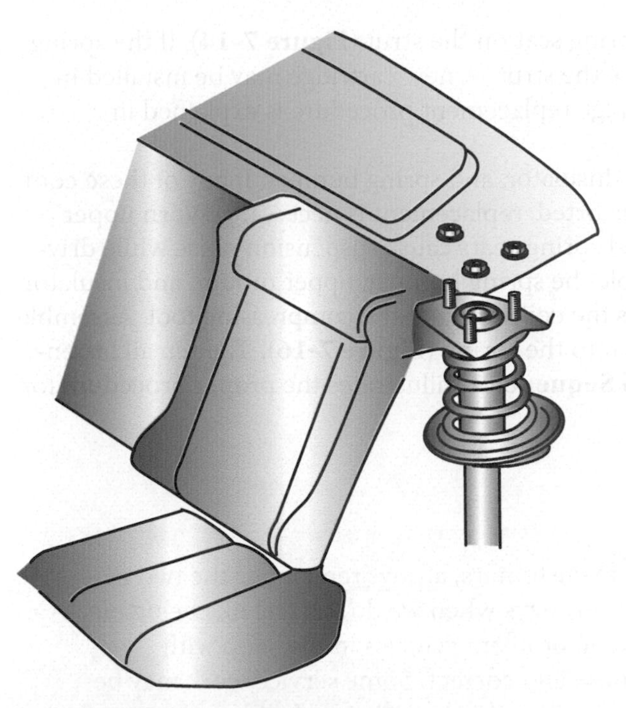
Figure 7-10 Removing upper mount nuts.

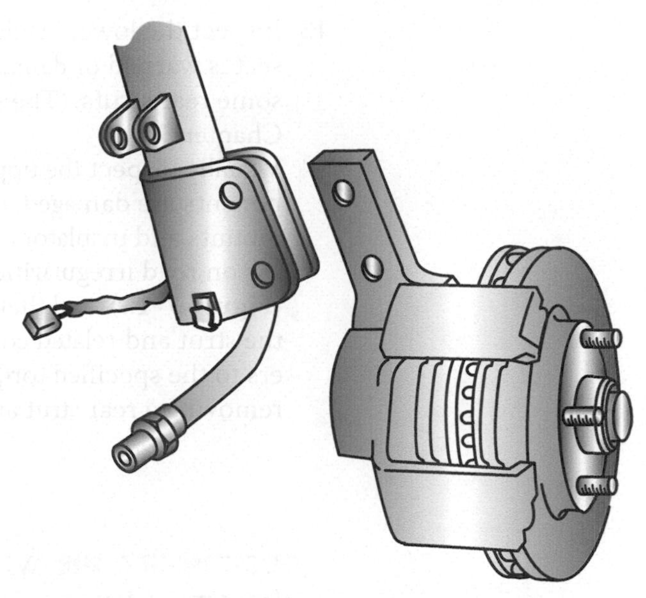
Figure 7-11 Removing the strut from the rear spindle.

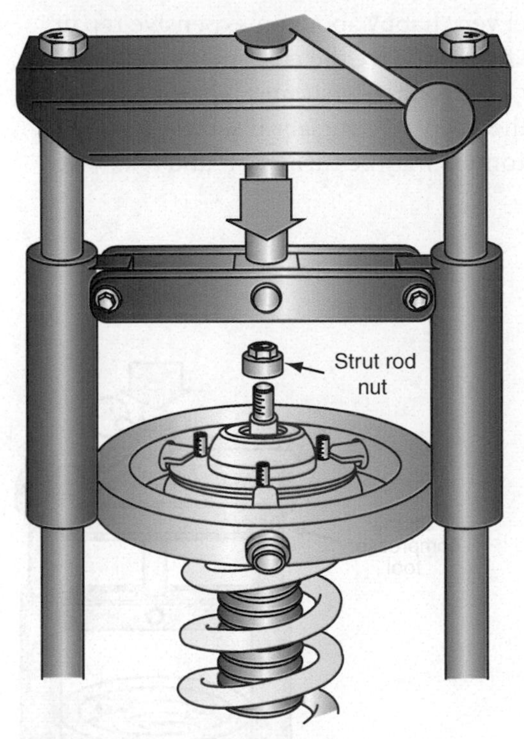

Figure 7-12 A spring compressing tool is used to compress the coil spring.

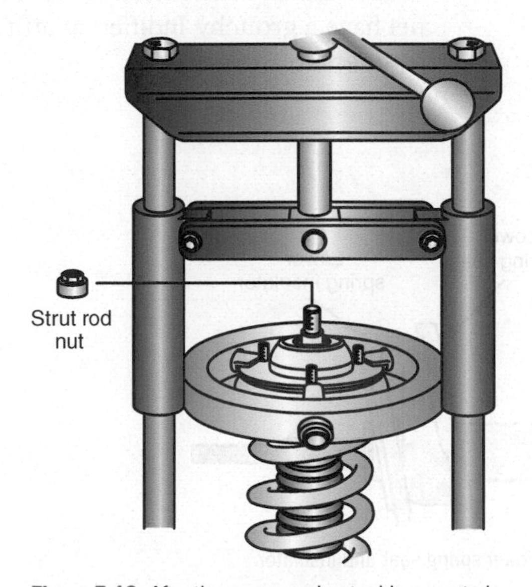

Figure 7-13 After the compressing tool is operated to remove the entire spring tension, remove the strut rod nut.

11. Remove the strut from the lower end of the spring.
12. If the spring is to be replaced, rotate the compressing tool handle until the entire spring tension is removed from the compressing tool, and then remove the spring from the tool.

Classroom Manual
Chapter 7, page 191

13. Inspect the lower insulator and spring seat on the strut (**Figure 7-14**). If the spring seat is warped or damaged, replace the strut. A new cartridge may be installed in some rear struts. (The strut cartridge replacement procedure is explained in Chapter 5.)
14. Visually inspect the upper mount, insulator, and spring bumper. If any of these components are damaged, worn, or distorted, replacement is necessary. Worn upper mounts and insulators or damaged spring seats cause suspension noise while driving on road irregularities. Assemble the spring bumper, upper mount, and insulator (**Figure 7-15**), and then compress the coil spring in the compressing tool. Assemble the strut and related components into the spring (**Figure 7-16**). Tighten all fasteners to the specified torque. **Photo Sequence 13** illustrates the proper procedure for removing a rear strut and spring.

CUSTOMER CARE When talking to customers, always remember the two *Ps*—*pleasant* and *polite.* There may be many days when we do not feel like being pleasant and polite. Perhaps we have several problem vehicles in the shop with symptoms that are difficult to diagnose and correct. Some service work may be behind schedule, and customers may be irate because their vehicles are not ready on time. However, we should always remain pleasant and polite with customers. Our attitude does much to make the customer feel better and realize their business is appreciated. A customer may not feel very happy about an expensive repair bill, but a pleasant attitude on our part may help improve the customer's feelings. When the two *Ps* are remembered by service personnel, customer relations are enhanced, and the customer will return to the shop. Conversely, if service personnel have a grouchy, indifferent attitude, customers may be turned off and take their business to another shop.

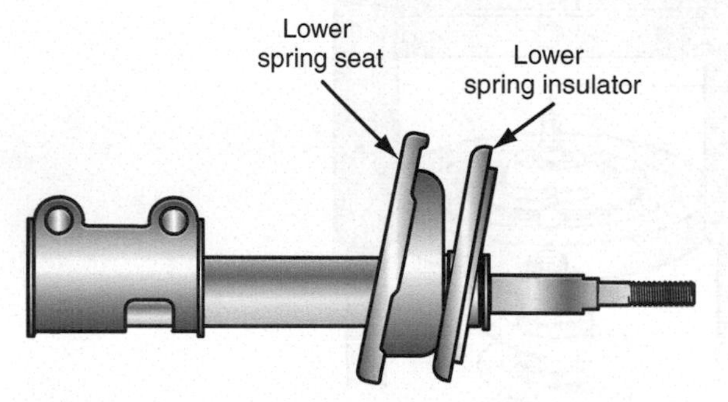

Figure 7-14 Inspecting lower spring seat and insulator.

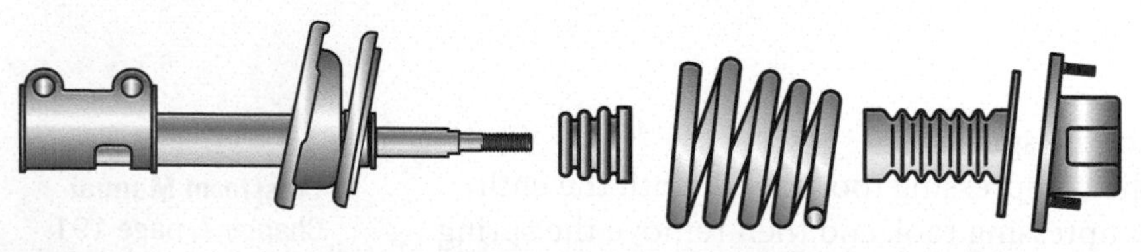

Figure 7-15 Assembly of rear strut, spring bumper, spring, upper mount, and insulator.

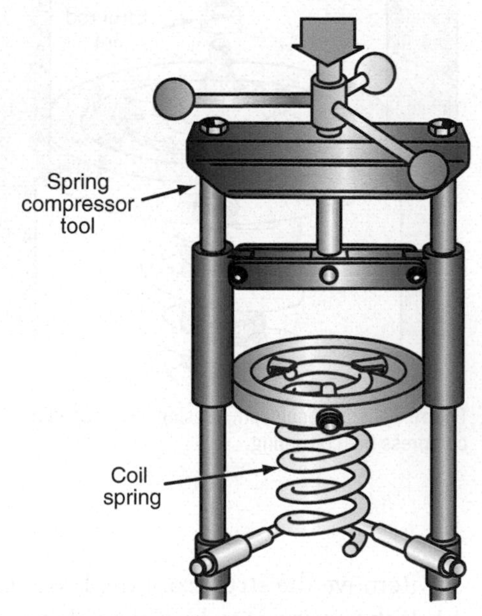

Figure 7-16 Compressing the coil spring in the compressing tool prior to strut-and-spring assembly.

PHOTO SEQUENCE 13

Typical Procedure for Removing a Rear Strut-and-Spring Assembly on a Front-Wheel-Drive Car

P13-1 Remove the left rear wheel cover, and loosen the wheel nuts on the left rear wheel.

P13-2 Raise the vehicle on a frame-contact lift and chalk-mark one of the left rear wheel studs in relation to the wheel, and remove the wheel nuts followed by the wheel-and-tire assembly.

P13-3 Remove the left rear brake caliper, and use a piece of wire to suspend the caliper from a chassis member. Do not allow the caliper to hang from the flexible brake hose.

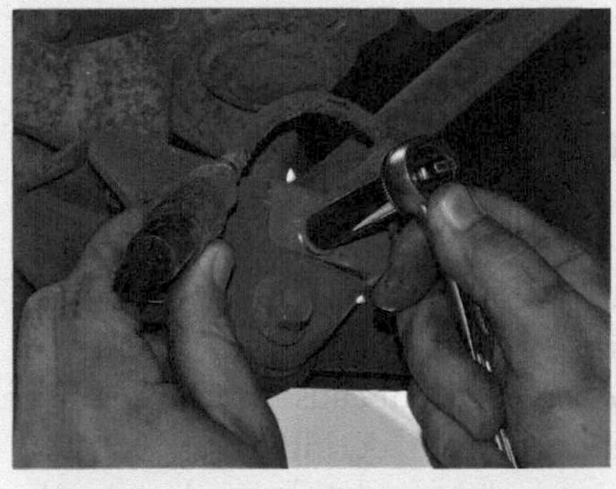

P13-4 If the vehicle is equipped with ABS, remove the left rear speed sensor wire, routing tube, and bracket from the trailing arm.

P13-5 Remove both forward and rearward lateral lower link-to-spindle attaching bolts by installing a thin open-end wrench on the hex of the attaching link stud to prevent the stud from turning and then remove the lateral link-to-spindle attaching bolts.

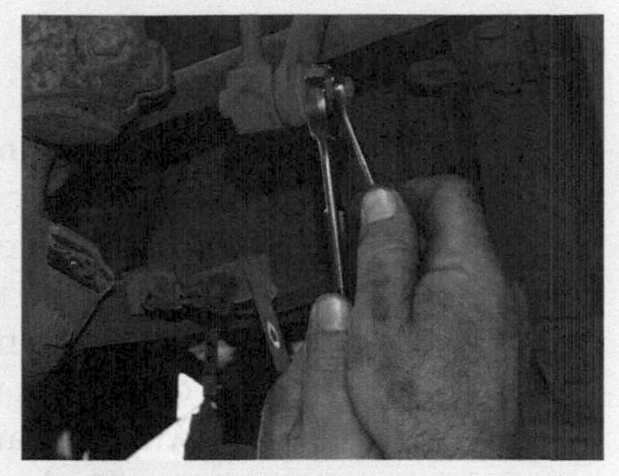

P13-6 Use an open-end wrench to hold the left rear stabilizer bar attaching link stud from turning, and remove the stabilizer bar attaching link nut. Disconnect the stabilizer bar from the attaching link stud.

P13-7 Remove the nut from the strut-to-spindle pinch bolt, and remove the pinch bolt.

P13-8 Tap a center punch into the slot in the lower end of the strut to spread the strut opening slightly. Be sure the inner end of the punch does not contact and puncture the strut.

P13-9 Use a hammer to tap the top of the spindle and drive it downward off the end of the strut. Allow the left rear spindle and assembled components to hang from the trailing arm.

PHOTO SEQUENCE 13 (CONTINUED)

P13-10 Lower the vehicle on the lift so the disconnected components on the left rear suspension and the other tires are a short distance off the shop floor. Open the trunk and remove the dust cap on top of the left rear strut opening. Loosen the left rear strut mounting nuts.

P13-11 Have a coworker hold the left rear strut-and-spring assembly and remove the strut mounting nuts. Remove the left rear strut from the vehicle.

Lower Control Arm and Ball Joint Diagnosis and Replacement Worn bushings on the lower control arms may cause incorrect rear wheel camber or toe, which results in rear tire wear and steering pull. Bent lower control arms must be replaced. When ball joints with **wear indicators** are in normal condition, there is 0.050 in. (1.27 mm) between the grease nipple shoulder and the cover. If the ball joint is worn, the grease nipple shoulder is flush with or inside the cover (**Figure 7-17**). A worn ball joint causes improper rear wheel toe and/or camber, which may result in tire tread wear or steering pull.

Wear indicators show ball joint wear by the position of the grease fitting in the ball joint.

The lower control arm removal and replacement procedure varies depending on the vehicle and the type of suspension. Always follow the vehicle manufacturer's recommended procedure in the service manual.

The following is a typical lower control arm removal procedure:

Classroom Manual
Chapter 7, page 193

1. Lift the vehicle on a hoist with the chassis supported on the hoist and the control arms dropped downward. The vehicle may be lifted with a floor jack, and the chassis supported on safety stands.
2. Remove the tire-and-wheel assembly.

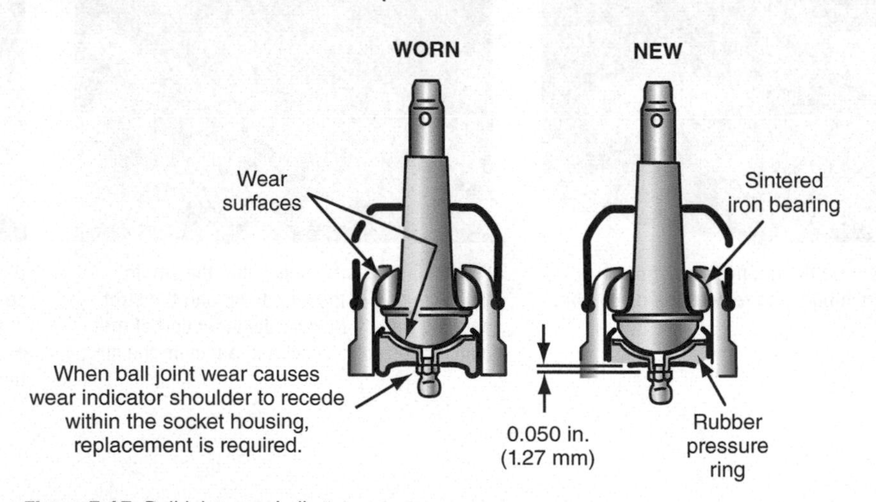

Figure 7-17 Ball joint wear indicator.

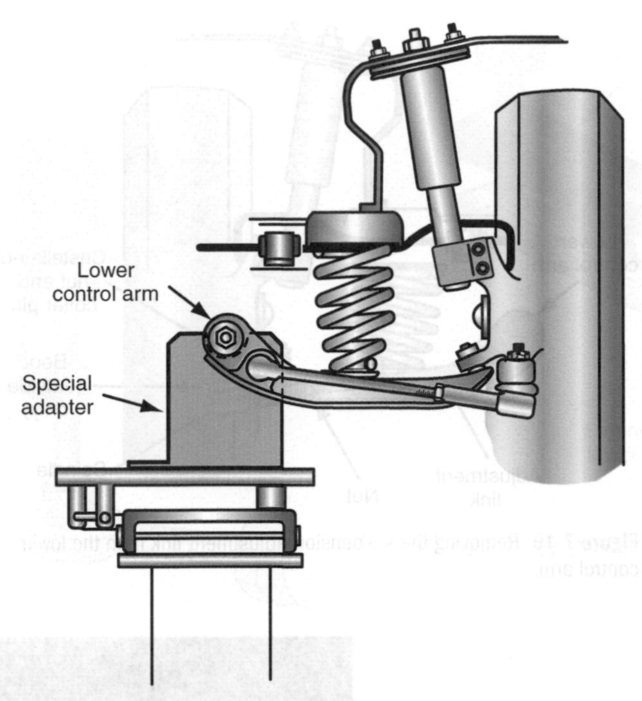

Figure 7-18 Special tool installed to support the inner end of the lower control arm.

3. Remove the stabilizer bar from the knuckle bracket.
4. Remove the parking brake cable retaining clip from the lower control arm.
5. If the car has electronic level control (ELC), disconnect the height sensor link from the control arm.
6. Install a special tool to support the lower control arm in the bushing areas (**Figure 7-18**).
7. Place a transmission jack under the special tool and raise the jack enough to remove the tension from the control arm bushing retaining bolts. If the car was lifted with a floor jack and supported on safety stands, place a floor jack under the special tool.
8. Place a safety chain through the coil spring and around the lower control arm.
9. Remove the bolt from the rear control arm bushing.
10. Be sure the jack is raised enough to relieve the tension on the front bolt in the lower control arm and remove this bolt.
11. Lower the jack slowly and allow the control arm to pivot downward. When all the tension is released from the coil spring, remove the safety chain, coil spring, and insulators.

Check the coil spring for distortion and proper free length. If the spring has a vinyl coating, check this coating for scratches and nicks. Check the spring insulators for cracks and wear.

After the coil spring is removed, follow these steps to remove the lower control arm:

1. Remove the nut on the inner end of the **suspension adjustment link**, and disconnect this link from the lower control arm (**Figure 7-19**).
2. Remove the cotter pin from the ball joint nut, and loosen, but do not remove, the nut from the ball joint stud.
3. Use a special ball joint removal tool to loosen the ball joint in the knuckle.
4. Remove the ball joint nut and the lower control arm (**Figure 7-20**).

Lower control arm support tool

A **suspension adjustment link** may be connected from the rear knuckle to the lower control arm to adjust rear wheel toe.

7. Install the stabilizer-bar-to-knuckle bracket and tighten the fasteners to the specified torque.
8. Install the parking brake retaining clip.
9. If the vehicle has ELC, install the height sensor link, and tighten the fastener to the specified torque.
10. Install the suspension adjustment link and tighten the retaining nuts to the specified torque. Install cotter pins as required.
11. Remove the transmission jack and install the tire-and-wheel assembly.
12. Lower the vehicle onto the floor and tighten the wheel hub nuts and lower control arm bolts and nuts to the specified torque.

REAR LEAF-SPRING DIAGNOSIS AND REPLACEMENT

Spring silencers are plastic spacers mounted between the spring leaves to reduce spring noise.

This leaf-spring discussion applies to multiple-leaf springs on rear suspension systems that have two springs mounted longitudinally in relation to the chassis. Many leaf springs have plastic **spring silencers** between the spring leaves. If these silencers are worn out, creaking and squawking noises are heard when the vehicle is driven over road irregularities at low speeds.

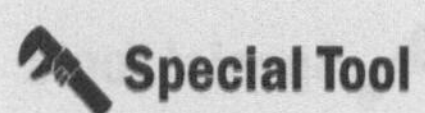

Special Tool

Suspension adjustment link removal tool

When the silencers require checking or replacement, lift the vehicle with a floor jack and support the frame on safety stands so the rear suspension moves downward. With the vehicle weight no longer applied to the springs, the spring leaves may be pried apart with a pry bar to remove and replace the silencers.

Caution

Never loosen a ball joint nut to install a cotter pin, because this action causes improper torquing of the nut.

Worn shackle bushings, brackets, and mounts cause excessive chassis lateral movement and rattling noises. With the normal vehicle weight resting on the springs, insert a pry bar between the rear outer end of the spring and the frame. Apply downward pressure on the bar and observe the rear shackle for movement. Shackle bushings, brackets, or mounts must be replaced if there is movement in the shackle. The same procedure may be followed to check the front bushing in the main leaf. A broken spring center bolt may allow the rear axle assembly to move rearward on one side. This movement changes rear wheel tracking, which results in handling problems, tire wear, and reduced directional stability. Sagged rear springs reduce the curb riding height. Spring replacement is necessary if the springs are sagged.

When rear leaf-spring replacement is necessary, proceed as follows:

Caution

The pivot bolts and nuts in the inner ends of the lower control arm must be tightened to the specified torque with the vehicle weight supported on the wheels and the suspension at normal curb height. Failure to follow this procedure may adversely affect ride quality and steering characteristics.

1. Lift the vehicle with a floor jack and place safety stands under the frame. Lower the vehicle weight onto the safety stands, and leave the floor jack under the differential housing to support the rear suspension weight.
2. Remove the nuts from the spring U-bolts, and remove the U-bolts and lower spring plate (**Figure 7-26**). The spring plate may be left on the rear shock absorber and moved out of the way.

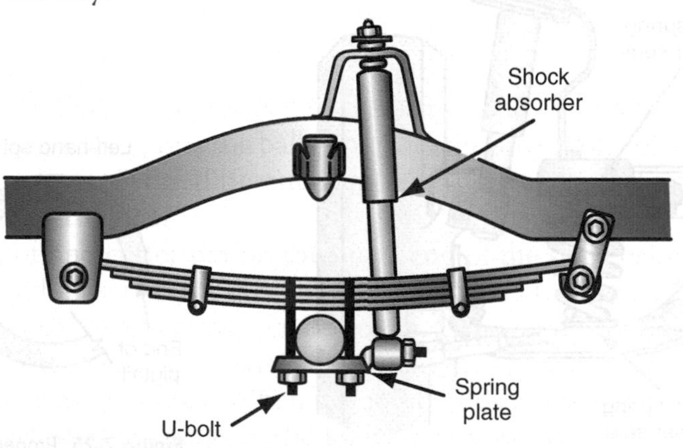

Figure 7-26 Leaf-spring rear suspension.

3. Be sure the floor jack is lowered sufficiently to relieve the vehicle weight from the rear springs.
4. Remove the rear shackle nuts, plate, shackle, and bushings.
5. Remove the front spring mounting bolt and remove the spring from the chassis. Check the spring center bolt to be sure it is not broken.
6. Check the front hanger, bushing, and bolt, and replace as necessary.
7. Check the rear shackle, bushings, plate, and mount; replace the worn components.
8. Reverse Steps 1 through 5 to install the spring. Tighten all bolts and nuts to the specified torque.

Classroom Manual
Chapter 7, page 185

Directional stability refers to the tendency of the vehicle steering to remain in the straight-ahead position when driven straight ahead on a reasonably smooth, level road surface.

TRACK BAR DIAGNOSIS AND REPLACEMENT

Some rear suspension systems have a track bar to control **lateral chassis movement.** Rubber mounting bushings insulate the track bar from the chassis components. Worn track bar mounts and bushings may cause rattling and excessive lateral chassis movement.

Lateral chassis movement refers to sideways movement.

When the track bar is inspected, lift the vehicle on a hoist or floor jack with the rear suspension in the normal riding height position. If the vehicle is lifted with a floor jack, place safety stands under the rear axle to support the vehicle weight. Grasp the track bar firmly and apply vertical and horizontal force. If there is movement in the track bar mountings, track bar, or bushing, replacement is essential (**Figure 7-27**).

Another track bar checking method is to leave the vehicle on the shop floor and observe the track bar mounts as an assistant applies side force to the chassis or rear bumper. If there is lateral movement in the track bar bushings or brackets, replace the bushings and check the bracket bolts. Bent track bars must be replaced.

When the track bar is replaced, remove the mounting bolts, bushings, grommets, and track bar. Inspect the mounting bolt holes in the chassis for wear. After the track bar is installed with the proper grommets and bushings, tighten the mounting bolts to the specified torque.

STABILIZER BAR DIAGNOSIS AND SERVICE

Worn stabilizer bar mounting bushings, grommets, or mounting bolts cause a rattling noise as the vehicle is driven on irregular road surfaces. A weak stabilizer bar or worn bushings and grommets cause harsh riding and excessive body sway while driving on irregular road

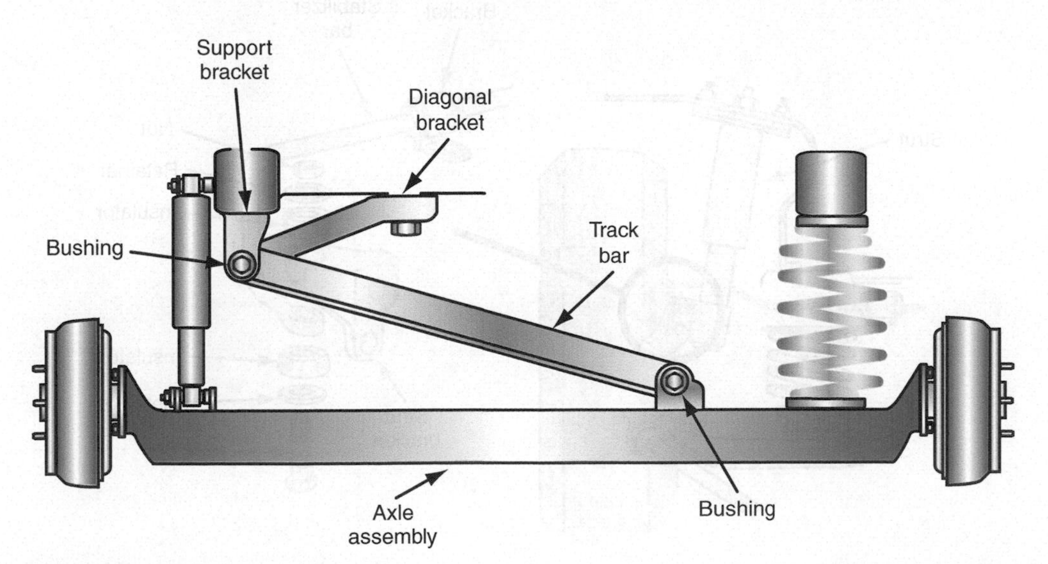

Figure 7-27 Checking track bar bushings.

A stabilizer bar may be referred to as a sway bar.

surfaces. Worn or very dry stabilizer bar bushings may cause a squeaking noise on irregular road surfaces. All stabilizer bar components should be visually inspected for wear. Stabilizer bar removal and replacement procedures vary depending on the vehicle. Always follow the vehicle manufacturer's recommended procedure in the service manual.

SERVICE TIP On rear suspension systems with an inverted U-channel, the stabilizer bar inside the U-channel sometimes breaks away where it is welded to the end plate in the U-channel. This results in a rattling, scraping noise when the car is driven over road irregularities.

Following is a typical rear stabilizer bar removal and replacement procedure:

1. Lift the vehicle on a hoist and allow both sides of the rear suspension to drop downward as the vehicle chassis is supported on the hoist.
2. Remove the mounting bolts at the outer ends of the stabilizer bar and remove the bushings, grommets, brackets, or spacers (**Figure 7-28**).
3. Remove the mounting bolts in the center area of the stabilizer bar.
4. Remove the stabilizer bar from the chassis.
5. Visually inspect all stabilizer bar components, such as bushings, bolts, and spacer sleeves. Replace the stabilizer bar, grommets, bushings, brackets, or spacers as required. Split bushings may be removed over the stabilizer bar. Bushings that are not split must be pulled from the bar.
6. Reverse Steps 2 through 4 to install the stabilizer bar. Make sure all stabilizer bar components are installed in the original position, and tighten all fasteners to the specified torque.

REAR SUSPENSION TIE ROD INSPECTION AND REPLACEMENT

Rear tie rods should be inspected for worn grommets, loose mountings, and bent conditions. Loose tie rod bushings or a bent tie rod will change the rear wheel tracking and result in reduced directional stability. Worn tie rod bushings also cause a rattling noise on

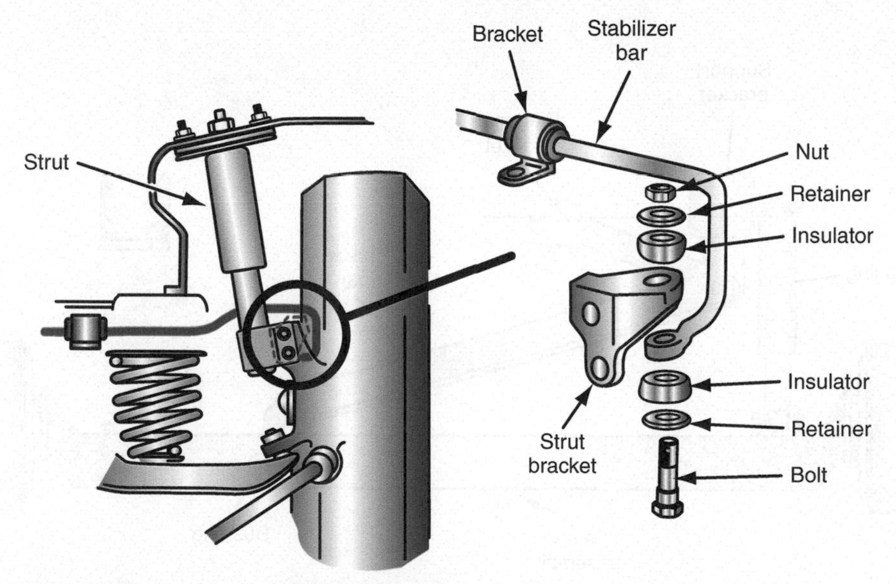

Figure 7-28 Stabilizer bar, bushings, grommets, and brackets.

TABLE 7-1 REAR SUSPENSION DIAGNOSIS

Problem	Symptoms	Possible Causes
Low riding height	Harsh ride quality, worn strikeout bumpers	Weak springs, worn control arm bushings, bent trailing arms or control arms
Steering pull	Steering pulls to the right or left when driving straight ahead	Improper rear wheel toe
Rear tire wear	Excessive rear tire tread wear	Improper rear wheel toe or camber
Rear chassis vibration	Rear chassis vibration when driving at a certain speed	Improper rear wheel, rotor, or drum balance; improper drive shaft balance or angles on rear-wheel-drive vehicles
Rear suspension noise	Rattling or squeaking noise when driving on road irregularities	Worn or dry stabilizer bar links and bushings, worn shock absorber or strut rod bushings, broken tubular rod in the rear axle inverted U-channel
Excessive rear suspension vertical oscillations	Excessive rear suspension bouncing when driving on road irregularities	Worn out shock absorbers or struts
Excessive rear chassis waddle	Excessive rear chassis lateral oscillations	Worn track bar bushings, loose track bar or brace

road irregularities. When the rear tie rod is replaced, remove the front and rear rod mounting nuts. The lower control arm or rear axle may have to be pried rearward to remove the tie rod. Inspect the tie rod grommets and mountings for wear, and replace parts as required. When the tie rod is reinstalled, tighten the mounting bolts to specifications, and measure the rear wheel toe.

Classroom Manual
Chapter 7, page 201

CASE STUDY

A customer complained about steering pull to the left on a Chevrolet Impala. The customer also said the problem had just occurred in the last few days. The technician road tested the car and found the customer's description of the complaint to be accurate. A careful inspection of the tires indicated there was no abnormal wear on the front or rear tires. The vehicle was inspected for recent collision damage, but there was no evidence of this type of damage. After lifting the vehicle on a hoist, the technician checked all the front suspension components, including ball joints, control arms, control arm bushings, and wheel bearings. However, none of these front suspension components indicated any sign of wear or looseness.

Realizing that improper rear wheel tracking causes steering pull, the technician inspected the rear suspension. A pry bar was used to apply downward and rearward force to the trailing arms on the rear suspension. When this action was taken on the right rear trailing arm, the technician discovered the trailing arm bushing was very loose. This defect had allowed the right side of the rear axle to move rearward a considerable amount, which explained why the steering pulled to the left.

After installing a new trailing arm bushing, the alignment was checked on all four wheels, and the wheel alignment was within specifications. A road test indicated no evidence of steering pull or other steering problems.

Task Completed

8. Following the vehicle and equipment manufacturers' recommended procedures, install a spring compressor on the coil spring, and tighten the spring compressor until the entire spring tension is removed from the upper support. ☐

   Is the spring compressor properly installed on the coil spring? ☐ Yes ☐ No

   Is the spring compressor tightened so all tension is removed from the upper support?

   ____

   Instructor check ______________________________

   ____

9. Install a bolt in the upper strut-to-spindle bolt hole and tighten the two nuts on the end of this bolt. One nut must be on each side of the strut bracket. ☐

   Clamp this bolt in a vise to hold the strut, coil spring, and spring compressor.

10. Using the special tool to hold the [illegible] loosen the strut rod nut.

    Is the entire coil spring tension removed from the upper strut mount by the spring compressor? ☐ Yes ☐ No

    Is the strut rod nut loosened? ☐ Yes ☐ No

    Instructor check ______________________________

11. Remove the strut rod nut, upper support, and upper insulator. ☐

12. Remove the strut from the lower end of the spring. ☐

13. If the spring is to be replaced, rotate the compressor bolt until the entire spring tension is removed from the compressing tool. Then remove the spring from the tool. ☐

14. Inspect the lower insulator and spring seat on the strut. If the spring seat is warped or damaged, replace the strut. A new cartridge may be installed in some rear struts.

    Lower insulator condition: ☐ Satisfactory ☐ Unsatisfactory

    Spring seat condition: ☐ Satisfactory ☐ Unsatisfactory

    Strut condition: ☐ Satisfactory ☐ Unsatisfactory

    List all the components that require replacement and explain the reasons for your diagnosis.

    ____

    ____

15. [illegible]

    [illegible]

    ____

    ____

Name ______________________ Date ______________

JOB SHEET 35

REMOVE REAR SUSPENSION LOWER CONTROL ARM AND BALL JOINT ASSEMBLY

Upon completion of this job sheet, you should be able to remove rear suspension lower control arm and ball joint assemblies.

ASE Education Foundation Task Correlation:

This job sheet addresses the following **MLR** tasks:

B.9. Inspect upper and lower control arms, bushings, and shafts. **(P-1)**

B.12. Inspect upper and lower ball joints (with or without wear indicators). **(P-1)**

B.18. Inspect rear suspension system lateral links/arms (track bars), control (trailing) arms. **(P-1)**

B.20. Inspect, remove, and/or replace shock absorbers; inspect mounts and bushings. **(P-1)**

This job sheet addresses the following **AST/MAST** tasks:

C.3. Inspect, remove, and/or replace upper and lower control arms, bushings, shafts, and rebound bumpers. **(P-3)**

C.5. Inspect, remove, and/or replace upper and/or lower ball joints (with or without wear indicators). **(P-2)**

We Support ASE Education Foundation

Tools and Materials

Floor jack
Control arm removing tool
Safety stands
Transmission jack
Hoist
Ball joint/bushing removal and replacement press

Describe the Vehicle Being Worked on:

Year ______________ Make ______________ Model ______________

VIN ______________________ Engine type and size ______________________

Procedure **Task Completed**

1. Lift the vehicle on a hoist with the chassis supported in the hoist and control arms dropped downward. The vehicle may be lifted with a floor jack and the chassis supported on safety stands. ☐
2. Remove the tire-and-wheel assembly. ☐
3. Remove the stabilizer bar from the knuckle bracket. ☐
4. Remove the parking brake cable retaining clip from the lower control arm. ☐
5. If the car has electronic level control (ELC), disconnect the height sensor link from the control arm. ☐
6. Install a special tool to support the lower control arm in the bushing areas. ☐

Task Completed

7. Place a transmission jack under the special tool and raise the jack enough to remove the tension from the control arm bushing retaining bolts. If the car was lifted with a floor jack and supported on safety stands, place a floor jack under the special tool.

   Is the special control arm support tool properly installed and supported?
   ☐ Yes ☐ No

   Instructor check ________________

8. Place a safety chain through the coil spring and around the lower control arm.

   Is the safety chain properly installed? ☐ Yes ☐ No

   Instructor check ________________

9. Remove the bolt from the rear control arm bushing. ☐

10. Be sure the jack is raised enough to relieve the tension on the front bolt in the lower control arm and remove this bolt. ☐

11. Lower the jack slowly and allow the control arm to pivot downward. When all the tension is released from the coil spring, remove the safety chain, coil spring, and insulators. ☐

12. Inspect the coil spring for distortion and proper free length. If the spring has a vinyl coating, check this coating for scratches or nicks. Check the spring insulators for cracks and wear.

    Coil spring condition: ☐ Satisfactory ☐ Unsatisfactory

    List all the components that require replacement and explain the reasons for your diagnosis.

    ____

    ____

    ____

13. Remove the nut on the inner end of the suspension adjustment link and disconnect this link from the lower control arm. ☐

14. Remove the cotter pin from the ball joint nut, and loosen, but do not remove, the nut from the ball joint stud.

    Is the ball joint nut loosened? ☐ Yes ☐ No

    Instructor check ________________

15. Use a special ball joint removal tool to loosen the ball joint in the knuckle. ☐

16. Remove the ball joint nut and the control arm. ☐

17. Inspect the lower control arm for bends, distortion, and worn bushings.

    Lower control arm condition: ☐ Satisfactory ☐ Unsatisfactory

Task Completed

List the control arm and related parts that require replacement and explain the reasons for your diagnosis.

____

____

____

Instructor check ______________________________

18. Use a special ball joint pressing tool to press the ball joint from the lower control arm. ☐

19. Use the same pressing tool with different adapters to press the new ball joint into the control arm. ☐

20. If control arm bushings are required, the same ball joint/bushing press as above may be used to remove and install bushings. ☐

Instructor's Response

____

____

____

Task Completed

Is the top of the spring properly positioned? ☐ Yes ☐ No

Instructor check ____

4. Install the special tool on the inner ends of the control arm, and place the transmission jack or floor jack under the special tool.

   Is the special tool properly supported on the control arm? ☐ Yes ☐ No

   Instructor check _______________

   Is the special tool properly supported on the transmission or floor jack?

   Instructor check _______________

5. Slowly raise the transmission jack until the control arm bushing openings are aligned with the openings in the chassis. ☐

6. Install the bolts and nuts in the inner ends of the control arm. Do not torque these bolts and nuts at this time. ☐

7. Install the stabilizer-bar-to-knuckle bracket fasteners to the specified torque.

   Specified torque on stabilizer-bar-to-knuckle bracket fasteners _______________

   Actual torque on stabilizer-bar-to-knuckle bracket fasteners _______________

8. Install the parking brake retaining clip. ☐

9. If the vehicle has ELC, install the height sensor link, and tighten the fastener to the specified torque. ☐

10. Install the suspension adjustment link and tighten the fastener to the specified torque. Install cotter pins as required.

    Specified adjustment link retaining nut torque _______________

    Actual adjustment link retaining nut torque _______________

    Are the cotter pins properly installed in adjustment link retaining nuts?
    ☐ Yes ☐ No

    Instructor check _______________

11. Remove the transmission jack or floor jack, and install the tire-and-wheel assembly. ☐

12. Lower the vehicle onto the floor. Tighten the wheel hub nuts and lower control arm bolts and nuts to the specified torque.

    Specified wheel nut torque _______________

    Actual wheel nut torque _______________

    Specified control arm retaining nut torque _______________

    Actual control arm retaining nut torque _______________

Caution

The pivot bolts and nuts in the inner ends of the lower control arm must be tightened to the specified torque with the vehicle supported on the wheels and the suspension at normal curb height. Failure to follow this procedure may adversely affect ride quality and steering characteristics.

Instructor's Response

____

____

____

Name ______________________________ Date ______________

JOB SHEET 37

STRUT ROD/RADIUS ARM REPLACEMENT

Upon completion of this job sheet, you should be able to inspect, remove, and/or replace track bar, strut rods/radius arms, and related mounts and bushings.

ASE Education Foundation Task Correlation

This job sheet addresses the following **MLR** task:

B.11. Inspect, remove, and/or replace track bar, strut rods/radius arms, and related mounts and bushings. **(P-1)**

This job sheet addresses the following **AST/MAST** task:

C.11. Inspect, remove, and/or replace track bar, strut rods/radius arms, and related mounts and bushings. **(P-3)**

We Support ASE Education Foundation

Tools and Materials

Hand tool

Tape measure

Describe the Vehicle Being Worked on:

Year ______________ Make ______________ Model ______________

VIN ______________________ Engine type and size ______________________

Procedure **Task Completed**

1. Lift the vehicle chassis on a hoist and allow the rear suspension to drop downward. ☐
2. Remove the rear wheel assembly. ☐
3. Remove the wheel speed sensor harness if necessary. ☐
4. Loosen and remove the strut rod/radius arm front mounting bolt and nut. ☐
5. Loosen and remove the strut rod/radius arm bolt(s) from the lower control arm. ☐
6. Remove the strut rod/radius arm from the subframe mount. ☐
7. Perform a visual inspection of the strut rod and bushings. Describe your findings.

   ____

   ____

8. Replace worn or faulty components. A ball joint/bushing press may be required if bushings are serviceable. ☐
9. Reverse Steps 3 through 6 to install the strut rod and bushing assembly. ☐
10. Torque strut rod/radius arm mounting bolts to manufacturer's specification.

    Torque Specification: ______________

11. Lower vehicle. ☐

Instructor's Response

Name ______________________ Date ______________

JOB SHEET 38

LEAF-SPRING REPLACEMENT

Upon completion of this job sheet, you should be able to inspect, remove, and/or replace rear suspension system leaf spring(s), spring insulators (silencers), shackles, brackets, bushings, center pins/bolts, and mount.

ASE Education Foundation Task Correlation

This job sheet addresses the following **MLR** task:

B.19. Inspect rear suspension system leaf spring(s), spring insulators (silencers), shackles, brackets, bushings, center pins/bolts, and mounts. **(P-1)**

This job sheet addresses the following **AST/MAST** task:

C.12. Inspect rear suspension system leaf spring(s), spring insulators (silencers), shackles, brackets, bushings, center pins/bolts, and mounts. **(P-1)**

We Support ASE | Education Foundation

Tools and Materials

Hand tool

Tape measure

Describe the Vehicle Being Worked on:

Year ______________ Make ______________ Model ______________

VIN ______________ Engine type and size ______________

Procedure **Task Completed**

1. Lift the vehicle chassis on a hoist and allow the rear suspension to drop downward. ☐
2. Remove the rear wheel assembly. ☐
3. Support the rear axle with a safety stand in order to relieve the tension on the leaf spring. ☐

   Refer to manufacturer's service information for specific removal and installation steps.
4. Remove the mounting U-bolts from the spring mounts. The rear shock absorber may be mounted to the spring plate. If this is the case the mounting plate and rear shock absorber may be moved out of the way. ☐
5. Remove the rear shackle mounting bolts. ☐
6. Remove the front spring mounting bolt. ☐
7. Remove the leaf-spring assembly from the vehicle. ☐
8. Visually inspect all components, such as bushings, bolts, and spacer sleeves and leaf centering bolt. Describe your findings.

   ____

   ____
9. Replace the leaf spring, grommets, bushings, brackets, or spacers as required. ☐
10. Reverse Steps 2 through 7 to install bar and torque all fasteners. ☐

AIR BAG DEPLOYMENT MODULE, STEERING WHEEL, AND CLOCK SPRING ELECTRICAL CONNECTOR REMOVAL AND REPLACEMENT

Prior to working on an air bag system, always disconnect the negative battery cable and wait 1 minute before proceeding with the diagnostic or service work. Many air bag systems have a **backup power supply** circuit designed into the air bag computer or located in a separate module. This backup power supply provides power to deploy the air bag for a specific length of time after the battery power is disconnected in a collision. One minute after the negative battery terminal is disconnected, this backup power supply is powered down, and no power is available to deploy the air bag while the battery remains disconnected.

Caution

Always disconnect the negative battery terminal and wait one minute before diagnosing or servicing any air bag system component. Failure to observe this precaution may result in accidental air bag deployment. Air bag service precautions vary depending on the vehicle. Always follow the vehicle manufacturer's air bag service precautions in the service manual.

Classroom Manual
Chapter 8, page 207

CUSTOMER CARE Before disconnecting the negative battery cable, note how the customer has the radio stations programmed, and reset the radio and clock after the negative battery cable is reconnected.

An air bag warning light in the instrument panel indicates the status of the air bag system. On some vehicles, this warning light should be illuminated for a few seconds when the ignition switch is turned on. The warning light should remain off while cranking the engine, and it should be on for a few seconds after the engine starts. After the engine has been running for a few seconds, the air bag warning light should remain off. On other vehicles, the air bag warning light flashes seven to nine times when the ignition switch is turned on and after the engine is started. The air bag warning light operation varies depending on the make and model year of the vehicle. Always check the vehicle manufacturer's service manual for the exact air bag warning light operation. When the air bag warning light does not operate as specified by the vehicle manufacturer, the air bag system is defective, and the air bag or bags will probably not deploy if the vehicle is involved in a collision.

Air bag module and steering wheel removal and replacement procedures vary depending on the vehicle. Always follow the vehicle manufacturer's recommended procedure in the service manual.

The following is a typical air bag module and steering wheel removal and replacement procedure:

1. Turn the ignition switch to the Lock position and place the front wheels facing straight ahead.
2. Remove the negative battery terminal and wait for the specified time period.
3. Loosen the three air bag-retaining Torx screws under the steering wheel (**Figure 8-1**).
4. Loosen the other two air bag-retaining Torx screws under the steering wheel (**Figure 8-2**). Loosen all five Torx screws until the groove along the screw circumference catches on the screw case. Some air bag deployment modules are retained on top of the steering wheel with a spring-loaded clip. To release this clip, insert a flat-tipped screw driver into the slot at the lower edge of the steering wheel and turn the screw driver one-quarter turn to release the clip.

WARNING When an air bag deployment module is temporarily stored on the workbench, always place this module face upward. If the air bag deployment module accidentally deployed when facing downward, the module would become a projectile, and personal injury might result.

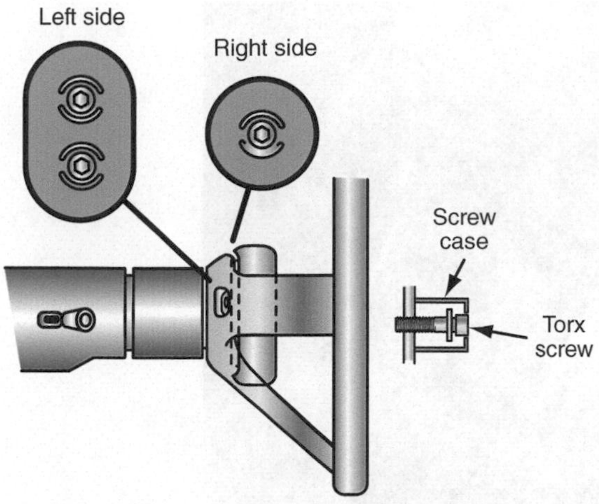
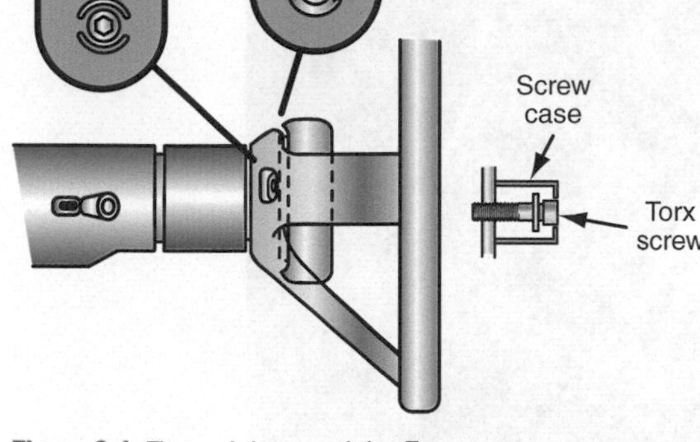

Figure 8-1 Three air bag-retaining Torx screws.

Figure 8-2 Two air bag-retaining Torx screws.

5. Pull the **air bag deployment module** from the steering wheel and disconnect the air bag module electrical connector (**Figure 8-3**). Do not pull on the air bag wires in the steering column. Place the air bag deployment module face upward on the workbench.
6. Disconnect the air bag wiring retainer in the steering wheel (**Figure 8-4**).
7. Use the proper size socket and a ratchet to remove the steering wheel retaining nut.
8. Observe the matching alignment marks on the steering wheel and the steering shaft. If these alignment marks are not present, place alignment marks on the steering wheel and steering shaft with a center punch and a hammer.

An air bag deployment module may be referred to as a steering wheel pad.

The **air bag deployment module** contains the air bag, inflation chemicals, and an igniting device.

WARNING Photo Sequence 14 shows a typical procedure for removing a steering wheel. Do not pull on the steering wheel in an attempt to remove it from the steering shaft. The steering wheel may suddenly come off, resulting in personal injury, or the steering wheel may be damaged by the pulling force.

9. Install a steering wheel puller with the puller bolts threaded into the bolt holes in the steering wheel. On some vehicles, the steering wheel has slots for the pulley adapters to fit into rather than threaded bolt holes. This design prevents the possibility of the technician turning the bolts too far into the steering wheel and damaging components below the steering wheel. Tighten the puller nut to remove the steering wheel (**Figure 8-5**). Visually check the steering wheel condition. If the steering wheel is bent or cracked, replace the wheel.

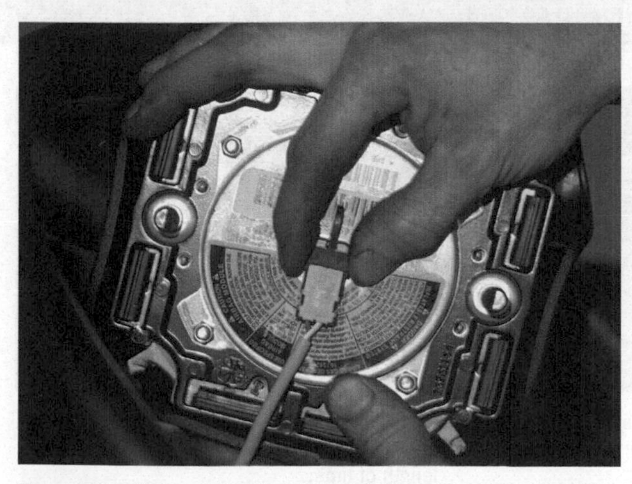

Figure 8-3 Disconnecting the air bag module electrical connector.

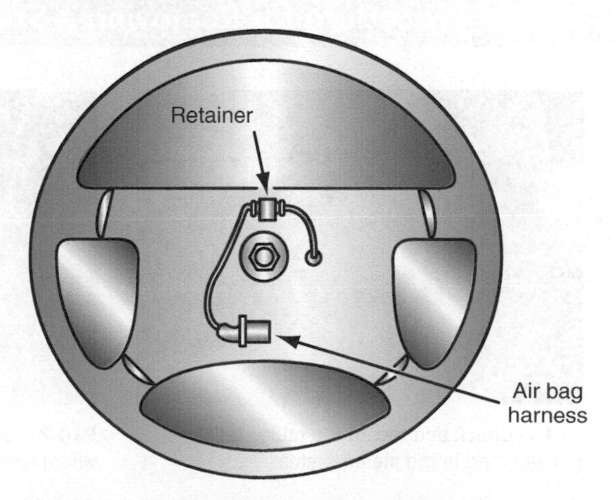

Figure 8-4 Disconnecting the air bag wiring retainer.

Caution

On an air bag-equipped vehicle, the wait time prior to servicing electrical components after the negative battery terminal is disconnected varies depending on the vehicle make and model year. Always follow the wait time and all other service precautions recommended in the vehicle manufacturer's service manual.

Figure 8-5 Removing steering wheel with the proper steering wheel puller.

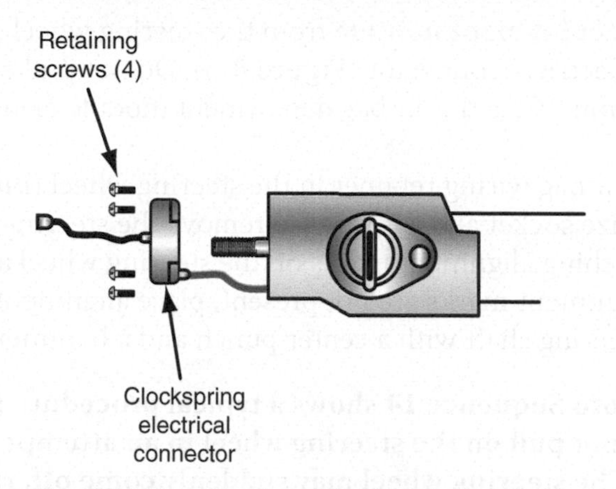

Figure 8-6 Removing clock spring electrical connector.

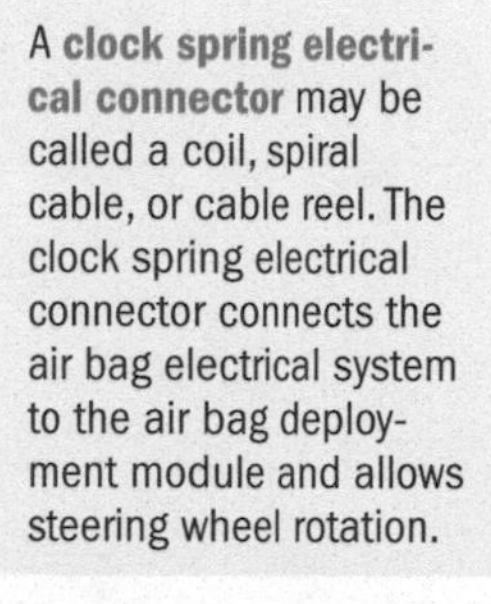

A **clock spring electrical connector** may be called a coil, spiral cable, or cable reel. The clock spring electrical connector connects the air bag electrical system to the air bag deployment module and allows steering wheel rotation.

10. Disconnect the four retaining screws and remove the **clock spring electrical connector** (**Figure 8-6**).
11. Be sure the front wheels are facing straight ahead. Turn the clock spring electrical connector counterclockwise by hand until it becomes harder to turn as it becomes fully wound in that direction.

PHOTO SEQUENCE 14
Typical Procedure for Removing a Steering Wheel

P14-1 Check and record the radio stations programmed in the stereo system.

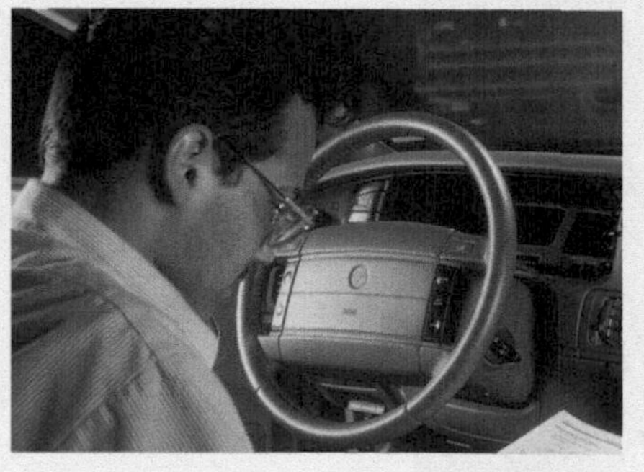

P14-2 Look up the car manufacturer's steering wheel removal procedure in the service manual.

P14-3 Disconnect the negative battery cable, and wait for the car manufacturer's specified length of time.

PHOTO SEQUENCE 14 (CONTINUED)

P14-4 Remove the air bag deployment module retaining screws under the steering wheel.

P14-5 Lift the air bag deployment module upward from the steering wheel, and disconnect the module wiring connector.

P14-6 Set the air bag deployment module face upward on the workbench.

P14-7 Loosen and remove the steering wheel retaining nut.

P14-8 Observe the alignment marks on the steering wheel and shaft.

P14-9 Connect the proper steering wheel puller to the steering wheel.

P14-10 Turn the puller screw to loosen the steering wheel on the shaft.

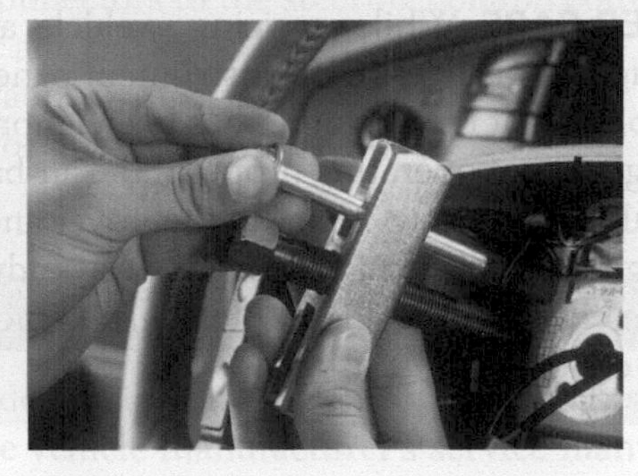

P14-11 Remove the puller from the steering wheel.

P14-12 Lift the steering wheel off the shaft.

12. Turn the clock spring electrical connector clockwise three turns and align the red mark on the center part of the spring face with the notch in the cable circumference (**Figure 8-7**). This action centers the clock spring electrical connector.
13. Install the clock spring electrical connector and tighten the four retaining screws to the specified torque.
14. Align the marks on the steering wheel and the steering shaft, and install the steering wheel on the shaft.

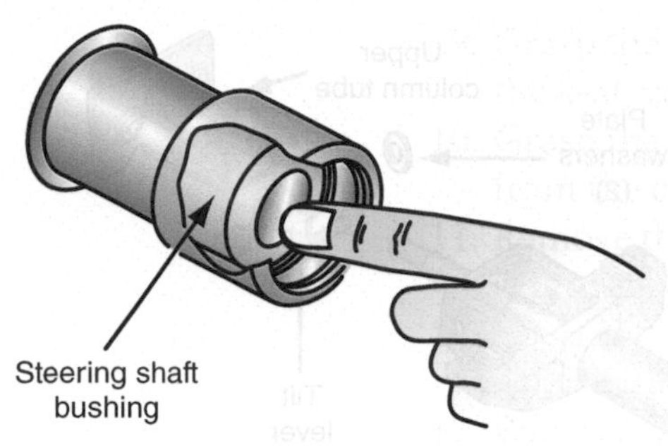

Figure 8-28 Inspecting upper steering shaft bearing.

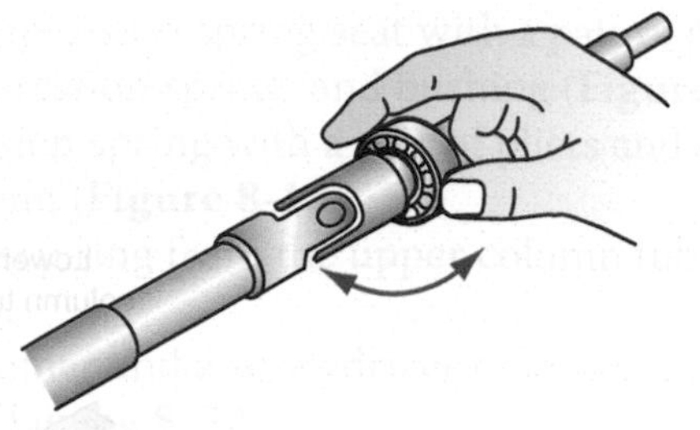

Figure 8-29 Inspecting lower steering shaft bearing.

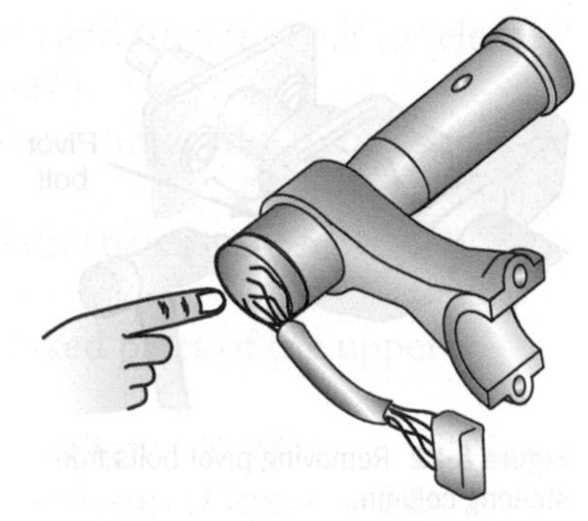

Figure 8-30 Inspecting ignition key interlock solenoid.

2. Rotate the upper steering shaft bearing and inspect for noise, looseness, and wear (**Figure 8-28**). If any of these conditions are present, replace the upper tube.
3. Inspect the lower steering shaft bearing for noise, looseness, and wear (**Figure 8-29**). Replace this bearing if necessary.
4. Inspect the ignition key interlock solenoid for damaged wires and loose mounting screws (**Figure 8-30**). Repair or replace this solenoid as required.

TILT STEERING COLUMN ASSEMBLY

Molybdenum disulphide lithium-based grease is a special grease that is applied to contacting parts in a tilt steering column.

1. Coat all rubbing parts with **molybdenum disulphide lithium-based grease.** Install the steering shaft bushing and the bushing snapring (**Figure 8-31**).
2. Install the steering shaft in the lower tube (**Figure 8-32**).
3. Install the upper column tube, tilt lever mechanism, and two plate washers (**Figure 8-33**). Install the two pivot bolts.
4. Install the pivot bolt nuts and tighten these nuts to the specified torque (**Figure 8-34**).
5. Use a brass bar and a hammer to tap the steering shaft into the upper column tube (**Figure 8-35**).
6. Stake the upper column tube with a pin punch and a hammer (**Figure 8-36**).
7. Install the snapring in the top of the upper column tube (**Figure 8-37**).
8. Install the tension spring (**Figure 8-38**). Assemble the compression spring, bushing, plate, and seat. Use a vise to install the compression spring and related components (**Figure 8-39**).
9. Install the upper bracket with two new tapered-head bolts. Tighten the tapered-head bolts until the tapered head breaks off (**Figure 8-40**).

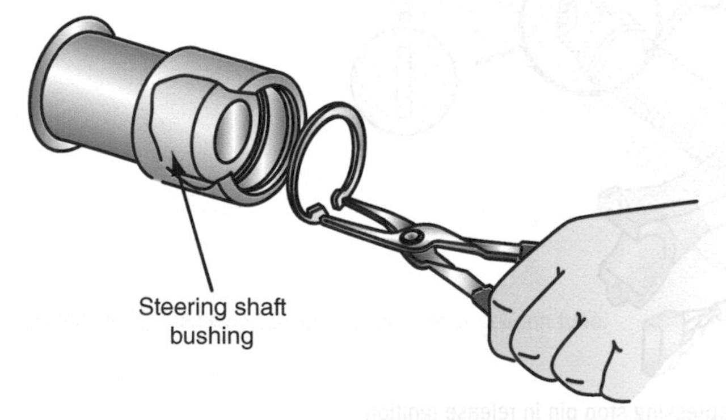

Figure 8-31 Installing steering shaft bushing and snapring.

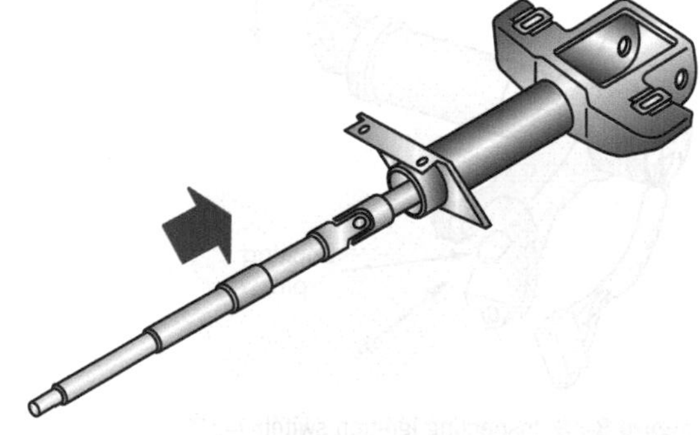

Figure 8-32 Installing the steering shaft in the lower tube.

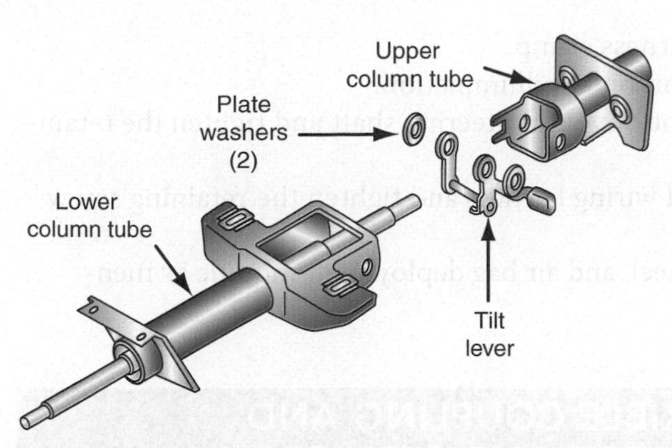

Figure 8-33 Installing upper column tube, tilt lever mechanism, and two plate washers.

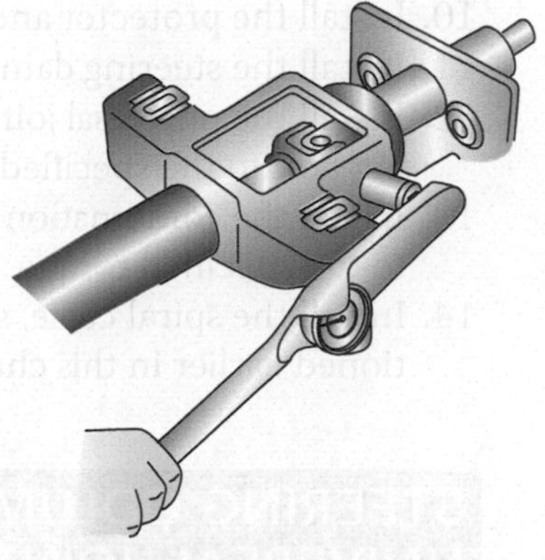

Figure 8-34 Tightening pivot bolt nuts.

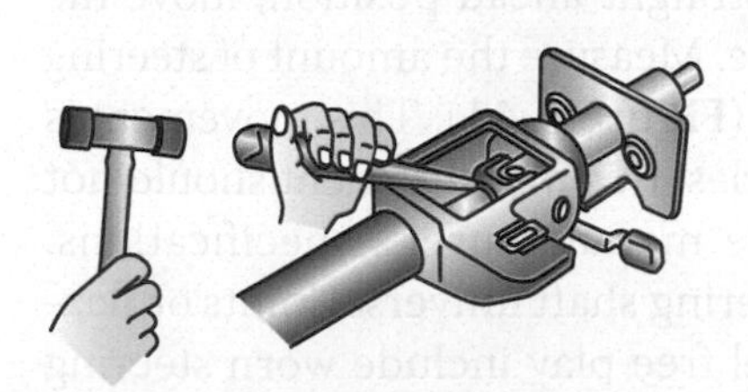

Figure 8-35 Tapping steering shaft into upper column tube.

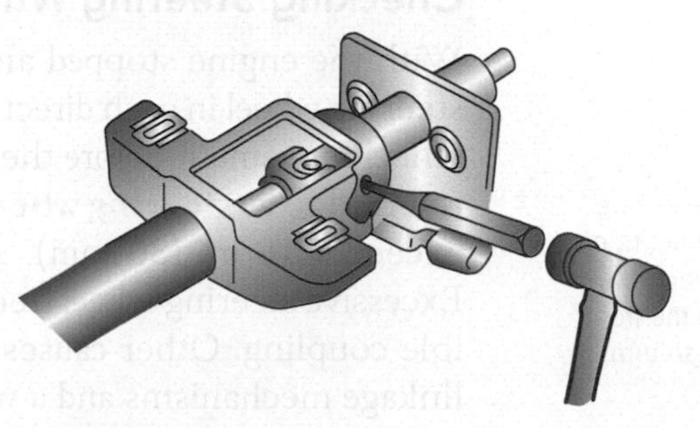

Figure 8-36 Staking upper column tube.

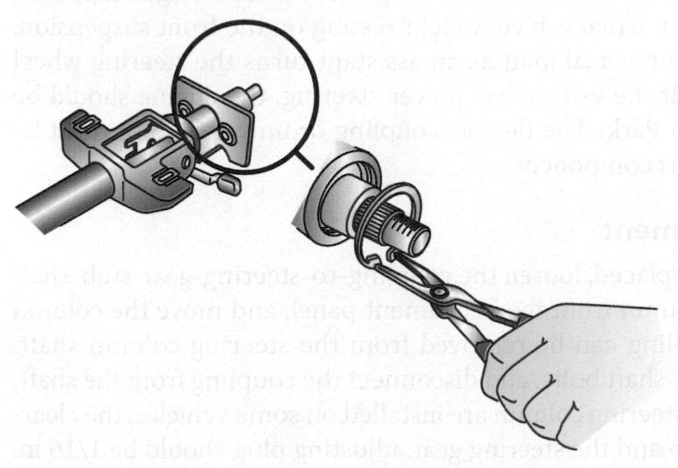

Figure 8-37 Installing snapring in upper column tube.

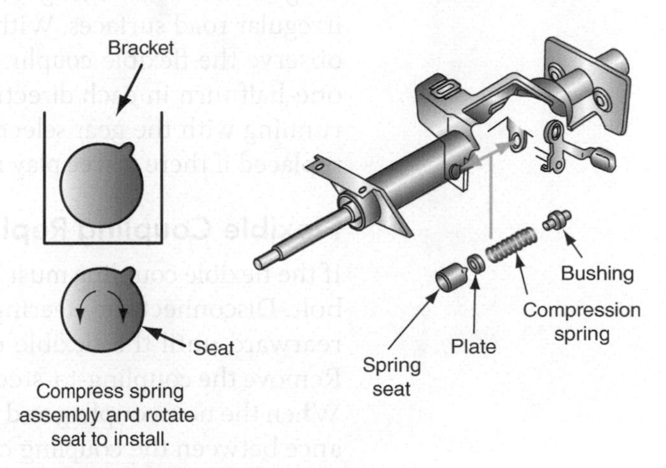

Figure 8-38 Installing tension spring.

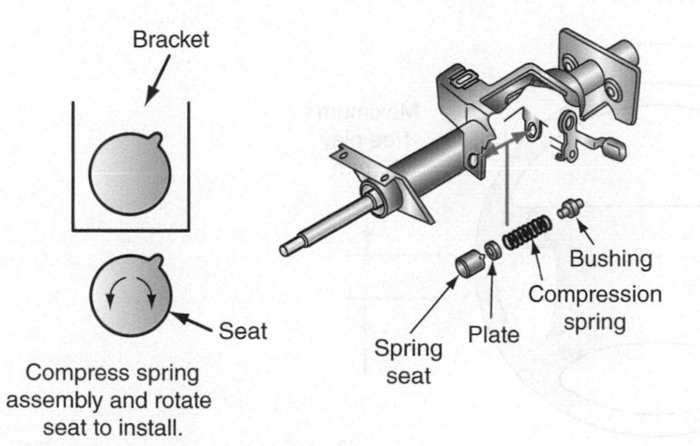

Figure 8-39 Installing compression spring.

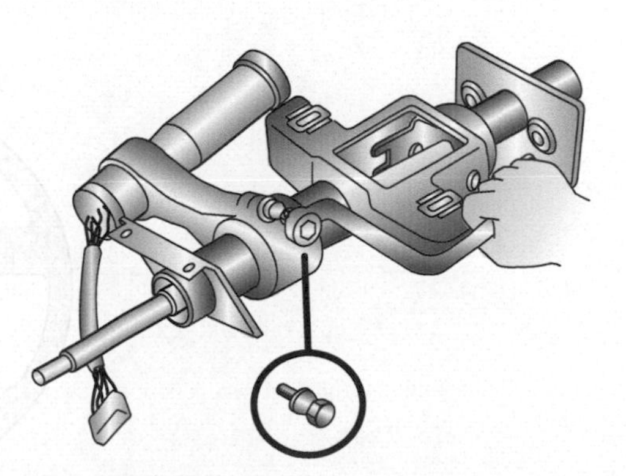

Figure 8-40 Tightening tapered-head bolts in upper bracket.

10. Install the protector and wiring harness clamp.
11. Install the steering damper and ignition key illumination.
12. Install the universal joint on the bottom of the steering shaft and tighten the retaining bolt to the specified torque.
13. Install the combination switch and wiring harness and tighten the retaining screw to the specified torque.
14. Install the spiral cable, steering wheel, and air bag deployment module as mentioned earlier in this chapter.

STEERING COLUMN FLEXIBLE COUPLING AND UNIVERSAL JOINT DIAGNOSIS AND SERVICE

Checking Steering Wheel Free Play

Steering wheel free play is the amount of steering wheel movement before the front wheels start to turn.

With the engine stopped and the front wheels in the straight-ahead position, move the steering wheel in each direction with light finger pressure. Measure the amount of steering wheel movement before the front wheels begin to turn (**Figure 8-41**). This movement is referred to as **steering wheel free play**. On some vehicles, this measurement should not exceed 1.18 in. (30 mm). Always refer to the vehicle manufacturer's specifications. Excessive steering wheel free play is caused by worn steering shaft universal joints or flexible coupling. Other causes of excessive steering wheel free play include worn steering linkage mechanisms and a worn or out of adjustment steering gear.

A worn universal joint or flexible coupling in the steering column may also cause rattling noises. The rattling noises may occur while driving the vehicle straight ahead on irregular road surfaces. With the normal vehicle weight resting on the front suspension, observe the flexible coupling or universal joint as an assistant turns the steering wheel one-half turn in each direction. If the vehicle has power steering, the engine should be running with the gear selector in Park. The flexible coupling or universal joint must be replaced if there is free play in this component.

Flexible Coupling Replacement

If the flexible coupling must be replaced, loosen the coupling-to-steering-gear-stub-shaft bolt. Disconnect the steering column from the instrument panel, and move the column rearward until the flexible coupling can be removed from the steering column shaft. Remove the coupling-to-steering-shaft bolts, and disconnect the coupling from the shaft. When the new coupling and the steering column are installed on some vehicles, the clearance between the coupling clamp and the steering gear adjusting plug should be 1/16 in. (1.5 mm) (**Figure 8-42**). This specification may vary depending on the vehicle. Always use the vehicle manufacturer's specifications in the service manual.

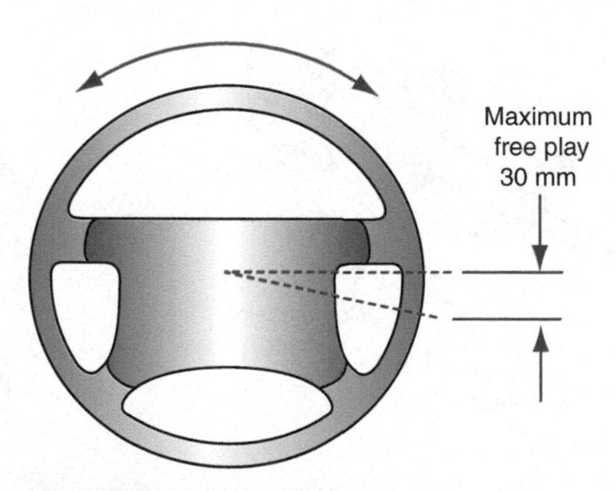

Figure 8-41 Measuring steering wheel free play.

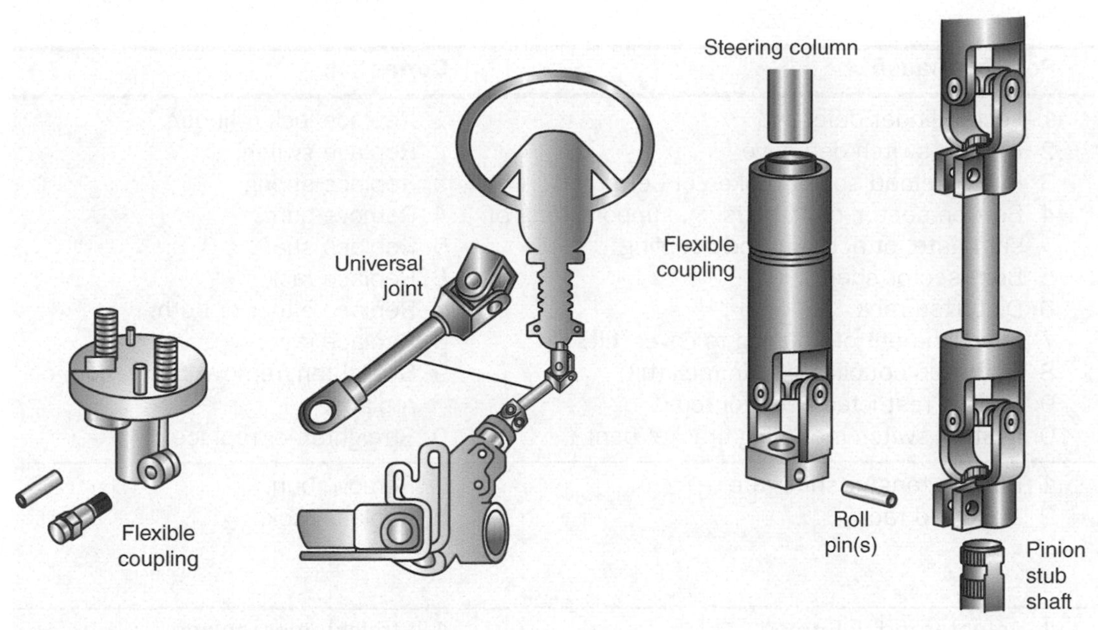

Figure 8-42 Flexible couplers.

STEERING COLUMN DIAGNOSIS

There are variations in steering columns depending on the vehicle, type of transmission, and the transmission gear selector position. Thus, different column diagnostic procedures may be required. See **Table 8-1**, **Table 8-2**, and **Table 8-3** for a typical steering column diagnosis.

TABLE 8-1 AUTOMATIC TRANSMISSION–STEERING COLUMN DIAGNOSIS

Condition	Possible Cause	Correction
Lock system—will not unlock	1 Lock bolt damaged 2 Defective lock cylinder 3 Damaged housing 4 Damaged or collapsed sector 5 Damaged rack 6 Shear flange on sector shaft collapsed	1 Replace lock bolt. 2 Replace or repair lock cylinder. 3 Replace housing. 4 Replace sector. 5 Replace rack. 6 Replace.
Lock system—will not lock	1 Lock bolt spring broken or defective 2 Damaged sector tooth, or sector installed incorrectly 3 Defective lock cylinder 4 Burr on lock bolt or housing 5 Damaged housing 6 Transmission linkage adjustment incorrect 7 Damaged rack 8 Interference between bowl and coupling (tilt) 9 Ignition switch stuck 10 Actuator rod restricted or bent	1 Replace spring. 2 Replace, or install correctly. 3 Replace lock cylinder. 4 Remove burr. 5 Replace housing. 6 Readjust. 7 Replace rack. 8 Adjust or replace as necessary. 9 Readjust or replace. 10 Readjust or replace.

(Continued)

TABLE 8-1 *(Continued)*

Condition	Possible Cause	Correction
Lock system—high effort	1 Lock cylinder defective 2 Ignition switch defective 3 Rack preload spring broken or deformed 4 Burr on sector, rack, housing, support, tang of shift gate, or actuator rod coupling 5 Bent sector shaft 6 Distorted rack 7 Misalignment of housing to cover (tilt) 8 Distorted coupling slot in rack (tilt) 9 Bent or restricted actuator rod 10 Ignition switch mounting bracket bent	1 Replace lock cylinder. 2 Replace switch. 3 Replace spring. 4 Remove burr. 5 Replace shaft. 6 Replace rack. 7 Replace either or both. 8 Replace rack. 9 Straighten, remove restriction, or replace. 10 Straighten or replace.
Lock cylinder—high effort between Off and Off-lock positions	1 Burr on tang of shift gate 2 Distorted rack	1 Remove burr. 2 Replace rack.
Sticks in Start position	1 Actuator rod deformed 2 Any high effort condition	1 Straighten or replace. 2 Check items under high effort section.
Key cannot be removed in Off-lock position	1 Ignition switch not set correctly 2 Defective lock cylinder	1 Readjust ignition switch. 2 Replace lock cylinder.
Lock cylinder can be removed without depressing retainer	1 Lock cylinder with defective retainer 2 Lock cylinder without retainer 3 Burr over retainer slot in housing cover	1 Replace lock cylinder. 2 Replace lock cylinder. 3 Remove burr.
Lock bolt hits shaft lock in Off and Park positions	Ignition switch not set correctly	Readjust ignition switch.
Ignition system—electrical system does not function	1 Defective fuse in "accessory" circuit 2 Connector body loose or defective 3 Defective wiring 4 Defective ignition switch 5 Ignition switch not adjusted properly	1 Replace fuse. 2 Tighten or replace. 3 Repair or replace. 4 Replace ignition switch. 5 Readjust ignition switch.
Switch does not actuate mechanically	Defective ignition switch	Replace ignition switch.
Switch cannot be set correctly	1 Switch actuator rod deformed 2 Sector to rack engaged in wrong tooth (tilt)	1 Repair or replace switch actuator rod. 2 Engage sector to rack correctly.
Noise in column	1 Coupling bolts loose 2 Column not correctly aligned 3 Coupling pulled apart 4 Sheared intermediate shaft plastic joint 5 Horn contact ring not lubricated 6 Lack of grease on bearings or bearing surfaces	1 Tighten pinch bolts to specified torque. 2 Realign column. 3 Replace coupling and realign column. 4 Replace or repair steering shaft and realign column. 5 Lubricate with Lubriplate. 6 Lubricate bearings.

Condition	Possible Cause	Correction
	7 Lower shaft bearing tight or frozen 8 Upper shaft tight or frozen 9 Shaft lock plate cover loose 10 Lock plate snapring not seated 11 Defective buzzer dog cam on lock cylinder 12 One click when in Off-lock position and the steering wheel is moved	7 Replace bearing. Check shaft and replace if scored. 8 Replace housing assembly. 9 Tighten three screws or, if missing, replace. Caution: Use specified screws (15 in.-lbs). 10 Replace snapring. Check for proper seating in groove. 11 Replace lock cylinder. 12 Normal condition: lock bolt is seating.
Steering shaft—high effort	1 Column assembly misaligned in vehicle 2 Improperly installed or deformed dust seal 3 Tight or frozen, upper or lower bearing 4 Flash on ID of shift tube from plastic joint	1 Realign. 2 Remove and replace. 3 Replace affected bearing or bearings. 4 Replace shift tube.
High shift effort	1 Column not aligned correctly in car 2 Improperly installed dust seal 3 Lack of grease on seal or bearing areas 4 Burr on upper or lower end of shift tube 5 Lower bowl bearing not assembled properly (tilt) 6 Wave washer with burrs (tilt)	1 Realign. 2 Remove and replace. 3 Lubricate bearings and seals. 4 Remove burr. 5 Reassemble properly. 6 Replace wave washer.
Improper transmission shifting	1 Sheared shift tube joint 2 Improper transmission linkage adjustment 3 Loose lower shift lever 4 Improper gate plate 5 Sheared lower shift lever weld	1 Replace shift tube assembly. 2 Readjust linkage. 3 Replace shift tube assembly. 4 Replace with correct part. 5 Replace tube assembly.
Lash in mounted column assembly	1 Instrument panel mounting bolts loose 2 Broken weld nuts on jacket 3 Instrument panel bracket capsule sheared 4 Instrument panel to jacket mounting bolts loose 5 Loose shoes in housing (tilt) 6 Loose tilt head pivot pins (tilt) 7 Loose shoe lock pin in support (tilt)	1 Tighten to specifications (20 ft.-lbs). 2 Replace jacket assembly. 3 Replace bracket assembly. 4 Tighten to specifications (15 ft.-lbs). 5 Replace. 6 Replace. 7 Replace.
Miscellaneous	1 Housing loose on jacket noticed with ignition in Off-lock position and a torque applied to the steering wheel 2 Shroud loose on shift bowl	1 Tighten four mounting screws (60 in.-lbs). 2 Bend tabs on shroud over lugs on bowl.

TABLE 8-2 MANUAL TRANSMISSION—STEERING COLUMN DIAGNOSIS

Condition	Possible Cause	Correction
Shift lever sticking	1 Defective upper shift lever 2 Defective shift lever gate 3 Loose relay lever on shift tube 4 Wrong shift lever	1 Replace shift lever. 2 Replace shift lever gate. 3 Replace shift tube assembly. 4 Replace with current lever.
High shift effort	1 Column not aligned correctly 2 Lower bowl bearing not assembled correctly 3 Improperly installed seal 4 Wave washer in lower bowl bearing defective 5 Improper adjustment of lower shift levers 6 Lack of grease on seal, bearing areas, or levers 7 Damaged shift tube in bearing areas	1 Realign column. 2 Reassemble correctly. 3 Remove and replace. 4 Replace wave washer. 5 Readjust. 6 Lubricate seal, levers, and bearings. 7 Replace shift tube assembly.
Improper transmission shifting	Loose relay lever on shift tube	Replace shift tube assembly.

TABLE 8-3 MANUAL TRANSMISSION—TILT COLUMN DIAGNOSIS

Condition	Possible Cause	Correction
Housing scraping on bowl	Bowl bent or not concentric with hub	Replace bowl.
Steering wheel loose	1 Excessive clearance between holes in support or housing and pivot pin diameters 2 Defective or missing antilash spring in spheres 3 Upper bearing seat not seating in bearing 4 Upper bearing inner race seat missing 5 Loose support screws 6 Bearing preload spring missing or broken	1 Replace either or both. 2 Add spring or replace both. 3 Replace both. 4 Install seat. 5 Tighten to 60 in.-lbs. 6 Replace preload spring.
Steering wheel loose every other tilt position	Loose fit between shoe and shoe pivot pin	Replace both.
Noise when tilting column	1 Upper tilt bumper worn 2 Tilt spring rubbing in housing or dirt	1 Replace tilt bumper. 2 Lubricate.
Steering column not locking in any tilt position	1 Shoe seized on its pivot pin 2 Shoe grooves might have burrs 3 Shoe lock spring weak or broken	1 Replace shoe and pivot pin. 2 Replace shoe. 3 Replace lock spring.
Steering wheel fails to return to top tilt position	1 Pivot pins bound up 2 Wheel tilt spring defective 3 Turn signal switch wires too tight	1 Replace pivot pins. 2 Replace tilt spring. 3 Reposition wires.

RACK AND PINION STEERING GEAR TIE ROD SERVICE

Follow these steps for rack and pinion steering inner tie rod replacement:

1. Remove the outer tie rod end from the steering arm.
2. Insert a short piece of wire through each outer tie rod end cotter pin hole and connect a pull scale to this wire. Pull upward on the scale to check the tie rod articulation effort (**Figure 8-43**). If this effort is not within specifications, replace the inner tie rod end.
3. Mark the outer tie rod and jam nut position with masking tape wrapped around the tie rod threads next to the jam nut, or place a dab of paint on the jam nut and tie rod threads (**Figure 8-44**). Loosen the jam nuts and remove the outer tie rod ends.
4. Remove the inner and outer bellows boot clamps and pull the bellows boots from the tie rods.
5. Straighten the lock plate where it is bent over the inner tie rod end (**Figure 8-45**). Hold the rack with a wrench, and use the proper size wrench to remove the inner tie rod end from the rack (**Figure 8-46**). A jam nut is used in place of the lock ring on some inner tie rod ends, and a roll pin retains some inner tie rod ends to the rack. Various inner tie rod socket removal procedures are required depending on the socket design.

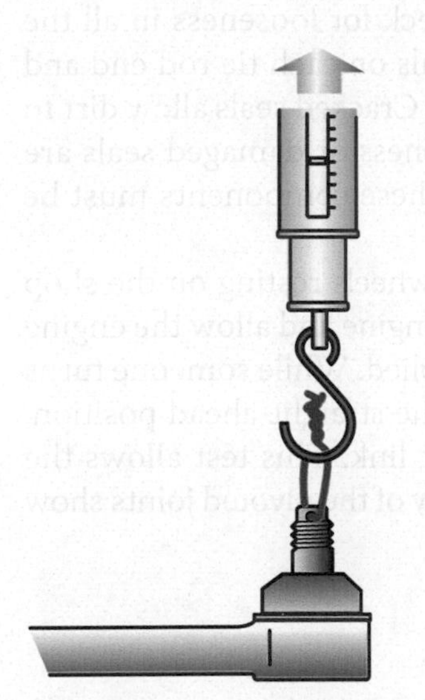

Figure 8-43 Measuring inner tie rod articulation effort.

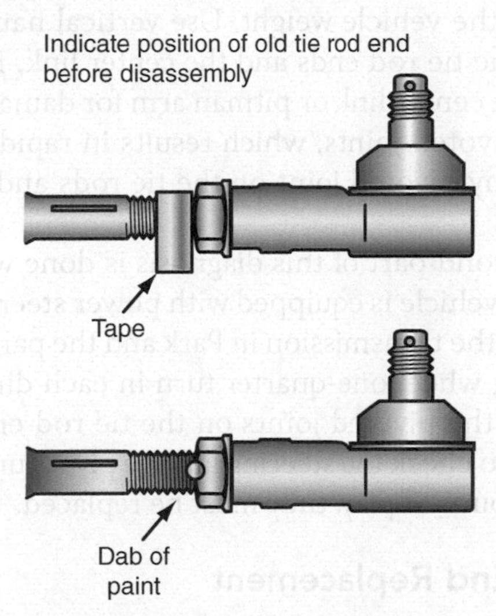

Figure 8-44 Before removing outer tie rod, mark the jam nut location or count thread turns to remove.

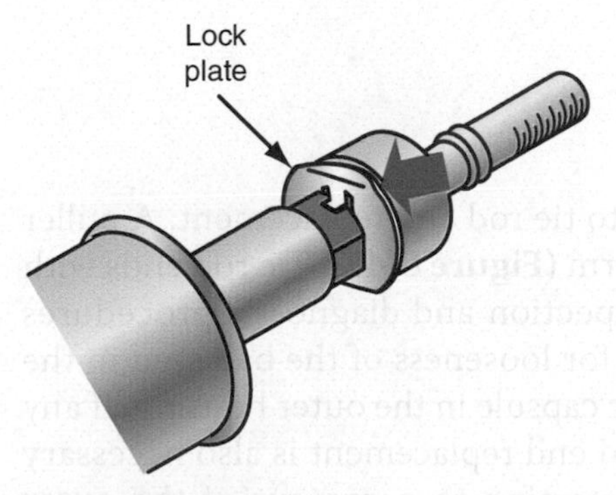

Figure 8-45 Straighten the lock plate to allow removal of inner tie rod end.

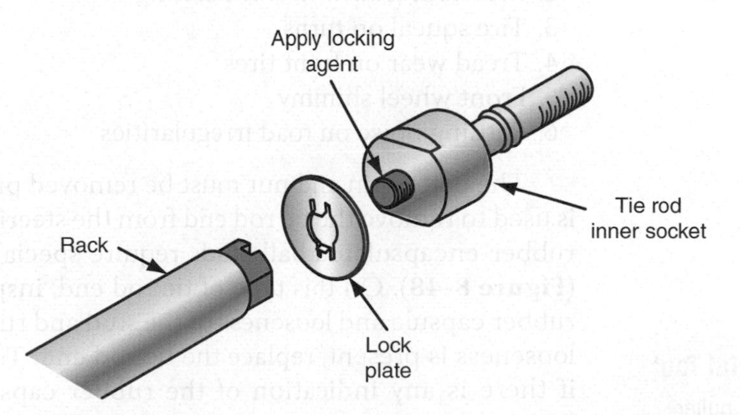

Figure 8-46 Removing inner tie rod end from rack.

6. Rotate the inner tie rod ends onto the rack until they bottom. Install the inner tie rod pins, or stake these ends as required. Always use a wooden block to support the opposite side of the rack and inner tie rod end while staking. If jam nuts are located on the inner tie rod ends, be sure they are tightened to the specified torque.
7. Place a large bellows boot clamp over each end of the gear housing. Install the bellows boots, and be sure the boots are seated in the housing and the tie rod undercuts. Install and tighten the large inner boot clamps.
8. Install the outer bellows boot clamps on the tie rods, but do not install these clamps on the boots until the steering gear is installed and the toe is adjusted.
9. Install the jam nuts and the outer tie rod ends. Align the marks placed on these components during the disassembly procedure. Leave the jam nuts loose until the steering gear is installed and the toe is adjusted.
10. Install the steering gear in the vehicle and check the front wheel toe. Tighten the outer bellows boot clamps and the outer tie rod end jam nuts.

Classroom Manual
Chapter 8, page 216

STEERING LINKAGE DIAGNOSIS AND SERVICE

Front wheel shimmy may be defined as a consistent, fast, side-to-side movement of the front wheels and steering wheel. This movement is usually experienced at speeds above 40 mph (25 km/h), and it may occur more frequently while driving on irregular road surfaces.

Diagnosis of Center Link, Pitman Arm, and Tie Rod Ends

The vehicle should be raised, and safety stands positioned under the lower control arms to support the vehicle weight. Use vertical hand force to check for looseness in all the pivots on the tie rod ends and the center link. Inspect the seals on each tie rod end and pivot on the center link or pitman arm for damage and cracks. Cracked seals allow dirt to enter the pivoted joints, which results in rapid wear. If looseness or damaged seals are found on any pivoted joint on the tie rods and center link, these components must be replaced.

The second part of this diagnosis is done with the front wheels resting on the shop floor. If the vehicle is equipped with power steering, start the engine and allow the engine to idle with the transmission in Park and the parking brake applied. While someone turns the steering wheel one-quarter turn in each direction from the straight-ahead position, observe all the pivoted joints on the tie rod ends and center link. This test allows the technician to check the steering linkage pivots under load. If any of the pivoted joints show a slight amount of play, they must be replaced.

Tie Rod End Replacement

Worn tie rod ends result in these problems:

1. Excessive steering wheel free play
2. Incorrect front wheel toe setting
3. Tire squeal on turns
4. Tread wear on front tires
5. Front wheel shimmy
6. Rattling noise on road irregularities

The cotter pin and nut must be removed prior to tie rod end replacement. A puller is used to remove the tie rod end from the steering arm (**Figure 8-47**). Tie rod ends with rubber-encapsulated ball studs require special inspection and diagnostic procedures (**Figure 8-48**). On this type of tie rod end, inspect for looseness of the ball stud in the rubber capsule and looseness of the stud and rubber capsule in the outer housing. If any looseness is present, replace the tie rod end. Tie rod end replacement is also necessary if there is any indication of the rubber capsule starting to come out of the outer housing.

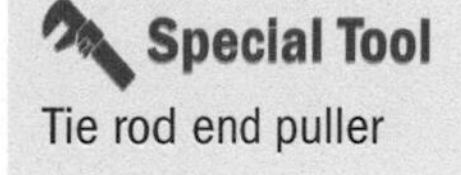

Tie rod end puller

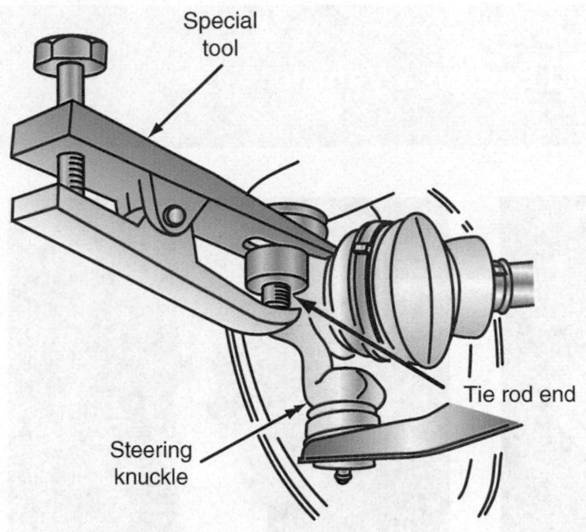

Figure 8-47 Removing tie rod end.

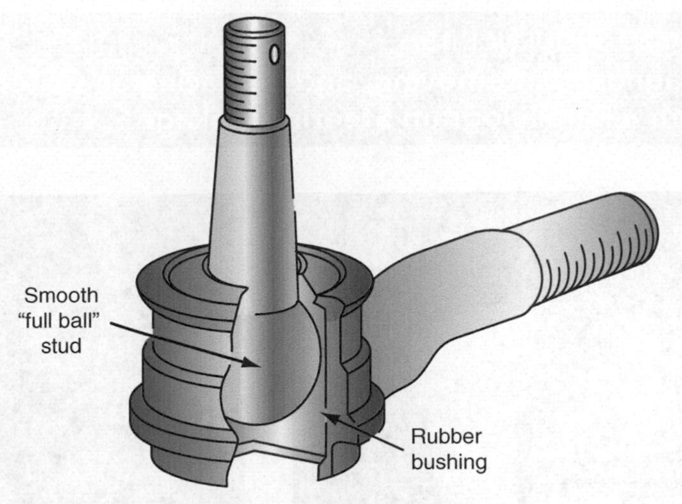

Figure 8-48 Tie rod end with rubber-encapsulated ball stud.

After the tie rod end is removed from the steering arm, install two nuts on the stud threads and tighten these nuts against each other. Use the proper size of socket and a torque wrench to rotate the ball stud through a 40-degree arc (**Figure 8-49**). If the ball stud turning torque is less than 20 ft.-lb. (27 N•m), replace the tie rod end.

The tie rod clamp must be loosened before the tie rod end is removed from the sleeve. Count the number of turns required to remove the tie rod end from the sleeve, and install the new tie rod with the same number of turns. Even when this procedure is followed, the toe must be checked after the steering linkage components are replaced. Before the new tie rod end is installed, center the stud in the tie rod end. When the tie rod end stud is installed in the steering arm opening, only the threads should be visible above the steering arm surface. If the machined surface of the tie rod end stud is visible above the steering arm surface, or if the stud fits loosely in the steering arm opening, this opening is worn, or the tie rod end is not correct for that application. The tie rod end nut must be torqued to the manufacturer's specifications, and the cotter pin must be installed through the tie rod end and nut openings (**Figure 8-50**). **Photo Sequence 15** illustrates the procedure for removing and replacing an outer tie rod end.

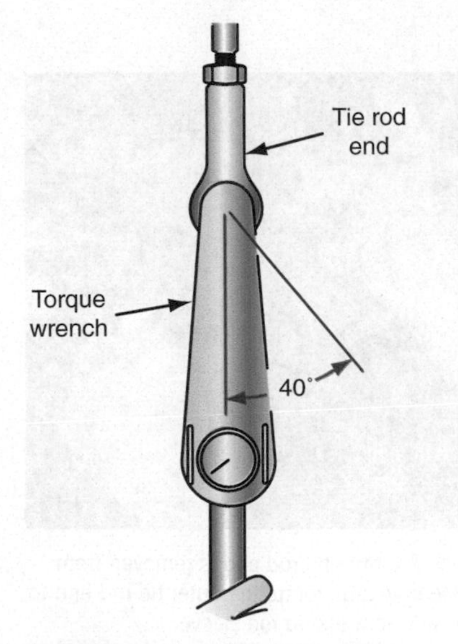

Figure 8-49 Tie rod end with rubber-encapsulated ball stud.

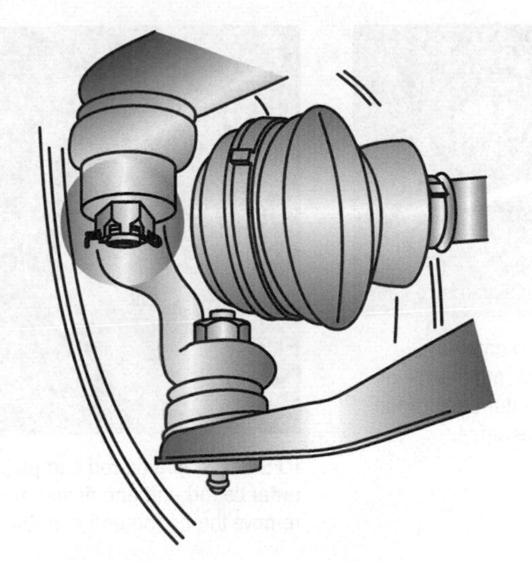

Figure 8-50 Tie rod nut and cotter pin installation.

PHOTO SEQUENCE 15

Diagnosing, Removing, and Replacing an Outer Tie Rod End on a Vehicle with a Parallelogram Steering Linkage

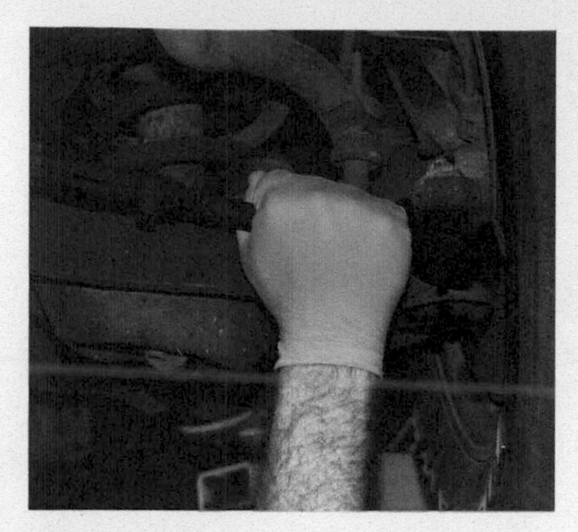

P15-1 Raise the vehicle on a lift and check for excessive vertical movement in the left outer tie rod end.

P15-2 Check for lateral movement in the left outer tie rod end.

P15-3 Visually inspect the left outer tie rod end seal for cracks and damage.

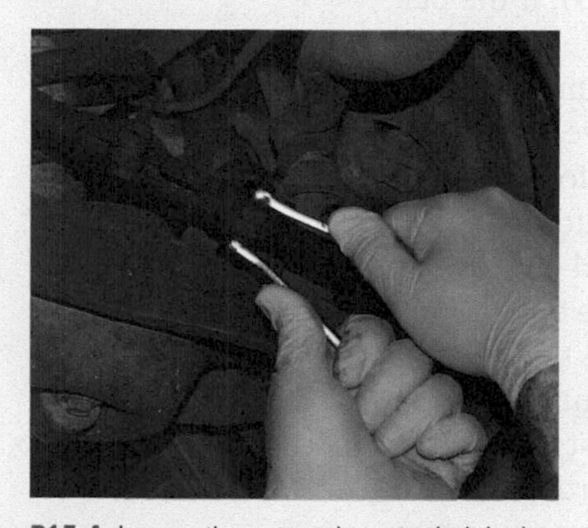

P15-4 Loosen the nut on the outer bolt in the left tie rod sleeve.

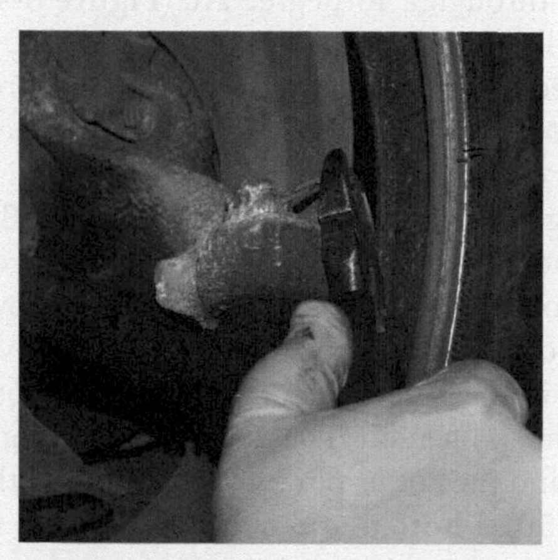

P15-5 Use a pair of side cutters to remove the cotter pin in the left outer tie rod end retaining nut.

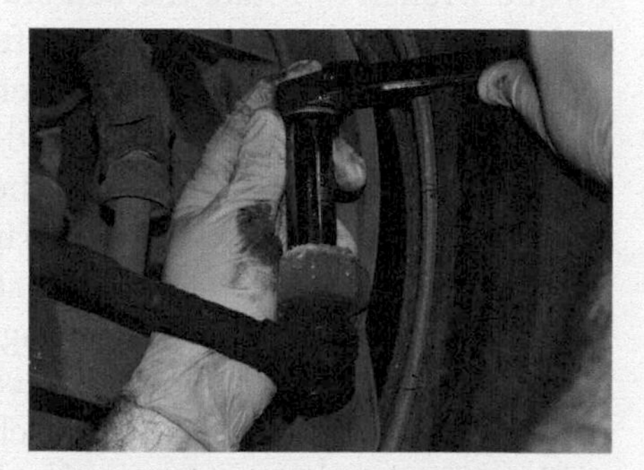

P15-6 Use the proper size socket and ratchet to remove the retaining nut on the outer tie rod end.

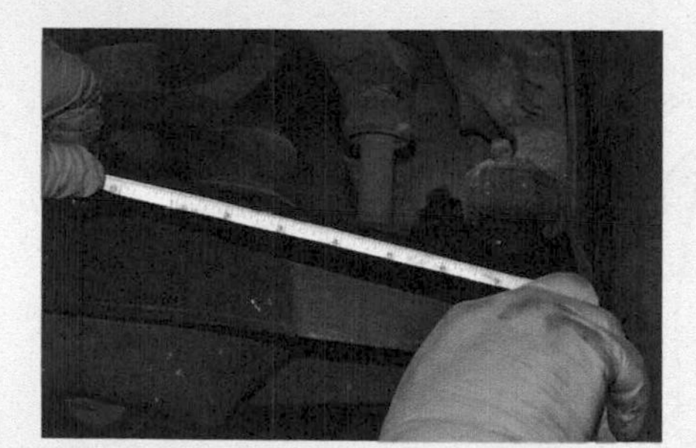

P15-7 Use a tape measure to carefully measure the distance from the outer edge of the left tie rod sleeve to the outer edge of the tie rod end, and record this distance.

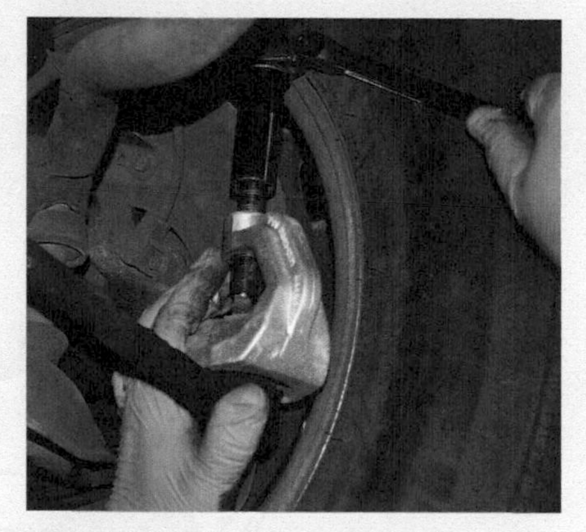

P15-8 Install a tie rod end puller on the left outer tie rod end and tighten the puller bolt to remove the tie rod end from the steering arm.

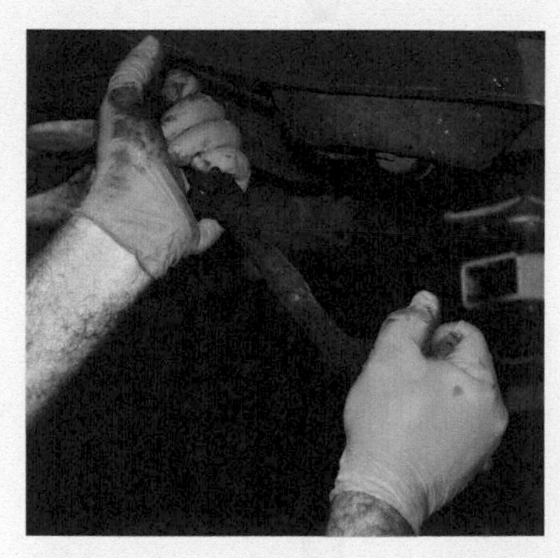

P15-9 After the tie rod end is removed from the steering arm, rotate the outer tie rod end to remove it from the tie rod sleeve.

PHOTO SEQUENCE 15 (CONTINUED)

P15-10 Inspect the outer tie rod sleeve for thread damage.

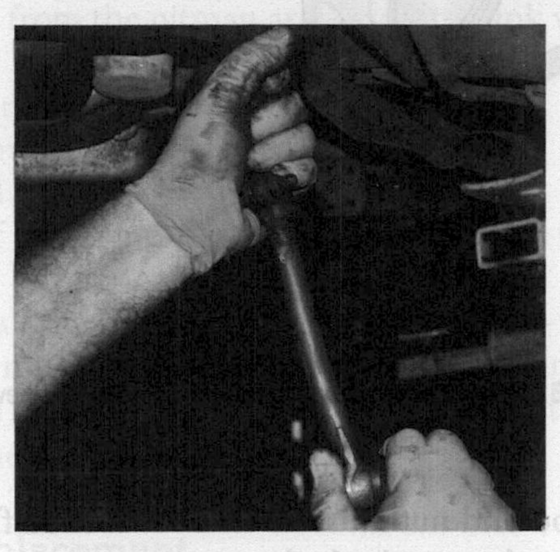

P15-11 Thread the new tie rod end into the outer tie rod sleeve until the distance from the outer edge of the sleeve to the outer edge of the tie rod end is exactly the same as the original distance measured and recorded previously.

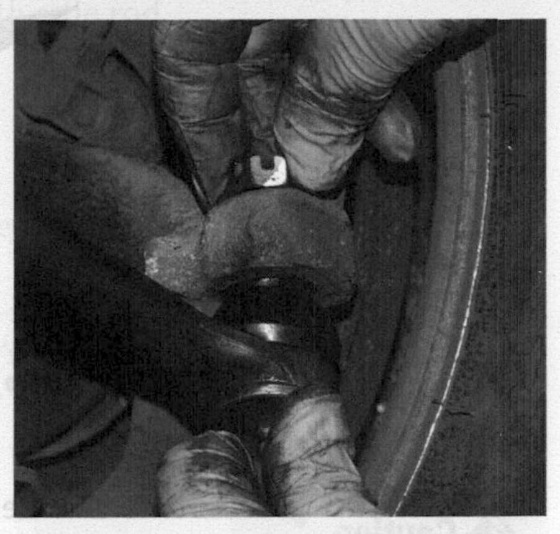

P15-12 Be sure the seal is properly installed on the new tie rod end, and install the outer tie rod end in the steering arm opening and install the nut to retain the tie rod end.

P15-13 Use a torque wrench and the proper size socket to tighten the left outer tie rod nut to the specified torque. Be sure the serrated openings in the nut are aligned with the hole in the tie rod stud.

P15-14 Install the cotter pin in the left outer tie rod nut and bend the ends of this cotter pin around the nut.

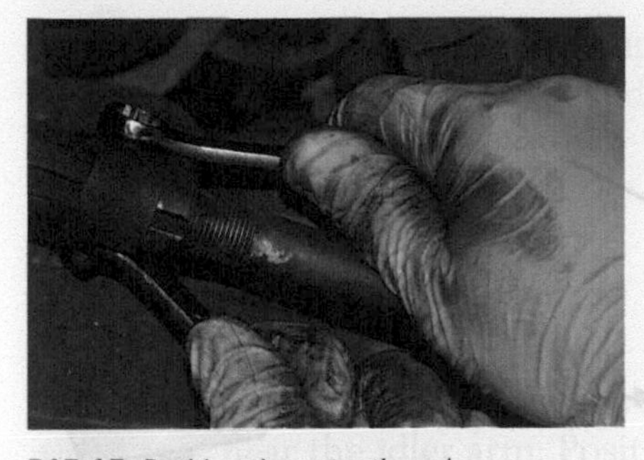

P15-15 Position the outer tie rod sleeve clamp so the clamp opening is positioned away from the slots in the tie rod sleeve, and tighten the clamp bolt nut to the specified torque.

P15-16 Inflate all the vehicle tires to the specified pressure, and position the vehicle properly on a wheel alignment ramp. Use the wheel aligner to check the front wheel toe. Adjust the front wheel toe if necessary.

TABLE 8-4 STEERING LINKAGE DIAGNOSIS

Condition	Possible Cause	Correction
Excessive play or looseness in steering system	1 Front wheel bearings loosely adjusted 2 Worn steering shaft couplings 3 Worn upper ball joints 4 Steering wheel loose on shaft or loose pitman arm, tie rods, steering arms, or steering linkage ball studs 5 Steering gear thrust bearings loosely adjusted 6 Excessive over-center lash in steering gear 7 Worn intermediate rod or tie rod sockets	1 Adjust bearings to obtain proper end play. 2 Replace part. 3 Check and replace if necessary. 4 Tighten to specified torque, or replace if necessary. 5 Adjust preload to specifications. 6 Adjust preload to specifications per shop manual. 7 Replace worn part.
Excessive looseness in tie rod or intermediate rod pivots, or excessive vertical lash in idler support	Seal damage and leakage resulting in loss of lubricant, corrosion, and excessive wear	Replace damaged parts as necessary. Properly position upon reassembly.
Hard steering—excessive effort required at steering wheel	1 Low or uneven tire pressure 2 Steering linkage or bolt joints need lubrication 3 Tight or frozen intermediate rod, tie rod, or idler socket 4 Steering gear-to-column misalignment 5 Steering gear adjusted too tightly 6 Front wheel alignment incorrect (manual gear)	1 Inflate to specified pressures. 2 Lube with specified lubricant. 3 Lube, replace, or reposition as necessary. 4 Align column. 5 Adjust over-center and thrust bearing preload to specification. 6 Check alignment and correct as necessary.
Poor returnability	1 Steering linkage or ball joints need lubrication 2 Steering gear adjusted too tightly 3 Steering gear-to-column misalignment 4 Front wheel alignment incorrect (caster)	1 Lube with specified lubricant. 2 Adjust over-center and thrust bearing preload to specifications. 3 Align columns. 4 Check alignment and correct as necessary.

CASE STUDY

A customer complained about excessive steering effort and poor steering wheel returnability after a turn on a 2009 General Motors Sierra truck with power steering. The technician road tested the car and found the customer's description of the problems to be accurate except for one point. The steering continually required excessive steering effort, and the steering wheel did not return properly after a turn. However, the customer did not mention, or possibly did not notice, that a squawking and creaking noise was sometimes heard during a turn.

The technician checked the power steering fluid level and condition, and found this fluid to be in good condition and at the proper level. Next, the technician checked the power steering belt condition and the power steering pump pressure. The belt tension and condition were satisfactory, and the power steering pump pressure was normal. A check of the power steering pump mounting bolts and brackets indicated they were in good condition. The technician disconnected the center link from the pitman arm and rotated the steering wheel with the engine running. Under this condition, the steering wheel turned very easily. Therefore, the technician concluded the excessive steering effort and poor returnability problems were not in the steering column or steering gear. Next, the technician disconnected the idler arm from the center link. When the technician attempted to move the idler arm back and forth, the idler arm had a severe binding problem. After an idler arm replacement, a road test revealed the excessive steering effort and poor returnability problems had disappeared.

ASE-STYLE REVIEW QUESTIONS

1. While servicing a steering column on an air bag-equipped vehicle, the system must be disarmed by:
   A. Locking the crash sensor in the disarmed position.
   B. Rewinding the clock spring.
   C. Using a scan tool to disable the system.
   D. Disconnecting the battery for manufacturer specified amount of time.
2. A steering wheel on an air bag-equipped vehicle moves from side to side in the column. Which of the following is the most likely cause?
   A. A loose rack mounting bushing.
   B. A worn upper steering shaft bearing.
   C. A worn pitman (sector) shaft.
   D. Loose steering gear box mounting bolts.
3. On a rack and pinion type steering system, the best way to inspect the inner tie rod for wear is with the wheels:
   A. Removed from the vehicle and the outer tie rod disconnected.
   B. Resting on the alignment rack turn plates.
   C. With the vehicle lifted off the ground by the frame or unibody and the suspension in the full rebound position.
   D. Steering the wheel to the left.
4. While discussing steering wheel removal:
   Technician A says a steering wheel should be marked in relation to the steering shaft.
   Technician B says the steering wheel should be grasped firmly with both hands and pulled from the shaft.
   Who is correct?
   A. A only C. Both A and B
   B. B only D. Neither A nor B
5. While discussing collapsible steering column damage:
   Technician A says if the vehicle has been in a frontal collision, the injected plastic in the column jacket, gearshift tube, and steering shaft may be sheared.
   Technician B says after a collision, some steering columns may be shifted on the bracket.
   Who is correct?
   A. A only C. Both A and B
   B. B only D. Neither A nor B
6. On a rack and pinion steering system when the inner tie rod is replaced the technician should:
   A. Set the steering rack preload.
   B. Center the steering rack piston prior to removing the inner tie rod.
   C. After the repair the power steering system should be bled.
   D. The rack piston should be prevented from rotating.
7. A rattling noise in the steering column and linkage may be caused by:
   A. A bent center link.
   B. A binding idler arm.
   C. A worn steering shaft U-joint.
   D. Loose tie rod sleeve clamps.
8. When diagnosing and servicing tie rod ends:
   A. Rubber-encapsulated tie rod ends should be tightened with the wheels turned fully to the right or left.
   B. The machined part of the tie rod stud should be visible above the surface of the steering arm.
   C. The nut on the tie rod stud may be loosened to install the cotter pin.
   D. If the turning torque on a rubber-encapsulated tie rod end is less than specified, the tie rod end must be replaced.

Task Completed

5. Pull the air bag deployment module from the steering wheel, and disconnect the air bag module electrical connector. Do not pull on the air bag wires in the steering column. Place the air bag deployment module face upward on the workbench. ☐

6. Disconnect the air bag wiring retainer in the steering wheel. ☐

7. Use the proper size socket and a ratchet to remove the steering wheel retaining nut. ☐

8. Observe the matching alignment marks on the steering wheel and the steering shaft. If these alignment marks are not present, place alignment marks on the steering wheel and steering shaft with a center punch and a hammer.

   Alignment marks on steering wheel and steering shaft:
   ☐ Satisfactory ☐ Unsatisfactory

WARNING Do not pull on the steering wheel in an attempt to remove it from the steering shaft. The steering wheel may suddenly come off, resulting in personal injury, or the steering wheel may be damaged by the pulling force.

9. Install a steering wheel puller with the puller bolts threaded into the bolt holes in the steering wheel. Tighten the puller nut to remove the steering wheel. Visually check the steering wheel condition. If the steering wheel is bent or cracked, replace the wheel.

   Steering wheel condition: ☐ Satisfactory ☐ Unsatisfactory

10. Disconnect the four retaining screws, and remove the clock spring electrical connector. ☐

11. Be sure the front wheels are facing straight ahead. Turn the clock spring electrical connector counterclockwise by hand until it becomes harder to turn as it becomes fully wound in that direction. ☐

12. Turn the clock spring electrical connector clockwise three turns, and align the red mark on the center part of the spring face with the notch in the cable circumference. This action centers the clock spring electrical connector.

    Is the clock spring centered? ☐ Yes ☐ No

    Instructor check ______________________________

    Explain why the clock spring electrical connector must be centered before it is installed.

    ____
    ____
    ____

13. Install the clock spring electrical connector, and tighten the four retaining screws to the specified torque. ☐

14. Align the marks on the steering wheel and the steering shaft, and install the steering wheel on the shaft.

    Are the marks on the steering wheel and steering shaft aligned? ☐ Yes ☐ No

    Instructor check ______________________________

15. Install the steering wheel retaining nut, and tighten this nut to the specified torque.

    Specified steering wheel retaining nut torque ______________________________

    Actual steering wheel retaining nut torque ______________________________

Caution

Do not hammer on the top of the steering shaft to remove the steering wheel. This action may damage the shaft.

Task Completed

16. Install the air bag wiring retainer in the steering wheel. ☐

17. Hold the air bag deployment module near the top of the steering wheel, and connect the air bag module connector. ☐

18. Install the air bag deployment module in the top of the steering wheel, and tighten the retaining screws. ☐

19. Reconnect the negative battery cable. ☐

20. Reset the clock and radio. ☐

21. Turn on the ignition switch and start the vehicle. Check the operation of the air bag system warning light.

    Explain the air bag system warning light operation that indicates normal air bag system operation.

    ____

    ____

    ____

Caution

Failure to center a clock spring prior to installation may cause a broken conductive tape in the clock spring.

Instructor's Response

____

____

____

State the necessary repairs and explain the reasons for your diagnosis.

____

____

____

____

3. While an assistant turns the steering wheel about one-half turn in both directions, watch for movement of the steering gear housing in the mounting bushings.

Looseness in steering gear mounting bushings: ☐ Satisfactory ☐ Unsatisfactory

State the necessary repairs and explain the reasons for your diagnosis.

____

____

____

____

4. Grasp the pinion shaft extending from the steering gear and attempt to move it vertically. If there is steering shaft vertical movement, a pinion bearing preload adjustment may be required. When the steering gear does not have a pinion bearing preload adjustment, replace the necessary steering gear components.

Excessive pinion shaft vertical end play: ☐ Satisfactory ☐ Unsatisfactory

State the necessary repairs and explain the reasons for your diagnosis.

____

____

____

____

5. Road test the vehicle and check for excessive steering effort. A bent steering rack, tight rack bearing adjustment, or damaged front drive axle joints in front-wheel-drive cars may cause excessive steering effort.

Steering effort: ☐ Satisfactory ☐ Excessive

State the necessary repairs and explain the reasons for your diagnosis.

____

____

____

____

6. Visually inspect the bellows boots for cracks, splits, leaks, and proper clamp installation. Replace any boot that indicates any of these conditions.

Condition of tie rod bellows boots and clamps:

Left side boot: ☐ Satisfactory ☐ Unsatisfactory

Right side boot: ☐ Satisfactory ☐ Unsatisfactory

State the necessary repairs and explain the reasons for your diagnosis.

____

____

____

____

7. Loosen the inner bellows boot clamps and move each boot toward the outer tie rod end until the inner tie rod end is visible. Push outward and inward on each front tire, and watch for movement in the inner tie rod end. An alternate method of checking the inner tie rod ends is to squeeze the bellows boots and grasp the inner tie rod end socket. Movement in the inner tie rod end is then felt as the front wheel is moved inward and outward. Hard plastic bellows boots may be found on some applications. With this type of bellows boot, remove the ignition key from the switch to lock the steering column, and push inward and outward on the front tire while observing any lateral movement in the tie rod.

   Condition of inner tie rod ends:

   Left side inner tie rod end: ☐ Satisfactory ☐ Unsatisfactory

   Right side inner tie rod end: ☐ Satisfactory ☐ Unsatisfactory

   State the necessary repairs and explain the reasons for your diagnosis.

   ____

   ____

   ____

   ____

WARNING Bent steering components must be replaced. Never straighten steering components, because this action may weaken the metal and result in sudden component failure, serious personal injury, and vehicle damage.

8. Grasp each outer tie rod end and check for vertical movement. While an assistant turns the steering wheel one-quarter turn in each direction, watch for looseness in the outer tie rod ends. Check the outer tie rod end seals for cracks and proper installation of the nuts and cotter pins. Cracked seals must be replaced. Inspect the tie rods for a bent condition.

   Tie rod and outer tie rod end condition:

   Left side tie rod and outer tie rod end condition: ☐ Satisfactory ☐ Unsatisfactory

   Right side tie rod and outer tie rod end condition: ☐ Satisfactory ☐ Unsatisfactory

   List the necessary repairs and explain the reasons for your diagnosis.

   ____

   ____

   ____

   ____

Instructor's Response

____

____

____

Task Completed

9. Install the new tie rod end in the tie rod sleeve using the same number of turns recorded in Step 8. ☐

10. Push the tie rod end ball stud fully into the steering arm opening. Only the threads on the tie rod end ball stud should be visible above the steering arm surface.

Is the taper on the tie rod end ball stud visible above the steering arm surface?
☐ Yes ☐ No

If the answer to this question is yes, state the necessary repairs required to correct this problem.

____

____

Instructor check

11. With the tie rod end ball stud pushed fully upward into the steering arm, measure the distance from the center of the ball stud to the outer end of the tie rod sleeve.

Distance from the center of the tie rod end ball stud to the outer end of the tie rod sleeve _______________

12. If this distance is not the same as recorded in Step 3, remove the tie rod end ball stud from the steering arm, and rotate that the tie rod end until this distance is the same as recorded in Step 3. ☐

13. Install the tie rod end ball stud retaining nut, and tighten this nut to the specified that torque.

Specified ball stud nut torque _______________

Actual ball stud nut torque _______________

WARNING Never loosen a tie rod end ball stud nut to align the openings in the ball stud and nut to allow cotter pin installation. This action may cause the nut to loosen during operation of the vehicle, and this will result in excessive steering free play. Complete loss of steering control will occur if the nut comes off the tie rod end ball stud, and this may cause a collision.

14. Install a new cotter pin through the tie rod end ball stud and nut openings, and bend the cotter pin legs separately around this nut. ☐

15. Tighten the nut on the outer tie rod sleeve clamp bolt to the specified torque. Be sure the slot in the tie rod sleeve is positioned away from the opening in the clamp. ☐

16. Measure the front wheel toe with a toe bar.

Specified front wheel toe _______________

Actual front wheel toe _______________

Explain the type of front tire wear that is caused by improper front wheel toe.

____

Instructor's Response

____

____

____

CHAPTER 9

FOUR-WHEEL ALIGNMENT PROCEDURE, PART 1

Upon completion and review of this chapter, you should be able to:

- Perform a prealignment inspection.
- Diagnose wheel alignment and tire wear problems.
- Position a vehicle on the alignment rack and connect the wheel units on a computer wheel aligner to the wheel rims.
- Select the specifications in the computer wheel aligner for the vehicle being aligned.
- Perform a preliminary wheel alignment inspection with the preliminary inspection screen.
- Perform a ride height inspection and measurement with the ride height screen.
- Perform an automatic or a manual wheel runout compensation procedure.
- Measure front wheel camber and caster.
- Recognize the symptoms of improper rear wheel alignment.
- Measure rear wheel camber and toe.
- Diagnose the causes of improper rear wheel alignment.
- Perform rear wheel camber adjustments.
- Perform rear wheel toe adjustments.
- Adjust front wheel camber on various front suspension systems.
- Adjust front wheel caster on various front suspension systems.
- Adjust front wheel toe.
- Center steering wheel.
- Measure toe change during front wheel jounce and rebound to check steering linkage height.

Basic Tools

Basic technician's tool set

Service manual

Terms To Know

Advanced vehicle handling (AVH)
Alignment ramp
Brake pedal depressor
Bump steer
Camber adjustment
Caster adjustment
Caster offset
Caster trail
Control arm movement monitor
Digital adjustment photos
Digital signal processor (DSP)
Eccentric cams
Front and rear wheel alignment angle screen
High-frequency transmitter
Included angle
Memory steer
Part-finder database
Prealignment inspection
Preliminary inspection screen
Rear wheel camber
Rear wheel toe
Ride height
Rim clamps
Road test
Setback
Shim display screen
Slip plates
Thrust line
Toe-in
Toe-out
Torque steer
Total toe

Proper front and rear wheel alignment is extremely important because it affects directional stability, tire tread wear, and vehicle safety! Technicians must know how to check front and rear wheel alignment angles and diagnose the causes of steering and alignment problems. It is also essential for technicians to know how to adjust front and rear

Prealignment Inspection

The technician should identify the exact suspension or steering complaint before a wheel alignment or other suspension work is performed. The technician must diagnose the cause or causes of the complaint and correct this problem during the wheel alignment (**Figure 9-3**). Because collision damage may affect wheel alignment angles, an inspection for collision damage on the vehicle should be completed prior to a wheel alignment (**Figure 9-4**).

Worn suspension and steering components must be replaced before a wheel alignment is performed.

Worn suspension and steering components must be replaced before a wheel alignment is performed. A wheel alignment should be performed after suspension components such as struts have been replaced.

Condition	Probable Cause
Tire wear:	
Outer shoulder	Too much positive camber
Inner shoulder	Too much negative camber
Sawtooth pattern:	
Sharp edge toward center	Too little toe-in
Sharp edge away from center	Too much toe-in
Both shoulders	Underinflation, high-speed cornering, overloading
Tread roll under	High speed in curves
Cupping, dishing	Wheel imbalance, radial runout
Diagonal wipe	Incorrect rear toe front-wheel-drive vehicle
Shimmy:	
With or without vehicle instability	Too much positive caster, unequal caster between wheels
Without instability	Tire imbalance, tire runout, driveline vibration
Vibration:	
Caster/camber not a probable cause. The condition is listed because it is sometimes misdiagnosed as shimmy.	Driveline misalignment, driveline imbalance, vehicle shake (accompanied by a characteristic moan), tire runout, unequal weight distribution between wheels
Steering wander/pull	Incorrect camber Unequal caster Overload, which elevates front end
Brake pull	Too much negative caster Unequal tire pressure, brake line damage that impedes hydraulic action on one side
Hard steering	Incorrect caster Damaged steering linkage, worn steering linkage, damaged spindles, rear-end overload, bent steering arm causing incorrect turning angle

Figure 9-3 Diagnosis of wheel alignment problems.

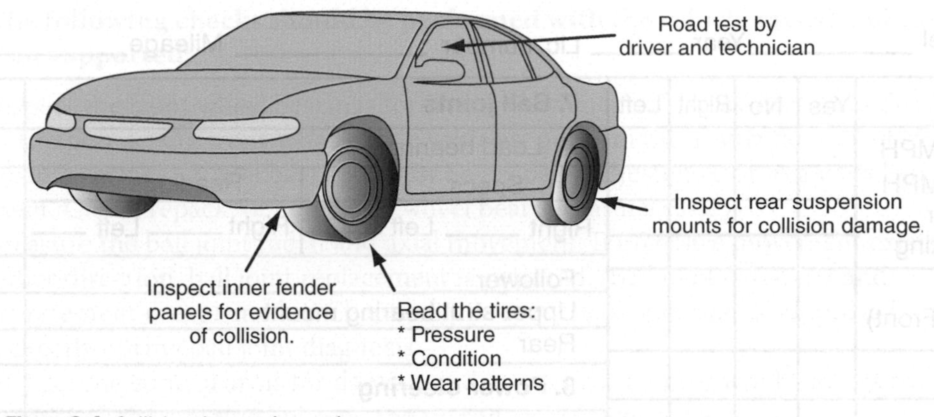
Look at the complete vehicle.
If you suspect collision damage to the inner body panels, measure the reference points or body and frame alignment as described in the service manual.
Road test by driver and technician
Inspect rear suspension mounts for collision damage.
Inspect inner fender panels for evidence of collision.
Read the tires:
* Pressure
* Condition
* Wear patterns

Figure 9-4 Collision damage inspection.

The following components and measurements should be checked in a prealignment inspection:

1. Curb weight
2. Tires
3. Suspension height
4. Steering wheel free play
5. Shock absorbers or struts
6. Wheel bearing adjustment
7. Ball joint condition
8. Control arms and bushings
9. Steering linkages and tie rod ends
10. Stabilizers and bushings
11. Full fuel tank

After suspension components such as struts have been replaced, a wheel alignment should be performed.

Components to be checked in a **prealignment inspection** and wheel alignment measurements are summarized in **Figure 9-5**.

A **prealignment inspection** of all steering and suspension components must be completed before a wheel alignment procedure.

These checks should be completed with the car on the floor:

1. Inspect for excessive mud adhered to the chassis. Remove heavy items from the trunk and passenger compartment that are not considered in the vehicle curb weight. If heavy items such as tool display cases or merchandise are normally carried in the vehicle, these items should be left in the car during a wheel alignment.
2. Inflate tires to the recommended pressure and note any abnormal tread wear or damage on each tire. Be sure all tires are the same size.
3. Check the front tires and wheels for radial runout.
4. Check the suspension **ride height**. If this measurement is not within specifications, check for broken or sagged springs. On a torsion bar suspension system, check the torsion bar adjustment.
5. Inspect front and rear suspension bumpers. Worn bumpers may indicate weak springs or worn-out shock absorbers or struts.
6. When the wheels are in the straight-ahead position, rotate the steering wheel back and forth to check for play in the steering column, steering gear, or linkages.
7. Inspect the shock absorbers or struts for loose mounting bushings and bolts. Examine each shock absorber or strut for leakage.

Ride height is the distance from a specified location on the front or rear suspension to the road surface.

Special Tool
Computer wheel aligner

An **alignment ramp** is a special steel ramp on which the vehicle is positioned for a wheel alignment.

A **turntable** is placed under each front wheel during a wheel alignment. The turntables allow the wheels to turn, and they also allow the wheels to move during alignment adjustments.

Slip plates under the rear wheels allow the wheels to move during alignment adjustments.

Rim clamps mount each wheel sensor to the rim.

A **digital signal processor (DSP)** in each wheel sensor provides digital signals indicating wheel alignment angles.

A **high-frequency transmitter** in each wheel sensor processes data from the DSP and transmits this data to the computer wheel aligner.

Classroom Manual
Chapter 9, page 353

FOUR-WHEEL ALIGNMENT WITH COMPUTER ALIGNMENT SYSTEMS

Preliminary Procedure

WARNING When using a computer wheel aligner, always follow the equipment manufacturer's recommended procedures to provide accurate readings and to avoid equipment and vehicle damage or personal injury.

The vehicle should be on an **alignment ramp** with a *turntable* under each front tire and conventional **slip plates** under the rear tires (**Figure 9-7**). Slip plates under the rear tires allow unrestricted movement of the rear wheels during rear wheel camber and toe adjustments (**Figure 9-8**). If suspension adjustments are made with the tires contacting the alignment ramp or the shop floor, the tires cannot move when the adjustment is completed. This action causes inaccurate suspension adjustments. The center of the front wheel spindles should be in line with the zero mark on the turning plates. The locking pins must be left in these plates and the parking brake applied.

Mount the **rim clamps** on each wheel. It may be necessary to remove the chrome discs or hub caps before these clamps can be installed. The adjustment knob on the rim clamp should be pointing toward the top of the wheel, and the bubble on the wheel unit should be centered (**Figure 9-9**). A set knob on the wheel unit allows the service technician to lock the wheel unit in this position. Some computer wheel aligners have a display on the monitor that indicates if any of the wheel sensors are not level (**Figure 9-10**). These sensors may be leveled by pressing a control knob on each wheel sensor. Some **digital signal processor (DSP)** wheel sensors contain a microprocessor and a **high-frequency transmitter** that acquire measurements and process data and then send these data to a *receiver* mounted on top of the wheel alignment monitor. This type of wheel sensor does not require any cables connected between the sensors and the computer wheel aligner. The data from this type of wheel sensor signal are virtually uninterruptible, even by solid objects. When these wheel sensors are stored on the computer wheel aligner, a "docking station" feature charges the batteries in the wheel sensors.

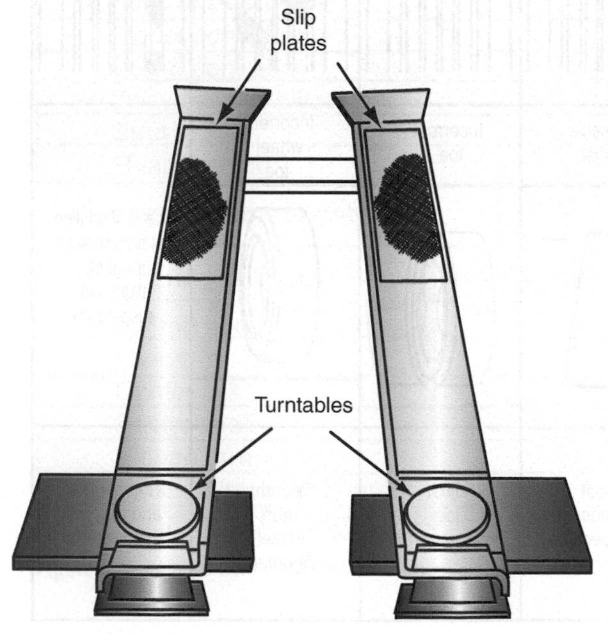

Figure 9-7 Alignment ramp with turntables.

Figure 9-8 Rear wheel slip plates.

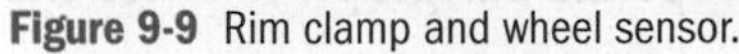

Figure 9-9 Rim clamp and wheel sensor.

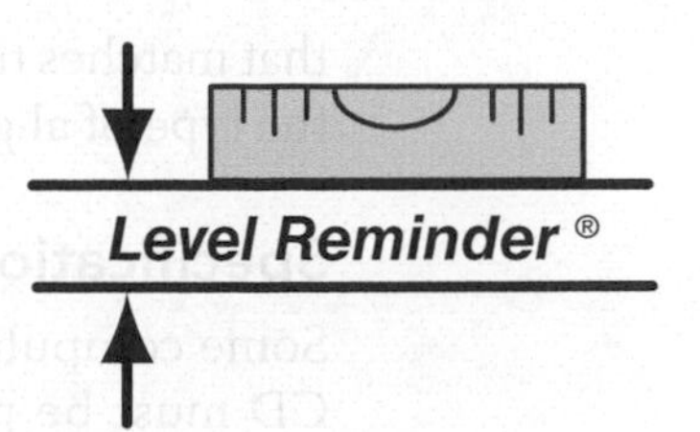

Figure 9-10 Wheel sensor level indicator on the monitor screen.

Some wheel sensors contain infrared emitter/detectors or LEDs between the front and rear wheel sensors to perform alignment measurements. Older wheel sensors have strings between the front and rear wheel sensors. Each wheel sensor contains a microprocessor with a preheat circuit that stabilizes the readings in relation to temperature. Some wheel sensors have touch keypads on the wheel units to allow the entry of commands from each unit. The front wheel units have arms that project toward the front of the vehicle to transmit signals between these units. When a blocked beam prompt appears on any screen, the beam between two wheel units is blocked. This could be caused by a person standing in the beam, an open car door, or a suspension jack. The blocked beam prompt must be eliminated before any tests are completed.

A **receiver** mounted on or in the computer wheel aligner receives data from the high-frequency transmitters in the wheel sensors.

The manufacturers of various types of computer alignment systems publish detailed operator's manuals for their specific equipment. Our objective here is to discuss some of the general screens that a technician uses while measuring wheel alignment angles with a typical computer aligner.

Main Menu

One of the first items displayed on the computer aligner screen is the *main menu* screen. From this screen, the technician makes a selection by touching the desired selection on the monitor screen with a mouse (**Figure 9-11**) or by pressing the number on the keypad

Figure 9-11 Making screen selections with a mouse.

that matches the number beside the desired procedure. A cursor may also be used to select the type of alignment.

Specifications Menu

Some computer wheel aligners have specifications contained on CDs. The appropriate CD must be placed in the aligner. Several methods may be used to enter the vehicle specifications. The technician may scroll through the *specifications menu* and then press Enter when the appropriate vehicle is highlighted on the screen (**Figure 9-12**). The technician may also select the proper vehicle specifications by pointing with the light pen or by typing in the vehicle identification number (VIN) or the first letter in the vehicle make.

On other computer wheel aligners, the specifications are contained on a floppy disk and the technician selects the appropriate make, model, and year of vehicle with the cursor on the screen.

On early computer alignment systems, the vehicle specifications were contained on bar graphs. The technician moved a wand across the appropriate bar graph to enter the specifications for the vehicle being aligned.

Preliminary Inspection Screen

The **preliminary inspection screen** allows the technician to enter the condition of all front and rear suspension components that must be checked during a prealignment inspection. On some computer alignment systems, the technician may enter checked, marginal, and repair or replace for each component (**Figure 9-13**). These entered conditions may be printed out at any time or with the complete wheel alignment results.

Caution

The ride height must be within specifications before proceeding with the wheel alignment. Improper curb riding height affects many of the other suspension angles.

Ride Height Screen

Some computer wheel aligners have optical encoders in the wheel sensors that measure ride height when this selection is displayed on the screen and the ride height attachment on each wheel sensor is lifted until it touches the lower edge of the fender (**Figure 9-14**). The ride height measurement is displayed on the screen. If the screen display is green, the ride height is within specifications. A red ride height display indicates this measurement is not within specifications. Improper ride height may be caused by sagged or broken springs, bent components such as control arms, or worn components such as control arm bushings.

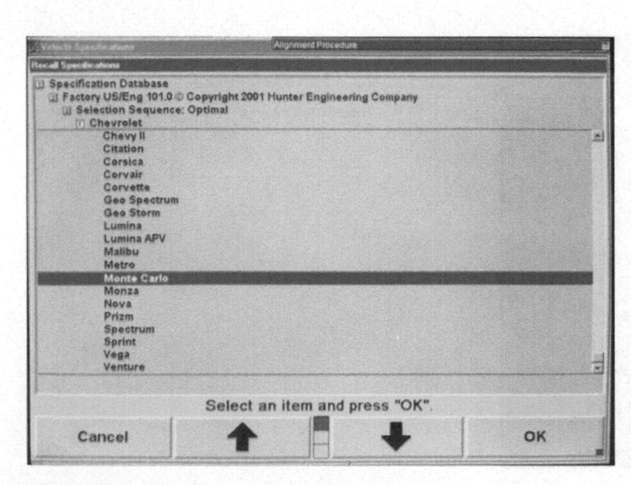

Figure 9-12 Entering vehicle specifications.

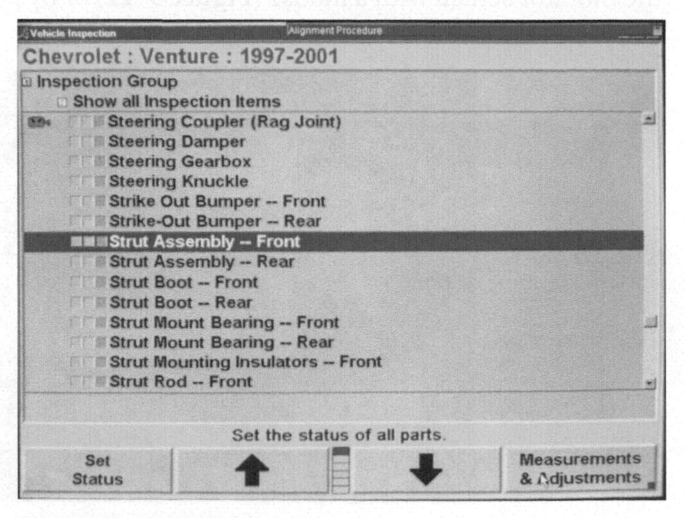

Figure 9-13 Preliminary inspection screen.

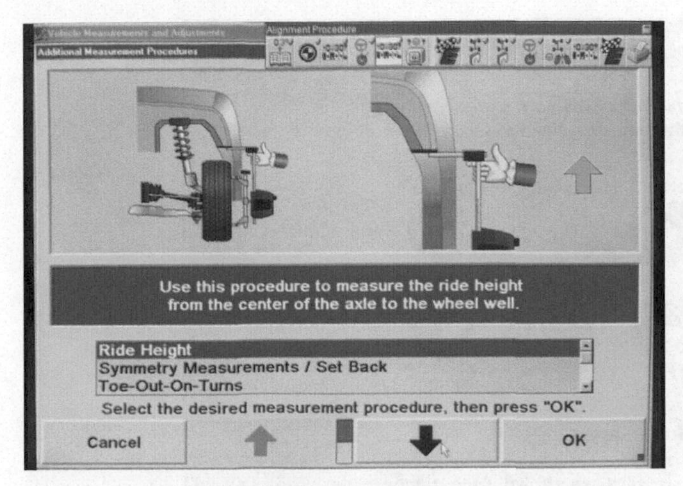

Figure 9-14 Ride height screen and ride height attachment in wheel sensor.

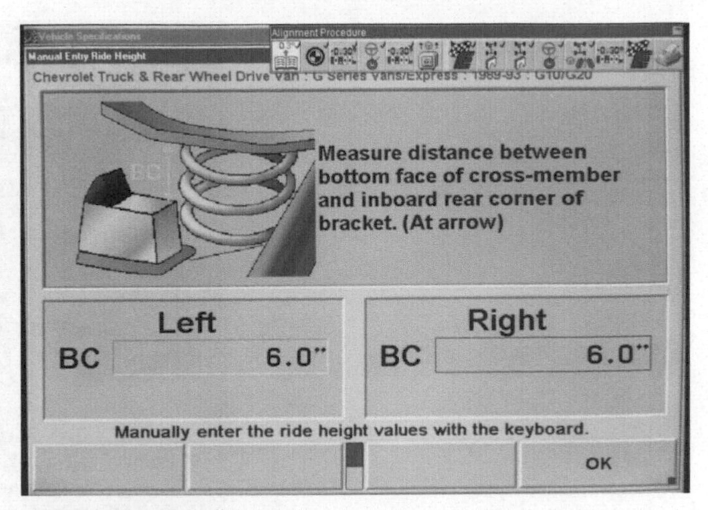

Figure 9-15 Ride height screen.

Some computer wheel aligners provide a **ride height screen** with a graphic display indicating the exact location where the ride height should be measured (**Figure 9-15**). The computer aligner compares the ride height measurements entered by the technician to the specifications.

Tire Condition Screen

When the **tire inspection screen** is displayed, the technician may enter various tire wear conditions for each tire (**Figure 9-16**). The cursor on the screen is moved to the tire being inspected. The tire conditions are printed out with the preliminary inspection results and the alignment report. On other computer wheel aligners, tire condition is included in the preliminary inspection screens.

Wheel Runout Compensation

Classroom Manual
Chapter 9, page 256

As mentioned previously, screen indicators on some computer wheel aligners inform the technician if any of the wheel sensors require leveling or compensating for wheel runout. Runout compensation is accomplished by pressing the appropriate button on each wheel sensor. Some wheel sensors provide continuous compensation. This feature provides accurate alignment angles even when a wheel is rotated after the compensation

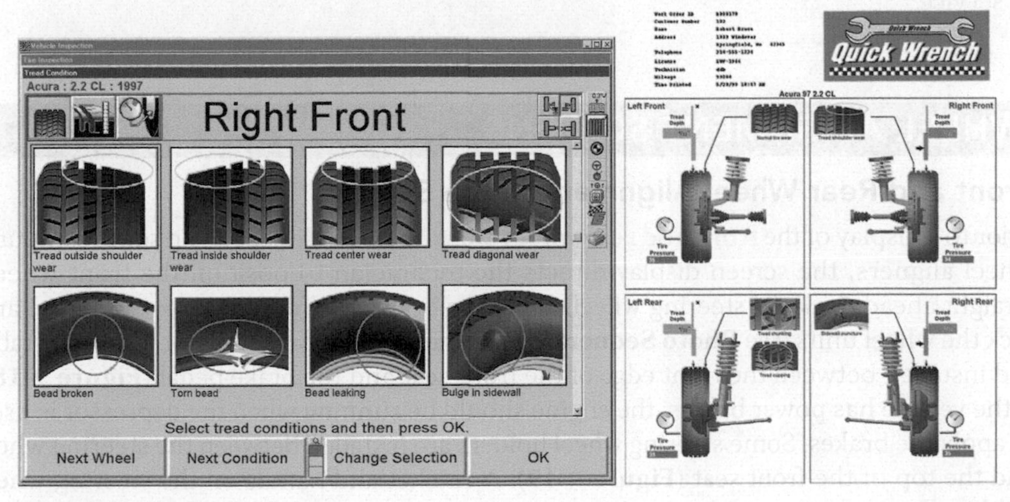

Figure 9-16 Tire inspection screen.

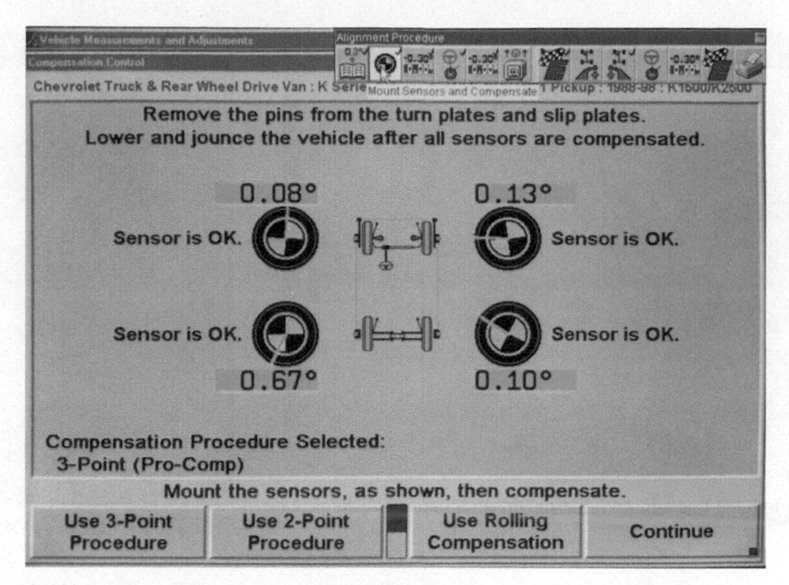

Figure 9-17 Wheel runout compensation screen.

button on the wheel sensor has been pressed. When the wheel runout screen is displayed, the technician is directed to level and lock each wheel unit and then press the compensation button on each wheel sensor to provide automatic wheel runout compensation (**Figure 9-17**).

If the computer aligner does not have automatic wheel runout compensation, a manual wheel runout procedure must be followed. This type of computer aligner displays a wheel runout measurement screen. During this procedure, the wheel being checked for runout is lifted with the hydraulic jack on the alignment rack and the wheel is rotated until the rim clamp knob faces downward. Level and lock the wheel unit in this position; then push Yes on the wheel unit as instructed on the screen. Rotate the wheel until the rim clamp knob faces upward; then level and lock the wheel unit. After this procedure, press Yes on the wheel unit. This same basic procedure is followed at each wheel.

SERVICE TIP Many computer wheel aligners can print out any single screen. It may be helpful to print out the wheel alignment angle screen before and after the alignment angles are adjusted to the vehicle manufacturer's specifications. These printouts may be presented to the customer; many customers appreciate this service.

WHEEL ALIGNMENT SCREENS

Front and Rear Wheel Alignment Angle Screen

Prior to a display of the **front and rear wheel alignment angle screen** on some computer wheel aligners, the screen display directs the technician to position the front wheels straight ahead, lock the steering wheel, apply the **brake pedal depressor**, and level and lock the wheel units (see **Photo Sequence 16**). The brake pedal depressor is an adjustable rod installed between the front edge of the front seat and the brake pedal (**Figure 9-18**). If the vehicle has power brakes, the engine should be running when the depressor is used to apply the brakes. Some steering wheel holders are installed between the steering wheel and the top of the front seat (**Figure 9-19**). A ratchet and handle on the steering wheel holder allow extension of this holder.

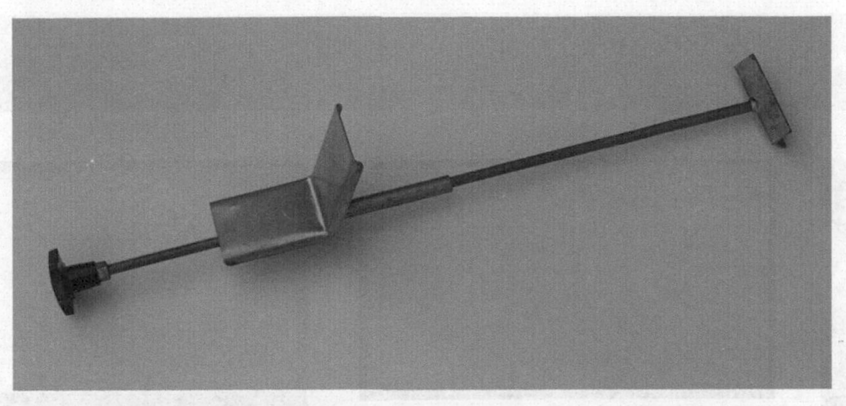

Figure 9-18 Brake pedal depressor.

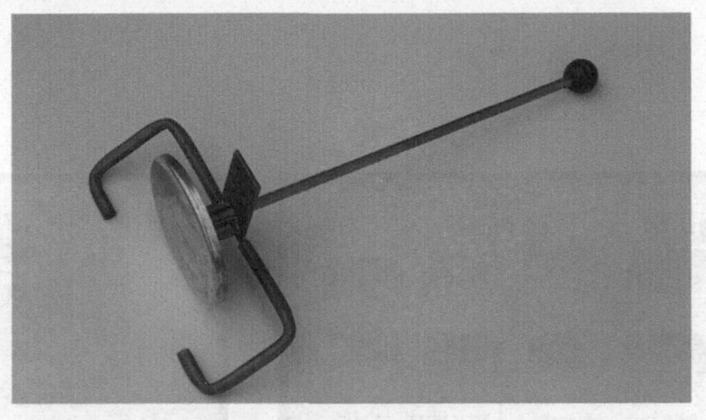

Figure 9-19 Steering wheel holder.

PHOTO SEQUENCE 16

Typical Procedure for Performing Four-Wheel Alignment with a Computer Wheel Aligner

P16-1 Position the vehicle on the alignment ramp.

P16-2 Be sure the front tires are positioned properly on the turntables.

P16-3 Position the rear wheels on slip plates.

PHOTO SEQUENCE 16 (CONTINUED)

P16-4 Attach the wheel units.

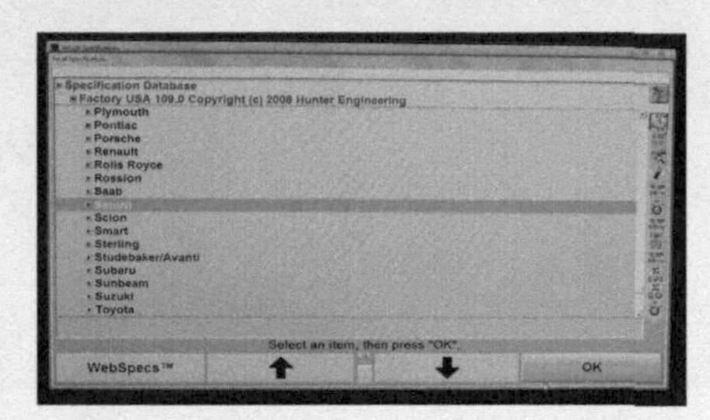

P16-5 Select the vehicle make and model year.

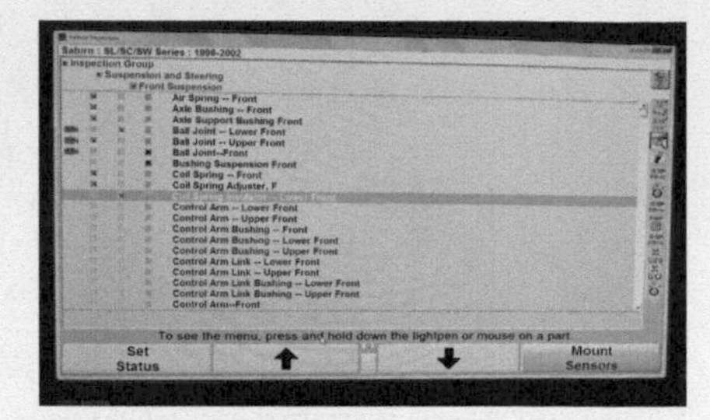

P16-6 Check items on the screen during preliminary inspection.

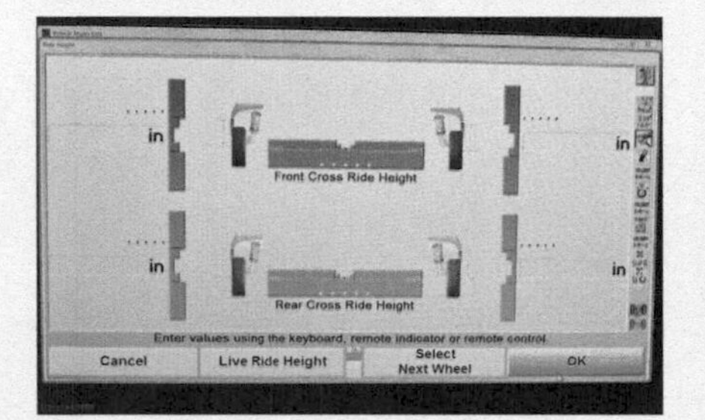

P16-7 Display the ride height screen.

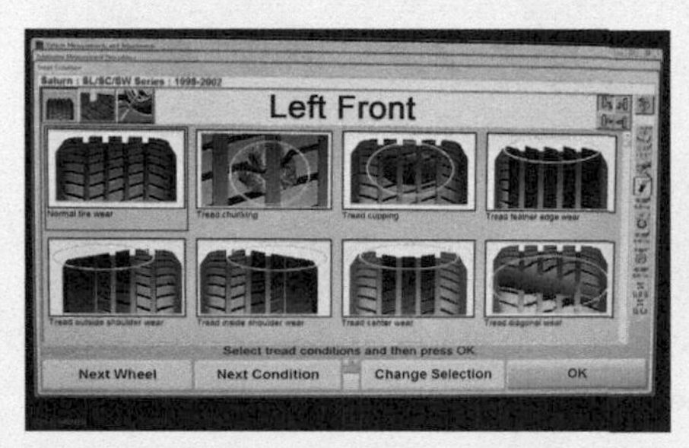

P16-8 Check the tire condition for each tire on the tire condition screen.

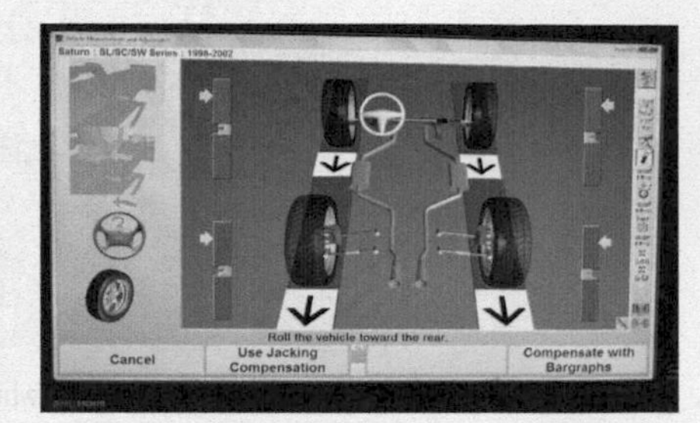

P16-9 Display the wheel runout compensation screen.

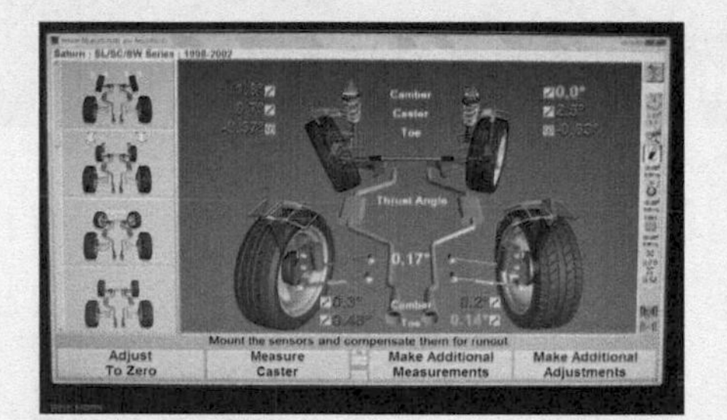

P16-10 Display the front and rear wheel alignment angle screen.

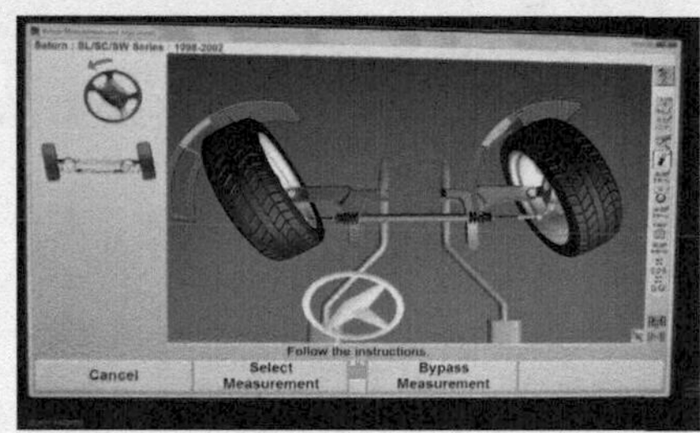

P16-11 Display the turning angle screen and perform the turning angle check.

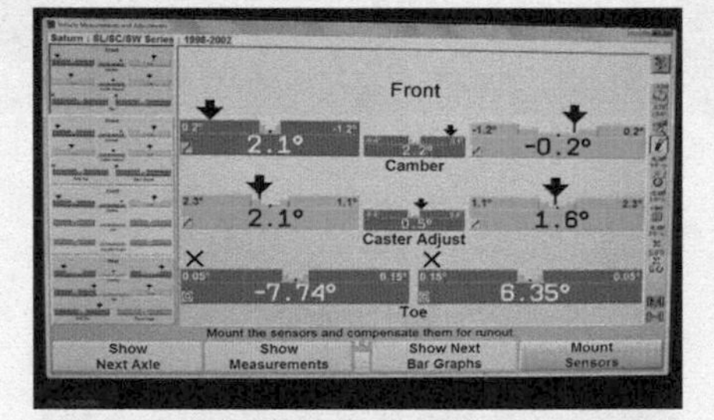

P16-12 Display the adjustment screens.

Camber is the tilt of a line through the tire and wheel centerline in relation to the true vertical centerline of the tire and wheel.

Total toe is the sum of the toe settings on the front wheels.

Setback is the distance that one front or rear wheel is moved rearward in relation to the opposite front or rear wheel.

After the wheel runout compensation procedure is completed at each wheel sensor, some computer wheel aligners automatically display *camber*, toe-in or toe-out, **total toe**, **setback**, and **thrust line** measurements on the front and rear suspension. After this procedure, the screen directs the technician to swing the front wheels outward a specific amount (**Figure 9-20**). The steering wheel holder must be released, and the lock pins removed from the front turntables before the wheel swing procedure. This front wheel swing may be referred to as a caster/SAI swing. The amount of front wheel swing varies depending on the make of the computer aligner. Older computer wheel aligners required the technician to read the degrees of wheel swing on the turntable degree scales. On some newer computer wheel aligners, the required amount of wheel swing is illustrated on the screen (**Figure 9-21**).

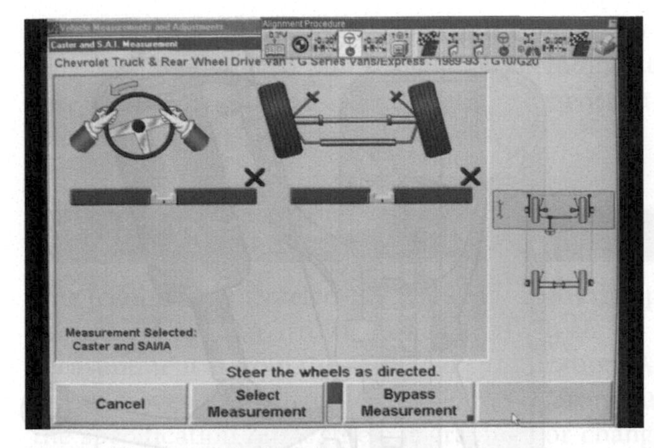

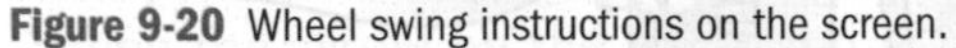

Figure 9-20 Wheel swing instructions on the screen.

Figure 9-21 Wheel swing procedure illustrated on the screen.

After the front wheel swing procedure, all the front and rear wheel alignment angles are displayed, including *caster, steering axis inclination (SAI),* and **included angle** (**Figure 9-22**), which are described in more detail in Chapter 10 of the Classroom Manual. Some computer wheel aligners highlight any wheel alignment angles that are not within specifications. If cross camber and cross caster are displayed on the screen, these readings indicate the maximum difference allowed between the right and left side readings. Alignment angles within specifications are highlighted in green; alignment angles that are not within specifications are highlighted in red.

The **thrust line** is positioned at 90 degrees in relation to the rear axle, and this line projects forward.

Caster is the tilt of a line through the centers of the lower ball joint and upper strut mount in relation to a vertical line through the center of the wheel and spindle as viewed from the side.

Toe-out is present when the distance between the front edges of the front or rear wheels is more than the distance between the rear edges of the front or rear wheels.

The **steering axis inclination (SAI)** line is an imaginary line through the centers of the upper and lower ball joints or through the center of the lower ball joint and the upper strut mount.

The **included angle** is the sum of the SAI angle and the camber angle if the camber is positive. If the camber is negative, the camber setting must be subtracted from the SAI angle to obtain the included angle.

The SAI angle is the angle between the SAI line and the true vertical centerline of the tire and wheel.

WinAlign Software with Advanced Vehicle Handling (AVH)

Advanced vehicle handling (AVH) is a standard feature of WinAlign computer alignment software. AVH allows the technician to locate the cause of hidden suspension, chassis, and body problems prior to alignment adjustments. All chassis and suspension components should be inspected during the preliminary alignment inspection, and any worn or defective components must be replaced prior to using the AVH alignment software. AVH measurements include the following:

1. **Caster offset**—Caster offset is the distance from the caster line to the center of the front spindle (**Figure 9-23**). Caster offset may be called **caster trail**. Assuming the right and left caster angles are equal, if the right and left caster offset measurements

Vehicle Measurements and Adjustments

Chevrolet Truck & Rear Wheel Drive Van : K Series/Silverado (4X4) 1972-2001 Pickup : 1999-2000 : Silverado (GMT800) : 1500

Front	Left		Right
Camber	0.4°		-0.2°
Cross Camber		0.6°	
Caster	4.0°		4.3°
Cross Caster		-0.4°	
SAI	10.3°		11.3°
Cross SAI		-1.0°	
Toe	-0.04°		0.10°
Total Toe		0.06°	
Rear	**Left**		**Right**
Camber	-0.1°		-0.5°
Cross Camber		0.4°	
Toe	0.15°		-0.05°
Total Toe		0.10°	
Thrust Angle		0.10°	

Save the "before" alignment measurements.

Show Virtual View | Show Secondary Measurements | Save "Before" Measurements

Figure 9-22 Front and rear wheel alignment screen.

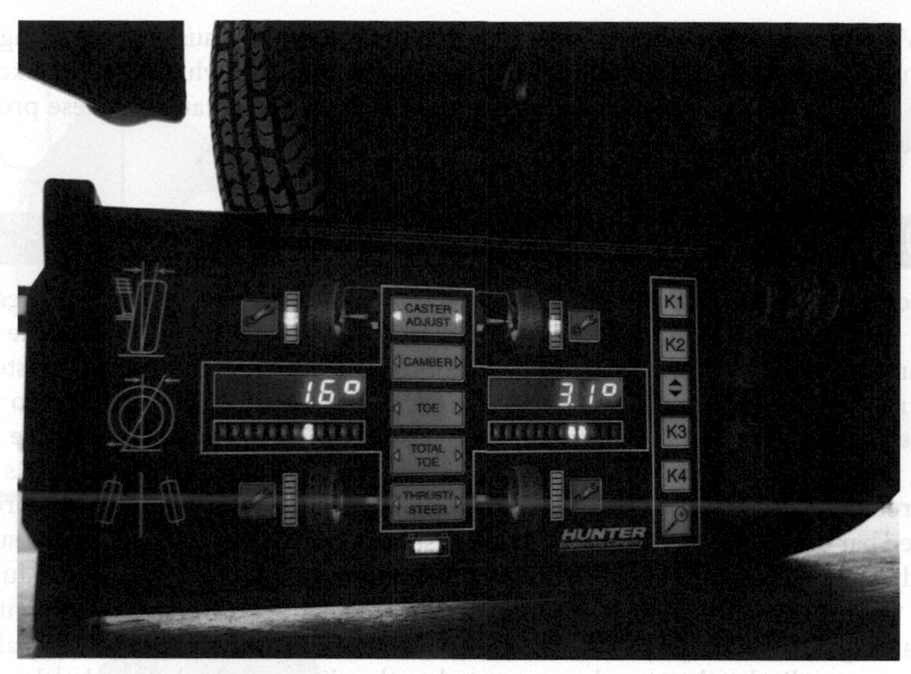

Figure 9-27 Remote display for computer wheel aligner.

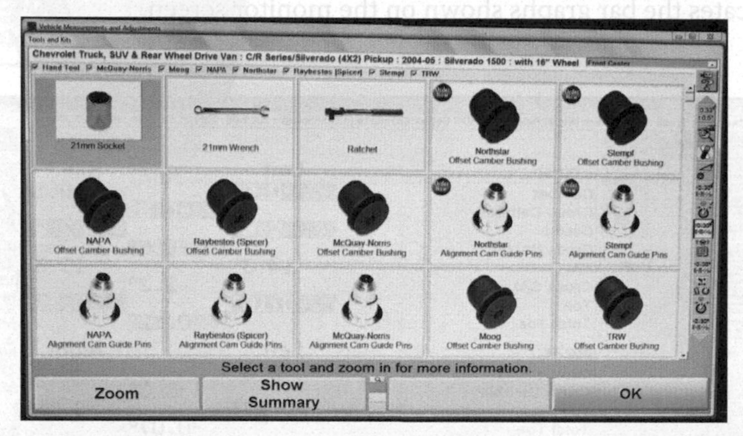

Figure 9-28 Wheel alignment tools and kits display.

DIAGNOSTIC DRAWING AND TEXT SCREENS

The technician may select tools and kits for the vehicle being serviced from the tools and kits database. When this feature is selected, the monitor screen displays the necessary wheel alignment adjustment tools for the vehicle selected at the beginning of the alignment (**Figure 9-28**). The kits displayed on the monitor screen are special components such as adjustment shims that are available for alignment adjustments on the car being serviced.

The technician may select **digital adjustment photos** that indicate how to perform wheel alignment adjustments (**Figure 9-29**). These digital photos include photos for cradle inspection and correction of cradle-to-body alignment (**Figure 9-30**). Live-action videos can also be selected. These CD videos provide suspension component inspection procedures (**Figure 9-31**).

A **part-finder database** is available in some computer wheel aligners. This database allows the technician to access part numbers and prices from many under-car parts manufacturers.

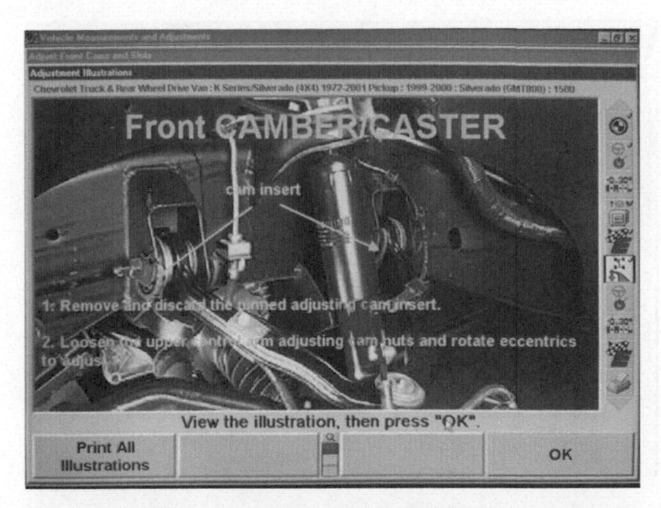

Figure 9-29 Wheel alignment adjustment photos.

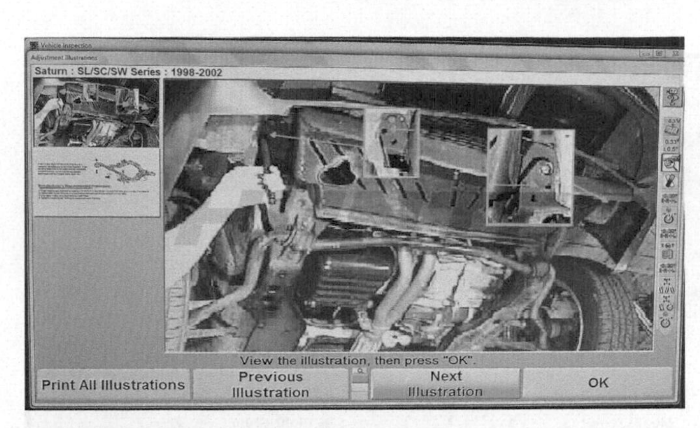

Figure 9-30 Cradle inspection and correction photos.

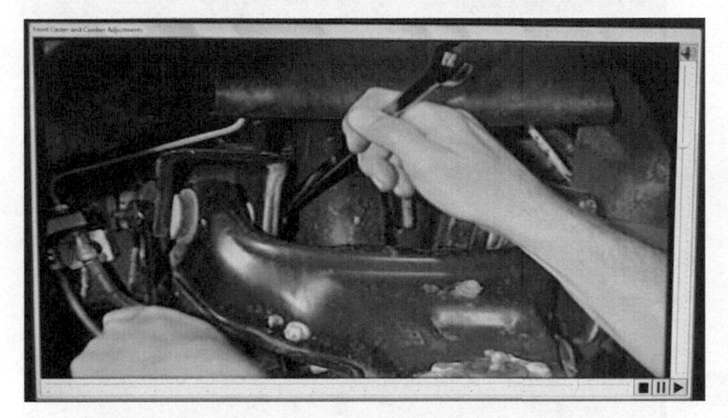
Figure 9-31 Live-action videos of suspension inspection and service.

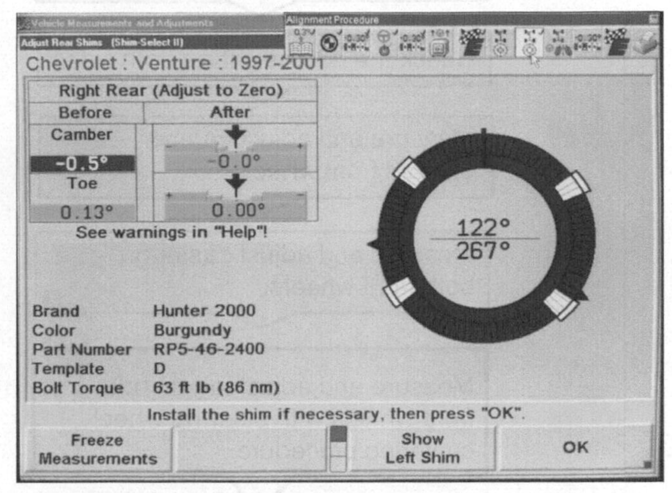

Figure 9-32 Shim selection screen.

On many front-wheel-drive cars, the rear wheel camber and toe are adjusted with shims. Some computer aligners have a **shim display screen** that indicates the thickness of shim required and the proper position of the shim. On some computer wheel aligners, the technician may use the light pen to change the orientation angle of the shim on the monitor screen while observing the resulting change in camber and toe (**Figure 9-32**). Some computer wheel aligners provide an automatic bushing calculator screen (**Figure 9-33**). This screen shows the required bushing and the proper bushing position to obtain the specified camber and caster on twin I-beam front suspension systems.

A **control arm movement monitor** is available on some computer wheel aligners. On short-and-long arm front suspensions, this feature indicates the required shim thickness to provide the specified camber and caster (**Figure 9-34**).

On some computer wheel aligners, the technician may select Print at any time, and print out the displayed screen, including diagnostic drawings. Another optional procedure is to print out the wheel alignment report before and after the adjustment of wheel alignment angles.

Wheel Alignment Procedure

The proper procedure for front and rear wheel alignment is important because adjusting one wheel alignment angle may change another angle. For example, adjusting front wheel caster changes front wheel toe. The wheel alignment adjustment procedure is especially

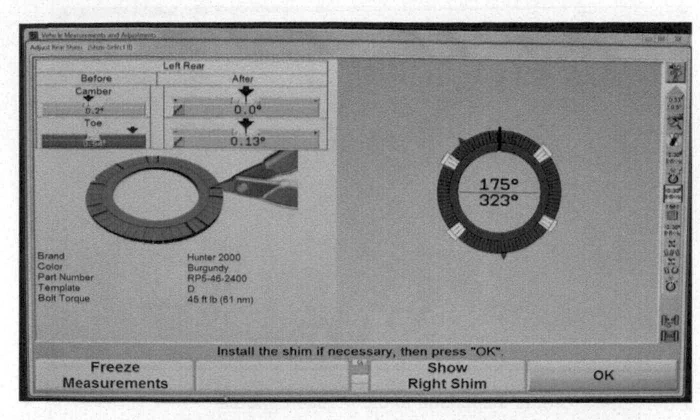

Figure 9-33 Automatic bushing calculation screen.

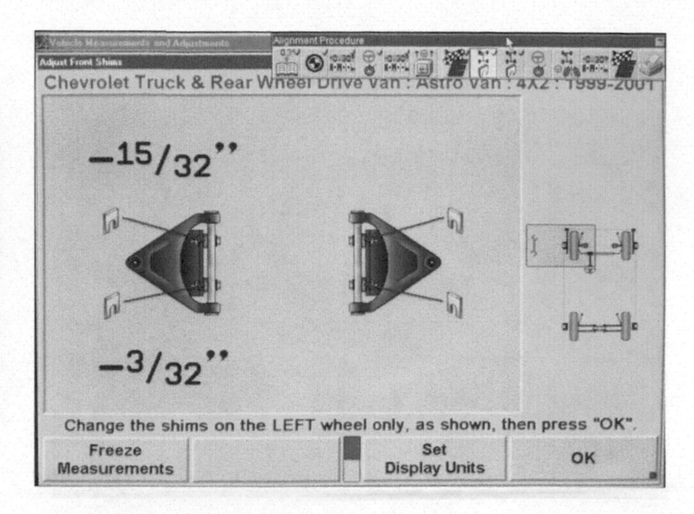

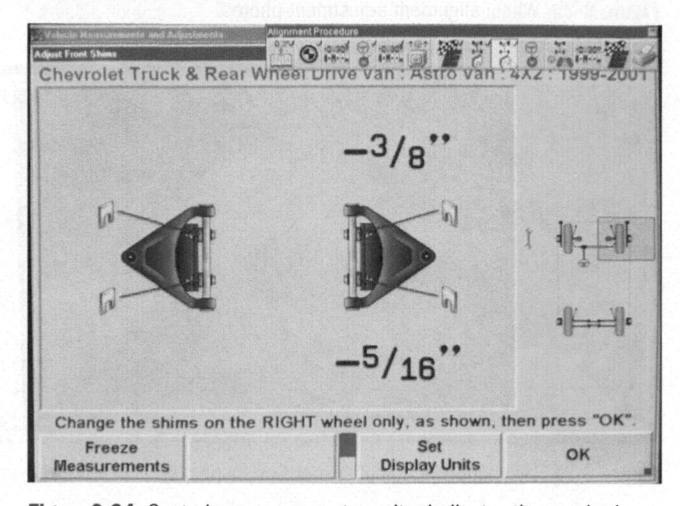

Figure 9-34 Control arm movement monitor indicates the required shim thickness to provide specified camber and caster.

Check tire pressure on both sides.

Check the ride height on both sides.

Measure and adjust camber on both front wheels.

Measure and adjust caster on both front wheels.

Measure and adjust toe on both front wheels with steering wheel centering procedure.

Figure 9-35 Front wheel alignment adjustment procedure.

critical on four-wheel independent suspension systems. A front wheel adjustment procedure is provided in **Figure 9-35**, and a typical rear wheel adjustment procedure is given in **Figure 9-36**. Always follow the wheel alignment procedure in the vehicle manufacturer's service manual.

On vehicles that have a provision for camber and toe adjustments in the rear of the vehicle, a four-wheel alignment may be performed. The alignment would begin at the rear of the vehicle to bring rear camber and toe into specifications. If you recall from the Classroom Manual rear toe errors will impact the thrust angle of the vehicle that is used for setting the front wheel toe angle. If you were to start in the front of the vehicle and then move to the rear, the changes you make to the rear toe would also change your front toe measurements proportionally. It is critical that you always begin by making all necessary adjustments to the rear alignment first and then move to the front.

SERVICE TIP Because rear wheel alignment plays a significant role in aiming or steering the vehicle, the rear wheel alignment should be corrected before adjusting front suspension angles. The rear wheel camber, toe, and thrust line must be within specifications before adjusting the front suspension angles.

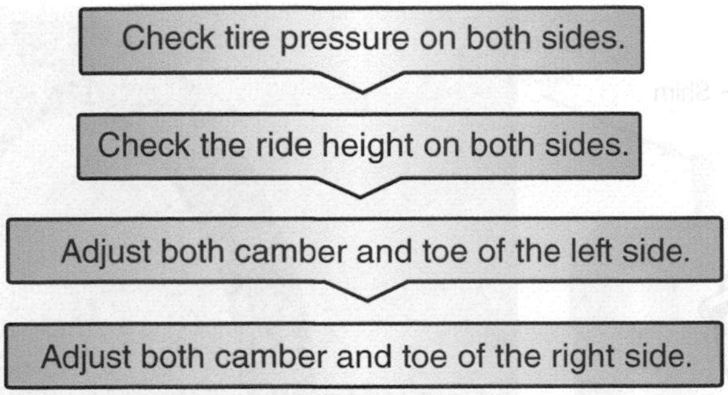

Check tire pressure on both sides.

Check the ride height on both sides.

Adjust both camber and toe of the left side.

Adjust both camber and toe of the right side.

Figure 9-36 Rear wheel alignment adjustment procedure.

CAUSES OF IMPROPER REAR WHEEL ALIGNMENT

The following suspension or chassis defects will cause incorrect rear wheel alignment:

1. Collision damage that results in a bent frame or distorted unitized body
2. A leaf-spring eye that is unwrapped or spread open
3. Leaf-spring shackles that are broken, bent, or worn
4. Broken leaf springs or leaf-spring center bolts
5. Worn rear upper or lower control arm bushings
6. Worn trailing arm bushings or dislocated trailing arm brackets
7. Bent components such as radius rods, control arms, struts, and rear axles

Caution

Always follow the vehicle manufacturer's recommended rear suspension adjustment procedure in the service manual to avoid improper alignment angles and reduced steering control.

REAR SUSPENSION ADJUSTMENTS

Rear Wheel Camber Adjustment

Improper **rear wheel camber** causes excessive wear on the edges of the rear tire treads.

Rear wheel camber is the tilt of a line through the center of a rear tire and wheel in relation to the true vertical centerline of the tire and wheel.

Cornering stability may be affected by improper rear wheel camber, especially while cornering at high speeds. Improper rear wheel camber on a front-wheel-drive vehicle may change the understeer or oversteer characteristics of the vehicle. Because a tilted wheel rolls in the direction of the tilt, rear wheel camber may cause steering pull. For example, excessive positive camber on the right rear wheel causes the rear suspension to drift to the right and steering to pull to the left.

Rear suspension adjustments vary depending on the type of suspension system. Some manufacturers of wheel alignment equipment and tools provide detailed diagrams for front and rear wheel alignment adjustments on all makes of domestic and imported vehicles. Our objective is to show a few of the common methods of adjusting rear wheel alignment angles on various types of rear suspension systems.

Special Tool

Rear suspension camber tool

On some semi-independent rear suspension systems, camber and toe are adjusted by inserting different sizes of shims between the rear spindle and the spindle mounting surface. These shims are retained by the spindle mounting bolts. The shim thickness is changed between the top and bottom of the spindle to adjust camber (**Figure 9-37**).

Many rear camber shims are now circular and are available in a wide variety of configurations to fit various rear wheels. Some computer wheel aligners indicate the thickness of shim required and the proper shim position. These same shims also adjust rear wheel toe (**Figure 9-38**).

On some transverse leaf rear suspension systems such as those on a later model Oldsmobile Cutlass Supreme, the spindle must be removed from the strut, and the lower strut bolt hole elongated to allow a camber adjustment (**Figure 9-39**). Once this bolt hole is elongated, assemble the knuckle and strut and leave the retaining bolts loose. The top of the tire

The **thrust line** is a line at 90 degrees to the rear axle and projecting forward.

Geometric centerline is the true centerline between the front and rear axles.

the left rear wheel moves the thrust line to the right of the geometric centerline. When toe-in or toe-out is excessive on the right rear wheel, it has the opposite effect on the thrust line compared to the left rear wheel. When the thrust line is moved to the left of the geometric centerline, the steering pulls to the right, whereas a thrust line positioned to the right of the *geometric centerline* results in steering pull to the left. **Photo Sequence 17** shows rear wheel alignment adjustment procedures.

Improper rear wheel toe adjustment on a nondriving rear axle causes diagonal wipe tire tread wear (**Figure 9-45**). The toe may be adjusted on many rear suspension systems used on front-wheel-drive cars by changing the tapered shim thickness between the front and rear edges of the rear spindle mounting flange (**Figure 9-46**). As mentioned previously, these shims are also used to adjust rear wheel camber.

On some transverse leaf rear suspension systems, a special toe adjusting tool is inserted in openings in the jack pad and the rear lower control rod (**Figure 9-47**). This tool has a turnbuckle in the center of the tool. When the nut on the inner end of the rear lower control rod is loosened, the turnbuckle is rotated to lengthen or shorten the tool and move the rear lower control rod to adjust the rear wheel toe (**Figure 9-48**). After the

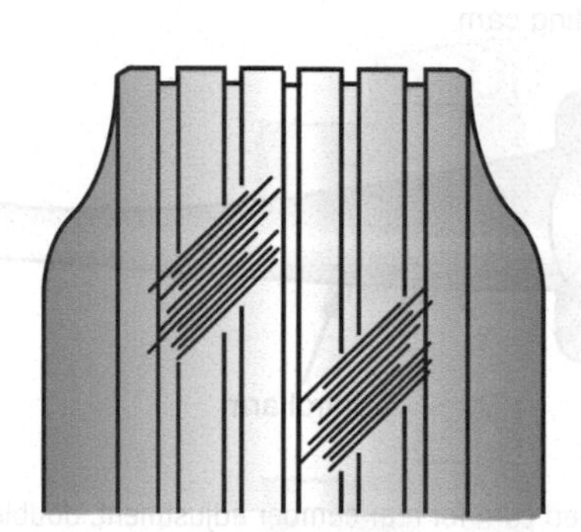

Figure 9-45 Diagonal wipe tire tread wear caused by improper rear wheel toe on a front-wheel-drive vehicle.

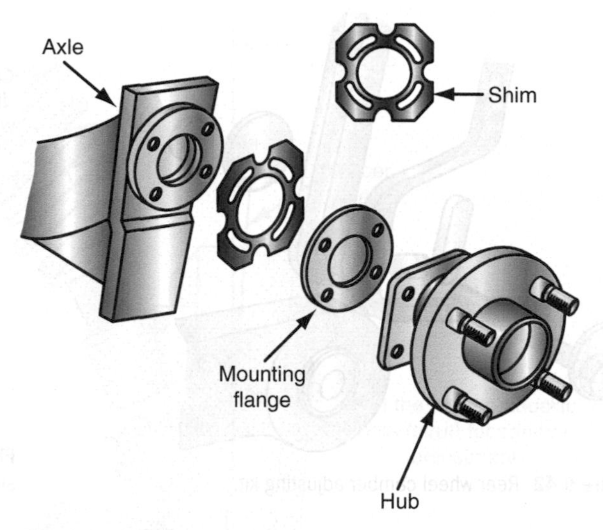
Axle
Shim
Mounting flange
Hub

Figure 9-46 Adjusting rear wheel toe by changing shim thickness between rear spindle and mounting flange.

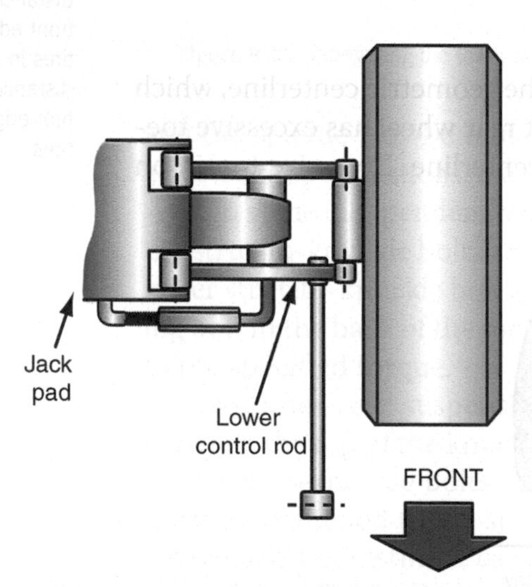
Jack pad
Lower control rod
FRONT

Figure 9-47 Special toe adjusting tool for transverse leaf-spring rear suspension.

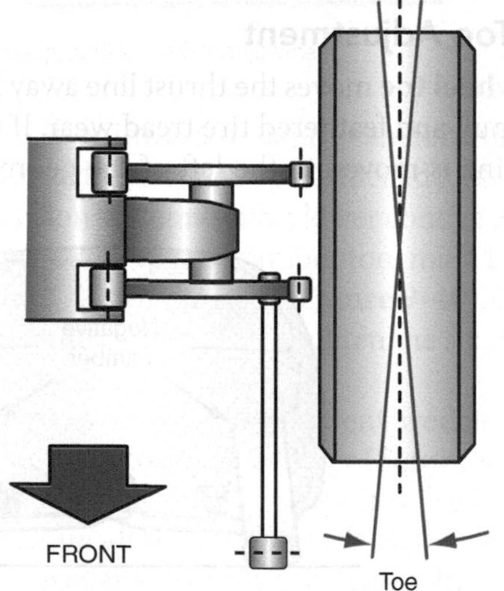
Check and set alignment with a full fuel tank.
Vehicle must be jounced three times before checking alignment to eliminate false readings.
Toe on the left and right side to be set separately per wheel to achieve specified total toe and thrust angle.
FRONT
Toe

Figure 9-48 Moving lower rear control rod position to adjust rear wheel toe.

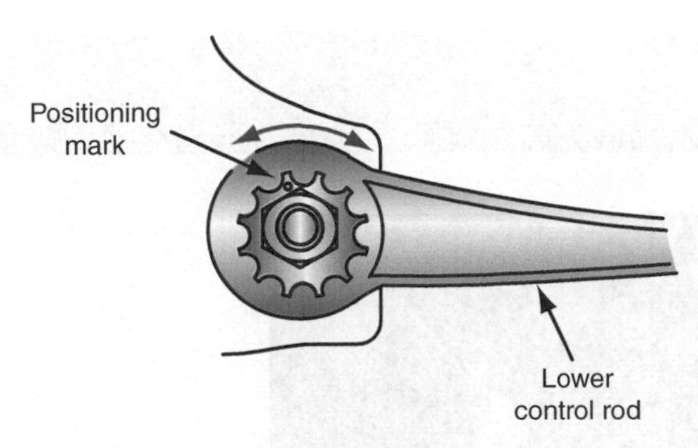

Figure 9-49 Eccentric star wheel on inner end of lower control rod for toe adjustment.

toe is adjusted to specification, the nut on the rear lower control rod bolt must be tightened to the specified torque. **Photo Sequence 18** illustrates front and rear suspension adjustments.

On some independent rear suspension systems, an eccentric star wheel on the rear lower control rod may be rotated to adjust the rear wheel toe (**Figure 9-49**).

PHOTO SEQUENCE 17

Typical Procedure for Adjusting Rear Wheel Camber and Toe

P17-1 Be sure the vehicle is properly positioned on an alignment ramp.

P17-2 Perform a prealignment inspection.

P17-3 Be sure the wheel aligner wheel sensors are properly connected to the vehicle.

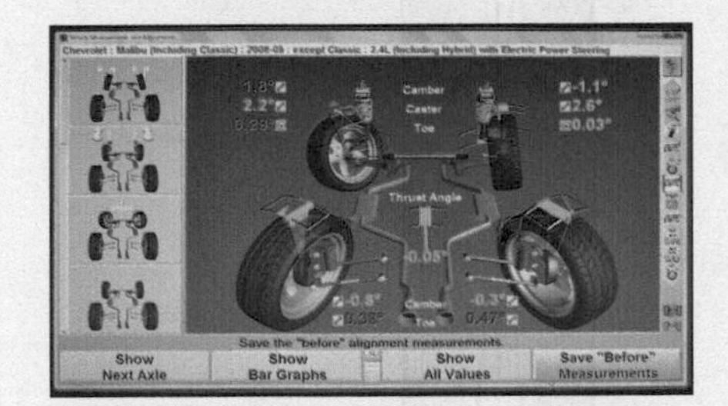

P17-4 Obtain the front and rear alignment angles using the procedures recommended by the equipment manufacturer. Display the front and rear wheel alignment angles on the aligner monitor.

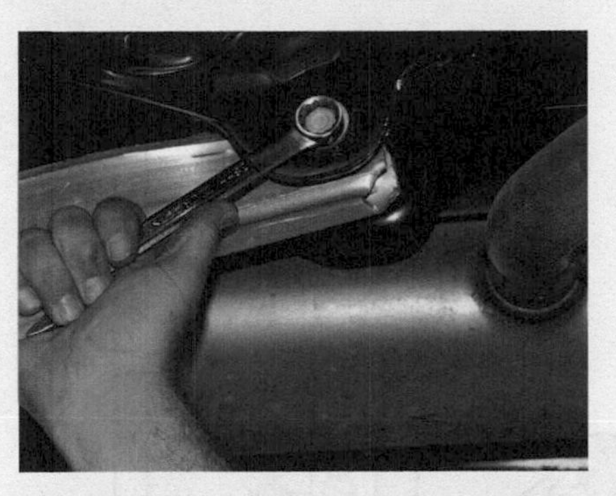

P17-5 Adjust the camber on the right rear wheel by loosening the cam bolt lock nut on the inner end of the right rear lower control arm and turning the eccentric cam until the camber setting is within specifications. Maintain the cam setting and tighten the lock nut to the specified torque.

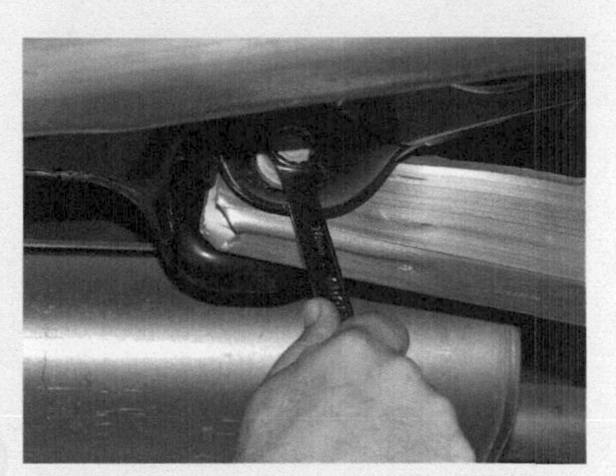

P17-6 Adjust the camber on the left rear wheel by loosening the cam bolt lock nut on the inner end of the right rear lower control arm and turning the eccentric cam until the camber setting is within specifications. Maintain the cam setting and tighten the lock nut to the specified torque.

PHOTO SEQUENCE 17 (CONTINUED)

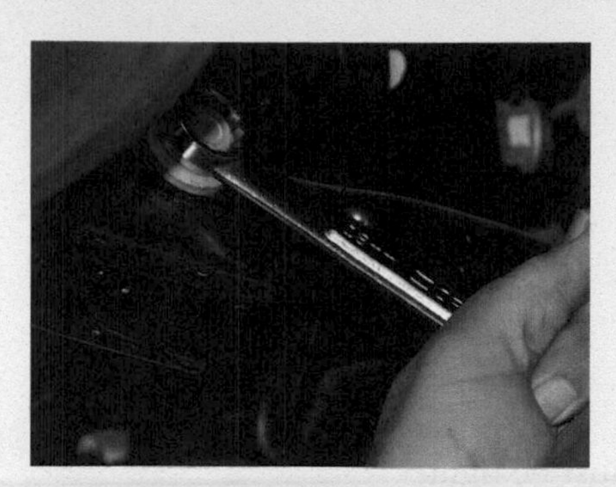

P17-7 Loosen the cam bolt nut on the inner end of the left rear toe link, and rotate the cam bolt to the required toe specification. Maintain the cam bolt position, and tighten the cam bolt nut to the specified torque.

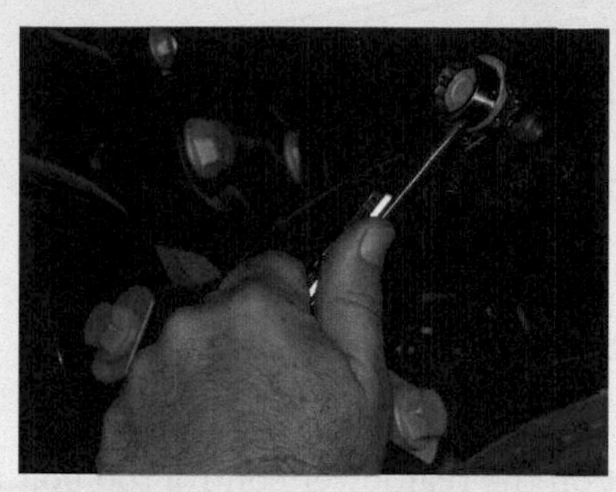

P17-8 Loosen the cam bolt nut on the inner end of the right rear toe link, and rotate the cam bolt to the required toe specification. Maintain the cam bolt position, and tighten the cam bolt nut to the specified torque.

On some rear suspension systems with a lower control arm and ball joint, the nut on the tie rod is loosened, and the tie rod is rotated to adjust the rear wheel toe (**Figure 9-50**). After the toe is adjusted to specifications, the tie rod nut must be tightened to the specified torque.

On some front-wheel-drive cars, the rear wheel toe is adjusted by loosening the toe link bolt and moving the wheel in or out to adjust the toe (**Figure 9-51**). After the rear wheel toe is properly adjusted, the toe link bolt must be tightened to the specified torque. On other front-wheel-drive cars, the rear wheel toe is adjusted by loosening the nut on

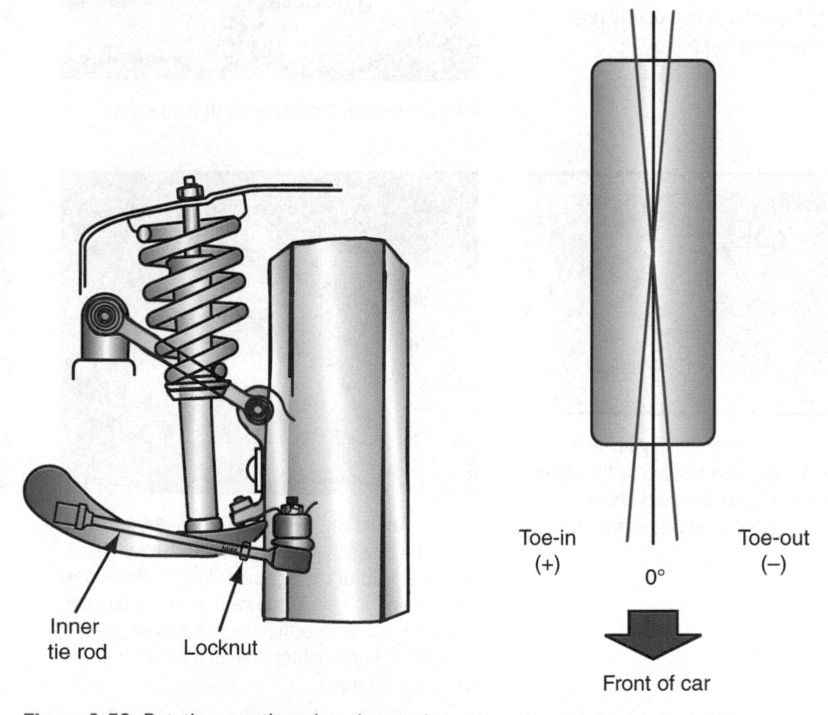

Figure 9-50 Rotating rear tie rod to change tie rod length and adjust rear wheel toe.

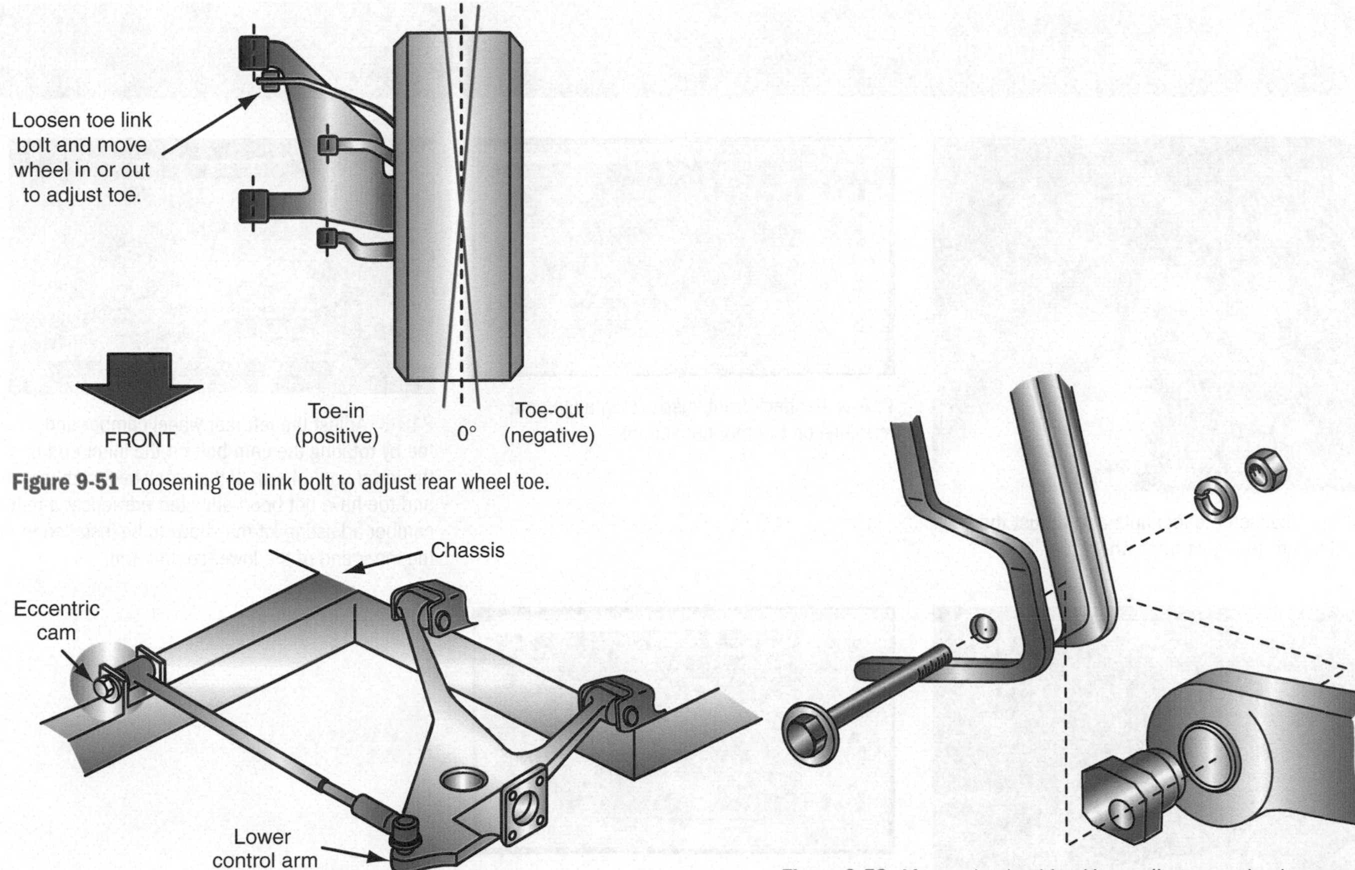

Figure 9-51 Loosening toe link bolt to adjust rear wheel toe.

Figure 9-52 Cam bolt for rear wheel toe adjustment.

Figure 9-53 Aftermarket bushing kit to adjust rear wheel toe and camber.

the inner adjustment link cam nut (**Figure 9-52**). The cam bolt is rotated to adjust the rear wheel toe. After the rear wheel toe is properly adjusted, the adjustment link cam nut must be tightened to the specified torque. Photo Sequence 18 illustrates front and rear suspension adjustments.

A variety of aftermarket bushing kits are available to adjust rear wheel toe. Some of these kits also adjust rear wheel camber (**Figure 9-53**).

PHOTO SEQUENCE 18
Typical Procedure for Performing Front and Rear Suspension Alignment Adjustments

P18-1 Position vehicle properly on alignment rack with front wheels centered on turntables and rear wheels on slip plates.

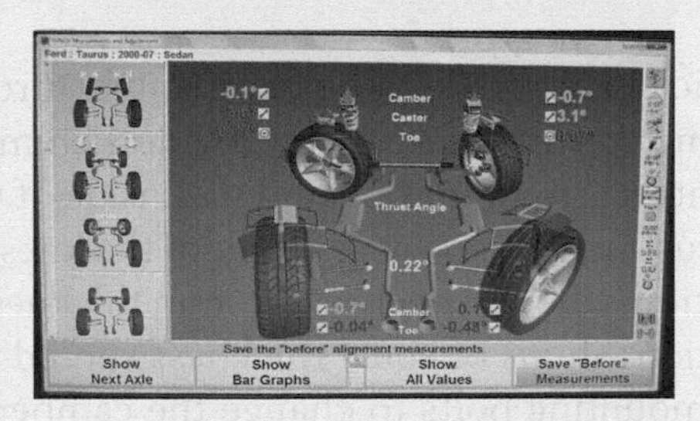

P18-2 After prealignment inspection and front and rear ride height measurement, measure front and rear suspension angles with computer wheel aligner.

P18-3 Drill spot welds to remove alignment plate on top of the strut tower. After strut upper mount retaining nuts and alignment plate are removed, shift the strut inward or outward to adjust front wheel camber. Shift the upper strut mount forward or rearward to adjust front wheel caster.

PHOTO SEQUENCE 18 (CONTINUED)

P18-4 Loosen the locknuts and adjust the tie rod length to adjust front wheel toe.

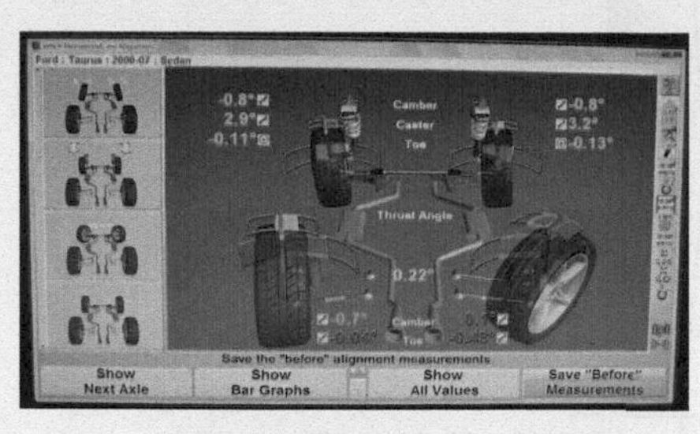

P18-5 Recheck front suspension alignment readings on the monitor screen.

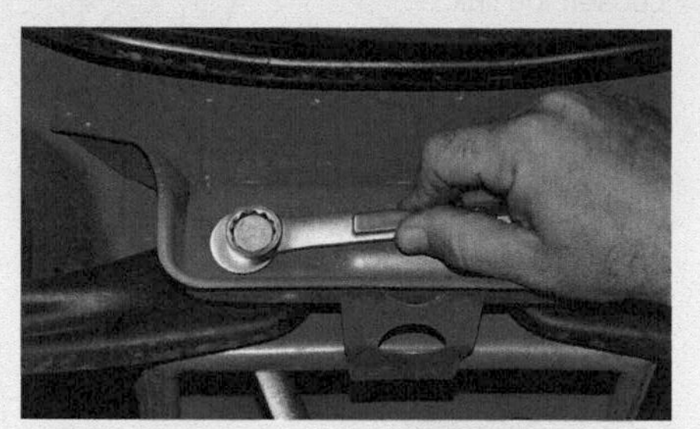

P18-6 Adjust the left rear wheel camber and toe by rotating the cam bolt on the inner end of the lower control arm. If the rear wheel camber and toe have not been adjusted previously, a rear camber adjusting kit may have to be installed in the inner end of the lower control arm.

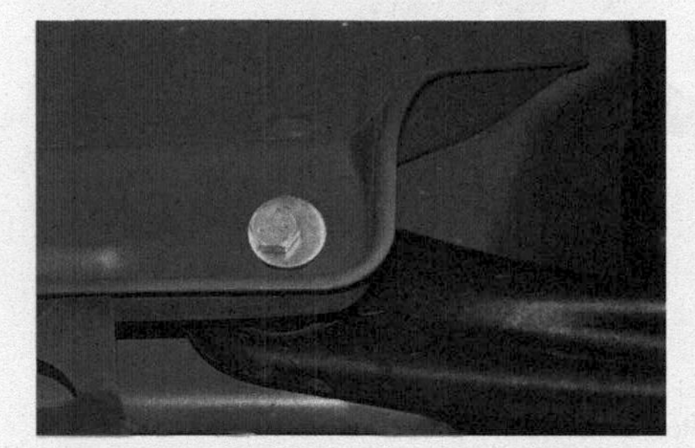

P18-7 Adjust the right rear wheel camber and toe by rotating the cam bolt on the inner end of the lower control arm. If the rear wheel camber and toe have not been adjusted previously, a rear camber adjusting kit may have to be installed in the inner end of the lower control arm.

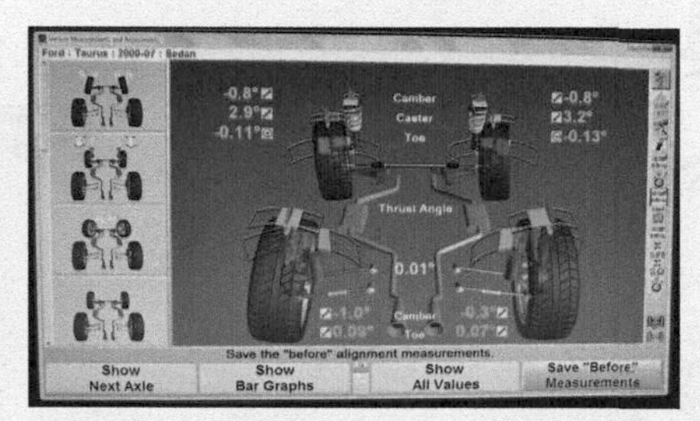

P18-8 Recheck all front and rear alignment readings on the monitor screen. Be sure all readings are within specifications, including the thrust angle.

FRONT CAMBER ADJUSTMENT

Shims

A **camber adjustment** is usually a shim-type or eccentric cam-type mechanism that is used to correct the camber angle on front or rear wheels.

Various methods are provided by car manufacturers for camber adjustment. On older platforms some car manufacturers provide a shim-type **camber adjustment** between the upper control arm mounting and the inside of the frame (**Figure 9-54**). In this type of camber adjustment, increasing the shim thickness moves the camber setting to a more negative position, whereas decreasing shim thickness changes the camber toward a more positive position. Shims of equal thickness should be added or removed on both upper control arm mounting bolts to change the camber setting without affecting the caster setting.

Other vehicle manufacturers provide a shim-type camber adjustment between the upper control arm mounting and the outside of the frame (**Figure 9-55**). On this shim-type camber adjustment, increasing the shim thickness increases positive camber.

Older vehicles have shim-type camber adjustments, but newer vehicles have eccentric cams for this adjustment.

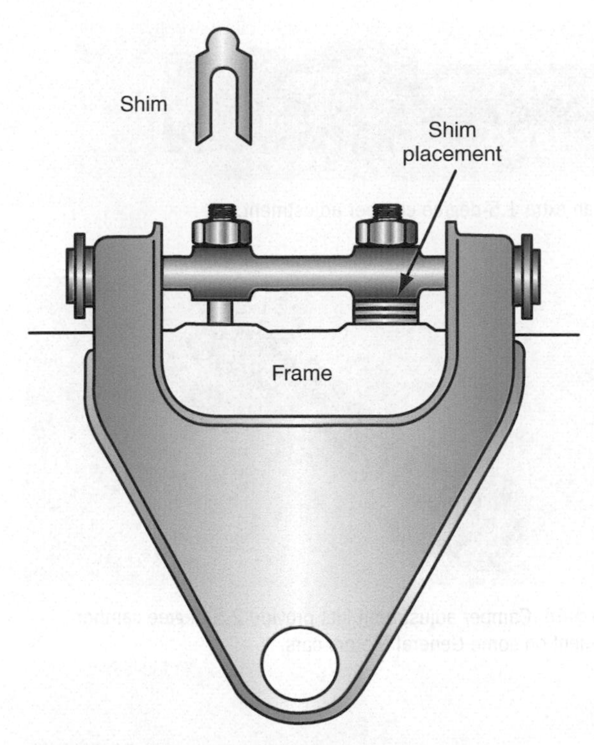

Figure 9-54 Shim-type camber adjustment between upper control arm mounting and the inside of the frame.

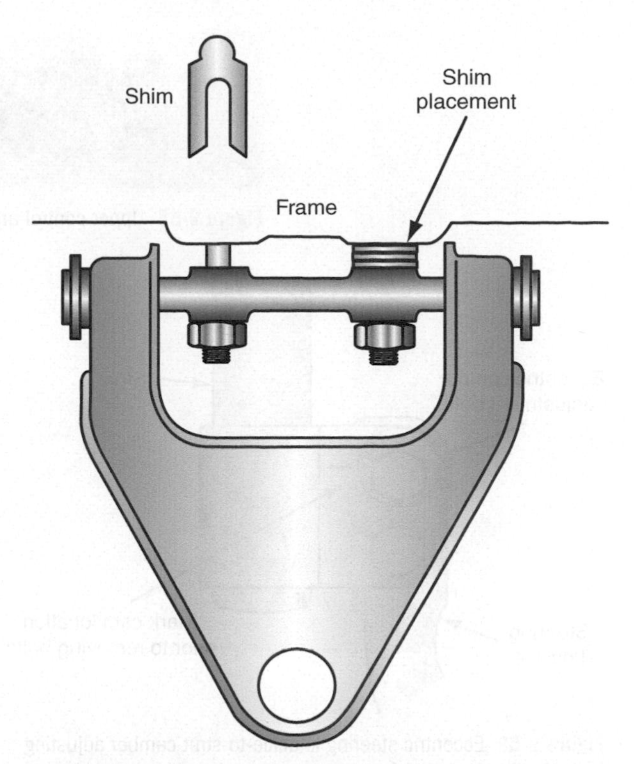

Figure 9-55 Shim-type camber adjustment between upper control arm mounting and the outside of the frame.

Eccentric Cams

Some vehicles have **eccentric cams** on the inner ends of the upper control arm to provide camber adjustment (**Figure 9-56**), whereas other suspension systems have eccentric cams on the inner ends of the lower control arms (**Figure 9-57**). If the original cam adjustment on the inner ends of the upper control arms does not provide enough camber adjustment, aftermarket upper control arm shaft kits are available that provide an extra 1.5 degrees of camber adjustment (**Figure 9-58**). The suspension adjustments shown in Figure 9-54 through Figure 9-58 are used on rear-wheel-drive cars with short-and-long arm front suspension systems.

Eccentric cams are out-of-round pieces of metal mounted on a retaining bolt with the shoulder of the cam positioned against a component. When the cam is rotated, the component contacting the cam shoulder is moved.

Some MacPherson strut front suspension systems on front-wheel-drive cars have a cam on one of the steering-knuckle-to-strut bolts to adjust camber (**Figure 9-59**). Aftermarket camber adjustment kits that provide 2.5 degrees of camber adjustment are

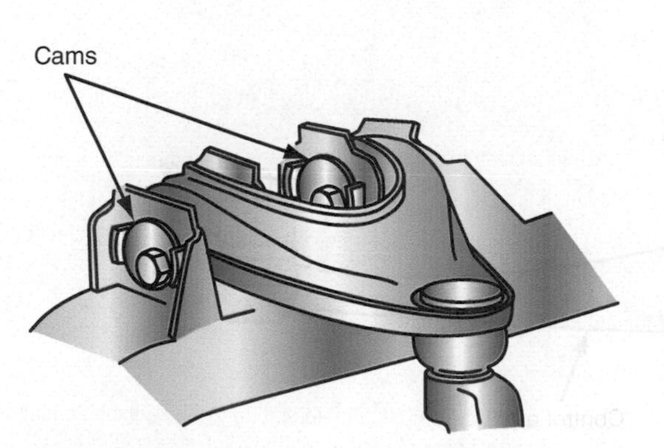

Figure 9-56 Cam adjustment on the inside ends of the upper control arm.

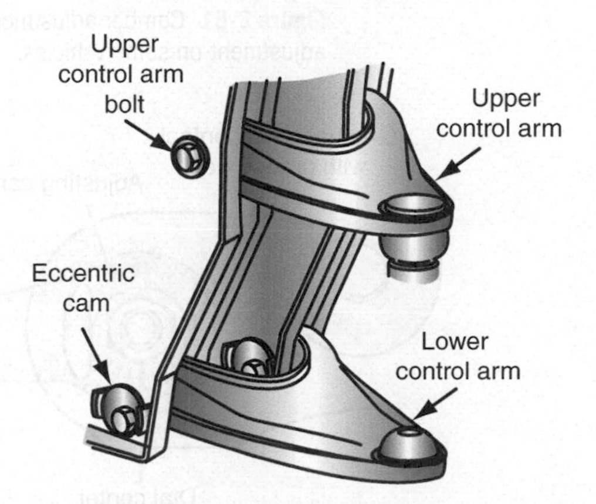

Figure 9-57 Cam adjustment on the inner ends of the lower control arms.

Figure 9-58 Upper control arm shaft kit provides an extra 1.5-degree camber adjustment.

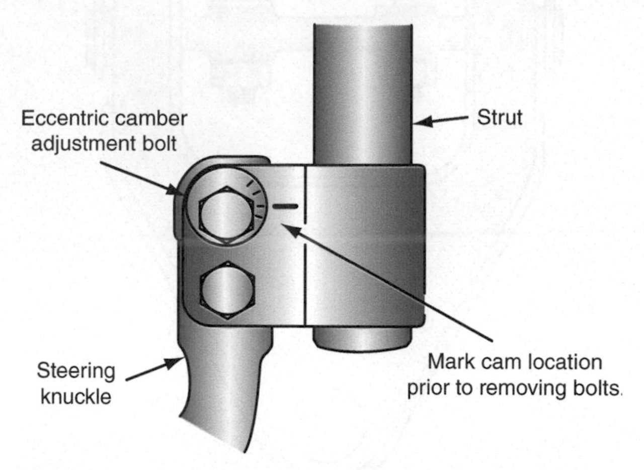

Figure 9-59 Eccentric steering knuckle-to-strut camber adjusting bolt, MacPherson strut front suspension.

Figure 9-60 Camber adjustment kits provide 2.5-degree camber adjustment on some General Motors cars.

available for many MacPherson strut suspension systems (**Figure 9-60** and **Figure 9-61**). Similar camber adjustment kits are available for many cars with MacPherson strut front suspension systems. On some double-wishbone front suspension systems, a graduated cam on the inner end of the lower control arm provides a camber adjustment (**Figure 9-62**).

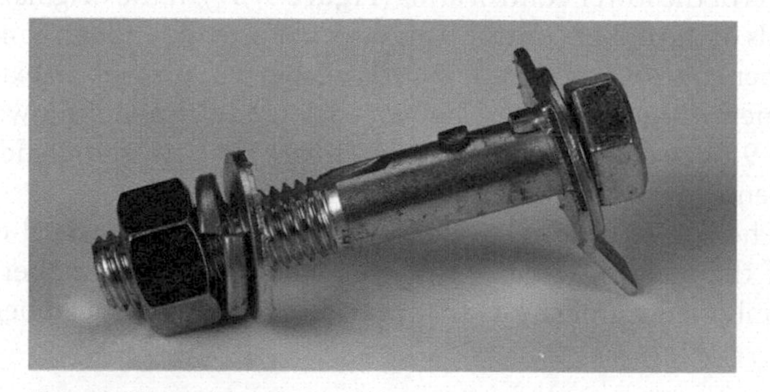

Figure 9-61 Camber adjustment kits provide 2.5-degree camber adjustment on some vehicles.

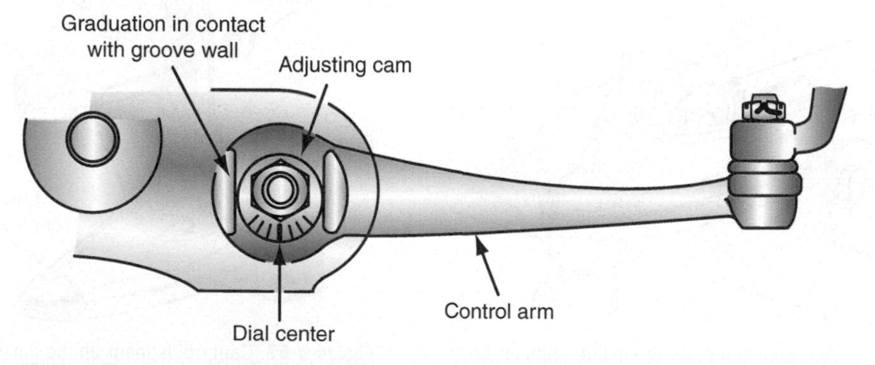

Figure 9-62 Camber adjustment double-wishbone front suspension.

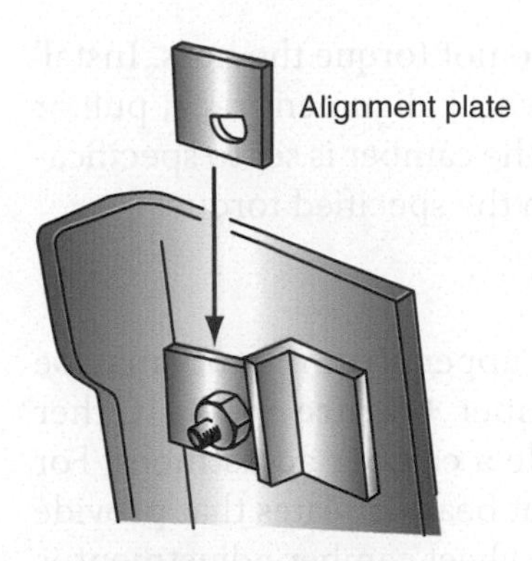

Figure 9-63 Alignment plate on upper control arm attaching nut.

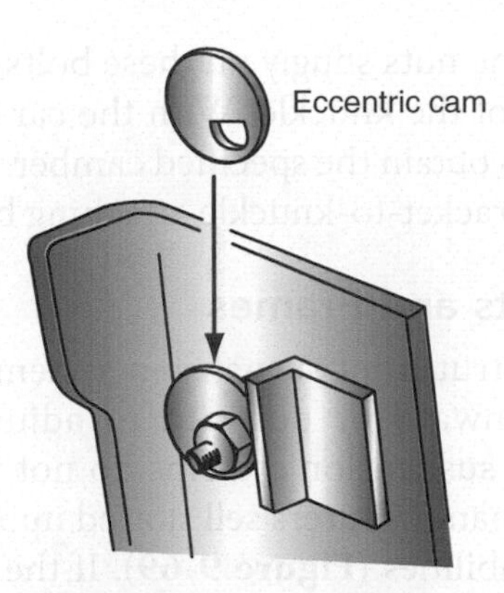

Figure 9-64 Eccentric cam installed behind upper control arm attaching nut.

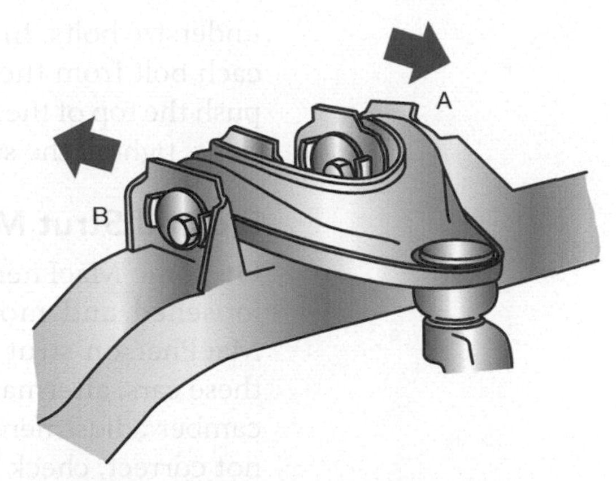

Figure 9-65 To increase the positive caster, move the front of the upper control arm (A) outboard and move the rear of the upper control arm (B) inboard.

On late-model Ford F-150 light-duty truck 4 × 2 front suspension systems, an alignment plate is installed on the upper control arm attaching nuts when the vehicle is manufactured (**Figure 9-63**). If a camber or caster adjustment is required, remove the control arm attaching nuts one at a time and install eccentric cams behind the adjusting nuts (**Figure 9-64**). Turn the eccentric cams equally on the front and rear upper control arm attaching nuts to adjust the camber. To increase the positive caster, move the front of the upper control arm outboard and move the rear of the upper control arm inboard (**Figure 9-65**). If it is necessary to decrease the positive caster, move the front of the upper control arm inboard and move the rear of the upper control arm outboard (**Figure 9-66**). These front suspension systems also have eccentric cams on the lower control arm attaching bolts (**Figure 9-67**). However, these eccentric cams are only for factory use, and they must not be adjusted in service. When removing the lower control arms, always mark the eccentric cams on the attaching bolts and reinstall these cams in their original position.

In some front suspension systems such as on the PT Cruiser, eccentric cams are not installed to provide a camber adjustment. In these suspension systems, the manufacturer supplies slightly undersize strut bracket-to-knuckle attaching bolts that allow a 2-degree camber adjustment on each side of the front suspension. If a camber adjustment is required, remove the strut bracket-to-knuckle attaching bolts one at a time and install the

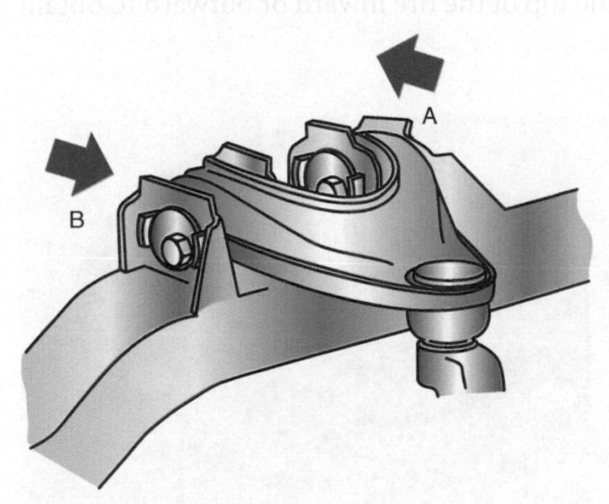

Figure 9-66 To decrease the positive caster move the front of the upper control arm (A) inboard and move the rear of the upper control arm (B) outboard.

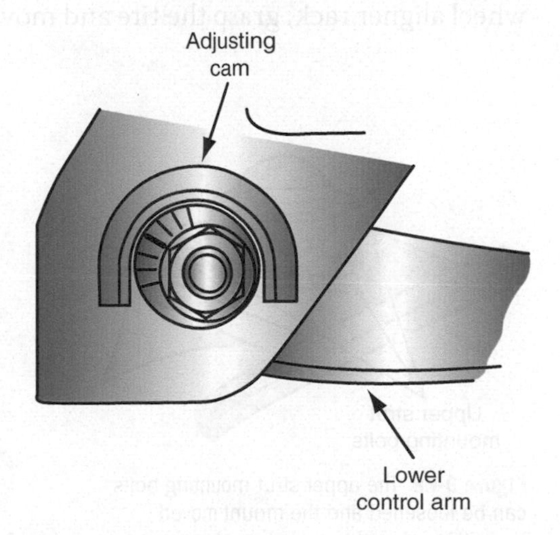

Figure 9-67 Factory-set eccentric cam on lower control arm attaching bolts.

undersize bolts. Install the nuts snugly on these bolts, but do not torque the nuts. Install each bolt from the rear of the knuckle. With the car on a wheel alignment rack, pull or push the top of the tire to obtain the specified camber. After the camber is set to specifications, tighten the strut bracket-to-knuckle attaching bolts to the specified torque.

Slotted Strut Mounts and Frames

On some MacPherson strut front suspension systems, the upper strut mounts may be loosened and moved inward or outward to adjust camber (**Figure 9-68**). Other MacPherson strut front suspension systems do not provide a camber adjustment. For these cars, aftermarket manufacturers sell slotted upper strut bearing plates that provide camber adjustment capabilities (**Figure 9-69**). If the front wheel camber adjustment is not correct, check for worn components such as ball joints or upper strut mounts. Also check the engine cradle/subframe mounts. Because the inner ends of the lower control arms are attached to the engine cradle on many front-wheel-drive cars, a cradle that is out of position may cause improper camber readings.

For example, if the vehicle has been impacted on the left front corner in a collision, the cradle may be shifted to the right. This action moves the bottom of the left front wheel inward and the bottom of the right front wheel outward. Therefore, the left front wheel has excessive positive camber and the right front wheel has too much negative camber. If the vehicle has been impacted on the left front corner, the cradle may be squashed to some extent with the left side being moved toward the right side without moving the right side of the cradle out of position. In this case the left front wheel has excessive positive camber, and the side-to-side cradle measurements are less than specified.

A camber plate is locked in place with a pop rivet on some upper strut mounts. The pop rivet may be removed to release the camber plate (**Figure 9-70**). After the upper strut mount bolts are loosened, the mount may be moved inward or outward to adjust the camber. When the camber is adjusted to the vehicle manufacturer's specification, the strut mount bolts should be tightened to the correct torque and a new pop rivet installed.

On some short-and-long arm front suspension systems, the bolt holes in the frame are elongated for the inner shaft on the upper control arm. When these bolts are loosened, the upper control arm may be moved inward or outward to adjust camber (**Figure 9-71**).

When a camber adjustment is required on some MacPherson strut front suspensions, the vehicle manufacturer recommends removing the strut-to-steering knuckle retaining bolts and removing the strut from the knuckle. A round file is then used to elongate the lower bolt opening in the strut (**Figure 9-72**). Reinstall the strut and retaining bolts, but leave the nuts slightly loose on the bolts. With the vehicle in the proper position on the wheel aligner rack, grasp the tire and move the top of the tire inward or outward to obtain

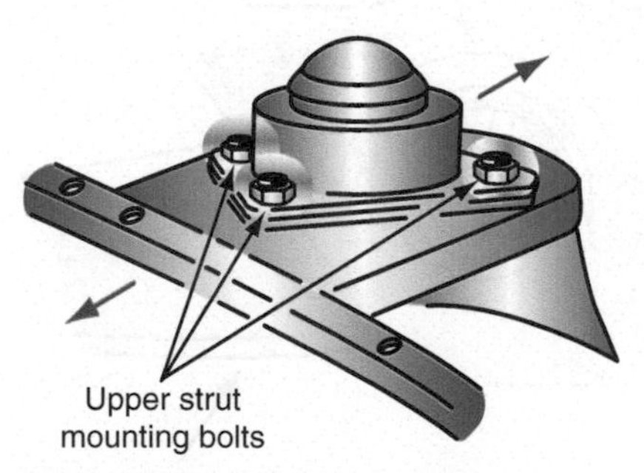

Figure 9-68 The upper strut mounting bolts can be loosened and the mount moved inward or outward to adjust camber on some MacPherson strut front suspension systems.

Figure 9-69 Slotted upper strut bearing plates provide camber adjustment on some MacPherson strut front suspension systems.

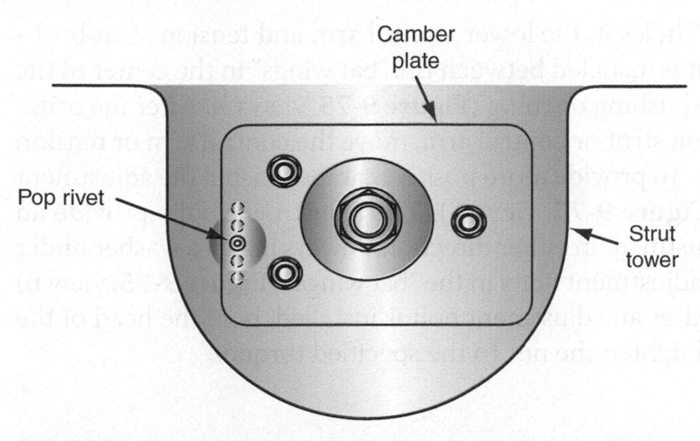

Figure 9-70 Upper strut mount with camber plate and rivet, modified MacPherson strut front suspension.

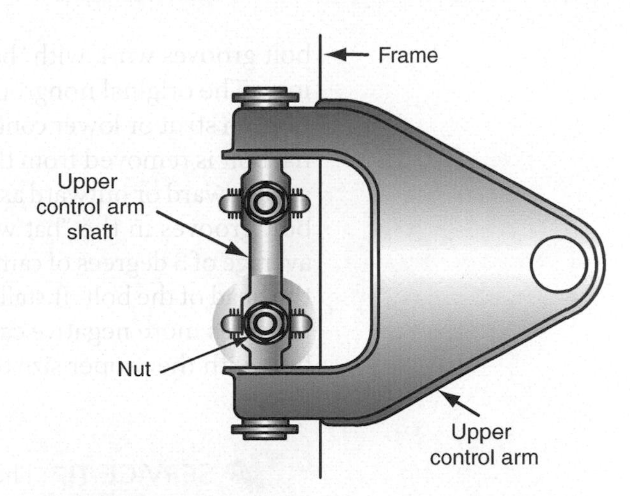

Figure 9-71 Elongated bolt holes for upper control arm shaft bolts provide camber adjustment on some short-and-long arm front suspension systems.

the desired camber adjustment, and then tighten the nuts on the strut-to-knuckle bolts to the specified torque.

Special Adjustment Bolts

On other vehicles such as the Dodge Magnum, the vehicle manufacturer recommends loosening the cradle/subframe mounts and shifting the cradle to provide minor camber or caster adjustments (**Figure 9-73**). If the desired camber adjustment cannot be obtained by shifting the cradle, the vehicle manufacturer provides an adjustment bolt package for camber and caster adjustment. These bolts are designed to replace the inboard bolts in the lower control arm and tension strut. To adjust camber use both bolts, and to adjust caster only use a replacement bolt in the tension strut.

The wheel alignment adjustment bolts have offset grooves cut into the length of the bolts (**Figure 9-74**). The grooves in the adjustment bolts force the bolt to be installed in one or two ways depending on whether more positive or negative camber is required. The

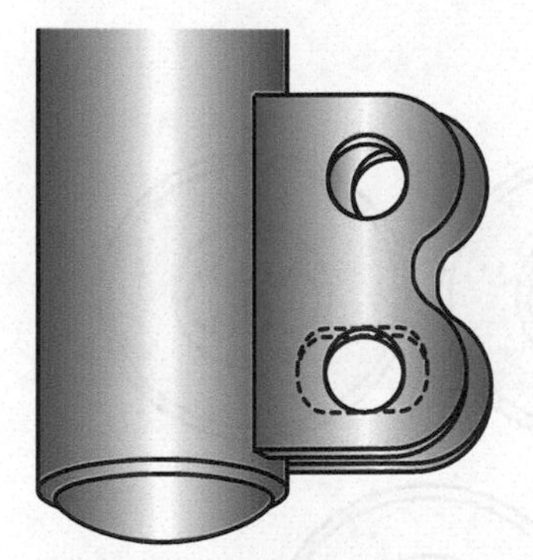

Figure 9-72 Elongated lower strut-to-knuckle bolt hole for camber adjustment.

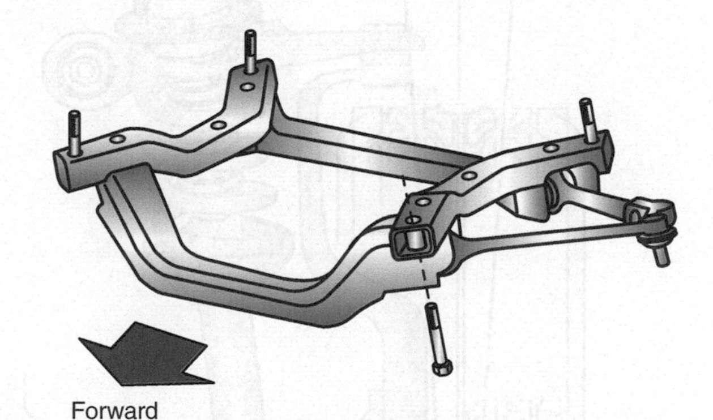

Figure 9-73 Loosening cradle mounting bolts and shifting the cradle to perform minor camber adjustments.

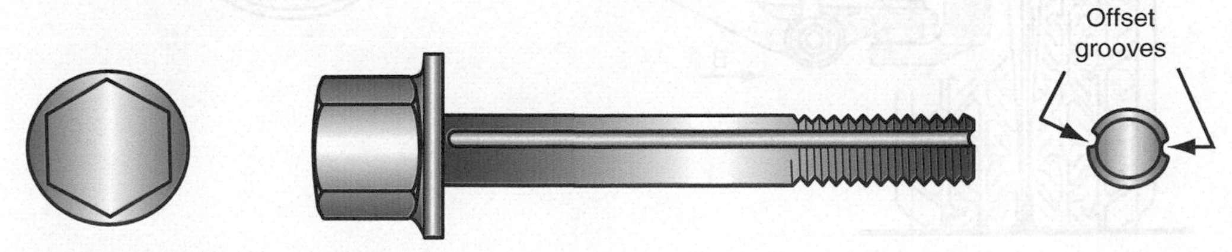

Figure 9-74 Camber/caster adjustment bolt with offset grooves.

bolt grooves work with "bat wing" holes in the lower control arm and tension strut bushings. The original nongrooved bolt is installed between the "bat wings" in the center of the tension strut or lower control arm bushing opening (**Figure 9-75**, view C). After the original bolt is removed from the tension strut or control arm, move the control arm or tension strut inward or outward as desired. To provide more positive camber, install the adjustment bolt grooves in the "bat wings" (**Figure 9-75**, view A). The adjustment bolts provide an average of 3 degrees of camber adjustment in either direction. Always install a washer under the head of the bolt. Installing the adjustment bolts in the "bat wings" (**Figure 9-75**, view B) provides more negative camber. After an adjustment bolt is installed, hold the head of the bolt with the proper size tool, and tighten the nut to the specified torque.

SERVICE TIP The bolt in the lower control arm bushing must be installed from the rear and the tension strut bushing bolt must be installed from the front to avoid contact with other suspension components.

Caution

After an adjustment bolt is installed, do not attempt to rotate the bolt. This action will damage the "bat wing" in the control arm or tension strut bushing.

To adjust the front wheel caster, follow the same procedure, but install the adjustment bolt only in the tension strut.

Special Tools for Camber Adjustment

Some vehicle manufacturers provide a special tool for adjusting front wheel camber and caster. This tool has threaded projections on each end extending at a right angle from a turnbuckle in the center of the tool. When the turnbuckle is rotated, the tool is lengthened

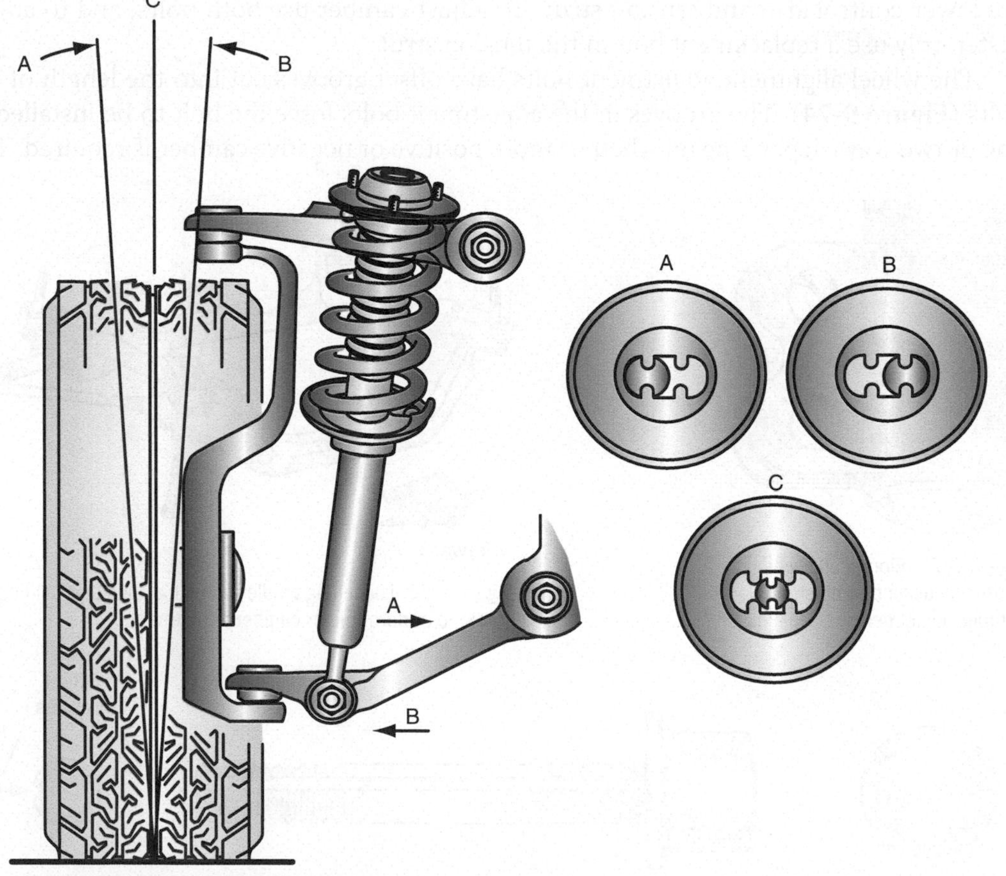

Figure 9-75 Adjustment bolt positions in tension strut or lower control arm bushing to provide camber adjustment.

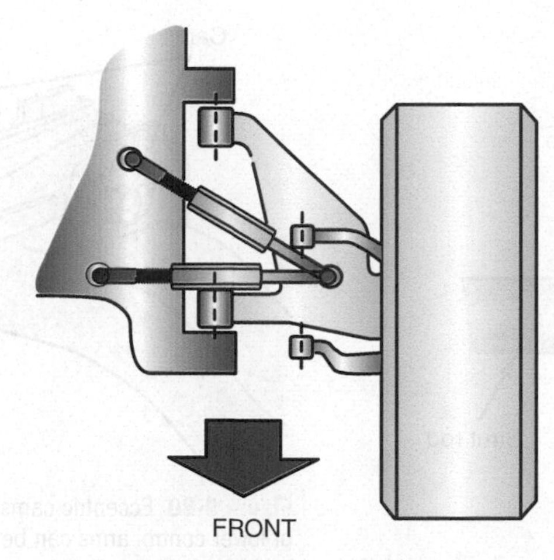

Figure 9-76 Tool installed in lower control arm and chassis openings to adjust camber/caster.

or shortened. The tool is installed into openings in the lower control arm and the chassis (**Figure 9-76**). After the bolt is loosened in the inner end of the control arm, the tool can be lengthened or shortened to adjust the camber or caster.

Caution

Some vehicle manufacturers do not recommend bending I-beams and other suspension components.

Camber Adjustment on Twin I-Beam Front Suspension System

On twin I-beam front suspension systems, a special clamp-type bending tool with a hydraulic jack is used to bend the I-beams and change the camber setting. On a twin I-beam front suspension, the clamp and hydraulic jack must be positioned properly to adjust the camber to a more positive setting (**Figure 9-77**) or a more negative setting (**Figure 9-78**).

FRONT CASTER ADJUSTMENT PROCEDURE

Strut Rod Length Adjustment

On some suspension systems, the nuts on the forward end of the strut rod may be adjusted to lengthen or shorten the strut rod, which changes the caster setting (**Figure 9-79**). Shorten the strut rod to increase the positive caster.

SERVICE TIP The vehicle manufacturer does not recommend bending the I-beams to adjust front wheel camber. However, this procedure is suggested by some equipment manufacturers.

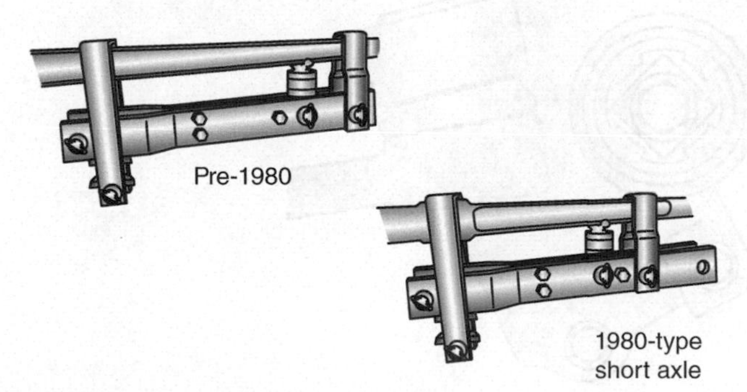

Figure 9-77 Adjusting camber to a more positive setting, twin I-beam front suspension.

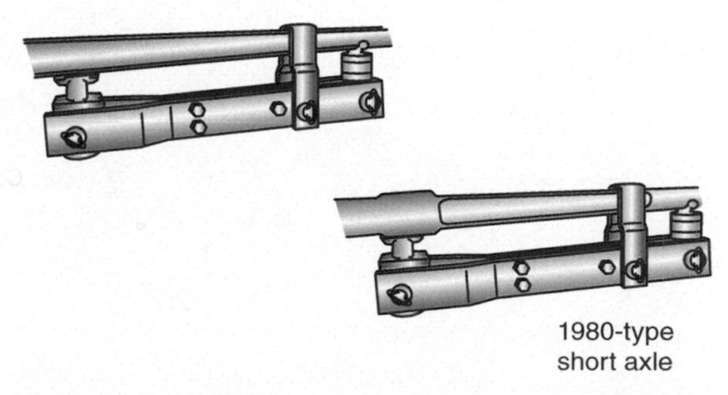

Figure 9-78 Adjusting camber to a more negative setting, twin I-beam front suspension.

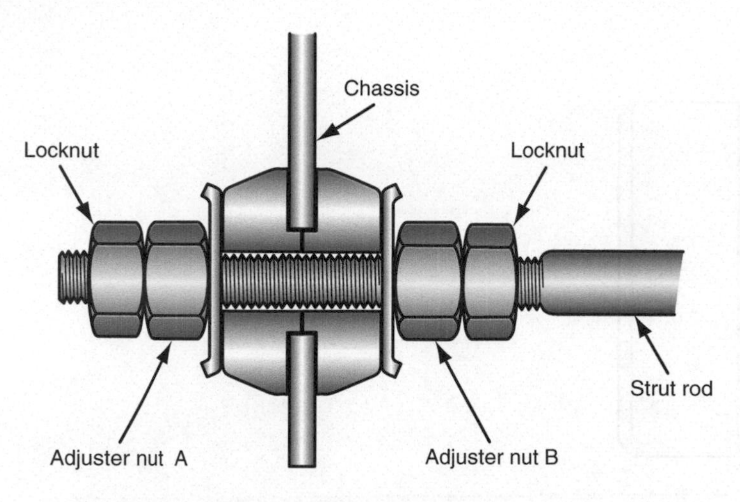

Unscrew A and then tighten B to increase caster.

Figure 9-79 Strut rod nuts can be adjusted to lengthen or shorten the strut rod to adjust caster.

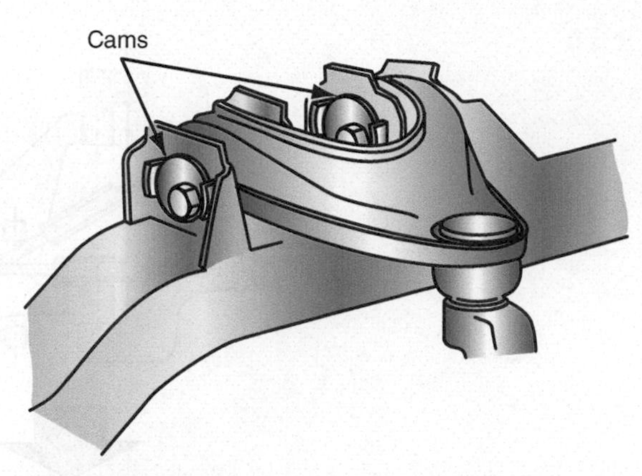

Figure 9-80 Eccentric cams on the inner ends of the upper or lower control arms can be used to adjust camber and caster.

Eccentric Cams

The same eccentric cams on the inner ends of the upper or lower control arms may be used to adjust camber or caster (**Figure 9-80**). If an eccentric bushing in the outer end of the upper control arm is rotated to adjust camber, this same eccentric also adjusts caster. On some double-wishbone front suspension systems, the pivot adjuster mounting bolt nuts must be loosened under the compliance pivot, and the graduated cam may then be rotated to adjust caster (**Figure 9-81**).

Slotted Strut Mounts and Frames

If a **caster adjustment** is required and the suspension does not have an adjustment for this purpose, check for worn components such as upper strut mounts and ball joints. Also check the engine cradle mounts. Because the inner ends of the lower control arms are attached to the engine cradle on many front-wheel-drive cars, a cradle that is out of position may cause improper caster readings.

The slots in the frame at the upper control arm shaft mounting bolts provide camber and caster adjustment. If the upper strut mount is adjustable on a MacPherson strut suspension, the mount retaining bolts may be loosened and the mount moved forward or rearward to adjust caster.

A strut rod may be called a radius rod.

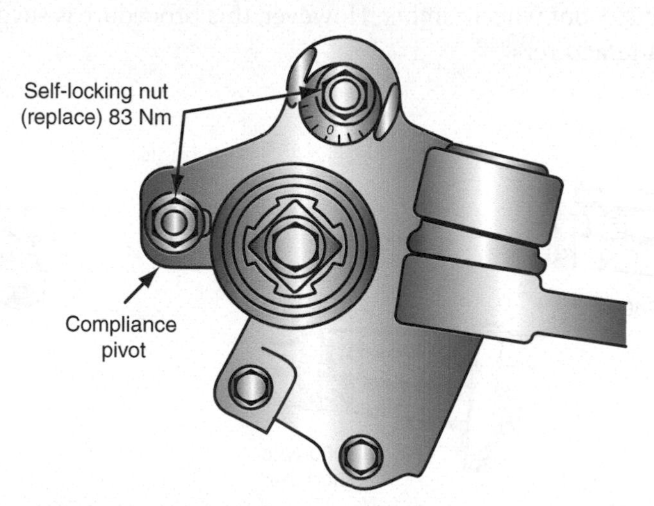

Figure 9-81 Graduated cam for caster adjustment, double-wishbone front suspension.

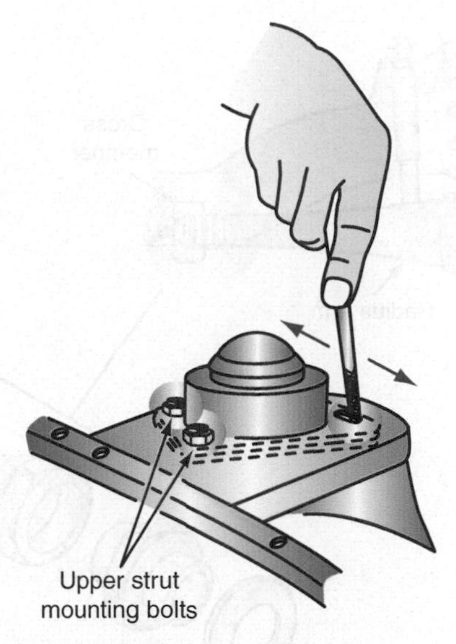

Figure 9-82 Elongating upper strut mount bolt openings to provide caster adjustment.

When a caster adjustment is necessary, some vehicle manufacturers recommend removing the upper strut mount bolts and elongating the bolt hole openings in the strut tower with a round file to provide a caster adjustment (**Figure 9-82**).

WARNING After any alignment angle adjustment, always be sure the adjustment bolts are tightened to the vehicle manufacturer's specified torque. Loose suspension adjustment bolts will result in improper alignment angles, steering pull, and tire wear. Loose suspension adjustment bolts may also cause reduced directional stability, which could result in a vehicle accident and possible personal injury.

Caster Adjustment on Twin I-Beam Front Suspension Systems

WARNING Improper use of bending equipment may result in personal injury. Always follow the equipment and vehicle manufacturer's recommended procedures.

The same bending tool and hydraulic jack may be used to adjust camber and caster on twin I-beam front suspension systems. This equipment is installed on the radius rods connected from the I-beams to the chassis to adjust caster. If the hydraulic jack is placed near the rear end of the radius rod and the clamp arm is positioned over the center area of this rod, the rod is bent downward in the center when the jack is operated (**Figure 9-83**). This bending action tilts the I-beam and caster line rearward to increase positive caster.

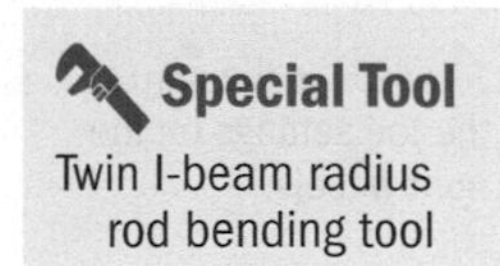
Special Tool
Twin I-beam radius rod bending tool

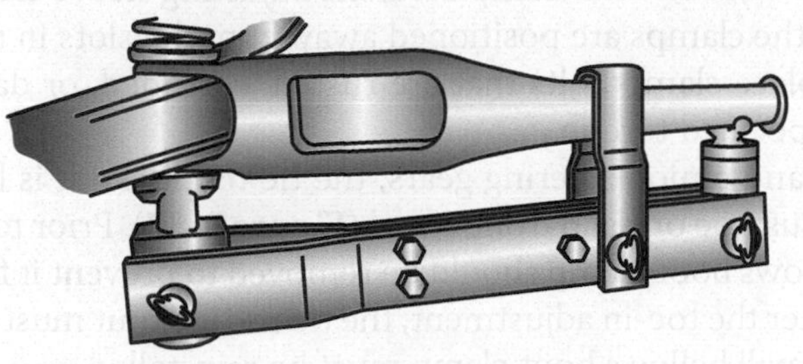

Figure 9-83 Hydraulic jack and bending tool positioned on the radius rod to increase positive caster on a twin I-beam front suspension.

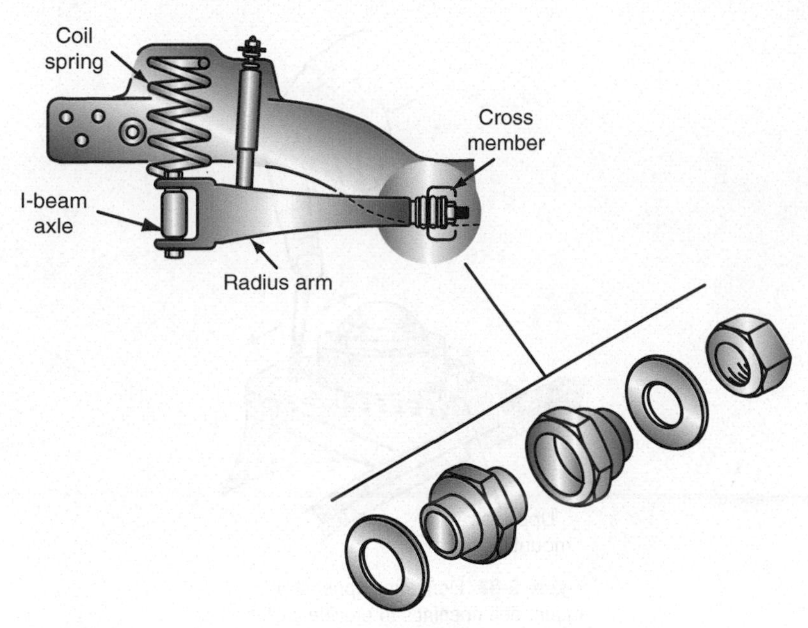

Figure 9-84 Radius arm bushing for caster adjustment on some twin I-beam suspension systems.

Aftermarket parts manufacturers supply a special radius arm bushing to provide a caster adjustment on some twin I-beam front suspension systems (**Figure 9-84**).

Special Tool
Tie rod sleeve rotating tool

TOE ADJUSTMENT

WARNING **Do not heat tie rod sleeves to loosen them. This action may weaken the sleeves, resulting in sudden sleeve failure, loss of steering control, personal injury, and vehicle damage.**

Classroom Manual
Chapter 9, page 242

WARNING **Do not use a pipe wrench to loosen tie rod sleeves. This tool may crush and weaken the sleeve, causing sleeve failure, loss of steering control, personal injury, and vehicle damage. A tie rod rotating tool must be used for this job.**

When a front or rear wheel toe adjustment is required, apply penetrating oil to the tie rod adjusting sleeves and sleeve clamp bolts. Loosen the tie rod adjusting sleeve clamp bolts enough to allow the clamps to partially rotate. One end of each tie rod sleeve contains a right-hand thread; the opposite end contains a left-hand thread. These threads match the threads on the tie rod and outer tie rod end. When the tie rod sleeve is rotated, the complete tie rod, sleeve, and tie rod end assembly is lengthened or shortened. Use a tie rod sleeve rotating tool to turn the sleeves until the toe-in on each front wheel is equal to one-half the total toe specification (**Figure 9-85**).

Total toe is the sum of the toe settings on the front wheels.

After the toe-in adjustment is completed, the adjusting sleeve clamps must be turned so the openings in the clamps are positioned away from the slots in the adjusting sleeves (**Figure 9-86**). Replace clamp bolts that are rusted, corroded, or damaged, and tighten these bolts to the specified torque.

On some rack and pinion steering gears, the tie rod locknut is loosened and the tie rod is rotated to adjust toe on each front wheel (**Figure 9-87**). Prior to rotating the tie rod, the small outer bellows boot clamp should be removed to prevent it from twisting during tie rod rotation. After the toe-in adjustment, the tie rod locknut must be torqued to specifications, and the small bellows boot clamp must be reinstalled.

On other rack and pinion steering gears, the tie rod and outer tie rod end have internal threads, and a threaded adjuster is installed in these threads. One end of the adjuster has

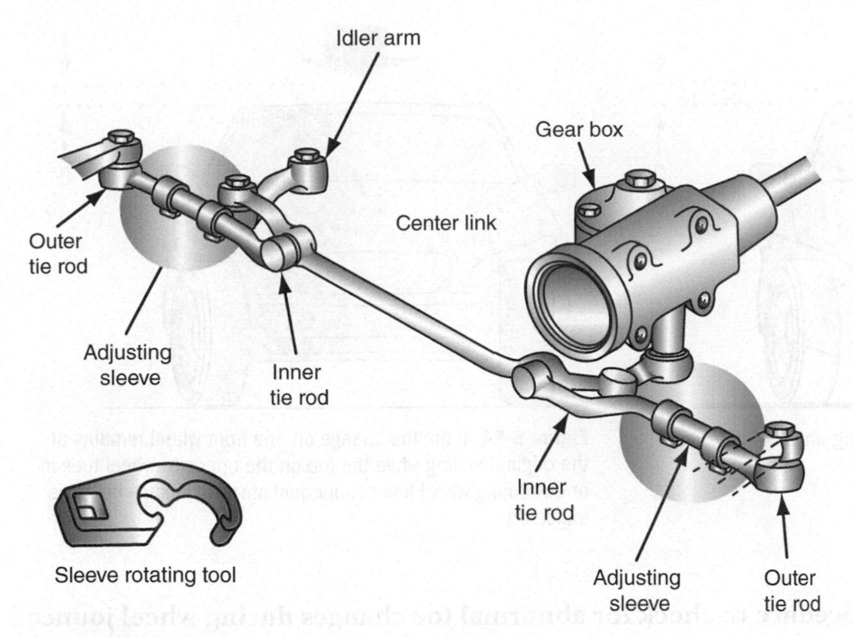

Figure 9-85 Rotating tie rod sleeves to adjust front wheel toe.

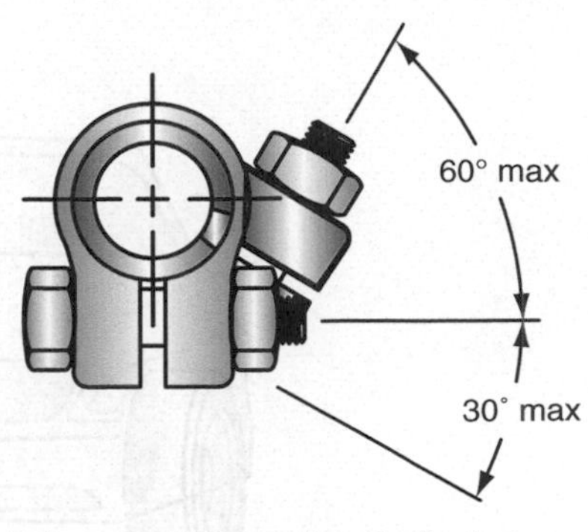

Figure 9-86 Proper tie rod adjusting sleeve and clamp position.

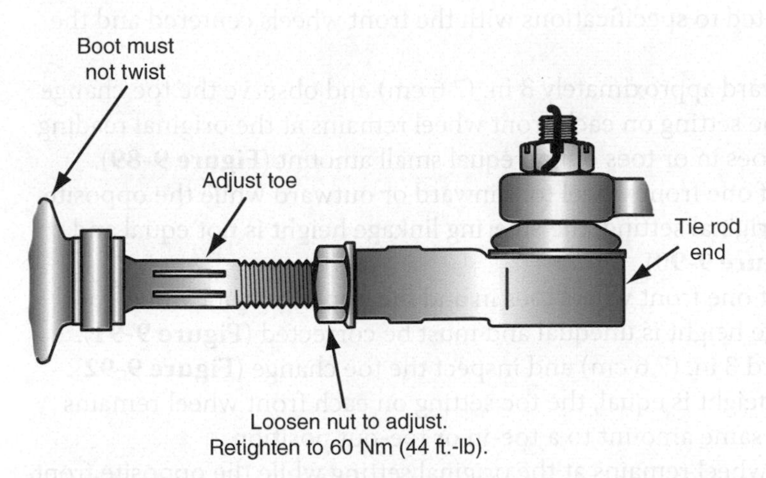

Figure 9-87 Rotating tie rod to adjust the toe on a rack and pinion steering gear.

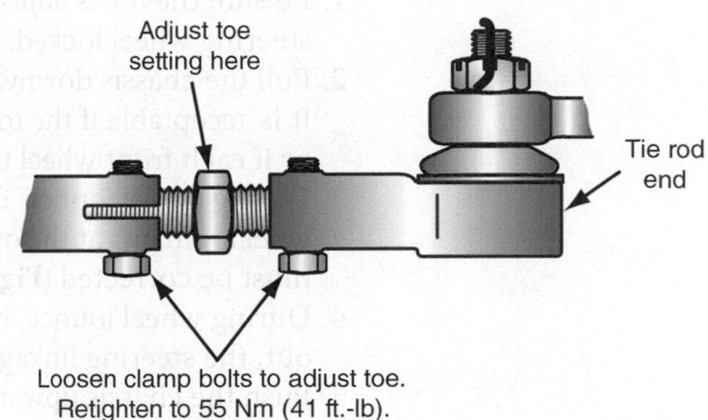

Figure 9-88 Rack and pinion steering gear with externally threaded toe adjuster and internal threads on the tie rod and outer tie rod end.

a right-hand thread, and the opposite end has a left-hand thread. Matching threads are located in the outer tie rod end and tie rod. Clamps on the outer tie rod end and tie rod secure these components to the adjuster. A hex-shaped nut is designed into the center of the adjuster (**Figure 9-88**). After the clamps are loosened, an open-end wrench is placed on this hex nut to rotate the adjuster and change the toe setting.

When the toe-in adjustment is completed, the toe must be set within the manufacturer's specifications with the steering wheel in the centered position. If the steering wheel is not centered, a centering adjustment is required.

CHECKING TOE CHANGE AND STEERING LINKAGE HEIGHT

If the steering linkage on each side is not the same height, a condition called bump steer may occur. When bump steer occurs, the steering suddenly swerves to one side when one of the front wheels strikes a road irregularity. When this steering linkage condition is present, abnormal toe changes occur during wheel jounce and rebound.

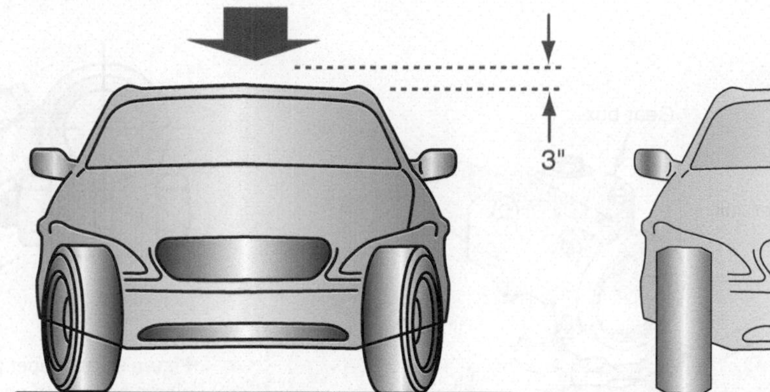

Figure 9-89 Normal toe change during wheel jounce.

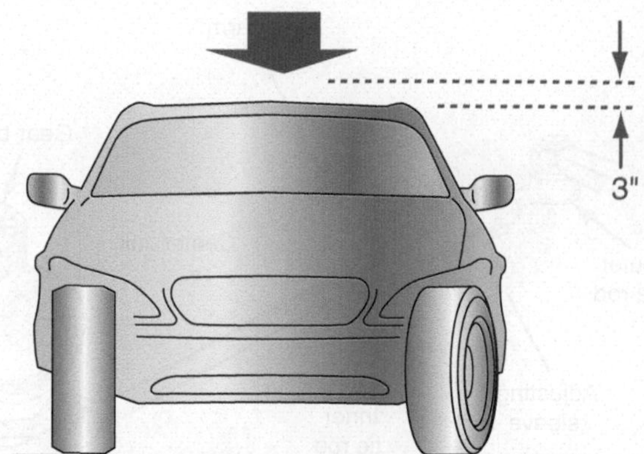

Figure 9-90 If the toe change on one front wheel remains at the original setting while the toe on the opposite wheel toes in or out during wheel jounce, unequal steering linkage height is indicated.

Follow this procedure to check for abnormal toe changes during wheel jounce and rebound:

1. Be sure the toe is adjusted to specifications with the front wheels centered and the steering wheel locked.
2. Pull the chassis downward approximately 3 in. (7.6 cm) and observe the toe change. It is acceptable if the toe setting on each front wheel remains at the original reading or if each front wheel toes in or toes out an equal small amount (**Figure 9-89**).
3. During wheel jounce, if one front wheel toes inward or outward while the opposite wheel remains at the original setting, the steering linkage height is not equal and must be corrected (**Figure 9-90**).
4. During wheel jounce, if one front wheel toes in and the opposite front wheel toes out, the steering linkage height is unequal and must be corrected (**Figure 9-91**).
5. Push the chassis upward 3 in. (7.6 cm) and inspect the toe change (**Figure 9-92**). If the steering linkage height is equal, the toe setting on each front wheel remains the same or moves the same amount to a toe-in or toe-out position.
6. If the toe on one front wheel remains at the original setting while the opposite front wheel toe changes to a toe-in or toe-out setting, the steering linkage height is unequal. When one front wheel moves to a toe-in position, and the opposite front wheel moves to a toe-out setting, unequal steering linkages are indicated.

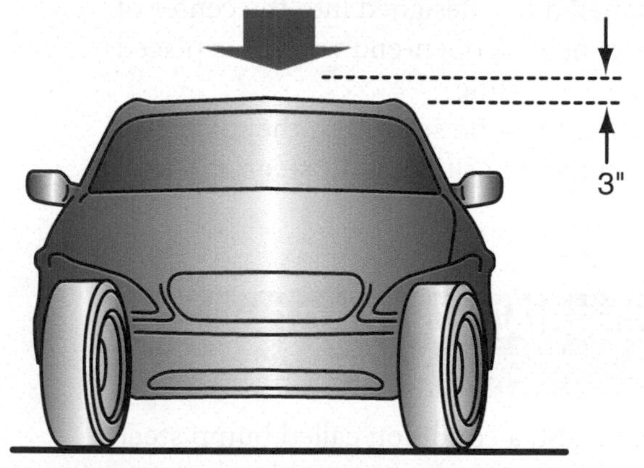

Figure 9-91 If the toe on one front wheel toes in and the toe on the opposite front wheel toes out during wheel jounce, unequal steering linkage height is indicated.

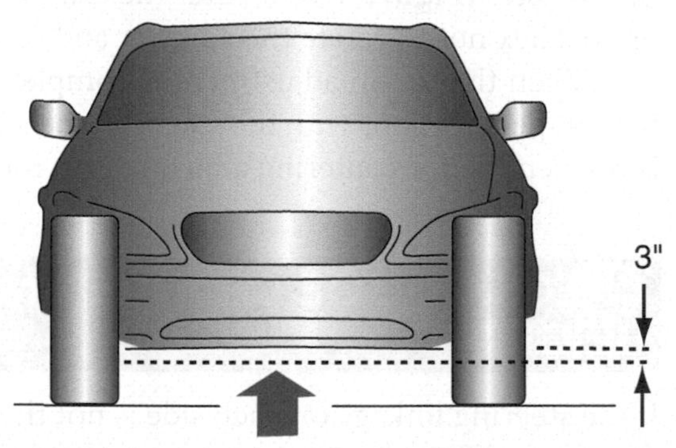

Figure 9-92 Pull the chassis upward 3 in. (7.6 cm) to check for improper toe change indicating unequal steering linkage height.

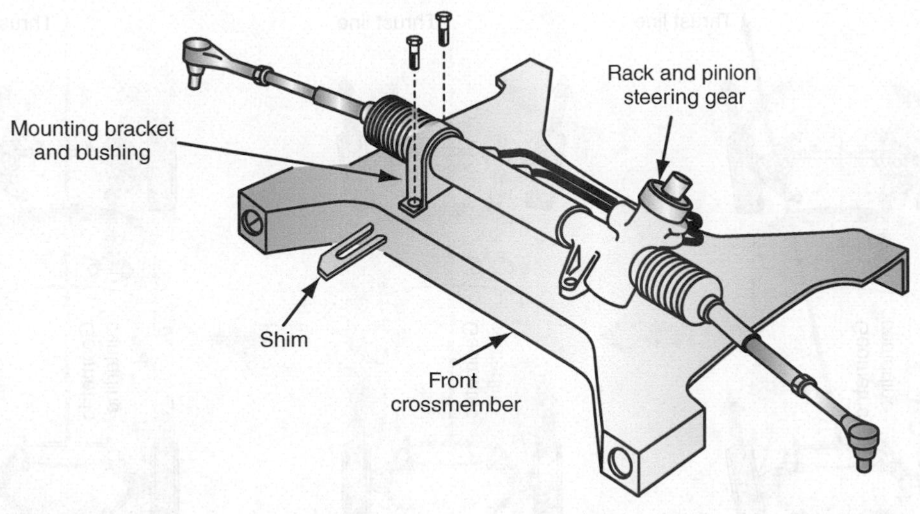

Figure 9-93 On some imported vehicles, shims between the rack and pinion steering gear and the chassis adjust the steering linkage height.

On some imported vehicles, the steering linkage height can be adjusted with shims between the rack and pinion steering gear and the chassis (**Figure 9-93**). The idler arm may be moved upward or downward on some domestic vehicles to adjust the steering linkage height. On other vehicles, if the steering linkage height is unequal, steering components are worn or bent. These components include tie rods, tie rod ends, idler arms, pitman arms, and rack and pinion steering gear mounting bushings.

Another method of inspecting for equal tie rod height on rack and pinion steering gears is to measure the distance from the center of the inner retaining bolt on the lower control arm to the alignment ramp on each side of the front suspension (distances A and B in **Figure 9-94**). If this distance is not equal on each side of the front suspension, one of the front springs may be sagged or such components as control arm bushings may be worn. When distances A and B are not equal, this problem must be corrected before performing the steering gear measurements. If distances A and B are equal, measure from the outer ends of the rack and pinion steering gear to the alignment ramp (distances C and D). If these distances are not equal, the rack and pinion steering gear mountings may be worn or distorted.

Steering Angle Sensor Reset

The last step on an alignment procedure that must be performed today on some vehicle platforms is a reset procedure for the steering angle sensor (SAS) and other related sensors to update the vehicle geometry information stored in the vehicle's software system after an alignment (**Figure 9-95**). On some vehicles, this is a simple procedure and on others

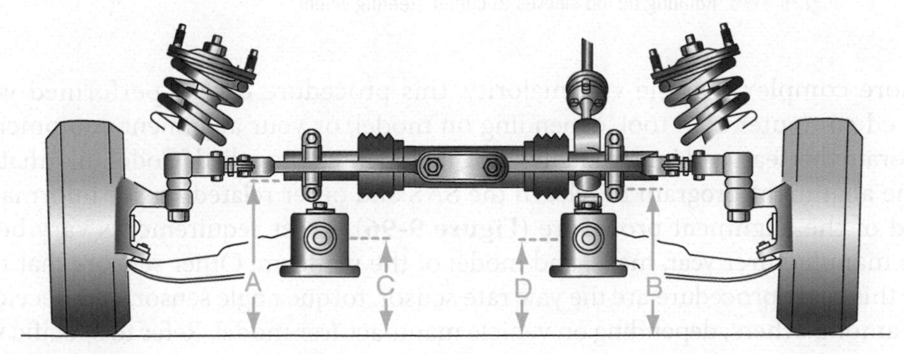

Figure 9-94 Measuring rack and pinion steering gear height at various locations.

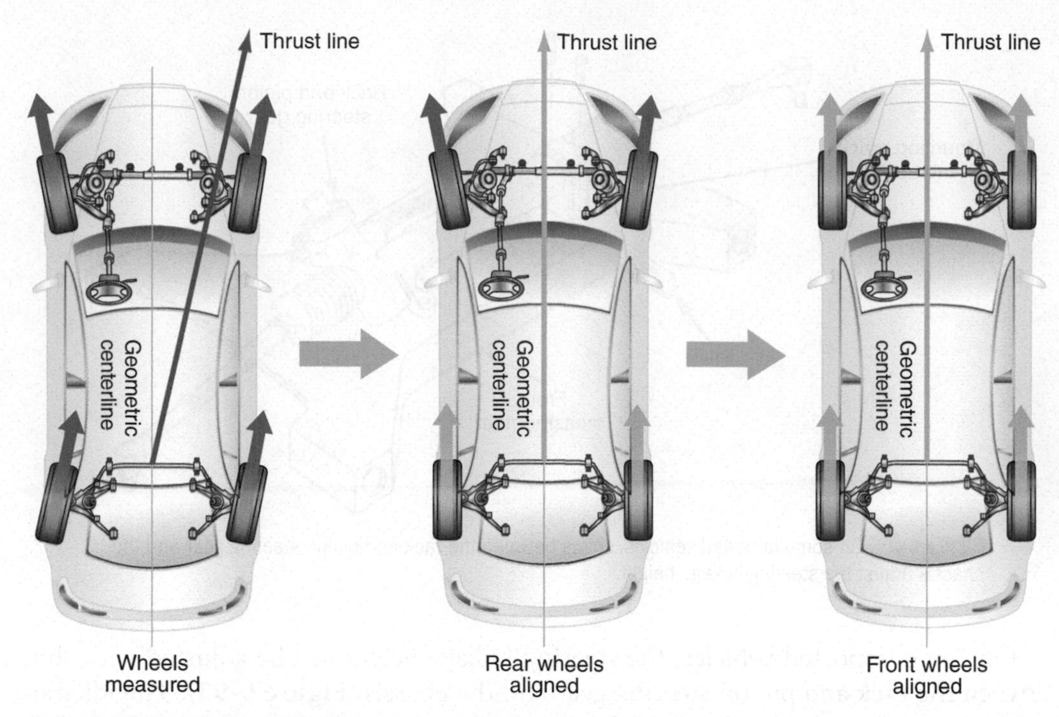

Figure 9-95 Marking tie rod sleeves in relation to the tie rods and outer tie rod ends.

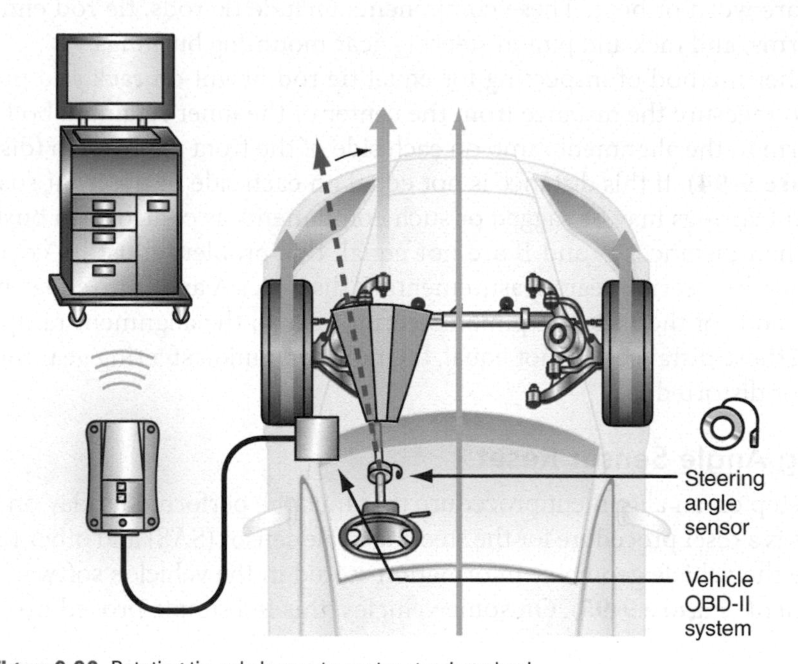

Figure 9-96 Rotating tie rod sleeves to center steering wheel.

it is more complex. On the vast majority, this procedure can be performed with an enhanced computer scan tool, depending on model, or your alignment equipment may incorporate this feature. Hunter Engineering has an interface called "CodeLink" that works with the alignment program to relearn the SAS and other related sensor information at the end of the alignment procedure (**Figure 9-96**). Reset requirements vary between vehicle manufacturer year, make, and model of the platform. Other sensors that may be part of this reset procedure are the yaw rate sensor, torque angle sensor, and deceleration sensor among others, depending on vehicle manufacturer model. Refer to specific vehicle information for specific steps and requirements.

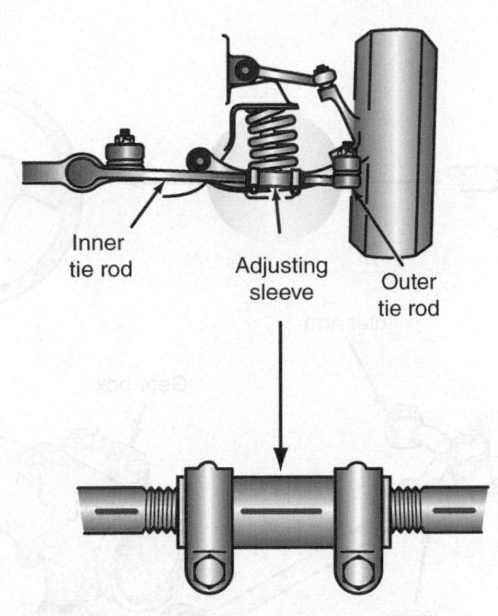

Figure 9-97 Marking tie rod sleeves in relation to the tie rods and outer tie rod ends.

MANUAL STEERING WHEEL CENTERING PROCEDURE

Road test the vehicle and determine if the steering wheel spoke is centered when the vehicle is driven straight ahead.

SERVICE TIP The most accurate check of a properly centered steering wheel is while driving the vehicle straight ahead during a road test.

If steering wheel centering is necessary, follow this procedure:

1. Lift the front end of the vehicle with a hydraulic jack and position safety stands under the lower control arms. Lower the vehicle onto the safety stands and place the front wheels in the straight-ahead position.
2. Use a piece of chalk to mark each tie rod sleeve in relation to the tie rod, and loosen the sleeve clamps (**Figure 9-97**).
3. Position the steering wheel spoke in the position it was in while driving straight ahead during the road test. Turn the steering wheel to the centered position and note the direction of the front wheels.
4. If the steering wheel spoke is low on the left side while driving the vehicle straight ahead, use a tie rod sleeve rotating tool to shorten the left tie rod and lengthen the right tie rod (**Figure 9-98**). One-quarter turn on a tie rod sleeve moves the steering wheel position approximately 1 in. Turn the tie rod sleeves the proper amount to bring the steering wheel to the centered position. For example, if the steering wheel spoke is 2 in. off-center, turn each tie rod sleeve one-half turn.
5. If the steering wheel spoke is low on the right side while driving the vehicle straight ahead, lengthen the left tie rod and shorten the right tie rod.
6. Mark each tie rod sleeve in its new position in relation to the tie rod. Be sure the sleeve clamp openings are positioned properly, as indicated earlier in this chapter. Tighten the clamp bolts to the specified torque.
7. Lift the front chassis with a floor jack and remove the safety stands. Lower the vehicle onto the shop floor, and check the steering wheel position during a road test.

Classroom Manual
Chapter 9, page 247

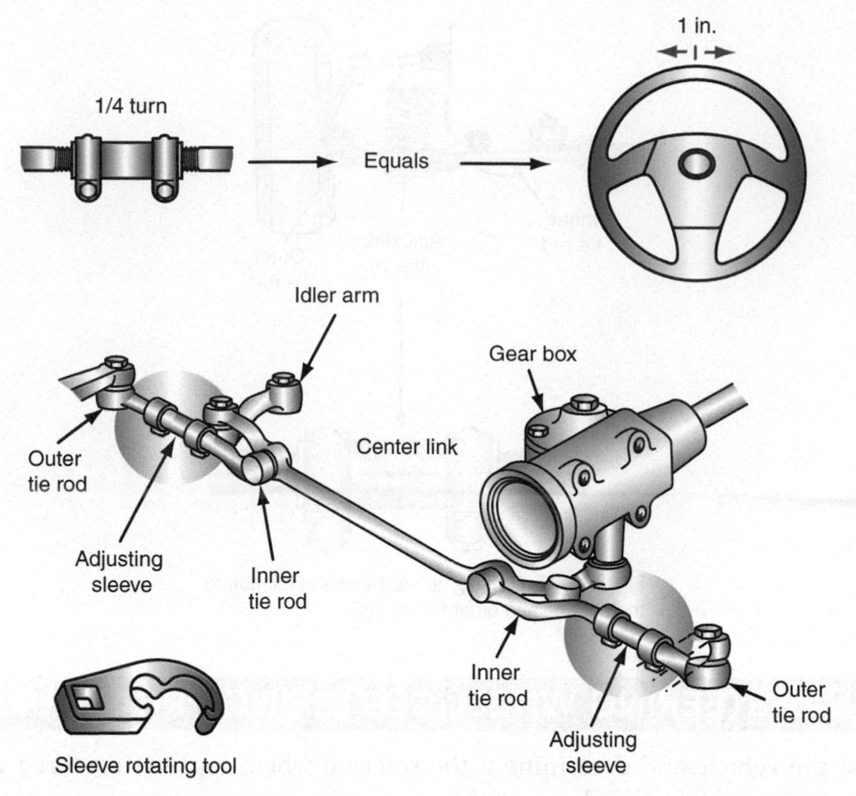

Figure 9-98 Rotating tie rod sleeves to center steering wheel.

SERVICE TIP Some computer wheel aligners have a steering wheel centering display that provides an easier method of steering wheel centering compared to the manual method.

When all the front suspension alignment angles are adjusted within the manufacturer's specifications, road test the vehicle. When the vehicle is driven on a relatively smooth, straight road, there should be directional stability with no drift to the right or left, and the steering wheel must be centered (**Table 9-1**).

CUSTOMER CARE Always concentrate on quality workmanship and customer satisfaction. Most customers do not mind paying for vehicle repairs if the work is done properly and their vehicle problem is corrected. To determine customer satisfaction, make follow-up phone calls a few days after repairing their vehicle. This indicates that you are interested in their vehicle and that you consider quality work and satisfied customers a priority.

TABLE 9-1 STEERING AND WHEEL ALIGNMENT DIAGNOSIS

Problem	Symptoms	Possible Causes
Tire tread wear, inside edge	Steering pull, premature tire replacement	Excessive negative camber
Tire tread wear, outside edge	Steering pull, premature tire replacement	Excessive positive camber
Tire tread wear, feathered	Premature tire replacement	Improper toe setting
Tire tread wear, cupped	Wheel vibration	Improper wheel balance
Steering wander	Reduced directional stability	Reduced positive caster or negative caster
Steering pull	Steering pull to the right when driving straight ahead	Reduced positive caster, right front wheel Excessive positive camber, right front wheel Excessive toe-out, left rear wheel
Steering pull	Steering pull to the left when driving straight ahead	Reduced positive caster, right front wheel Excessive positive camber, right front wheel Excessive toe-out, right rear wheel
Steering wheel return	Excessive steering wheel returning force after a turn	Excessive positive caster on front wheels
Steering wheel return	Steering wheel does not return properly after a turn	Binding column or linkage Reduced positive caster on front wheels
Harsh riding	Reduced ride quality when driving on road irregularities	Worn shock absorbers or struts Excessive positive caster, front wheels Reduced ride height

CASE STUDY

A customer complained about erratic steering on a front-wheel-drive Dodge Intrepid. A road test revealed the car steered reasonably well on a smooth road surface, but while driving on irregular road surfaces, the steering would suddenly swerve to the right or left.

The technician performed a preliminary wheel alignment inspection and found the right tie rod end was loose; all the other suspension and steering components were in satisfactory condition. The technician replaced the loose tie rod end, but a second road test indicated that the bump steer problem was still present. After advising the customer that a complete wheel alignment was necessary, the technician drove the vehicle on the wheel aligner and carefully checked all front and rear alignment angles. Each front and rear wheel alignment angle was within specifications. The technician realized that somehow he had not diagnosed this problem correctly.

While thinking about this problem, the technician remembered a general diagnostic procedure he learned while studying automotive technology. This procedure stated: Listen to the customer complaints, be sure the complaint is identified, think of the possible causes, test to locate the exact problem, and be sure the complaint is eliminated. The technician realized he had not thought much about the causes of the problem, and so he began to recall the wheel alignment theory he learned in college. He remembered that the tie rods must be parallel to the lower control arms, and if the tie rod height is unequal, this parallel condition no longer exists. The technician also recalled that unequal tie rod height causes improper toe changes during wheel jounce and rebound, which result in bump steer.

An inspection of the toe during front wheel jounce and rebound indicated the toe on the right front wheel remained the same during wheel jounce and rebound, but the toe on the left front wheel moved to a toe-out position. Because the tie rods had been inspected during the preliminary alignment inspection, the technician turned his attention to the rack and pinion steering gear mounting. He found the bushing on the right end of the steering gear was worn and loose. This bushing was replaced and all the steering gear mounting bolts were tightened to the specified torque. An inspection of the toe change during wheel jounce and rebound revealed a normal toe change.

From this experience the technician learned the following two things:

1. His understanding of wheel alignment theory was very important in diagnosing steering problems.
2. Always be thorough! During a prealignment inspection, check all suspension and steering components, including rack and pinion steering gear mountings.

ASE-STYLE REVIEW QUESTIONS

1. While discussing a front suspension height that is 1 in. (2.54 cm) less than specified:
   Technician A says the suspension height must be correct before a wheel alignment is performed.
   Technician B says the lower front suspension height may be caused by worn lower control arm bushings.
   Who is correct?
   A. A only C. Both A and B
   B. B only D. Neither A nor B
2. While performing a prealignment inspection:
   Technician A says improper front wheel bearing adjustment may affect wheel alignment angles.
   Technician B says worn ball joints have no effect on wheel alignment angles.
   Who is correct?
   A. A only C. Both A and B
   B. B only D. Neither A nor B
3. While discussing a front suspension system in which the right front wheel has 2 degrees positive camber and the left front wheel has 0.5 degree positive camber:
   Technician A says when the vehicle is driven straight ahead, the steering will pull to the left.
   Technician B says there will be excessive wear on the inside edge of the left front tire tread.
   Who is correct?
   A. A only C. Both A and B
   B. B only D. Neither A nor B
4. When discussing unsatisfactory steering wheel returnability:
   Technician A says the rack and pinion steering gear mounts may be worn.
   Technician B says this problem may be caused by interference between the dash seal and the steering shaft.
   Who is correct?
   A. A only C. Both A and B
   B. B only D. Neither A nor B
5. While discussing turning radius measurement:
   Technician A says a bent steering arm will cause the turning radius to be out of specification.
   Technician B says if the turning radius is not within specification, tire tread wear is excessive while cornering.
   Who is correct?
   A. A only C. Both A and B
   B. B only D. Neither A nor B
6. While measuring and adjusting front wheel toe:
   A. If the positive caster is increased on the right front wheel, this wheel moves toward a toe-in position.
   B. Improper front wheel toe setting causes steering wander and drift.
   C. The front wheel toe should be checked with the front wheels straight ahead.
   D. Improper front wheel toe setting causes wear on the inside edge of the tire tread.
7. All of these statements about front wheel toe change during wheel jounce and rebound are true EXCEPT:
   A. If one front wheel toes in and the opposite front wheel toes out during front wheel jounce and rebound, the tie rod height is unequal.
   B. If both front wheels toe in or toe out a small, equal amount during front wheel jounce and

rebound, the tie rods are parallel to the lower control arms.

C. The improper toe change during front wheel jounce and rebound may cause bump steer.

D. Improper toe change during front wheel jounce and rebound may be caused by a worn upper strut mount.

8. While using a computer wheel aligner:

A. The technician may select defective suspension components from a list on the screen.

B. It is not necessary to check or compensate for wheel runout.

C. If the camber bar graph is red, the camber setting is within specifications.

D. A wheel sensor containing a high-frequency transmitter requires a cable connected to the aligner computer.

9. While using computer wheel aligners:

A. On many computer wheel aligners, the technician may only print out the four-wheel alignment results.

B. Some wheel sensors have the capability to measure ride height and display this reading on the screen.

C. A front wheel swing is necessary before reading the front wheel camber.

D. If the computer aligner contains a control arm movement monitor, the technician has to estimate the necessary shim thickness.

10. While using computer wheel aligners:

A. Symmetry angle measurements display thrust angle, rear wheel toe, and rear wheel camber.

B. Setback is the angle formed by a line at 90 degrees to the vehicle centerline at the axle attachment point and a line through the left and right wheel centers.

C. Wheelbase difference is an angle created by a line through the rear wheel centers and the thrust line.

D. Right or left lateral offset is an angle between a line through the left front and left rear tires and a line between the right front and right rear tires.

ASE CHALLENGE QUESTIONS

1. While discussing turning radius:

Technician A says incorrect turning radius may often be noted by tire squeal while cornering.

Technician B says to adjust turning radius toe-out on turns, turn the inner wheel to stop.

Who is correct?

A. A only C. Both A and B

B. B only D. Neither A nor B

2. The customer says that sometimes her car suddenly swerves to one side on a bump. All of the following could cause this problem EXCEPT:

A. Loose steering gear.

B. Worn tie rods.

C. Sagging front springs.

D. Steering gear lash adjustment.

3. While discussing steering diagnosis:

Technician A says uneven half-shaft axle lengths may cause a vehicle to pull to one side when accelerating.

Technician B says abnormal toe changes can cause a vehicle to pull to one side on road irregularities.

Who is correct?

A. A only C. Both A and B

B. B only D. Neither A nor B

4. After you have completed a front end alignment, a customer returns to your shop and complains of continued vehicle drift. To correct this problem, you should:

A. Inspect the manual steering gear for possible miscentering of the sector gear.

B. Inspect rear wheel alignment.

C. Ask the customer to fill the fuel tank.

D. Inspect the steering column flex coupling.

5. While performing a prealignment inspection:

Technician A says a prealignment inspection should include checking the vehicle interior for heavy items.

Technician B says tools and other items normally carried in the vehicle should be included during an alignment.

Who is correct?

A. A only C. Both A and B

B. B only D. Neither A nor B

JOB SHEET 44

Name ______________________ Date ______________

ROAD TEST VEHICLE AND DIAGNOSE STEERING OPERATION

Upon completion of this job sheet, you should be able to road test a vehicle and diagnose steering operation.

ASE Education Foundation Task Correlation

This job sheet addresses the following **MLR** task:

C.1. Perform prealignment inspection; measure vehicle ride height. **(P-1)**

This job sheet addresses the following **AST/MAST** task:

E.1. Diagnose vehicle wander, drift, pull, hard steering, bump steer, memory steer, torque steer, and steering return concerns; determine needed action. **(P-1)**

E.2. Perform prealignment inspection; measure vehicle ride height; determine needed action. **(P-1)**

We Support
Education Foundation

Describe the vehicle being worked on:

Year ____________ Make ____________ Model ____________

VIN ____________ Engine type and size ____________

Procedure

1. Road test vehicle under such driving conditions as slow-speed driving, cornering, and normal cruising speed while driving on a straight, level road surface. Check for the following abnormal steering conditions:

   a. Vertical chassis oscillations: ☐ Satisfactory ☐ Unsatisfactory

   b. Chassis lateral waddle: ☐ Satisfactory ☐ Unsatisfactory

   c. Steering pull to right: ☐ Satisfactory ☐ Unsatisfactory

   d. Steering pull to left: ☐ Satisfactory ☐ Unsatisfactory

   e. Steering effort: ☐ Satisfactory ☐ Unsatisfactory

   f. Tire squeal while cornering: ☐ Satisfactory ☐ Unsatisfactory

   g. Bump steer: ☐ Satisfactory ☐ Unsatisfactory

   h. Torque steer: ☐ Satisfactory ☐ Unsatisfactory

   i. Memory steer: ☐ Satisfactory ☐ Unsatisfactory

   j. Steering wheel return: ☐ Satisfactory ☐ Unsatisfactory

   k. Steering wheel free play: ☐ Satisfactory ☐ Unsatisfactory

2. Return the vehicle to the shop, and inspect suspension and steering to determine the cause of abnormal conditions. List the necessary repairs or adjustments to correct all abnormal conditions that occurred during the road test.

____

____

Instructor's Response

____

____

____

Name ______________________________ Date ______________

JOB SHEET 45

MEASURE FRONT AND REAR WHEEL ALIGNMENT ANGLES WITH A COMPUTER WHEEL ALIGNER

Upon completion of this job sheet, you should be able to measure front and rear wheel alignment with a computer wheel aligner.

ASE Education Foundation Task Correlation

The following job sheet addresses the following **MLR** tasks:

C.1. Perform prealignment inspection; measure vehicle ride height. **(P-1)**

C.2. Describe alignment angles (camber, caster, and toe). **(P-1)**

The following job sheet addresses the following **AST/MAST** tasks:

E.2. Perform prealignment inspection; measure vehicle ride height; determine needed action. **(P-1)**

E.3. Prepare vehicle for wheel alignment on alignment machine; perform four-wheel alignment by checking and adjusting front and rear wheel caster, camber; and toe as required; center steering wheel. **(P-1)**

E.4. Check toe-out on turns (turning radius); determine needed action. **(P-2)**

E.5. Check steering axis inclination (SAI) and included angle; determine needed action. **(P-2)**

E.6. Check rear wheel thrust angle; determine needed action. **(P-1)**

E.7. Check for front wheel setback; determine needed action. **(P-2)**

E.8. Check front and/or rear cradle (subframe) alignment; determine needed action. **(P-3)**

We Support ASE | Education Foundation

Tools and Materials

Computer wheel aligner

Describe the vehicle being worked on:

Year ______________ Make ______________ Model ______________

VIN ______________________ Engine type and size ______________

Procedure

WARNING When you drive a vehicle onto an alignment ramp, be sure no one is standing in front of the vehicle to avoid causing personal injury.

1. Lock the front turntables and drive the vehicle onto the alignment ramp. Apply the parking brake.

   Are the front tires properly positioned on the front turntables? ☐ Yes ☐ No

   Are the rear tires properly positioned on the slip plates? ☐ Yes ☐ No

   Instructor check ______________________________

2. Install the rim clamps and wheel sensors. Perform wheel sensor leveling and wheel runout compensation procedures.

   Are the wheel sensors level? ☐ Yes ☐ No

   Is the wheel runout compensation procedure completed? ☐ Yes ☐ No

   Instructor check ______________________

3. Select the specifications for the vehicle being aligned.

   Are the specifications selected? ☐ Yes ☐ No

   Instructor check ______________________

4. Perform a prealignment inspection using the checklist in the computer wheel aligner. List any components that must be repaired or replaced, and explain the reasons for your diagnosis.

   ____
   ____
   ____
   ____
   ____

5. Measure the ride height.

Left front ride height ______	Specified ride height ______
Right front ride height ______	Specified ride height ______
Left rear ride height ______	Specified ride height ______
Right rear ride height ______	Specified ride height ______

   State the necessary repairs to correct ride height and explain the reasons for your diagnosis.

   ____
   ____
   ____
   ____

6. Measure the front and rear suspension alignment angles following the prompts on the computer wheel aligner screen.

Left front camber ______	Right front camber ______
Cross camber ______	Specified front wheel camber ______
Specified cross camber ______	
Left front caster ______	Right front caster ______
Cross caster ______	Specified front wheel caster ______
Specified cross caster ______	
Left front SAI ______	Right front SAI ______
Specified SAI ______	Included angle ______

Front wheel set-back +/– ______	
Thrust angle ______	
Specified thrust angle ______	
Left front toe ______	Right front toe ______
Total toe ______	Specified front wheel toe ______
Left rear camber ______	Right rear camber ______
Specified camber ______	
Left rear toe ______	Right rear toe ______
Total toe ______	Specified rear wheel toe ______

State the necessary adjustments and repairs to correct front and rear suspension alignment angles and explain the reasons for your diagnosis.

______
______
______
______
______

7. Measure the turning radius.

Left turn:	Turning radius right front wheel ______
	Turning radius left front wheel ______
	Specified turning radius ______
Right turn:	Turning radius left front wheel ______
	Turning radius right front wheel ______
	Specified turning radius ______

State the necessary repairs to correct the turning radius, and explain the reasons for your diagnosis.

______
______
______
______
______
______

Instructor's Response

______
______
______

Task Completed

4. Use a piece of chalk to mark each tie rod sleeve in relation to the tie rod, and loosen the sleeve clamps.

Are the tie rod sleeves marked in relation to the tie rods? ☐ Yes ☐ No

Instructor check ______________________________

5. Position the steering wheel spoke in the position it was in while driving straight ahead during the road test. Turn the steering wheel to the centered position and note the direction of the front wheels.

Direction the front wheels are turned with the steering wheel centered:
______________ right ______________ left

6. If the steering wheel spoke is low on the left side while driving the vehicle straight ahead, use a tie rod sleeve rotating tool to shorten the left tie rod and lengthen the right tie rod. A one-quarter turn on a tie rod sleeve moves the steering wheel position approximately 1 in. Turn the tie rod sleeves the proper amount to bring the steering wheel to the centered position. For example, if the steering wheel spoke is 2 in. off-center, turn each tie rod sleeve one-half turn.

Left tie rod sleeve: lengthened ______________ shortened ______________

Right tie rod sleeve: lengthened ______________ shortened ______________

Amount of left tie rod sleeve rotation ______________

Amount of right tie rod sleeve rotation ______________

Is the steering wheel centered with front wheel straight ahead? ☐ Yes ☐ No

Instructor check ______________________________

7. If the steering wheel spoke is low on the right side while driving the vehicle straight ahead, lengthen the left tie rod and shorten the right tie rod. ☐

8. Mark each tie rod sleeve in its new position in relation to the tie rod. Be sure the sleeve clamp openings are positioned properly, as indicated in this chapter. Tighten the clamp bolts to the specified torque.

Are the tie rod sleeves marked in relation to tie rods? ☐ Yes ☐ No

Are the tie rod sleeve clamps properly installed? ☐ Yes ☐ No

Tie rod sleeve bolt specified torque ______________

Tie rod sleeve bolt actual torque ______________

Instructor check ______________________________

9. Lift the front chassis with a floor jack, and remove the safety stands. Lower the vehicle onto the shop floor, and check the steering wheel position during a road test.

Steering wheel position while driving straight ahead during a road test:
☐ Satisfactory ☐ Unsatisfactory

Instructor's Response

____

____

____

Name ______________________ Date ____________

JOB SHEET
48

ADJUST FRONT WHEEL ALIGNMENT ANGLES

Upon completion of this job sheet, you should be able to adjust front wheel alignment angles.

ASE Education Foundation Task Correlation

This job sheet addresses the following **MLR** tasks:

C.1. Perform prealignment inspection; measure vehicle ride height. **(P-1)**

C.2. Describe alignment angles (camber, caster, and toe). **(P-1)**

This job sheet addresses the following **AST/MAST** tasks:

E.2. Perform prealignment inspection; measure vehicle ride height; determine needed action. **(P-1)**

E.3. Prepare vehicle for wheel alignment on alignment machine; perform four-wheel alignment by checking and adjusting front and rear wheel caster, camber; and toe as required; center steering wheel. **(P-1)**

E.4. Check toe-out on turns (turning radius); determine needed action. **(P-2)**

E.5. Check steering axis inclination (SAI) and included angle; determine needed action. **(P-2)**

E.6. Check rear wheel thrust angle; determine needed action. **(P-1)**

E.7. Check for front wheel setback; determine needed action. **(P-2)**

E.8. Check front and/or rear cradle (subframe) alignment; determine needed action. **(P-3)**

E.9. Reset steering angle sensor. **(P-2)**

We Support
ASE Education Foundation

Tools and Materials

A vehicle with improperly adjusted front wheel camber, caster, and toe.

Describe the vehicle being worked on:

Year ____________ Make ____________ Model ____________

VIN ________________ Engine type and size ________________

Procedure

1. Position the vehicle properly on the alignment ramp with the front wheels properly positioned on the turntables and the rear wheels on slip plates.

   Are the front wheels properly positioned on turntables? ☐ Yes ☐ No

   Are the rear wheels properly positioned on slip plates? ☐ Yes ☐ No

   Is the parking brake applied? ☐ Yes ☐ No

   Instructor check ______________________________

2. Perform a prealignment inspection and correct any defective conditions or components found during the inspection. List the defective conditions or components, and explain the reasons for your diagnosis.

   ____

   ____

   ____

3. Measure front and rear ride height and correct it if necessary.

Specified front suspension ride height ____________________

Actual front suspension ride height ____________________

Specified rear suspension ride height ____________________

Actual rear suspension ride height ____________________

List the necessary repairs to correct the front and rear suspension ride height and explain the reasons for your diagnosis.

____

____

____

4. Measure and adjust front wheel camber.

Specified front wheel camber ____________________

Actual front wheel camber: left ____________ right ____________

Front wheel camber after adjustment: left ____________ right ____________

Camber adjustment method ____________________

5. Measure SAI and included angle.

Specified SAI ____________________

Actual SAI: left ____________ right ____________

Specified included angle ____________________

Actual included angle: left ____________ right ____________

If the SAI and included angle are not within specifications, state the necessary repairs and explain the reasons for your diagnosis.

____

____

____

6. Measure and adjust front wheel caster.

Specified front wheel caster ____________________

Actual front wheel caster: left ____________________

right ____________________

Front wheel caster after adjustment: left ____________ right ____________

Caster adjustment method ____________________

7. Inspect camber measurements and adjust if necessary.

Camber adjustment: ☐ Satisfactory ☐ Unsatisfactory

Left front camber readjusted to ____________________

Right front camber readjusted to ____________________

8. Measure and adjust front wheel toe.

Specified front wheel toe ______________________

Actual front wheel toe: left ______________ right ______________

Total toe ______________________

Front wheel toe after adjustment: left ______________ right ______________

Total front wheel toe ______________________

9. Inspect steering wheel centering with front wheels straight ahead and center the wheel if necessary.

Is steering wheel centered? ☐ Yes ☐ No

Instructor check ______________________

10. Measure turning angle.

Specified turning angle ______________________

Actual turning angle left front wheel ______________________

Actual turning angle right front wheel ______________________

Turning angle correction required: ☐ Yes ☐ No

Left front wheel turned inward 20 degrees, turning angle on right front wheel ______________________

Right front wheel turned inward 20 degrees, turning angle on left front wheel ______________________

If the turning angle is not within specifications, state the necessary repairs and explain the reasons for your diagnosis.

____

____

____

____

11. Record alignment specifications both before and after the alignment.

BEFORE ALIGNMENT READINGS								
Left Front					**Right Front**			
Min.	**Preferred Spec.**	**Max.**	**Actual**		**Actual**	**Min.**	**Preferred Spec.**	**Max.**
				CAMBER				
				CASTER				
				TOE				
				SAI				
				INCLUDED ANGLE				
				TURNING ANGLE				

FRONT				
	Actual	Min.	Preferred Spec.	Max.
CROSS CAMBER				
CROSS CASTER				
TOTAL TOE				
SET BACK				

Left Rear					Right Rear			
Min.	Preferred Spec.	Max.	Actual		Actual	Min.	Preferred Spec.	Max.
				CAMBER				
				TOE				

REAR				
	Actual	Min.	Preferred Spec.	Max.
Total Toe				
Thrust Line				

AFTER ALIGNMENT READINGS								
Left Front					Right Front			
Min.	Preferred Spec.	Max.	Actual		Actual	Min.	Preferred Spec.	Max.
				CAMBER				
				CASTER				
				TOE				
				SAI				
				INCLUDED ANGLE				
				TURNING ANGLE				

FRONT				
	Actual	Min.	Preferred Spec.	Max.
CROSS CAMBER				
CROSS CASTER				
TOTAL TOE				
SET BACK				

Left Rear					Right Rear			
Min.	Preferred Spec.	Max.	Actual		Actual	Min.	Preferred Spec.	Max.
				CAMBER				
				TOE				

REAR				
	Actual	Min.	Preferred Spec.	Max.
TOTAL TOE				
THRUST LINE				

12. Steering Angle Sensor Reset: The last step of an alignment procedure that must be performed on some vehicle platforms is a reset procedure for the steering angle sensor (SAS) and other related sensors to update the vehicle geometry information stored in the vehicle's software system after an alignment. This procedure can be performed with an enhanced computer scan tool, depending on model, or your alignment equipment may incorporate this feature. Reset requirements vary with vehicle manufacturer year, make, and model of the platform. Other sensors that may be part of this reset procedure are the yaw rate sensor, torque angle sensor, and deceleration sensor among others depending on vehicle manufacturer model. Refer to specific vehicle information for specific steps and requirements.

Instructor's Response

____

____

____

Name ______________________ Date ____________

JOB SHEET 49

FOUR (ALL) WHEEL ALIGNMENT

Upon completion of this job sheet, you should be able to adjust rear wheel alignment angles.

ASE Education Foundation Task Correlation

This job sheet addresses the following **MLR** tasks:

C.1. Perform prealignment inspection; measure vehicle ride height. **(P-1)**

C.2. Describe alignment angles (camber, caster, and toe). **(P-1)**

This job sheet addresses the following **AST/MAST** tasks:

E.2. Perform prealignment inspection; measure vehicle ride height; determine needed action. **(P-1)**

E.3. Prepare vehicle for wheel alignment on alignment machine; perform four-wheel alignment by checking and adjusting front and rear wheel caster, camber; and toe as required; center steering wheel. **(P-1)**

E.4. Check toe-out on turns (turning radius); determine needed action. **(P-2)**

E.5. Check steering axis inclination (SAI) and included angle; determine needed action. **(P-2)**

E.6. Check rear wheel thrust angle; determine needed action. **(P-1)**

E.7. Check for front wheel setback; determine needed action. **(P-2)**

E.8. Check front and/or rear cradle (subframe) alignment; determine needed action. **(P-3)**

E.9. Reset steering angle sensor. **(P-2)**

We Support ASE Education Foundation

Tools and Materials

A vehicle with improperly adjusted rear wheel camber and toe.

Describe the vehicle being worked on:

Year ____________ Make ____________ Model ____________

VIN ____________ Engine type and size ____________

Procedure

1. Position the vehicle properly on the alignment ramp with the front wheels properly positioned on the turntables and the rear wheels on slip plates.

   Are the front wheels properly positioned on turntables? ☐ Yes ☐ No

   Are the rear wheels properly positioned on slip plates? ☐ Yes ☐ No

   Is the parking brake applied? ☐ Yes ☐ No

   Instructor check ____________

2. Perform a prealignment inspection, and correct any defective conditions or components found during the inspection. List the defective conditions or components, and explain the reasons for your diagnosis.

   ____

   ____

   ____

3. Measure front and rear ride height and correct if necessary.

   Specified front suspension ride height ______________________

   Actual front suspension ride height ______________________

   Specified rear suspension ride height ______________________

   Actual rear suspension ride height ______________________

   If the rear ride height is not within specifications, state the necessary repairs and explain the reasons for your diagnosis.

   ____

   ____

   ____

   ____

4. Record alignment specifications both before and after the alignment.

BEFORE ALIGNMENT READINGS								
Left Front					Right Front			
Min.	Preferred Spec.	Max.	Actual		Actual	Min.	Preferred Spec.	Max.
				CAMBER				
				CASTER				
				TOE				
				SAI				
				INCLUDED ANGLE				
				TURNING ANGLE				

FRONT				
	Actual	Min.	Preferred Spec.	Max.
CROSS CAMBER				
CROSS CASTER				
TOTAL TOE				
SET BACK				

Left Rear					Right Rear			
Min.	PREFERRED SPEC.	Max.	ACTUAL		ACTUAL	Min.	PREFERRED SPEC.	Max.
				CAMBER				
				TOE				

REAR				
	Actual	Min.	Preferred Spec.	Max.
Total Toe				
Thrust Line				

AFTER ALIGNMENT READINGS								
Left Front					Right Front			
Min.	Preferred Spec.	Max.	Actual		Actual	Min.	Preferred Spec.	Max.
				CAMBER				
				CASTER				
				TOE				
				SAI				
				INCLUDED ANGLE				
				TURNING ANGLE				

FRONT				
	Actual	Min.	Preferred Spec.	Max.
CROSS CAMBER				
CROSS CASTER				
TOTAL TOE				
SET BACK				

Left Rear					Right Rear			
Min.	PREFERRED SPEC.	Max.	ACTUAL		ACTUAL	Min.	PREFERRED SPEC.	Max.
				CAMBER				
				TOE				

REAR				
	Actual	Min.	Preferred Spec.	Max.
Total Toe				
Thrust Line				

5. Measure and adjust camber on left rear wheel.

Specified left rear wheel camber _______________

Actual left rear wheel camber _______________

Left rear wheel camber after adjustment _______________

Camber adjustment method _______________

If the turning angle is not within specifications, state the necessary repairs and explain the reasons for your diagnosis.

____

____

____

____

17. Steering Angle Sensor Reset: The last step of an alignment procedure that must be performed on some vehicle platforms is a reset procedure for the steering angle sensor (SAS) and other related sensors to update the vehicle geometry information stored in the vehicle's software system after an alignment. This procedure can be performed with an enhanced computer scan tool, depending on model, or your alignment equipment may incorporate this feature. Reset requirements vary with vehicle manufacturer year, make, and model of the platform. Other sensors that may be part of this reset procedure are the yaw rate sensor, torque angle sensor, and deceleration sensor among others depending on vehicle manufacturer model. Refer to specific vehicle information for specific steps and requirements.

18. Road test vehicle for satisfactory steering and suspension operation.

    Steering directional control: ☐ Satisfactory ☐ Unsatisfactory

    Complete steering operation: ☐ Satisfactory ☐ Unsatisfactory

    Suspension operation: ☐ Satisfactory ☐ Unsatisfactory

    If the steering or suspension operation is unsatisfactory, state the necessary repairs and explain the reasons for your diagnosis.

    ____

    ____

    ____

    ____

Instructor's Response

____

____

____

CHAPTER 10
DIAGNOSTIC ALIGNMENT ANGLES AND FRAME DIAGNOSIS

Upon completion and review of this chapter, you should be able to:

- Measure front and rear wheel setback.
- Measure steering axis inclination (SAI).
- Measure front wheel toe and turning radius.
- Diagnose bent front struts.
- Correct setback conditions.
- Check and correct front engine cradle position.
- Correct SAI angles that are not within specifications.
- Use a track gauge to measure rear wheel tracking.
- Diagnose rear wheel tracking problems from the track gauge measurements.
- Take the necessary precautions to avoid frame damage.
- Diagnose the causes of frame damage.
- Follow safety precautions when measuring and welding frames.
- Visually inspect frames.
- Measure frames with a plumb bob and diagnose frame damage.
- Measure frames with a tram gauge and diagnose frame damage.
- Perform unitized body measurements with a tram gauge.
- Perform unitized body measurements with a dedicated bench system.

Basic Tools

Basic technician's tool set
Service manual
Tape measure
Chalk
Plumb bob
Floor jack
Safety stands

Terms To Know

Axle offset
Datum line
Dedicated bench system
Frame flange
Frame web
Geometric centerline
Lateral axle sideset
Multipull
Plumb bob
Rear axle offset
Rear wheel tracking
Section modulus
Setback
Single-pull
Steering axis inclination (SAI)
Sunburst cracks
Symmetry angle measurements
Track gauge
Tram gauge
Turntable

Today's computerized wheel alignment system performs many vehicle geometric angle measurements beyond the standard alignment angles of camber, caster, and toe. What used to take many hours using mechanical measuring devices can now be done fast and efficiently with today's computerized alignment systems. While not all of these measurements are displayed as part of a routine wheel alignment, they are available as additional measurements in your computerized wheel alignment system, such as that produced by Hunter Engineering, to aid you in diagnosing vehicle geometry issues.

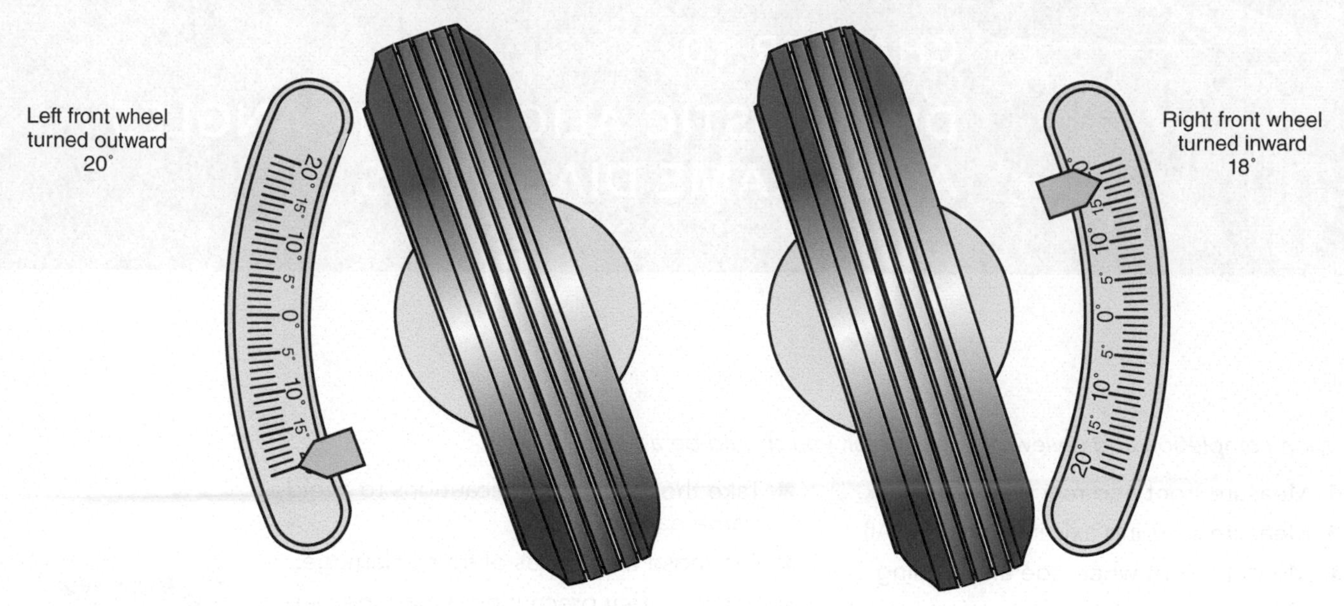

Figure 10-1 Turning angle screen.

Turning Angle Screen

Some computer wheel aligners have a turning angle screen. When this screen is displayed, the technician removes the locking pins from the turntables. Each front wheel must be turned outward a specified amount, and the turning angle on the opposite front wheel must be entered with the keypad as directed on the screen (**Figure 10-1**). A toe-out on turns option is available on some DSP wheel sensors. This option has optical encoders in the wheel sensors that measure the turning angle electronically rather than reading the degree scales on the front turntables. **Photo Sequence 19** shows a manual procedure for measuring turning angle.

PHOTO SEQUENCE 19

Typical Procedure for Front Wheel Turning Radius Measurement

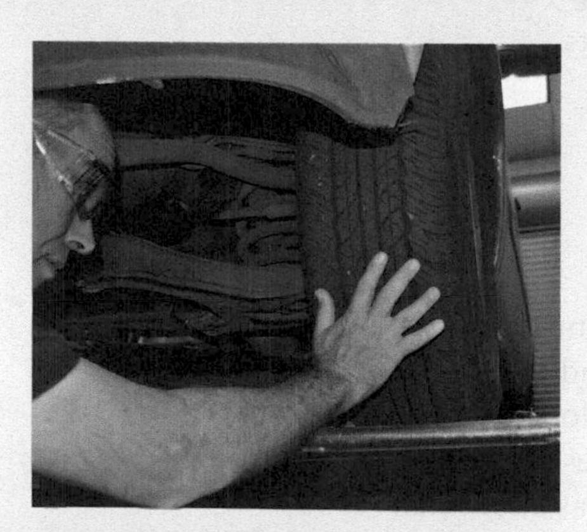

P19-1 Perform a prealignment inspection.

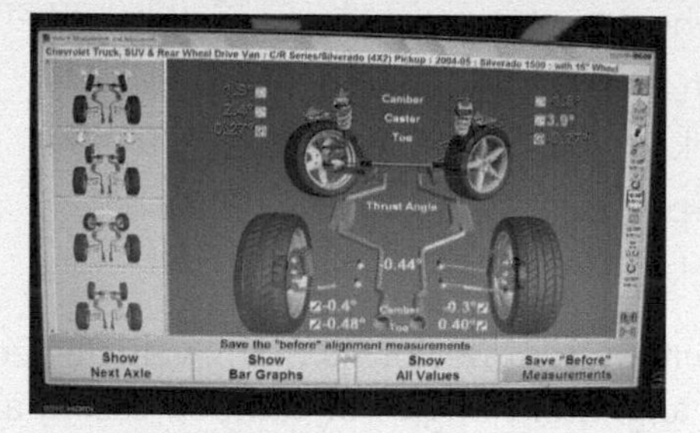

P19-2 Measure and adjust the other front and rear suspension angles.

P19-3 Be sure the front wheels are centered on the turntables and the brake pedal jack is installed to apply the brakes.

PHOTO SEQUENCE 19 (CONTINUED)

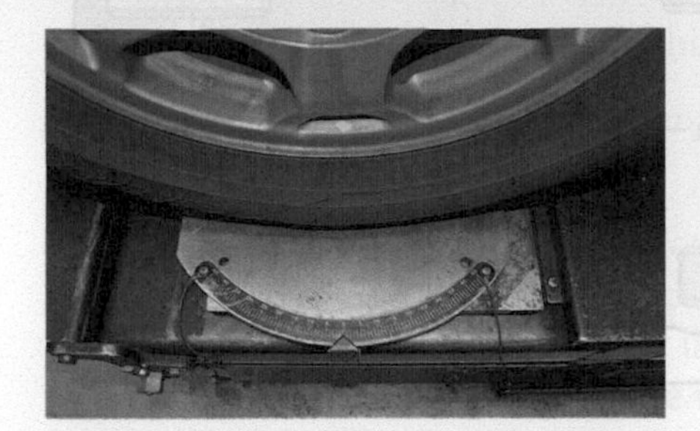

P19-4 Remove the turntable locking pins and be sure the turning radius gauges are in the zero position with the front wheels straight ahead.

P19-5 Turn the right front wheel inward toward the center of the vehicle until the turning radius gauge indicates 20 degrees.

P19-6 With the front wheels positioned as described in step 5, read and record the reading on the left turning radius gauge. The reading on the left turning radius gauge should be 22 degrees, or 2 degrees more than the reading on the right turning radius gauge.

P19-7 Turn the left front wheel inward toward the center of the vehicle until the turning radius gauge indicates 20 degrees.

P19-8 With the front wheels positioned as described in Step 7, read and record the reading on the right turning radius gauge. The reading on the right turning radius gauge should be 22 degrees, or 2 degrees more than the reading on the left turning radius gauge.

Additional Diagnostic Screens

Some computer wheel aligners provide **symmetry angle measurements** that help the technician determine if out-of-specification readings may have been caused by collision or frame damage. These symmetry angle measurements display **axle offset**, right or left lateral **axle sideset** and track-width difference, front and rear wheel setback, and wheelbase difference (**Figure 10-2**). Setback is an angle formed by a line drawn at a 90-degree

Axle offset occurs when the rear axle is rotated so it is no longer at a 90-degree angle to the vehicle centerline.

Lateral axle sideset occurs when the rear axle moves inward or outward in relation to the vehicle centerline, but the axle and vehicle centerline remain at a 90-degree angle.

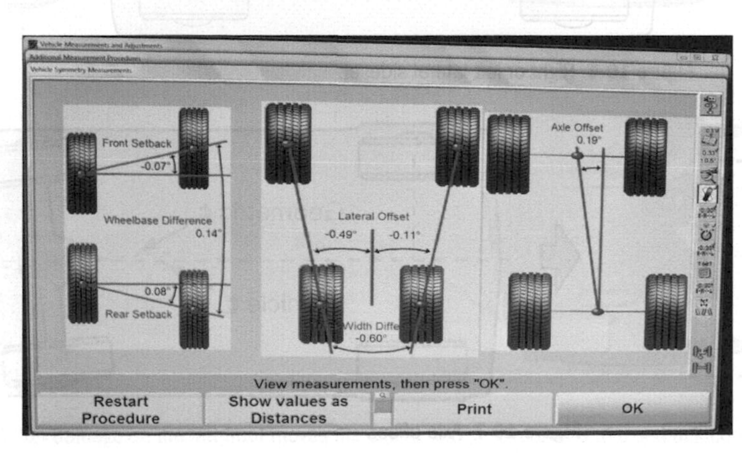

Figure 10-2 Symmetry angle measurements.

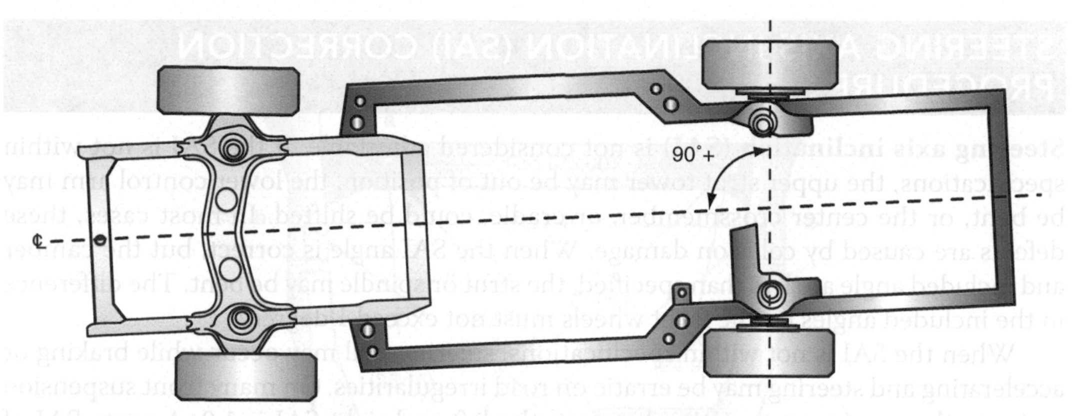

Figure 10-15 Collision damage causes front wheel setback and improper rear wheel tracking with rear wheel offset.

moved lengthwise on the track bar. The metal pointers may be adjusted inward or outward in the brackets.

To inspect rear wheel tracking, one of the track gauge pointers is positioned against the rear outside edge of the front wheel rim at spindle height. The other two pointers are positioned on the outside edges of the rear wheel rim at spindle height. The track gauge must be kept parallel to the side of the vehicle (**Figure 10-16**).

Rear Wheel Tracking Measurement Procedure

Follow these steps to measure rear wheel tracking:

1. Lift one side of the front suspension with a floor jack and lower the suspension onto a safety stand. Be sure the tire is lifted off the shop floor.
2. Place a lateral runout gauge against the outside edge of the rim lip and rotate the wheel to check wheel runout. (Refer to Chapter 3 for wheel runout measurement.) Place a chalk mark at the location of the maximum wheel runout. If the wheel runout is excessive, replace the wheel.
3. Rotate the chalk mark to the exact top or bottom of the wheel, and remove the safety stand and floor jack to lower the wheel onto the shop floor.

Figure 10-16 Track bar positioned to measure rear wheel tracking.

4. Repeat Steps 1, 2, and 3 at the other three wheels.
5. Hold the single pointer on the track gauge so it contacts the rear edge of the front rim.
6. Adjust the other two pointers so they contact the outer edges of the rear wheel rim.
7. Adjust the pointers on the rear wheel so the track gauge is parallel to the side of the vehicle. Be sure all three pointers are clamped securely.
8. Move the track gauge to the opposite side of the vehicle and position the pointers in the same location on these wheels. If the vehicle has proper tracking, all three pointers on the track gauge will contact the front and rear rims at exactly the same location on both sides of the vehicle, and the track gauge will be parallel to the sides of the vehicle.

Examples of Improper Rear Wheel Tracking Measured with a Track Gauge

Rear Axle Offset When the track gauge is positioned on the left side of the vehicle, the rear pointer on the rear wheel must be moved farther inward compared with the pointer at the front edge of the rear rim. On the right side of the vehicle, the front pointer is positioned forward and ahead of the rim edge when the two pointers are positioned on the edges of the rear rim. This indicates the wheelbase is less between the wheels on the right side of the vehicle compared with the left side. The front pointer on the rear rim is some distance from the rim face (**Figure 10-17**). These pointer positions indicate the right rear wheel has moved forward and the left rear wheel has moved rearward. This **rear axle offset** condition moves the thrust line to the left of the geometric centerline.

Rear axle offset occurs when the rear axle is turned so one rear wheel is moved rearward and the opposite rear wheel is moved forward.

Wheel Setback If a vehicle has been subjected to a severe right front sideways impact, the left front wheel may be driven outward and rearward. The track gauge and pointers are positioned on the right side of the vehicle and then moved to the left side. The front pointer is positioned ahead of the rim edge on the left side, and the track gauge is not parallel to the side of the vehicle (**Figure 10-18**).

The track gauge must be moved inward on the front pointer to position the bar parallel to the side of the vehicle. When this adjustment is made, the distance on the front pointer is less on the left side of the vehicle compared with the distance on the right side of the vehicle. These track gauge measurements indicate the left front wheel is moved rearward to a setback position as well as being outward.

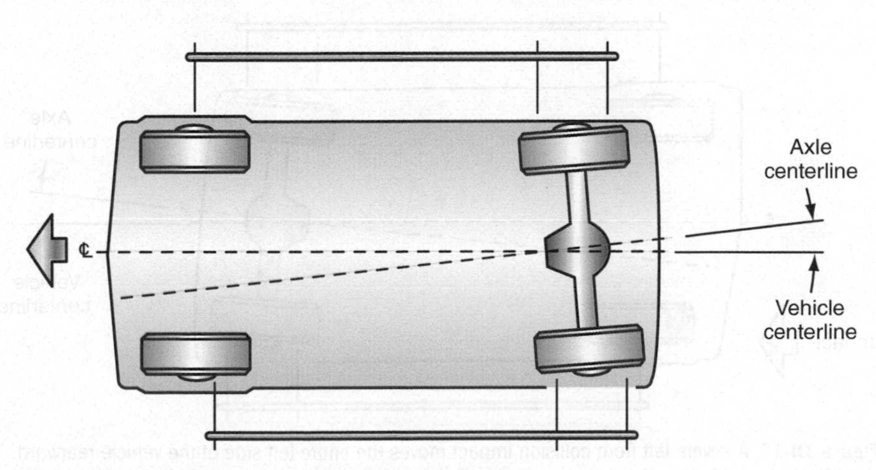

Figure 10-17 Track bar measurement of an offset rear axle.

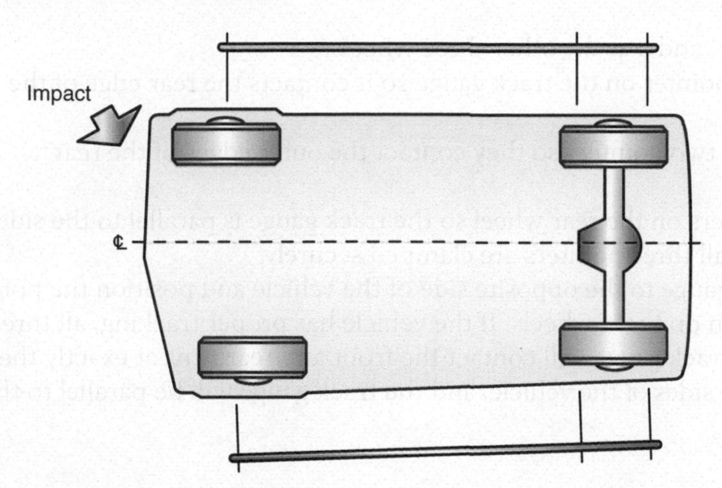

Figure 10-18 Track bar measurement of a left front wheel that is moved rearward to a setback position.

Diamond-Shaped Frame Condition When a vehicle experiences severe collision damage on the left front, the entire left side of the vehicle may be driven rearward in relation to the right side of the chassis. Under this condition, the left front wheel is moved to a setback condition, and the rear axle is offset, which moves the thrust line to the left of the geometric centerline. This particular type of damage is most likely to occur on a rear-wheel-drive vehicle with a frame and a one-piece rear axle.

Because the entire left side of the vehicle is driven rearward, the wheelbase is the same on both sides of the vehicle. Therefore, the track gauge pointers indicate the wheelbase is the same between the front and rear wheels on both sides of the vehicle. However, the two rear track gauge pointers must be adjusted to different lengths to contact the rear rim edges. On the left side of the vehicle, the rear track gauge pointer is some distance from the rear rim edge if these two pointers are set at the same measurement (**Figure 10-19**).

When the track gauge is positioned on the right side of the vehicle, there is some distance between the front pointer and the edge of the rear rim if both rear pointers are set at the same distance.

Measuring vehicle tracking with a track gauge is a two-person operation and requires a considerable amount of time to perform the wheel runout and track bar measurements. The track gauge also lacks precision accuracy.

Computer alignment systems perform the same checks by measuring the thrust line angle, setback, and all the other front and rear suspension alignment angles with speed and accuracy.

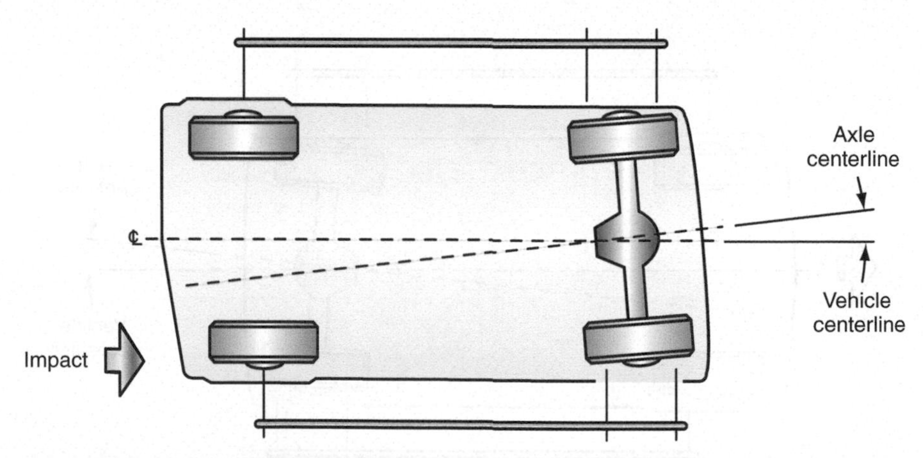

Figure 10-19 A severe left front collision impact moves the entire left side of the vehicle rearward, resulting in front wheel setback and rear axle offset, which provides a diamond-shaped chassis.

CUSTOMER CARE Those of us involved in the automotive service industry are like everyone else: we do make mistakes. If you make a mistake that results in a customer complaint, always be willing to admit your mistake and correct it. Do not try to cover up the mistake or blame someone else. Customers are usually willing to live with an occasional mistake that is corrected quickly and efficiently.

FRAME DIAGNOSIS AND SERVICE

Improper frame or unitized body alignment may cause wheel alignment defects, rapid tire tread wear, and reduced directional stability. After collision damage, a unitized body vehicle may be repaired so it has satisfactory cosmetic appearance, but body and wheel alignment defects are still present. These defects may contribute to reduced directional stability, especially during such extreme conditions as hard cornering or severe braking. Therefore, diagnosing and correcting frame and unitized body alignment is very important to restoring vehicle safety! Body shops have mechanical or electronic body and chassis alignment equipment that can help correct improper body alignment.

The most common cause of frame damage on cars and light-duty trucks is collision damage. In some cases, the collision damage may be repaired to make the cosmetic appearance of the vehicle satisfactory, but the frame damage may not always be correctable.

When driving the vehicle, some indications of frame damage are:

1. Excessive tire wear when the front suspension alignment angles are correct.
2. Steering pull when the front suspension alignment angles are correct.
3. Steering wheel not centered when driving straight ahead, but the steering wheel was centered in the shop.

Classroom Manual
Chapter 10, page 278

FRAME DIAGNOSIS

Preventing Frame Damage

Because frame problems affect wheel alignment, technicians must be able to diagnose frame defects so they are not confused with wheel alignment problems. Refer to **Table 10-1** for additional information.

Follow these precautions to minimize frame damage:

1. Do not overload the vehicle.
2. Place the load evenly in a vehicle.
3. Do not operate the vehicle on extremely rough terrain.
4. Do not mount equipment such as a snowplow on a vehicle unless the frame is strong enough to carry the additional load and force.

Yield strength is a measure of the strength of the material from which the frame is manufactured. Yield strength is the maximum load that may be placed on a material and still have the material retain its original shape. Yield strength is measured in pounds per square inch (psi) or kilopascals (kPa).

Diagnosis of Frame Problems

Side Sway The causes of frame side sway are:

1. Collision damage.
2. Fire damage.
3. Use of equipment on the vehicle for which the frame was not designed.

TABLE 10-1 DIAGNOSING FRAME AND BODY DAMAGE

Problem	Symptoms	Possible Causes
Excessive front tire tread wear	Front tire tread wear with correct front suspension alignment angles	Frame, front cradle damage
Steering pull when driving straight ahead	Steering pull with correct front and rear alignment angles	Frame, front cradle damage
Steering wheel not centered when driving straight ahead	Steering wheel centered in the shop, but it is not centered when driving straight ahead	Frame, front cradle damage
Improper strut tower position	Incorrect front suspension alignment angles, steering pull, tire tread wear	Collision damage
Frame buckle	Wrinkles in frame flanges	Collision damage, vehicle abuse
Improper cradle measurements or position	Incorrect front suspension alignment angles, steering pull, tire tread wear	Collision damage, worn or improperly positioned cradle mounts

Sag The causes of frame sag are:

1. Vehicle loads that exceed the load-carrying capacity of the frame.
2. Uneven load distribution.
3. Sudden changes in **section modulus**.
4. Holes drilled in the **frame flange**.
5. Too many holes drilled in the **frame web**.
6. Holes drilled too close together in the frame web.
7. Welds on the frame flange.
8. Cutting holes in the frame with a cutting torch.
9. Cutting notches in the frame rails.
10. A fire involving the vehicle.
11. Collision damage.
12. The use of equipment for which the frame was not designed.

Section modulus is a measure of the frame's strength based on its height, width, thickness, and the shape of the side rails. Section modulus does not account for the type of material in the frame.

Frame flange is the horizontal part of the frame on the top and bottom of the web.

Frame web refers to the vertical side of the frame.

Buckle The causes of frame buckle are:

1. Collision damage.
2. Using equipment such as a snowplow when the frame was not designed for this type of service.
3. A fire involving the vehicle.

Diamond-Shaped Diamond-shaped frame damage may be caused by:

1. Collision damage.
2. Towing another vehicle with a chain attached to one corner of the frame.
3. Being towed by another vehicle with a chain attached to one corner of the vehicle frame.

Twist Frame twist may be caused by:

1. An accident or collision, especially one involving a rollover.
2. Operating the vehicle on extremely rough terrain.

CHECKING FRAME ALIGNMENT

Safety Concerns

While servicing vehicle frames, always wear the proper safety clothing and safety items for the job being performed. This includes proper work clothing, safety goggles, ear protection, respirator, proper gloves, welding shield, and safety shoes (**Figure 10-20**).

If arc welding is necessary on a vehicle frame, follow these precautions:

1. Remove the negative battery cable before welding (**Figure 10-21**).
2. Remove the fuel tank before welding (**Figure 10-22**).
3. Protect the interior and exterior of the vehicle as necessary (**Figure 10-23**).

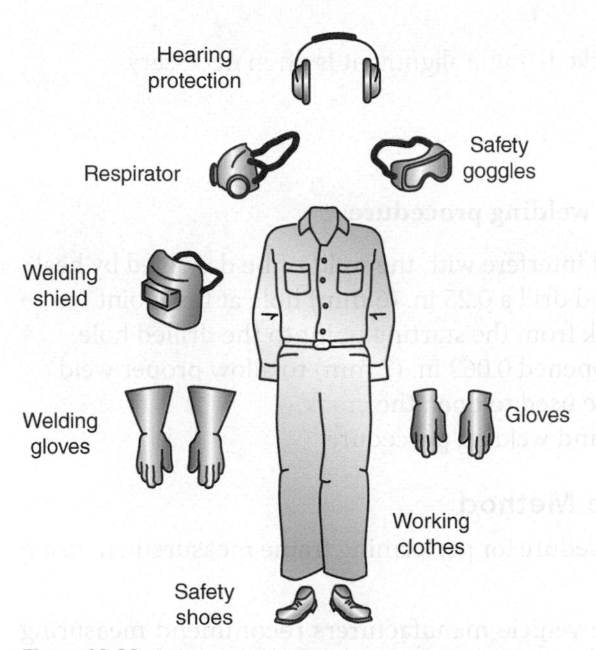

Figure 10-20 Safety items for frame service.

Figure 10-21 Remove the negative battery cable before arc welding on a vehicle.

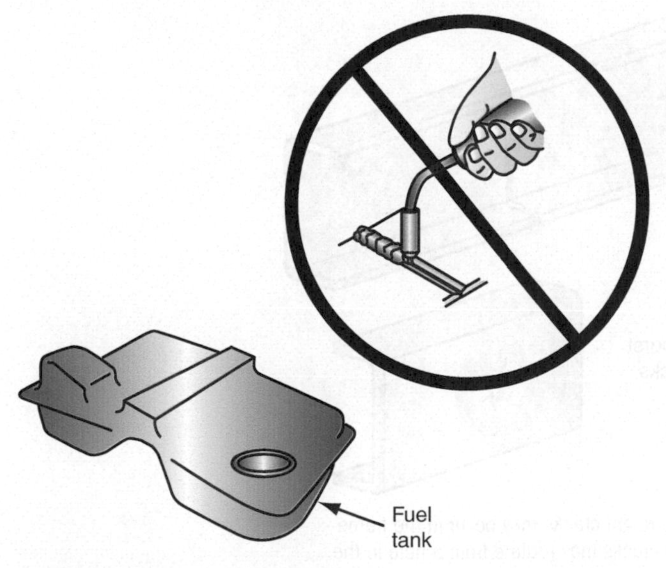

Figure 10-22 Remove the fuel tank before arc welding on a vehicle.

Figure 10-23 Protect the interior and exterior of the vehicle while arc welding.

Sunburst cracks radiate outward from a hole in a frame web or crossmember.

Caution

Always follow the vehicle manufacturer's recommendations in the service manual regarding welding or reinforcing the frame, or mounting additional equipment on the frame. If these recommendations are not followed, the frame may be weakened and damaged.

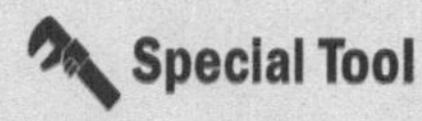

Special Tool

Plumb bob

A **plumb bob** is a weight with a sharp, tapered point that is suspended and centered on a string. Plumbers use this tool for locating pipe openings directly above each other in the tops and bottoms of partitions.

VISUAL INSPECTION

Prior to frame measurement, the frame and suspension should be visually inspected. Check for wrinkles on the upper flange of the frame, which indicate a sag problem. Visually inspect the lower flange for wrinkles, which are definite evidence of buckle. Because suspension or axle problems may appear as frame problems, the suspension components should be inspected for wear and damage. For example, an offset rear axle may appear as a diamond-shaped frame. Check all suspension mounting bushings and inspect leaf-spring shackles and center bolts.

The frame should be inspected for cracks, bends, and severe corrosion. Minor frame bends are not visible, but severe bends may be visible. Straight cracks may occur at the edge of the frame flange, and **sunburst cracks** may radiate from a hole in the frame web or crossmember (**Figure 10-24**).

SERVICE TIP If the frame is cracked, frame alignment is often necessary.

Frame Welding

Follow these steps for a typical frame welding procedure:

1. Remove any components that would interfere with the weld or be damaged by heat.
2. Find the extreme end of the crack and drill a 0.25 in. (6 mm) hole at this point.
3. V-grind the entire length of the crack from the starting point to the drilled hole.
4. The bottom of the crack should be opened 0.062 in. (2 mm) to allow proper weld penetration. A hacksaw blade may be used to open the crack.
5. Arc weld with the proper electrode and welding procedure.

Frame Measurement, Plumb Bob Method

Photo Sequence 20 shows a typical procedure for performing frame measurement using the plumb bob method.

Locating the Frame Centerline Some vehicle manufacturers recommend measuring the frame with the **plumb bob** method.

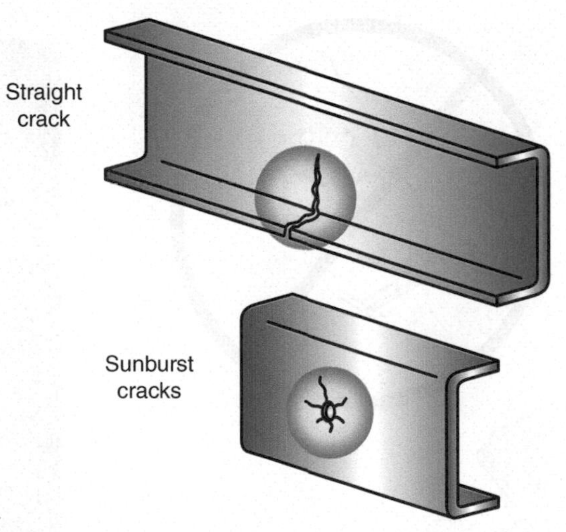

Figure 10-24 Straight cracks may occur in the frame flange; sunburst cracks may radiate from a hole in the frame web or crossmember.

Follow these steps to complete a plumb bob measurement for frame damage:

1. Place the vehicle on a level area of the shop floor and use a floor jack to raise the front and rear suspension off the floor. Support the chassis on safety stands at the manufacturer's recommended locations.

PHOTO SEQUENCE 20

Typical Procedure for Performing Frame Measurement, Plumb Bob Method

P20-1 Park the vehicle on a level area of the shop floor.

P20-2 Raise the front suspension with a floor jack and lower the chassis onto safety stands positioned at the manufacturer's recommended lifting points.

P20-3 Raise the rear suspension with a floor jack and lower the chassis onto safety stands positioned at the manufacturer's recommended lifting points.

P20-4 Suspend a plumb bob at the manufacturer's recommended frame measurement locations and place a chalk mark on the floor directly under the plumb bob.

P20-5 Use a floor jack to lift the vehicle, remove the safety stands, and lower the vehicle.

P20-6 Drive the vehicle away from the chalk-marked area.

P20-7 Use a tape measure to measure the vehicle's frame measurements between the chalk marks on the floor.

P20-8 Compare the frame measurements obtained with the vehicle manufacturer's specifications in the service manual.

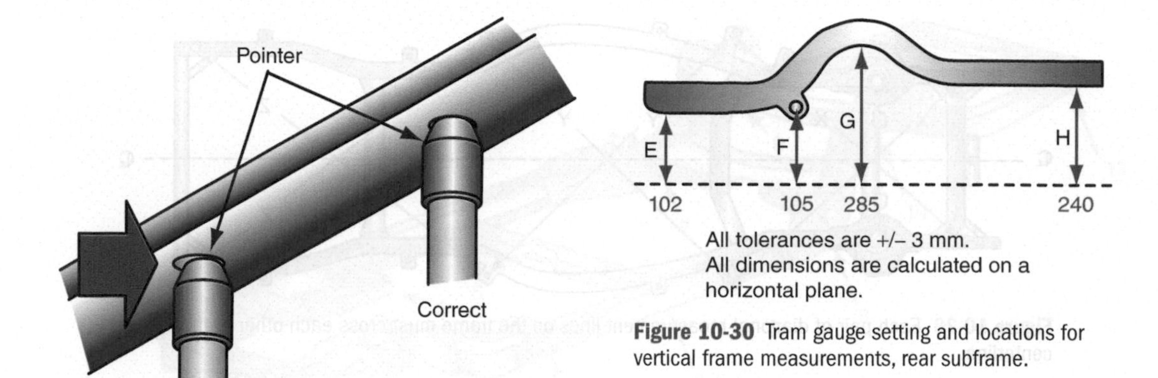

Figure 10-30 Tram gauge setting and locations for vertical frame measurements, rear subframe.

Figure 10-29 Correct and incorrect tram gauge pointer seating in the frame openings.

When vertical frame measurements are completed with the tram gauge, the gauge pointers must be set at the manufacturer's specified distance for each frame location (**Figure 10-30**). With the tram gauge pointers set at the specified height and the gauge properly installed across the frame in the recommended frame openings, the tram gauge should be level if the vertical frame measurements are within specifications. If the frame is twisted, the tram gauge is not level when adjusted to specifications and installed in the twisted area.

To measure frame sag, three tram gauges may be installed at various locations for vertical frame measurement. When viewed from the front or the rear, the tram gauges must be level with each other. If the tram gauge near the center of the frame is lower than the tram gauges at the front and rear of the frame, the frame is sagged.

SERVICE TIP When the tram gauge is installed on the frame, be sure the pointers are seated properly in the frame openings (**Figure 10-29**).

Tram gauge

Frame Straightening

Frame straightening is usually done with special hydraulically operated bending equipment. This equipment must be operated according to the equipment manufacturer's recommended procedures. Because frame straightening is usually done by experienced body technicians, we will not discuss this service in detail. All safety precautions must be observed while operating frame-straightening equipment. Never stand in front of pulling equipment when it is in operation (**Figure 10-31**).

Figure 10-31 Do not stand in front of pulling equipment while it is in operation.

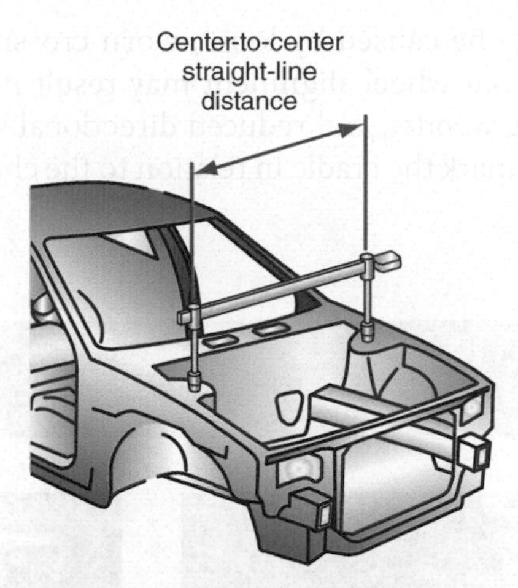

Figure 10-32 Performing an upper strut tower measurement with a tram gauge on a unitized body vehicle.

MEASURING UNITIZED BODY ALIGNMENT

Tram Gauge

On unitized bodies, a tram gauge may be used to perform such measurements as the upper strut towers (**Figure 10-32**) and the front crossmember (**Figure 10-33**).

If the underhood body measurements such as the strut tower measurements are not within specifications, the strut towers may not be in the original position. Collision damage may cause the strut towers to move out of the original position. If the strut towers are out of position, the front alignment angles are incorrect. **Photo Sequence 21** illustrates the underhood body measurements.

Because the lower control arms are attached to the crossmember or cradle and also to the lower end of the steering knuckle, if the crossmember is moved from the original intended position, front wheel alignment is adversely affected (**Figure 10-34**). Improper

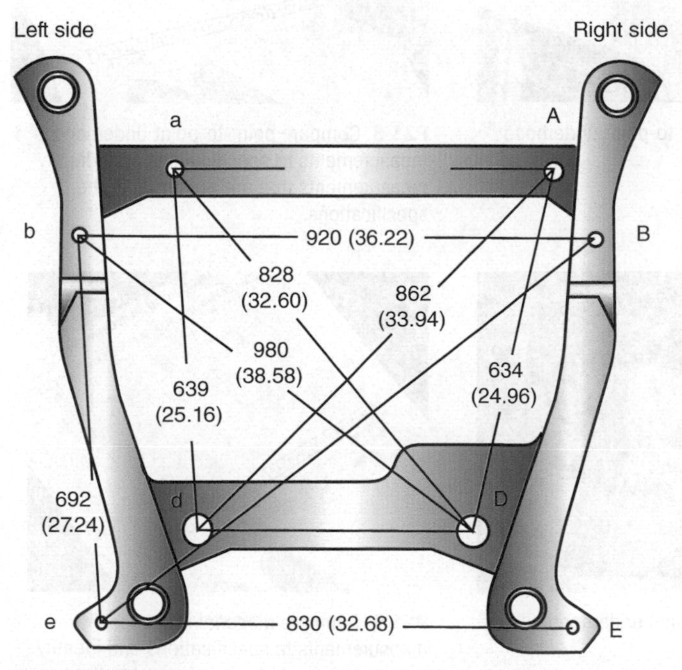

Figure 10-33 Tram gauge measurements on the front crossmember of a unitized body vehicle.

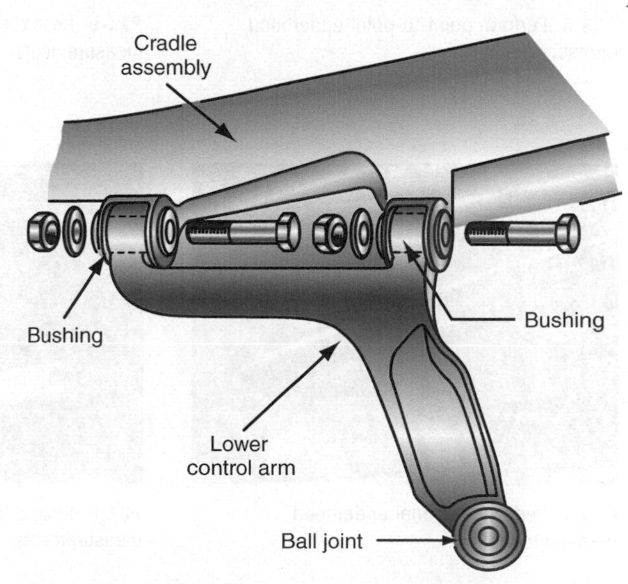

Figure 10-34 Lower control arm-to-cradle attachment.

crossmember position may be caused by loose, worn crossmember mounts or a bent crossmember. Improper front wheel alignment may result in excessive front tire tread wear, steering pull, steering wander, and reduced directional stability. When removing a front crossmember, always mark the cradle in relation to the chassis so it can be reinstalled in the original position.

PHOTO SEQUENCE 21
Performing Underhood Measurements

P21-1 Performing underhood measurements.

P21-2 Record the horizontal underhood measurements.

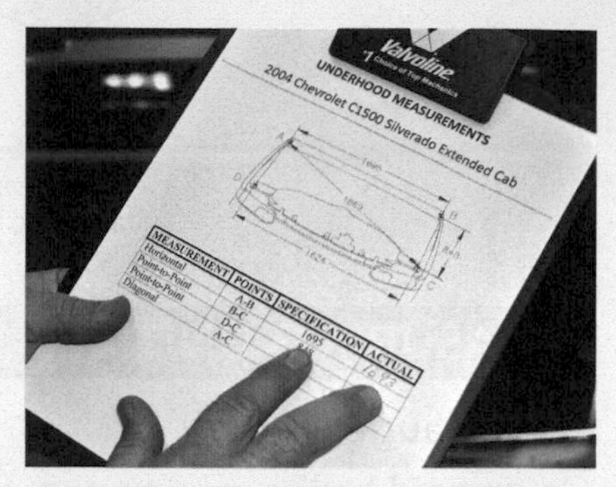

P21-3 Compare horizontal underhood measurements to specifications and identify measurements that are not equal to the specifications.

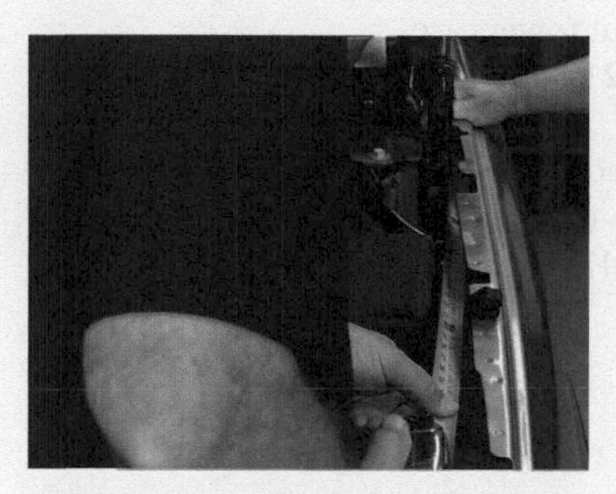

P21-4 Perform point-to-point underhood measurements.

P21-5 Record the point-to-point underhood measurements.

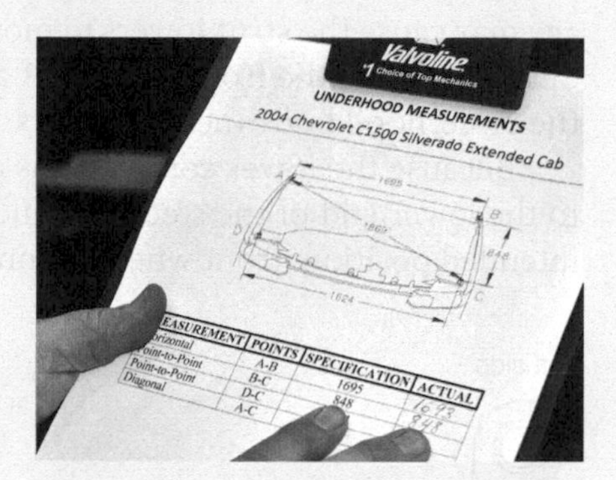

P21-6 Compare point-to-point underhood measurements to specifications and identify measurements that are not equal to the specifications.

P21-7 Perform diagonal underhood measurements.

P21-8 Record the diagonal underhood measurements.

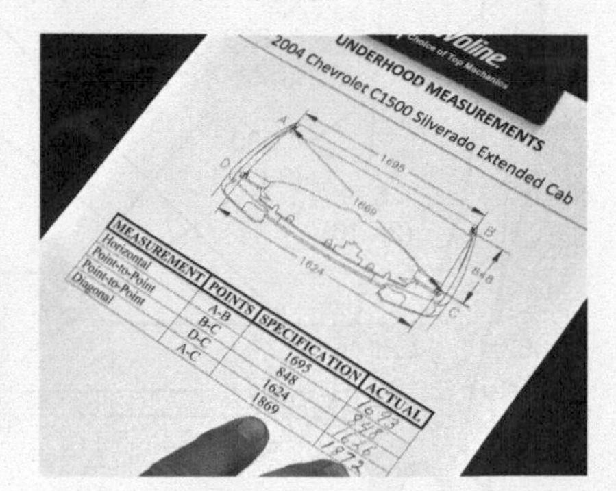

P21-9 Compare diagonal underhood measurements to specifications and identify measurements that are not equal to the specifications.

The engine and transaxle mounts are also connected to the front crossmember. Therefore, a bent crossmember or improperly positioned crossmember mounts may cause improper engine and transaxle position, which results in vibration problems. Therefore, when diagnosing and correcting engine, drive axle, or transaxle vibration problems, always be sure the front crossmember, engine mounts, transaxle mounts, and crossmember mounts are in satisfactory condition. The tram gauge may be used to perform vertical measurements on the front and rear subframes on a unitized body vehicle.

Special Tool
Dedicated bench system

Bench

A **dedicated bench system** is necessary to check many unitized body measurements after medium or heavy collision damage. The dedicated bench system has three main parts: the bench, transverse beams, and the dedicated fixtures. When used together, this equipment will perform many undercar unitized body measurements, including length, width, and height at the same time. The bench contains strong steel beams mounted on heavy casters (**Figure 10-35**). The top of the bench acts as a datum line.

Transverse Beams

The transverse beams are mounted perpendicular to the bench. There are a variety of holes in the bench for proper transverse beam attachment (**Figure 10-36**). Various holes in the transverse beams provide the correct dedicated fixture positions.

Dedicated Fixtures

The dedicated fixtures are specific to the body style being measured (**Figure 10-37**). These fixtures are bolted to the transverse beams, and the holes in the upper end of the fixtures must be aligned with specific body openings (**Figure 10-38**).

The unitized body must be straightened until the holes in the dedicated fixtures and the body openings are aligned. If the holes in the unitized body are behind and above the dedicated fixture openings, the unitized body must be pulled forward and downward in that area (**Figure 10-39**).

Universal benches are now available with computer measuring systems. Measuring a unitized body with a bench system and straightening these bodies with special body pulling equipment is usually done in an autobody shop rather than an automotive repair shop.

Electronic body and chassis alignment equipment using laser beam technology is used in many body shops during collision repair.

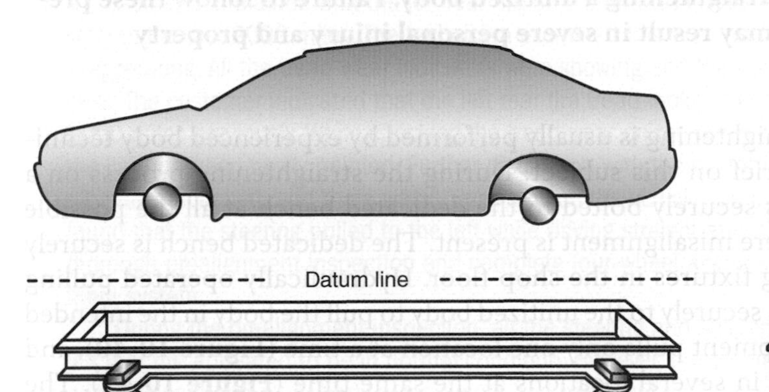

Figure 10-35 Bench from a dedicated bench system.

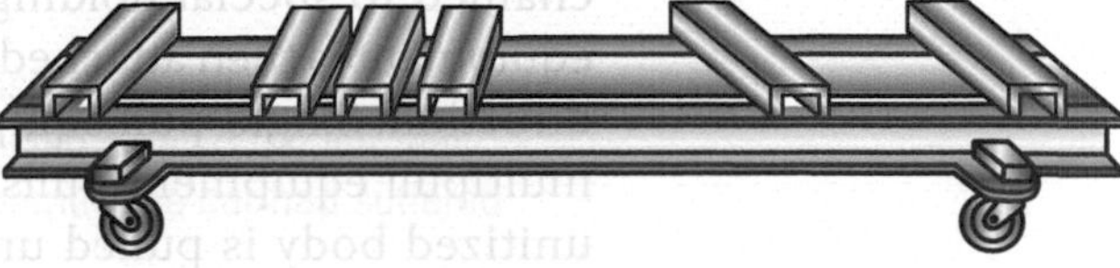

Figure 10-36 Transverse beams placed on the bench.

The four-wheel alignment results were satisfactory except the left rear wheel, which had a 5-degree toe-in and a thrust angle of 8 degrees. This severe rear toe-in condition had moved the thrust line to the right of the geometric centerline. When the technician referred to the manufacturer's service manual, this information stated the rear wheel toe is not adjustable, and if the rear wheel camber and toe are not within specifications, the rear underbody and suspension should be inspected for damage. The technician located the underbody dimension specifications in the service manual (**Figure 10-42** and **Figure 10-43**).

The technician began checking the rear suspension underbody measurements with a tram gauge. When dimension M was measured between the center of the bolt heads in the trailing arm brackets, this measurement was 1 in. (25.4 mm) less than specified. This measurement indicated the left rear subframe had been forced inward toward the center of the vehicle, which resulted in the improper toe and thrust angle. Further inspection of the rear axle channel with a long straightedge indicated this channel was bent rearward in the center. The car was sent to the body shop to pull the rear subframe and unitized body back to the original position and provide the specified distance between the center of bolt heads on the trailing arms. Because the rear axle channel is a critical component in providing rear suspension ride quality and alignment, the customer was advised this component should be replaced. A new rear tire was also installed.

When the rear suspension was reassembled, all the alignment angles were rechecked and the rear wheel toe and thrust line were within specifications. During a road test, there was no evidence of steering pull or any other steering problems.

From this experience, the technician learned the importance of four-wheel alignment. The technician also discovered that accurate underbody measurement specifications are absolutely essential.

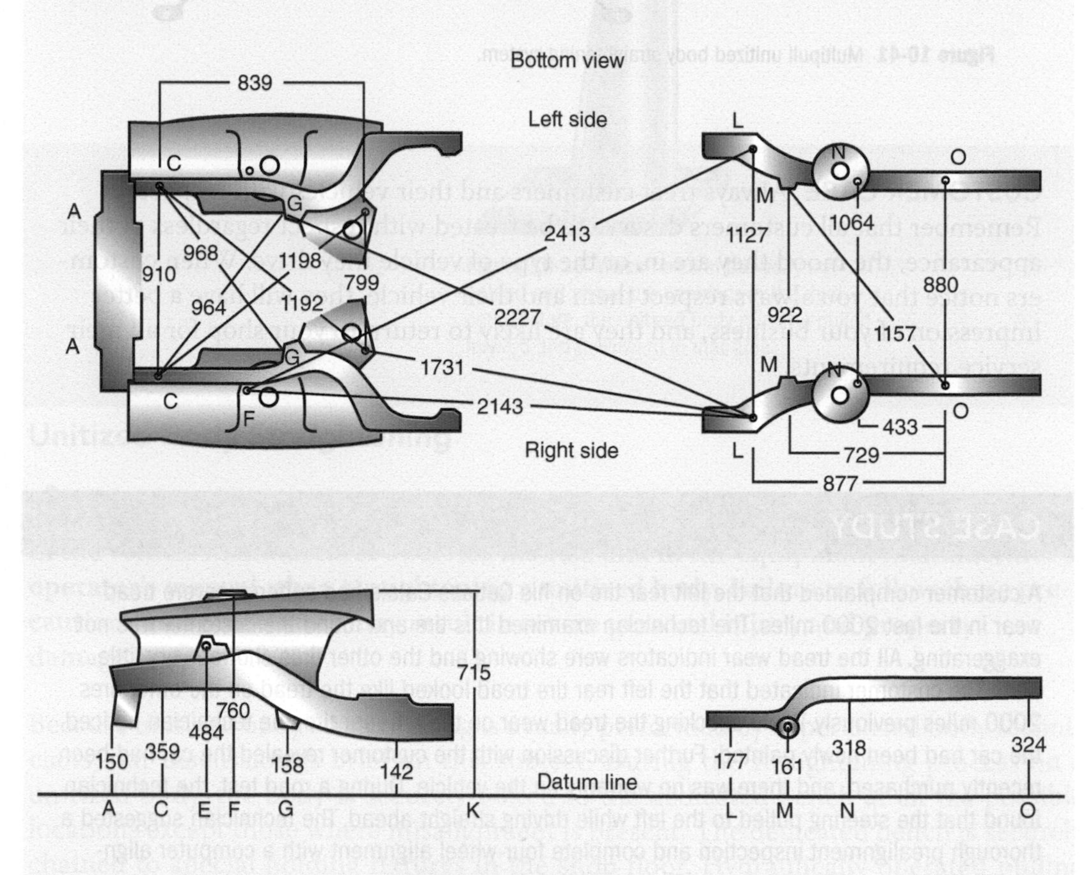

Figure 10-42 Underbody measurement specifications and locations.

REF	HORIZONTAL	VERTICAL	LOCATION
A	Leading edge of 24 mm gauge hole	Lower surface at gauge hole to datum line	Front end lower tie bar
B	Center of 16 mm gauge hole	None	Front end upper tie bar
C	Leading edge of 20 mm gauge hole	Lower surface at gauge hole to datum line	Motor compartment side rail, forward of transaxle anchor support plate on left side rail, and at rear of engine mounting support on right side rail. For access on right side with air conditioning, remove air compressor.
D	Center of 16 mm gauge hole	None	Front upper surface of motor compartment side rail
E	Center of lower attaching hole in transaxle support anchor plate	Datum line to horizontal centerline of transaxle support anchor plate lower attaching hole	Transaxle support anchor plate on left side rail
F	Center of shock tower strut front attaching hole	Upper surface at shock tower strut front attaching hole	Shock tower
G	Leading edge of oblong master gauge hole	Datum line to horizontal leading edge of oblong master gauge hole	Front suspension lower control arm mount support
H	Center of 16 mm gauge hole	None	Gauge hole in upper surface of motor compartment side rail
I	Center of 19 mm gauge hole	Datum line to lower surface at gauge hole	Outboard 19 mm gauge hole in front suspension control arm mounting reinforcement support
J	Center of hood hinge pivot pin head	None	Hood hinge
K	Center of front upper hinge pin hole	Upper surface at hinge pin hole	Front upper door hinge, body side
L	Leading edge of 22 mm gauge hole	Lower surface at leading edge of gauge hole to datum line	Compartment pan longitudinal rail, forward of rear suspension spring seat support
M	Center of attaching bolt head	Datum line to horizontal centerline of attaching bolt head of rear suspension control arm support.	Rear suspension control arm support forward of rear suspension spring seat support
N	Center of 16 mm gauge hole	Datum line to lower surface at gauge hole	Outboard gauge hole at rear suspension spring seat support
O	Leading edge of 21 mm gauge hole	Lower surface at gauge hole to datum line	Compartment pan longitudinal rail

Figure 10-43 Explanation of underbody horizontal and vertical measurement locations.

ASE-STYLE REVIEW QUESTIONS

1. While discussing a diamond-shaped frame condition:
   Technician A says this condition may be caused when the vehicle is involved in a fire resulting from an accident.
   Technician B says this condition may be caused by towing another vehicle with the chain attached to one corner of the frame.
   Who is correct?
   A. A only
   B. B only
   C. Both A and B
   D. Neither A nor B

2. When welding a vehicle frame crack:
   A. A 0.25-in. hole should be drilled in the center of the crack.
   B. The crack should be V-ground at both ends of the crack.
   C. Disconnect the negative battery cable before welding the crack.
   D. Use a cutting torch to open the crack for proper weld penetration.

3. A vehicle has a drift to one side, camber is negative on one side and positive on the other, but the cross camber is within specification and caster is within specifications. There are no adjustments available for either camber or caster. The technician suggests that the camber may be brought into specifications by:
   A. Rotating the front coil springs.
   B. Bending the lower control arms.
   C. Repositioning the cradle/subframe assembly.
   D. Replacing the front wheel hub assemblies.
4. While inspecting and measuring a front cradle on a front-wheel-drive car:
   A. Front wheel alignment angles may be changed by a bent front cradle.
   B. Engine mounting position is not affected by a bent front cradle.
   C. A bent front cradle has no effect on front drive axle vibration.
   D. A bent front cradle may cause premature front wheel bearing failure.
5. All of these statements about frame problems are true EXCEPT:
   A. Frame sag may be caused by overloading the vehicle.
   B. Frame sag may be caused by the use of equipment for which the frame was not designed.
   C. Frame buckle may be caused by using a snowplow on the front of the vehicle.
   D. A diamond-shaped frame may be caused by operating the vehicle on extremely rough terrain.
6. Unequal SAI angles on the left and right sides of the front suspension may cause:
   A. Tread wear on the front tires.
   B. Brake pull during sudden stops.
   C. Bump steer on a rough road.
   D. Steering wander while driving straight ahead.
7. On many front suspension systems, the maximum variation between the left and right SAI is:
   A. 0.75 degree
   B. 1 degree
   C. 1.5 degrees
   D. 2.5 degrees
8. A bent front crossmember affects all of the following alignments EXCEPT:
   A. Centerline.
   B. Suspension.
   C. Axle.
   D. Engine.
9. Referring to Figure 10-12:
   Technician A says the measurement between "e" and "b" should be no more than ±1.0 in. different from the measurement between "E" and "B."
   Technician B says frame damage problems, such as buckling, can be determined if the measurements between points "e" and "E" are more than ±0.25 in. different than between "b" and "B."
   Who is correct?
   A. A only
   B. B only
   C. Both A and B
   D. Neither A nor B
10. When making horizontal frame measurements to determine if the frame or unibody is straight, all of the measurements should be made with reference to which one of the following?
    A. Vertical
    B. Centerline
    C. Horizontal
    D. Body bushings

ASE CHALLENGE QUESTIONS

1. A car with a MacPherson strut front suspension has 1.5-degree positive camber on the left front wheel and 1.5-degree negative camber on the right front wheel. The most likely cause of this problem is:
   A. Worn upper strut mounts.
   B. Improperly positioned steering gear.
   C. Engine cradle/subframe shifted to the right.
   D. Bent strut rods.
2. A pickup truck with a MacPherson strut front suspension pulls to the left. Preliminary inspection shows that the SAI is correct, but the camber of the right front wheel is –1.25 degrees.
   Technician A says a bent spindle could be the cause.
   Technician B says the problem may be a worn-out strut.
   Who is correct?
   A. A only
   B. B only
   C. Both A and B
   D. Neither A nor B

3. While discussing frame damage:

   Technician A says a possible indication of frame damage is an uncentered steering wheel when driving straight ahead.

   Technician B says suspension alignment may be impossible if the frame is bent.

   Who is correct?

   A. A only
   B. B only
   C. Both A and B
   D. Neither A nor B

4. Which of the following makes this statement true? Measurement points of a frame or unibody using a tram gauge are:

   A. Seam notches.
   B. Calculated by computer.
   C. Based on a vertical datum plane.
   D. Manufacturer's specified frame openings.

5. The owner of an older pickup says the truck has begun to "wander and shimmy" in the past few months, especially when the bed of the truck is empty.

   Technician A says the twin I-beam front suspension caster is probably out of spec.

   Technician B says the eyes of the rear leaf springs are probably worn.

   Who is correct?

   A. A only
   B. B only
   C. Both A and B
   D. Neither A nor B

Name ______________________ **Date** ______________

JOB SHEET
50

FRAME MEASUREMENT, PLUMB BOB METHOD

Upon completion of this job sheet, you should be able to perform frame measurements with a plumb bob.

ASE Education Foundation Correlation

This task applies to the following **AST/MAST** task:

E.8. Check front and/or rear cradle (subframe) alignment; determine needed action. **(P-3)**

We Support | Education Foundation

Tools and Materials

Floor jack
Plumb bob
Safety stands
Chalk

Describe the vehicle being worked on:

Year ______________ Make ______________ Model ______________

VIN ______________________ Engine type and size ______________________

Procedure **Task Completed**

1. Place the vehicle on a level area of the shop floor and use a floor jack to raise the front and rear suspension off the floor. Support the chassis on safety stands at the manufacturer's recommended locations.

Is the vehicle chassis supported securely on safety stands? ☐ Yes ☐ No

Instructor check ______________________

2. Locate the frame measurement locations in the vehicle manufacturer's service manual. ☐

3. Suspend a plumb bob at each frame measurement location and place a chalk mark on the floor directly below the tip of the plumb bob.

Are all frame measurement locations chalk-marked on the shop floor? ☐ Yes ☐ No

Instructor check ______________________

4. Raise the vehicle with the floor jack, remove the safety stands, and lower the vehicle onto the floor. Move the vehicle away from the chalk-marked area. ☐

5. Measure the distance between the two chalk marks directly across from each other at the front of frame, and chalk-mark the exact halfway point in this distance. This mark is the frame centerline at the front.

Distance between the two chalk marks directly across from each other at the front of the frame ______________________

Midpoint in the distance between the two chalk marks directly across from each other at the front of the frame ______________________

SERVICE TIP The measurement points on the left and right frame webs must be at the same location on each web to obtain accurate measurements.

Diagonal frame measurements at the rear of the frame:

diagonal E ______________ diagonal F ______________

difference between diagonals E and F ______________

Necessary frame service: ______________________________

____

____

10. Use a long, straight steel bar to place a straight chalk mark between all frame diagonal measurement locations. These diagonal lines should cross each other at the frame centerline.

If all diagonal lines do not cross the frame centerline at the same location, state the required frame repairs. ______________________________

____

____

Instructor's Response ______________________________

____

____

Name ______________________ Date ______________

JOB SHEET 51

INSPECT AND MEASURE FRONT CRADLE

Upon completion of this job sheet, you should be able to inspect and measure front cradles.

ASE Education Foundation Task Correlation

This job sheet addresses the following **AST/MAST** task:

E.8. Check front and/or rear cradle (subframe) alignment; determine needed action. **(P-3)**

We Support ASE Education Foundation

Tools and Materials

Tram gauge

Describe the vehicle being worked on:

Year ______________ Make ______________ Model ______________

VIN ______________ Engine type and size ______________

Procedure

Task Completed

1. Raise the vehicle on a lift using the vehicle manufacturer's specified lifting points. ☐
2. Inspect front cradle mounts for looseness, damage, oil soaking, wear, and deterioration.

   Cradle mount condition: right front ______________ right rear

   left front ______________ left rear ______________
3. Inspect front cradle for visible bends and damage.

   Front cradle condition indicated in visible inspection:

   ____

   ____

   ____
4. Inspect front cradle alignment.

   Is the front cradle aligning hole(s) properly aligned with the matching hole in the chassis? ☐ Yes ☐ No

   Recommended cradle service:

   ____

   ____

   ____
5. Use a tram gauge to complete all measurements across the width of the cradle starting at the front of the cradle.

   Width measurement A ______________

   Specified width measurement A ______________

   Width measurement B ______________

   Specified width measurement B ______________

Name ______________________________ Date ____________

JOB SHEET
52

INSPECT AND WELD VEHICLE FRAME

Upon completion of this job sheet, you should be able to inspect and weld a vehicle frame.

Tools and Materials

Hand pump
Welding hammer
Arc welder
Welding rods

Describe the Vehicle being Worked On:

Year ________________ Make ________________ Model ________________

VIN ________________ Engine type and size ________________

Procedure

Task Completed

1. Raise the vehicle on a lift using the vehicle manufacturer's specified lifting points. ☐

2. Read the instructions for disconnecting the negative battery cable in the vehicle manufacturer's service manual. These instructions may include connecting a 12 V power source to the cigarette lighter socket to maintain power to computer memories, radio-stereo, and other electronic equipment.

   List the vehicle manufacturer's instructions for disconnecting the negative battery cable.

   ____

   ____

   Have these instructions been completed on the vehicle being serviced? ☐ Yes ☐ No

   Instructor check ________________________________

Caution

Always wear eye protection while working in the shop, and use the specified eye protection while arc welding to prevent eye injury.

3. If the vehicle is air bag-equipped, determine the specified waiting period after the negative battery cable is disconnected before performing service work.

   Waiting period ________________________________

4. Disconnect the negative battery cable.

   Is the negative battery cable disconnected? ☐ Yes ☐ No

   Has the specified waiting period been completed before servicing the vehicle?
   ☐ Yes ☐ No

   Instructor check ________________________________

5. Use a hand pump to pump all the fuel from the fuel tank. Place the fuel in the proper fuel safety containers and store this fuel away from the work area and other ignition sources.

   Has all the fuel been pumped from the tank? ☐ Yes ☐ No

   Is this fuel placed in proper fuel safety containers and stored away from the work area? ☐ Yes ☐ No

   Instructor check ________________________________

CHAPTER 11
COMPUTER-CONTROLLED SUSPENSION SYSTEM SERVICE

Upon completion and review of this chapter, you should be able to:

- Diagnose programmed ride control (PRC) systems.
- Diagnose electronic air suspension systems.
- Remove, replace, and inflate air springs.
- Adjust front and rear trim height on electronic air suspension systems.
- Service and repair nylon air lines.
- Diagnose vehicle dynamic suspension (VDS) systems.
- Remove and replace air springs on VDS systems.
- Perform an on-demand self-test on VDS systems.
- Diagnose electronic suspension control (ESC) systems.
- Display and interpret scan tool data on ESC systems.
- Use the output controls function to diagnose ESC systems.
- Diagnose vehicle networks.
- Post-alignment sensor adjustment procedures on driver assistance systems.
- Adjust adaptive cruise control radar sensor.
- Lane departure warning system sensor calibration.

Basic Tools

Basic technician's tool set

Service manual

Floor jack

Safety stands

Utility knife

Machinist's ruler or tape measure

Terms To Know

Antilock brake system (ABS)

Brake pressure modulator valve (BPMV)

Continuously variable road sensing suspension (CVRSS)

Data link connector (DLC)

Electronic brake and traction control module (EBTCM)

Electronic suspension control (ESC)

Programmed ride control (PRC)

Powertrain control module (PCM)

Scan tool

Splice pack

Stabilitrak®

Trim height

Vehicle dynamic suspension (VDS)

Each year more vehicles are equipped with computer-controlled suspension systems, and these systems are becoming increasingly complex. Therefore, technicians must understand the correct procedures for diagnosing and servicing these systems. When a technician understands computer-controlled suspension systems and the proper diagnostic procedures for these systems, diagnosis becomes faster and more accurate.

PRELIMINARY INSPECTION OF COMPUTER-CONTROLLED SUSPENSION SYSTEMS

Prior to diagnosing a computer-controlled suspension system, a preliminary inspection should be performed. The preliminary inspection may locate a minor defect that is the cause of the problem. If the preliminary inspection is not performed, a lot of time may be

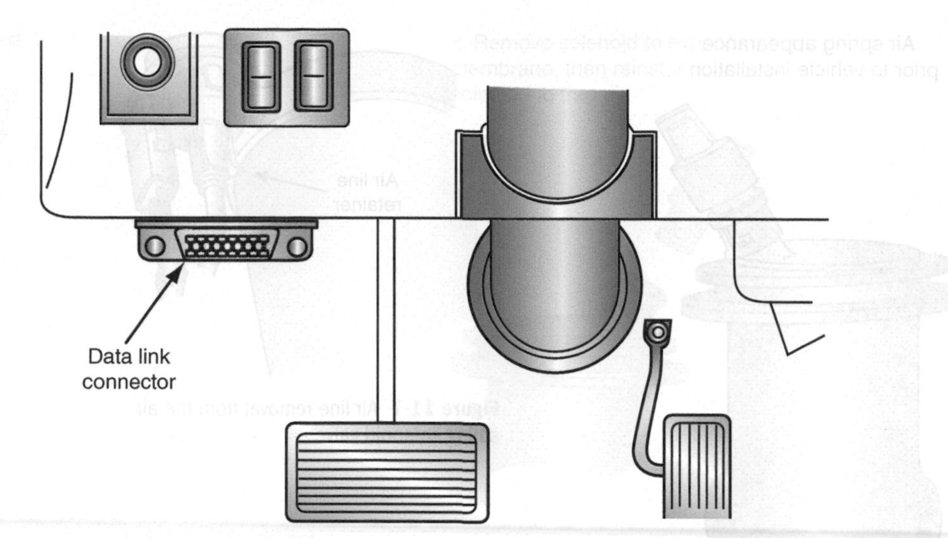

Figure 11-9 Data link connector (DLC), on-board diagnostic II (OBD-II) system.

When an air spring is being inflated, use this procedure:

1. With the vehicle chassis supported on a hoist, lower the hoist until a slight load is placed on the suspension. Do not lower the hoist until the suspension is heavily loaded.
2. Turn on the air suspension system switch.
3. Turn the ignition switch from Off to Run for 5 seconds with the driver's door open and the other doors shut. Turn the ignition switch off.
4. Ground the diagnostic lead.
5. Apply the brake pedal and turn the ignition to the Run position. The warning lamp will flash every 2 seconds to indicate the fill mode.
6. To fill a rear spring or springs, close and open the driver's door once. After a 6-second delay, the rear spring will fill for 60 seconds.
7. To fill a front spring or springs, close and open the driver's door twice. After a 6-second delay, the front spring will fill for 60 seconds.
8. When front and rear springs require filling, fill the rear springs first. Once the rear springs are filled, close and open the driver's door once to begin filling the front springs.
9. The spring fill mode is terminated if the diagnostic lead is disconnected from ground. Termination also occurs if the ignition switch is turned off or the brake pedal is applied.

Scan Tool Procedure for Air Spring Inflation

On some air suspension systems, the **scan tool** is used to activate the air compressor and energize the appropriate air spring solenoid valve to inflate an air spring. The following is a typical spring fill procedure using a scan tool on an automatic air suspension system:

A **scan tool** is a digital tester that can be used to diagnose various on-board computer systems.

The **data link connector (DLC)** is a 16-terminal connector positioned under the left side of the instrument panel. A scan tool is connected to the DLC to diagnose various on-board electronic systems.

1. Turn off the air suspension switch.
2. Be sure the proper module is installed in the scan tool for suspension diagnosis and service on the vehicle being serviced.
3. Connect the scan tool to the **data link connector (DLC)** on the vehicle.
4. Be sure the vehicle is positioned on a frame-contact lift with the suspension dropped downward and no load on the suspension.
5. Connect a battery charger to the battery terminals with the correct polarity, and set the charger at a low charging rate.

6. Select Function Test on the scan tool, and select the desired function test number for the air spring being filled as follows: test number 212—LH front air spring, test number 214—RH front air spring, test number 216—LH rear air spring, and test number 218—RH rear air spring. When the proper function test number is selected, the scan tool commands the compressor and the appropriate air spring solenoid valve on, to allow air spring inflation.
7. When the compressor shuts off, the air spring inflation is complete, and any further air spring inflation or deflation is done during normal operation of the air suspension system.
8. Lower the vehicle onto the shop floor and lower the lift completely. Be sure the lift arms are not contacting the vehicle.
9. Inspect the air spring(s) that were completely deflated, and be sure there are no folds and creases in these springs.
10. Turn on the suspension switch.
11. If the lower front strut-to-control arm bolt and nut were removed and reinstalled during the suspension service procedure, use your knee to push downward and release the front bumper. Repeat this procedure three times.
12. Tighten the lower front strut-to-control arm nut and bolt to the specified torque.
13. Road test the vehicle and verify proper air suspension operation and trim height.

Photo Sequence 22 shows the proper procedure for air spring inflation with a scan tool.

OBD-II computer systems became mandatory on cars and light trucks in the 1996 model year. Some vehicles prior to 1996 were equipped with OBD-II systems. OBD-II systems have some standardized features such as a 16-terminal DLC under the left side of the dash. OBD-II systems have increased monitoring capabilities in the PCM. Many engine control systems are monitored, and if any engine system is defective it allows the exhaust emissions to increase more than 1.5 times the normal limit for that vehicle, the malfunction indicator light (MIL) illuminates in the instrument panel.

PHOTO SEQUENCE 22
Inflate L/H Front Air Spring Using a Scan Tool

P22-1 Turn off the air suspension switch.

P22-2 Connect the scan tool to the data link connector (DLC) on the vehicle.

P22-3 Be sure the vehicle is positioned on a frame-contact lift with the suspension dropped downward and no load on the suspension.

P22-4 Connect a battery charger to the battery terminals with the correct polarity, and set the charger at a low charging rate.

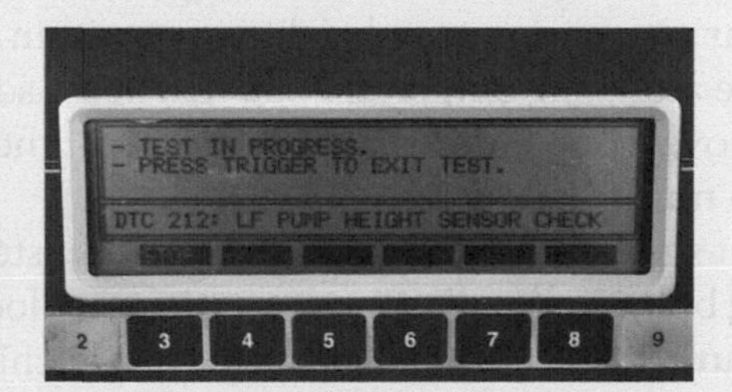

P22-5 Select Function Test on the scan tool, and select the desired Function Test number 212–LH front air spring. When the proper function test number is selected, the scan tool commands the compressor and the appropriate air spring solenoid valve on, to allow air spring inflation.

P22-6 When the compressor shuts off, the air spring inflation is complete. Inspect the L/H front air spring and be sure there are no folds and creases in the spring.

PHOTO SEQUENCE 22 (CONTINUED)

P22-7 Shut the battery charger off and disconnect the charger cables. Lower the vehicle onto the shop floor and completely lower the lift. Be sure the lift arms are not contacting the vehicle.

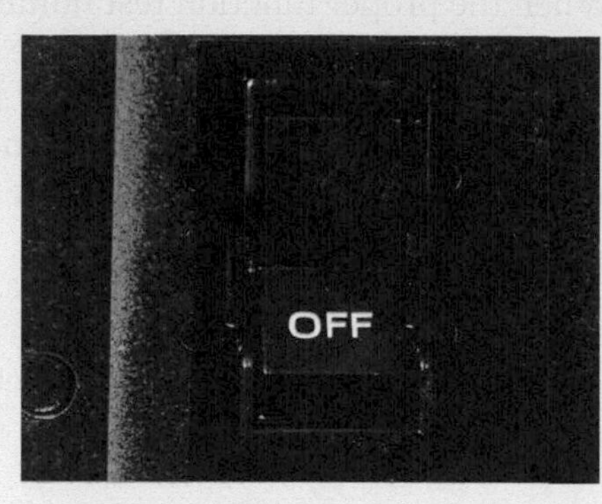

P22-8 Disconnect the scan tool, and turn on the suspension switch.

P22-9 If the lower front strut-to-control arm bolt and nut were removed and reinstalled during the suspension service procedure, use your knee to push downward and release the front bumper. Repeat this procedure three times. Tighten the lower front strut-to-control arm nut and bolt to the specified torque.

P22-10 Road test the vehicle and verify proper air suspension operation and trim height.

Trim Height Mechanical Adjustment

On some vehicles with air suspension, the trim height is adjusted by lengthening or shortening the front or rear height sensors. On other vehicles, the manufacturer recommends using the scan tool to adjust the trim height. Always follow the trim height adjustment procedure in the vehicle manufacturer's service manual.

Front and rear **trim height** is the distance between the vehicle chassis and the road surface measured at locations specified by the vehicle manufacturer.

The **trim height** should be measured on the front and rear suspension at the locations specified by the vehicle manufacturer (**Figure 11-10**).

If the rear suspension trim height is not within specifications, it may be adjusted by loosening the attaching bolt on the top height sensor bracket (**Figure 11-11**). When the bracket is moved one index mark up or down, the ride height is lowered or raised to 0.25 in. (6.35 mm).

The front suspension trim height may be adjusted by loosening the lower height sensor attaching bolt. Three adjustment positions are located in the lower front height sensor bracket (**Figure 11-12**). If the height sensor attaching bolt is moved one position up or down, the front suspension height is lowered or raised to 0.5 in. (12.7 mm).

Line Service

Nylon lines on the electronic air suspension system have quick disconnect fittings. These fittings should be released by pushing downward and holding the plastic release ring, and

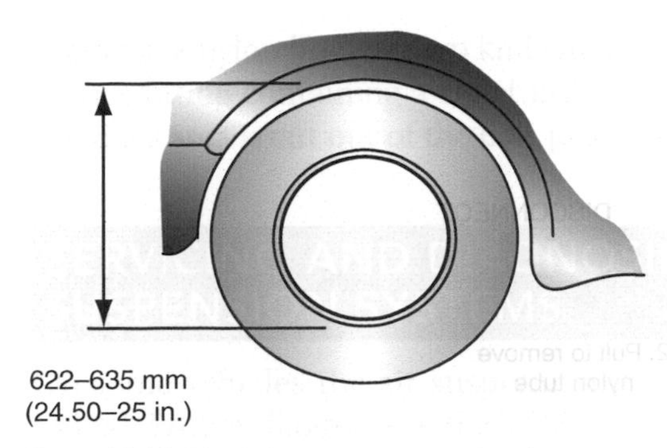

Figure 11-10 Trim height measurement location.

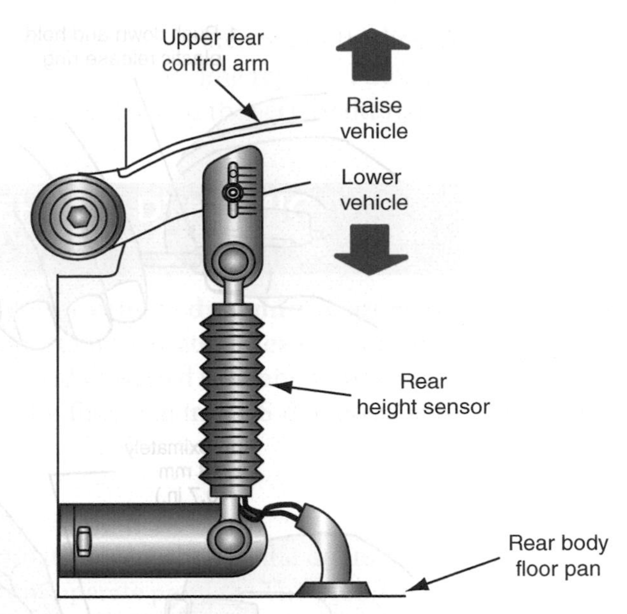

Figure 11-11 Rear trim height adjustment.

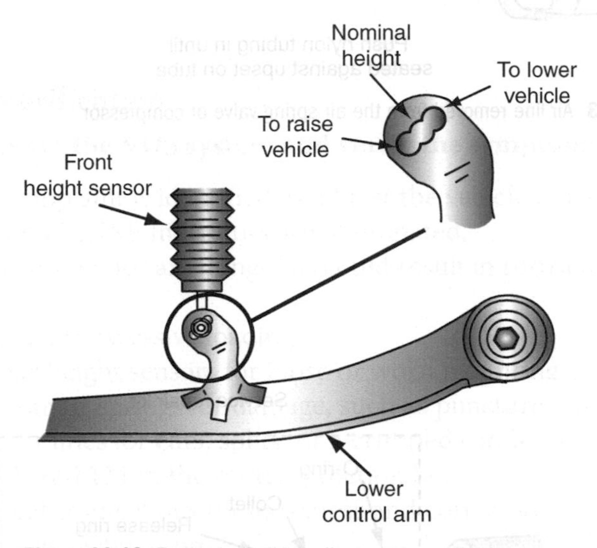

Figure 11-12 Front trim height adjustment.

SERVICE TIP Some on-board diagnostic I (OBD-I) vehicles have a separate suspension DLC positioned on the right-front strut tower. Other OBD-I vehicles have the suspension DLC in the trunk. **OBD-II computer systems** have a universal 16-terminal DLC mounted under the left side of the dash near the steering column (Figure 11-9). Data links are connected from various on-board computers including the suspension computer to the DLC to allow the on-board computers to communicate with the scan tool and vice versa.

then pulling outward on the nylon line (**Figure 11-13**). Simply push the nylon line into the fitting until it seats to reconnect an air line.

If a line fitting is damaged, it may be removed by looping the line around your fingers and pulling on the line without pushing on the release ring (**Figure 11-14**). When a new collet and release ring are installed, the O-ring under the collet must be replaced. If a leak

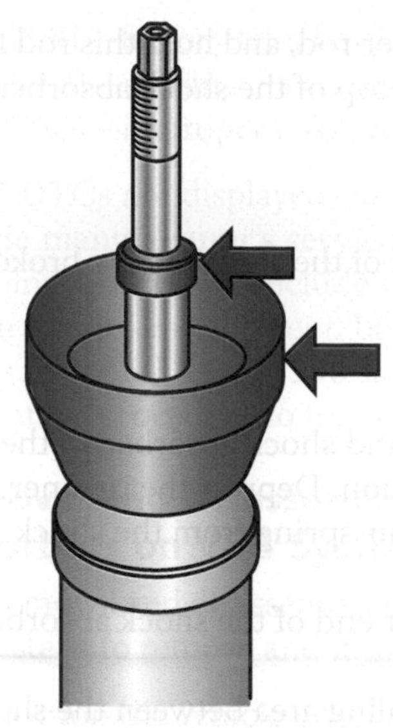

Figure 11-23 Seal and washer on the upper end of the shock absorber rod.

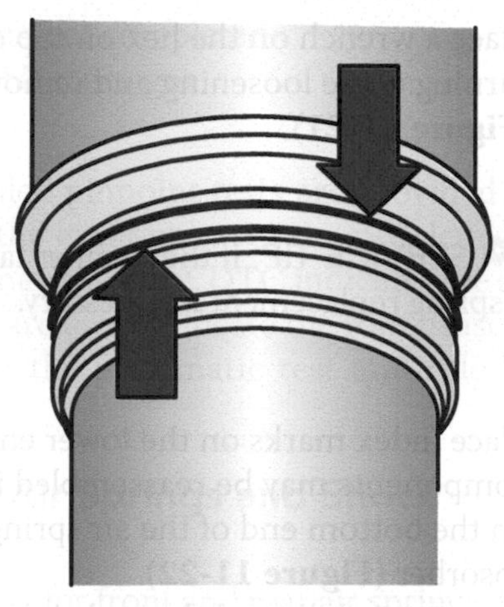

Figure 11-24 O-ring seals on the lower sealing area on the shock absorber.

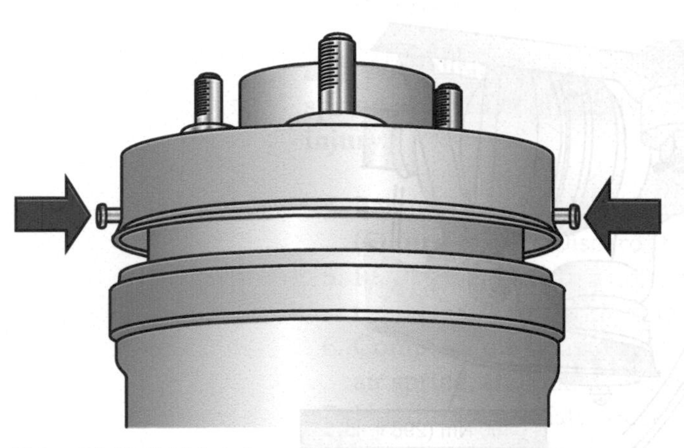

Figure 11-25 Retaining pins on the upper air spring mount.

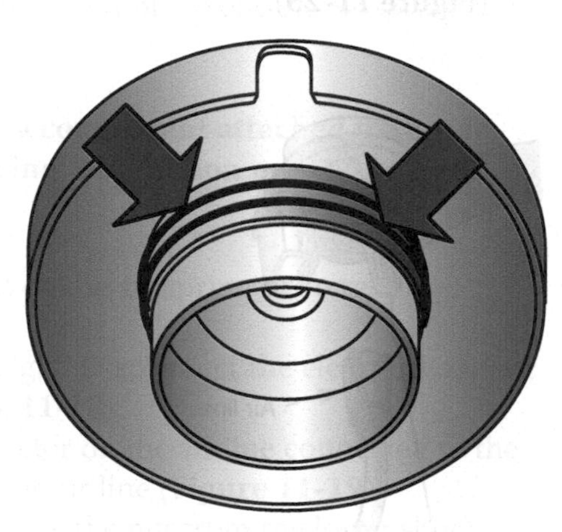

Figure 11-26 O-ring seals on the upper spring mount.

Caution

Do not supply voltage to or ground any circuit or component in a computer system unless instructed to do so in the vehicle manufacturer's service manual. This action may damage computer system components.

14. Remove and discard the O-ring seals on the upper mount area that provides a seal between the mount and the air spring (**Figure 11-26**).

When reassembling the air spring and shock absorber, replace all the O-rings and self-locking nuts, and place a light coating of chassis lubricant on all the O-rings.

DIAGNOSIS OF ELECTRONIC SUSPENSION CONTROL SYSTEMS

Diagnostic Trouble Code (DTC) Display

Vehicles equipped with the **electronic suspension control (ESC)** system may have a digital or an analog instrument panel cluster (IPC). If the vehicle has a digital IPC, the air-conditioning (A/C) control panel is contained in the IPC (**Figure 11-27**). A driver message center is located in the center of the IPC underneath the speedometer. This message center is a vacuum fluorescent (VF) display with twenty characters. The same

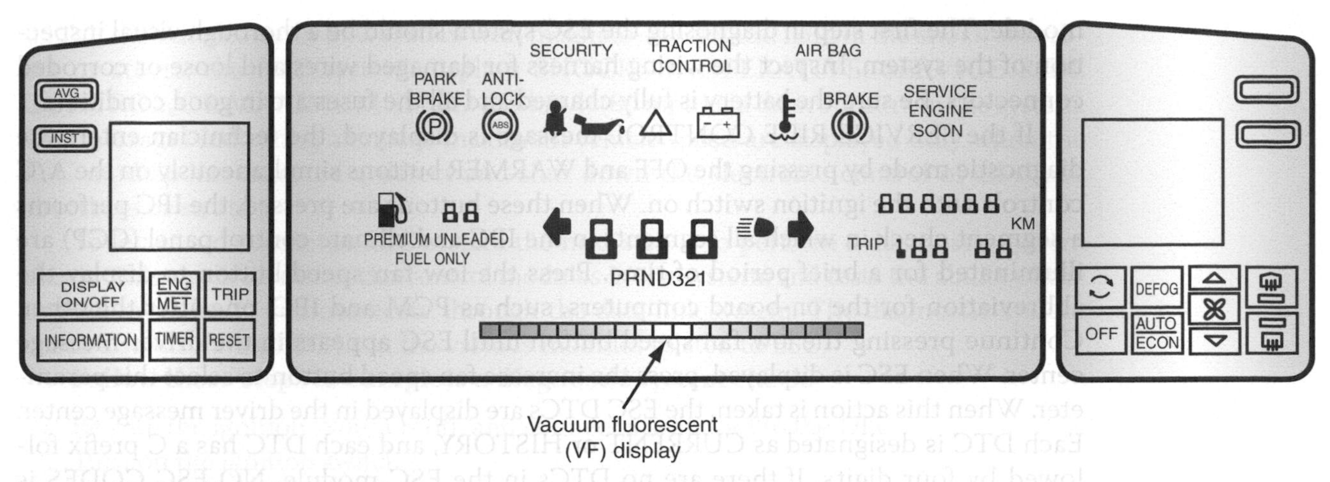

Figure 11-27 Digital instrument panel cluster.

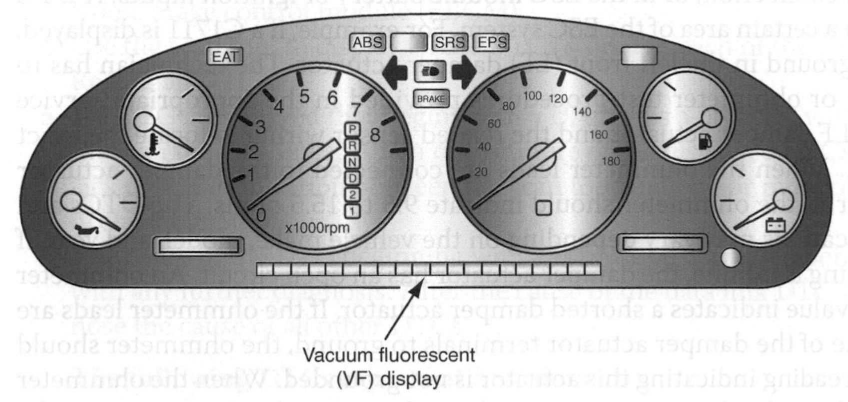

Figure 11-28 Analog instrument panel cluster.

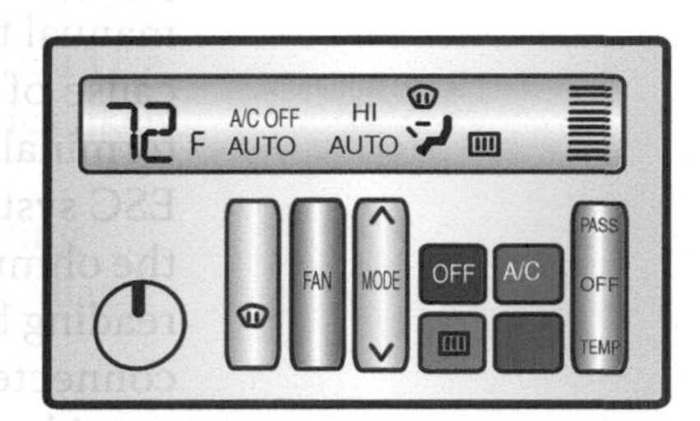

Figure 11-29 A/C controls used with the analog instrument panel cluster.

message center is mounted in an analog IPC (**Figure 11-28**). If the car has an analog IPC, the A/C controls are mounted in a separate display (**Figure 11-29**).

Early versions of the ESC system may be called **continuously variable road sensing suspension (CVRSS)** systems.

Caution

During computer system diagnosis, use only the test equipment recommended in the vehicle manufacturer's service manual to prevent damage to computer system components.

SERVICE TIP When removing, replacing, or servicing an electronic component on a vehicle, always disconnect the negative battery cable before starting the service procedure. If the vehicle is equipped with an air bag or bags, wait one minute after the battery negative cable is removed to prevent accidental air bag deployment.

Many air bag computers have a backup power supply capable of deploying the air bag for a specific length of time after the battery is disconnected.

When the information (INFO) button is pressed in the IPC, the driver message center cycles through a series of status messages: RANGE, AVG MPG, MPG INST, FUEL USED, OIL LIFE LEFT, ENGINE RPM, BATTERY VOLTS, COOLANT TEMP, ENGLISH/METRIC RESET. The parameters related to fuel usage are not displayed in the message center on a digital IPC, because these readings are displayed in the fuel data center that is contained in the IPC.

If an electrical defect occurs in the ESC system, a SERVICE RIDE CONTROL message is displayed in the driver information center, and a DTC is usually stored in the ESC

10. C0710—the ESC module does not detect a valid steering position signal from the EBCM for 5 seconds. The EBCM controls the ABS and the variable assist steering.
11. C0896—the battery voltage is not within the normal range of 9V to 15.5V.
12. P1652—the output driver module (ODM) in the PCM that controls suspension lift/dive has detected an improper circuit condition.

To set most DTCs in the ESC module, the module must sense a defect on three consecutive ignition cycles, or during the same ignition cycle after clearing the DTC with a scan tool.

Scan Tool Data Display

Select *Special Functions* on the scan tool to access the ESC system data. The ESC data is very helpful to pinpoint the exact cause of DTCs. The following ESC data may be displayed on the scan tool with the ignition switch on, the engine off, and the front wheels straight ahead:

1. Battery voltage signal—typical data 12.6 V.
2. DSP software version ID—numerical value.
3. EEPROM calibration ID—numerical value.
4. GM part number ESC module—numerical value.
5. ESC module software version—numerical value.
6. L/F damper actuator command—0 to 100 percent—this percentage indicates the commanded on-time of the damper actuator. If the ESC module commands the damper actuator On, the percentage increases.
7. L/F position sensor—0 V to 5 V—this voltage is the signal voltage sent from the position sensor to the ESC module. If downward force is supplied to the L/F of the chassis, the voltage reading should change. When the position sensor voltage does not change, the sensor and connecting wires must be tested to locate the exact cause of the problem. After repairs have been completed to correct a problem causing a DTC, the scan tool must be used to clear the DTC from the ESC module.
8. L/F normal force—20 to 80 percent—this percentage indicates a measured level of the road surface condition sent from the ESC module to the EBCM, and the EBCM uses this information to improve antilock braking on irregular road surfaces.
9. The damper actuator command and position sensor data are displayed on the scan tool for the other three corners of the chassis. The normal force data is available only on the front wheels.
10. Lift/Dive status—active, inactive—if there is no change in chassis pitch, the scan tool displays Inactive. When a change in chassis pitch occurs, the scan tool displays Active.
11. Lift/Dive change—changed, unchanged—if there is no change in chassis pitch, Unchanged is displayed on the scan tool. A change in chassis pitch causes the scan tool to display Changed.
12. Steering position PWM—0 to 10 ms—this PWM signal is sent from the EBCM to the ESC module when the steering wheel is turned while cornering. The ESC module commands a firmer shock absorber state when the vehicle is turning a corner.
13. Vehicle speed—0 to 159 mph (0 to 255 km/h)—the vehicle speed sensor signal is relayed on the data links from the PCM to the ESC module.

Scan Tool Output Controls

The output controls are accessed on the scan tool under the Special Functions menu. When output controls are selected on the scan tool, followed by an individual damper solenoid selection, the solenoid is cycled on and off. When an individual damper solenoid actuator is cycled on and off, the technician may perform voltage tests to check the damper actuator and related circuit. The scan tool may also be used to cycle all the damper solenoids simultaneously.

In the Output Controls mode, the scan tool may be used to command the ESC module to send the L/F and R/F normal force percentage readings to be sent to the EBCM, and this percentage is displayed on the scan tool.

ESC Module Recalibration may also be selected on the scan tool in the output controls mode. This recalibration is necessary after a module replacement.

Photo Sequence 23 illustrates the procedure for reading scan tool data on an electronic suspension control system.

Diagnosis of Stability Control System

The stability control system module is combined with the **antilock brake system (ABS)** and traction control system (TCS) modules. Therefore, it is not possible to completely separate the diagnosis of these systems. However, our discussion is mostly concerned with the stability control system, and this diagnostic information pertains to the **Stabilitrak®** system. The Stabilitrak® system may be called a vehicle stability enhancement system (VSES).

The **antilock brake system (ABS)** prevents wheel lockup during a brake application.

The **Stabilitrak®** system is a type of vehicle stability control.

PHOTO SEQUENCE 23
Reading Scan Tool Data on an Electronic Suspension Control System

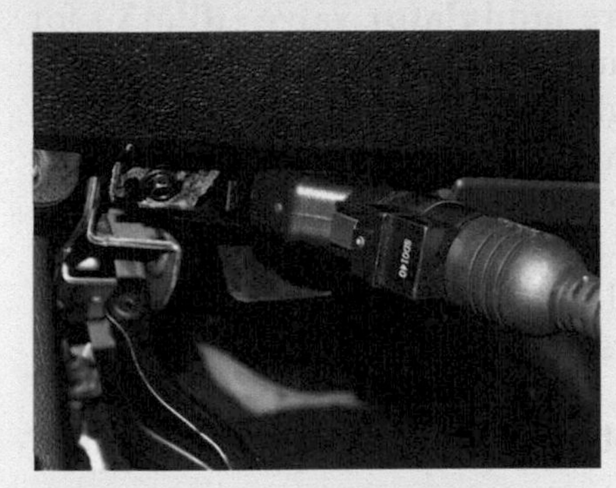

P23-1 Connect the scan tool to the DLC under the dash.

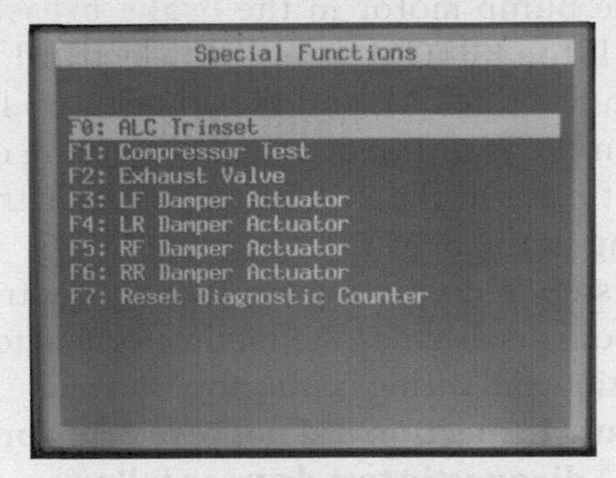

P23-2 Turn on the ignition switch, and select Special Functions in the ESC menu on the scan tool.

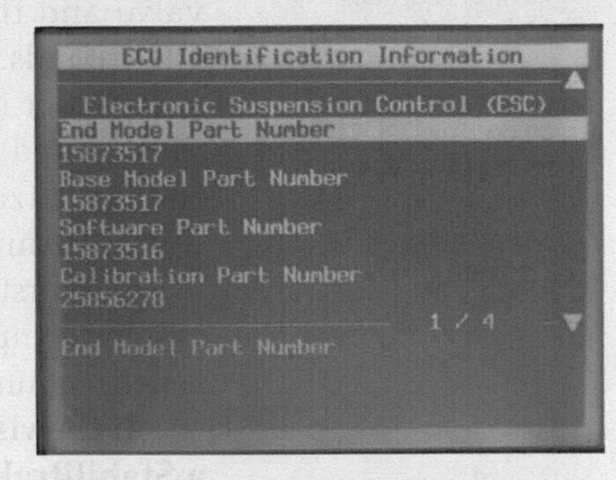

P23-3 Display and record the DSP software version ID, EEPROM Calibration ID, GM part number ESC module, and ESC module software version on the scan tool.

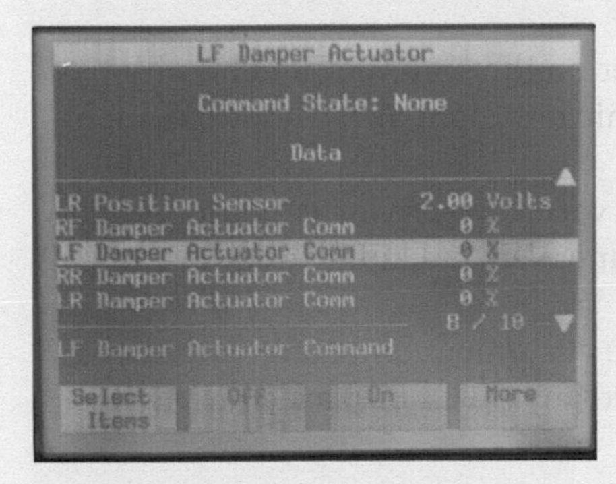

P23-4 Select the damper actuator command on each damper actuator. This percentage indicates the commanded on-time of the damper actuator.

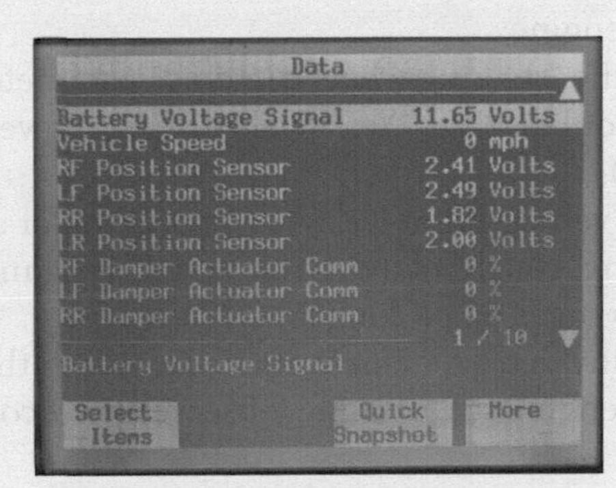

P23-5 Select the wheel position sensor voltage on each wheel position sensor. This voltage is the signal voltage sent from the position sensor to the ESC module.

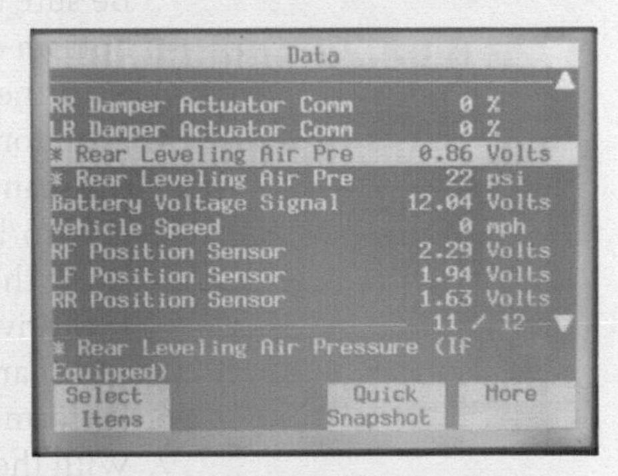

P23-6 Select the normal force display on the scan tool. This percentage indicates a measured level of the road surface condition sent from the ESC module to the EBCM, and the EBCM uses this information to improve antilock braking on irregular road surfaces.

PHOTO SEQUENCE 23 (CONTINUED)

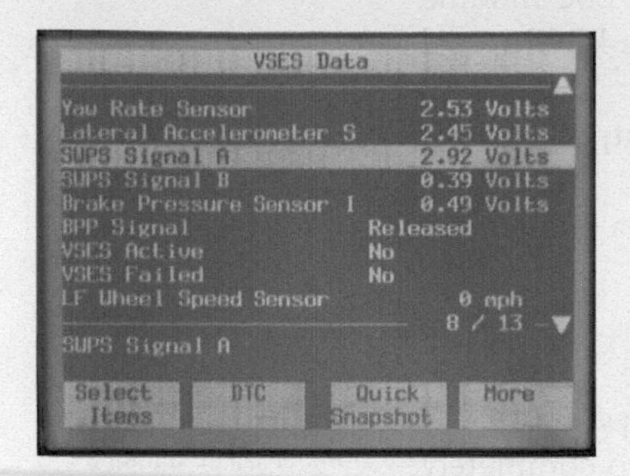

P23-7 Select the Lift/Dive status on the scan tool.

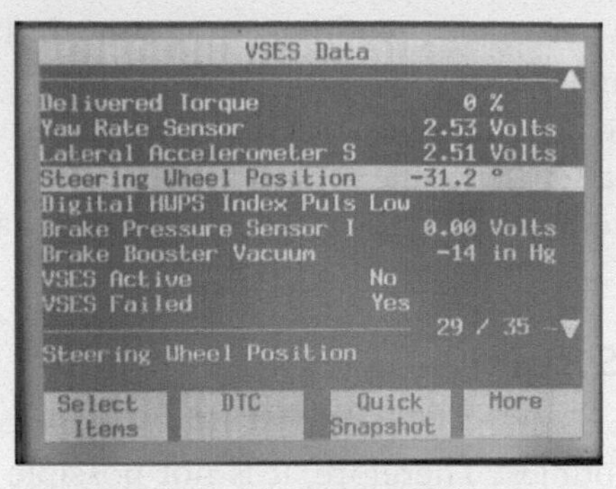

P23-8 Select Steering Position PWM on the scan tool.

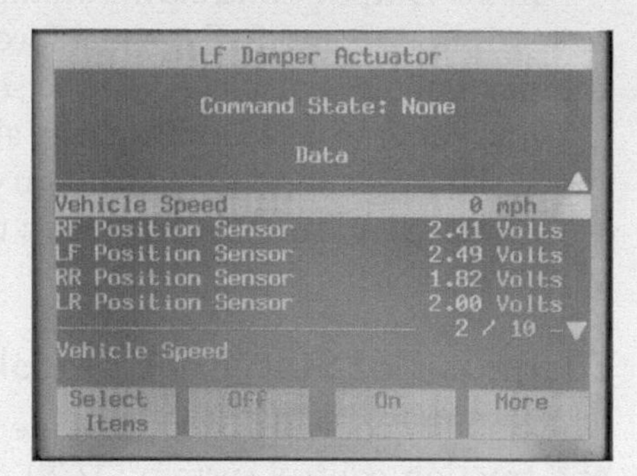

P23-9 Select Vehicle Speed on the scan tool.

The **electronic brake and traction control module (EBTCM)** controls ABS, TCS, and Stabilitrak® functions.

The **brake pressure modulator valve (BPMV)** controls brake fluid pressure to the wheel calipers or cylinders.

The **electronic brake and traction control module (EBTCM)** performs an initialization test each time the vehicle is started and the EBTCM does not receive a brake switch input. During the initialization test, the EBTCM cycles each solenoid valve and the pump motor in the **brake pressure modulator valve (BPMV)** for 1.5 seconds. If the EBTCM detects an electrical fault during the initialization test, a DTC is set in the EBTCM, and the amber ABS light and the red brake light may be illuminated in the instrument panel, depending on the severity of the defect. During the initialization test, the customer may hear the solenoids clicking and the pump motor turning in the BPMV.

The first step in diagnosing the stability control system is a visual inspection of all system components, such as the module connectors, fuses, relays, wiring harness, and sensor mounting and wiring connector.

If the visual inspection did not reveal any problems, the next step is to perform a Stabilitrak® diagnostic test drive as follows:

1. Turn the ignition switch off.
2. Connect a scan tool to the data link connector (DLC) under the instrument panel. Be sure the correct module for the system being tested is securely installed in the bottom of the scan tool.
3. Start the engine.
4. Monitor the "steering wheel centered" parameter on the scan tool, and be sure the scan tool displays "Yes" while driving the vehicle straight ahead above 15 mph (24 km/h).
5. Drive the vehicle for at least 10 minutes under a variety of driving conditions: highway driving, rough road driving, and turning maneuvers, including freeway ramps and sharp turns on parking lots.
6. Perform any driving maneuvers under which the customer complaint(s) occurred.
7. With the engine still running, observe and record any DTCs displayed on the scan tool.

If there is a defect in the Stabilitrak® system, "Stability Reduced" may be displayed in the driver information center. When "Display ABS/TCM/ICCS DTCs" is selected on the scan tool, any of the following DTCs may be displayed as indicated in **Figure 11-30** and

DTC	DESCRIPTION
C1211	ABS Indicator Lamp Circuit Malfunction
C1214	Solenoid Valve Relay Contact or Coil Circuit Open
C1217	BPMV Pump Motor Relay Contact Circuit Open
C1221	LF Wheel Speed Sensor Input Signal - 0
C1222	RF Wheel Speed Sensor Input Signal - 0
C1223	LR Wheel Speed Sensor Input Signal - 0
C1224	RR Wheel Speed Sensor Input Signal - 0
C1225	LF - Excessive Wheel Speed Variation
C1226	RF - Excessive Wheel Speed Variation
C1227	LR - Excessive Wheel Speed Variation
C1228	RR - Excessive Wheel Speed Variation
C1232	LF Wheel Speed Sensor Circuit Open or Shorted
C1233	RF Wheel Speed Sensor Circuit Open or Shorted
C1234	LR Wheel Speed Sensor Circuit Open or Shorted
C1235	RR Wheel Speed Sensor Circuit Open or Shorted
C1236	Low System Voltage
C1237	High System Voltage
C1238	Brake Thermal Model Limit Exceeded
C1241	Magna Steer® Circuit Malfunction
C1242	BPMV Pump Motor Ground Circuit Open
C1243	BPMV Pump Motor Stalled
C1251	RSS Steering Sensor Data Malfunction
C1252	ICCS2 Data Link Left Malfunction
C1253	ICCS2 Data Link Right Malfunction
C1255 xx	EBTCM Internal Malfunction (ABS/TCS/ICCS Disabled)
C1256 xx	EBTCM Internal Malfunction
C1261	LF Hold Valve Solenoid Malfunction

Figure 11-30 DTCs for ABS/TCS/ICCS systems.

Figure 11-31. The technician usually has to perform voltmeter or ohmmeter tests to locate the exact cause of a problem in the area indicated by the DTC.

Automated Test

An automated test may be selected on the scan tool. During this test, the EBTCM cycles all the solenoid valves and the pump motor in the BPMV, and DTCs are set in the EBTCM if there are defects in these components. This test is the same as the initialization test performed by the EBTCM when the engine is started. The scan tool also performs solenoid tests.

Valve Solenoid/Pressure Hold Test

To perform the valve solenoid/pressure hold test, follow this procedure:

1. Be sure the scan tool is properly connected to the DLC.
2. Turn on the ignition switch.
3. Raise the vehicle on a lift so all four wheels are at least 6 in. (15 cm) off the floor.
4. Select Valve Solenoid Test on the scan tool followed by Hold Pressure on a specific wheel.
5. Hold the brake pedal in the applied position.
6. Have a coworker try to turn the wheel being tested. If the hold pressure solenoid is operating properly, the coworker should be able to turn the wheel.
7. Repeat Steps 4 through 6 on the other wheels.

DTC	DESCRIPTION
C1262	LF Release Valve Solenoid Malfunction
C1263	RF Hold Valve Solenoid Malfunction
C1264	RF Release Valve Solenoid Malfunction
C1265	LR Hold Valve Solenoid Malfunction
C1266	LR Release Valve Solenoid Malfunction
C1267	RR Hold Valve Solenoid Malfunction
C1268	RR Release Valve Solenoid Malfunction
C1271	LF TCS Master Cylinder Isolation Valve Malfunction
C1272	LF TCS Prime Valve Malfunction
C1273	RF TCS Master Cylinder Isolation Valve Malfunction
C1274	RF TCS Prime Valve Malfunction
C1276	Delivered Torque Signal Circuit Malfunction
C1277	Requested Torque Signal Circuit Malfunction
C1278	TCS Temporarily Inhibited by PCM
C1281	Stabilitrak® Sensors Uncorrelated
C1282	Yaw Rate Sensor Bias Circuit Malfunction
C1283	Excessive Time to Center Steering
C1284	Lateral Accelerometer Sensor Self-Test Malfunction
C1285	Lateral Accelerometer Sensor Circuit Malfunction
C1286	Steering/Lateral Accelerometer Sensor Bias Malfunction
C1287	Steering Sensor Rate Malfunction
C1288	Steering Sensor Circuit Malfunction
C1291	Open Brake Lamp Switch During Deceleration
C1293	DTC C1291 Set In Current or Previous Ignition Cycle
C1294	Brake Lamp Switch Circuit Always Active
C1295	Brake Lamp Switch Circuit Open
C1297	PCM Indicated Brake Extended Travel Switch Failure
C1298	PCM Indicated Class 2 Serial Data Link Malfunction
U1016	Loss of PCM Communications
U1056	Loss of CVRSS Communications
U1255	Generic Loss of Communications
U1300	Class 2 Circuit Shorted to Ground
U1301	Class 2 Circuit Shorted to Battery +

Figure 11-31 DTCs for ABS/TCS/ICCS systems.

CUSTOMER CARE While discussing computer-controlled suspension systems with customers, remember that the average customer is not familiar with automotive electronics terminology. Always use basic terms that customers can understand when explaining electronic suspension problems. Most customers appreciate a few minutes spent by service personnel to explain their automotive electronic problems. It is not necessary to provide customers with a lesson in electronics, but it is important that customers understand the basic cause of the problem with their vehicle so they feel satisfied the repair expenditures are necessary. A satisfied customer is usually a repeat customer.

Valve Solenoid/Pressure Release Test

To perform a valve solenoid/pressure release test, follow this procedure:

1. Be sure the scan tool is properly connected to the DLC.
2. Turn on the ignition switch.
3. Raise the vehicle on a lift so all four wheels are at least 6 in. (15 cm) off the floor.

4. Hold the brake pedal in the applied position.
5. Select Valve Solenoid Test on the scan tool followed by Release Pressure on a specific wheel.
6. Have a coworker try to turn the wheel being tested. If the release pressure solenoid is operating properly, the coworker should be able to turn the wheel.
7. Repeat Steps 4 through 6 on the other wheels.

Classroom Manual
Chapter 11, page 289

If any of the hold or release solenoids do not operate properly, the BPMV requires replacement. The solenoids in the BPMV are not serviceable.

Diagnosis of Vehicle Networks

Vehicle networks have various voltages and operating characteristics. Therefore, the vehicle manufacturer's specific diagnostic procedure must be followed for each vehicle network. The most common equipment for diagnosing vehicle networks are a scan tool, a digital multimeter (DMM), and a lab scope. If a defect occurs in most networks, a DTC(s) is displayed on a scan tool connected to the DLC. The technician must follow the vehicle manufacturer's specified procedure to diagnose the cause of the DTC(s). When using a lab scope to diagnose vehicle networks, the technician must be familiar with the normal lab scope pattern for each network. We will discuss the lab scope patterns for two common networks and provide several case studies of network diagnosis.

A Class 2 network is a single-wire system. When the system is at rest with no communication taking place, the system voltage is 0 V. When communication takes place on this network, the data are transmitted by a variable pulse width signal of 0 V to 7 V. The Class 2 network is connected to terminal 2 in the DLC, and it also interconnects with various computers depending on the electronic equipment on the vehicle. To diagnose a Class 2 network with a lab scope, connect the Channel 1 positive lead of the scope to pin 2 in the DLC, and connect the ground lead on the scope to pin 4 in the DLC or to a good chassis ground. Adjust the scope voltage setting to 5 V and set the time base setting to 100 milliseconds (ms). With the ignition switch on, a normal network should display a voltage pattern as illustrated in **Figure 11-32**. A **splice pack** is used in the network on some vehicles. A parts locator manual will help the technician to locate the splice pack. This splice pack may be disconnected to isolate the PCM and truck body computer (TBC) from the rest of the network so the network cannot affect these computers. If the vehicle has a no-start condition, and the engine starts after the splice pack is disconnected, the defect is located in the network. Under this condition, the network may be shorted to ground, the network may be shorted to a 12 V source, or a module connected to the network may be shorted internally. The network modules may be disconnected one at a time. If the defect is corrected when one module is disconnected, the network problem is in that module. When none of the modules are causing the network problem, voltage and/or ohmmeter tests must be performed on the network wires to locate the shorted condition.

A **splice pack** is a special connector that may be disconnected to isolate specific components when diagnosing vehicle networks.

The Standard Corporate Protocol (SCP) network is used on many vehicles. The two wires in the SCP system are designated as Bus+ and Bus–. Bus+ is connected to pin 2 in the DLC, and Bus– is connected to pin 10 in the DLC. The network voltages at rest with

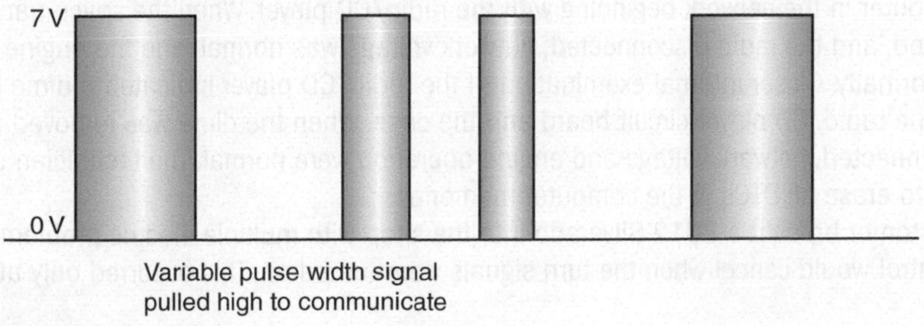

Figure 11-32 Class 2 network voltage signal.

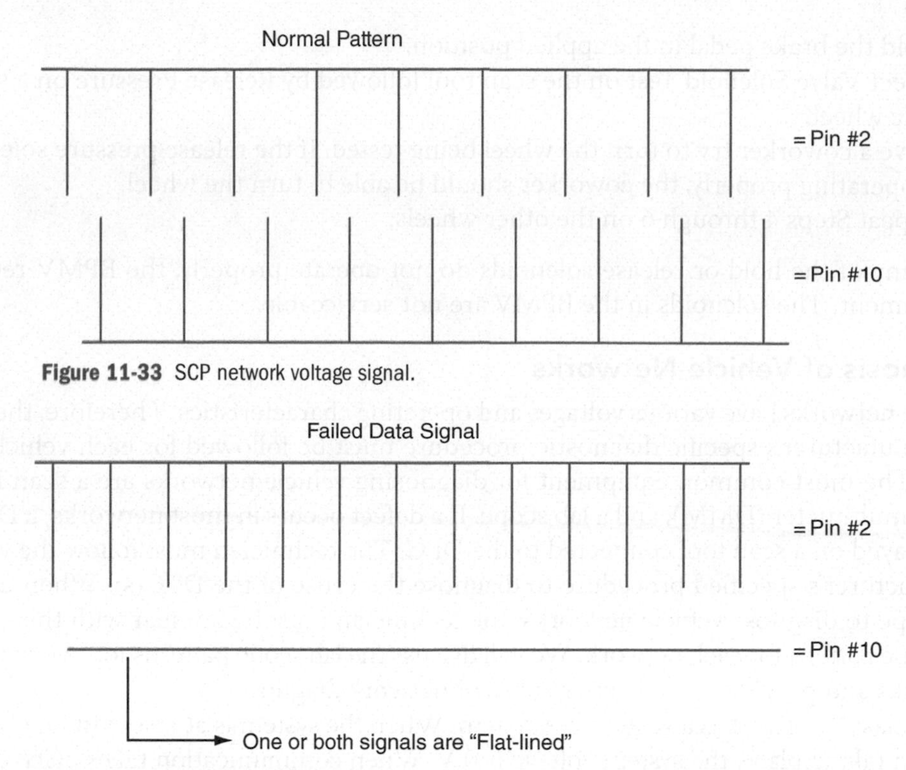

Figure 11-33 SCP network voltage signal.

Figure 11-34 SCP network defective voltage signal.

no communication are 5 V on Bus+, and 0 V–0.2 V on Bus–. A DMM may be used to read these voltages. To read the SCP voltages on a lab scope, connect the Channel 1 lead on the lab scope to pin 2 on the DLC, and connect the Channel 2 lead on the lab scope to pin 10 on the DLC. Connect the ground lead to pin 4 on the DLC or a good chassis ground. When the ignition switch is turned on, the normal SCP pattern shown in **Figure 11-33** should be displayed. If one network signal is displayed as a flat line, communication has failed between some of the computers and part of the network (**Figure 11-34**). When both network signals are displayed as a flat line, no communication is taking place through the network. Some networks remain functional with the ignition switch off, because certain module(s) require information under this condition. Always consult the vehicle manufacturer's specific diagnostic information when diagnosing networks.

CASE STUDY

A 2010 GM 1500 series truck was towed into the shop with a no-start condition. During a preliminary diagnosis, the technician discovered there was 12 V on the Class 2 vehicle network with the ignition switched on. The technician used a parts locator manual to locate the splice pack behind the radio. When the splice pack was disconnected, the engine started and ran normally. This proved that the PCM and TBC were not causing the problem, because disconnecting the splice pack isolated these computers from the network. Therefore, the defect must be in the Class 2 network or computers connected in the network. The technician decided to disconnect each computer in the network beginning with the radio/CD player. When the splice pack was reconnected, and the radio disconnected, network voltage was normal, and the engine started and ran normally. Closer internal examination of the radio/CD player indicated a dime lodged between the radio/CD player circuit board and the case. When the dime was removed and the radio reconnected, network voltage and engine operation were normal. The technician used a scan tool to erase all DTCs in the computer memories.

A customer brought a 2012 Silverado into the shop with multiple electric problems. The cruise control would cancel when the turn signals were turned on. This occurred only at night when the headlights were on. The customer said several instrument panel readings were randomly intermittent. When a scan tool was connected to the DLC, a U1041 was displayed indicating loss of electronic brake control module (EBCM) data on the network. The technician checked for service bulletins related to this problem and discovered this problem was detailed in a service bulletin. The bulletin indicated this problem was caused by high resistance in ground connection G110 on the vehicle frame below the driver's door. The ground connection was cleaned and tightened and the DTC erased, but the DTC reset again in a short time.

The technician considered the possibility of a defective EBCM. Prior to EBCM replacement, the technician checked the EBCM voltage supply and ground. The EBCM voltage supply was 12 V. When a pair of voltmeter leads was connected from the EBCM ground terminal to the battery ground, the voltmeter indicated 3V. The technician inspected all the wiring from the EBCM module to the battery, and discovered that the battery ground cable was connected to the radiator support rather than being connected to the specified location on the left front corner of the vehicle frame. The battery ground cable and the vehicle frame attaching location were thoroughly cleaned, and the ground cable was properly tightened. Now the voltage reading from the EBCM ground terminal to the battery ground was 0.2 V. All DTCs were erased with a scan tool, and after driving the vehicle on a road test the DTCs did not reset. All electronic systems operated normally during the road test.

CASE STUDY

A customer complained about the SERVICE RIDE CONTROL light being illuminated on his 2009 Cadillac XLR. When the technician visually inspected the ESC system, no defects were evident. During a diagnostic system check, the scan tool displayed DTC C0577. The detailed DTC explanation in the vehicle manufacturer's service manual indicated this DTC represented a short to ground in the L/F shock absorber damper solenoid circuit.

When the technician disconnected the wiring connector from the L/F damper solenoid and measured the solenoid resistance with an ohmmeter, he discovered the solenoid had the specified resistance of 2 ohms.

The technician disconnected the wiring connectors from the ESC module and identified the L/F damper solenoid wires connected to the module. When the technician connected a pair of ohmmeter leads across the wires from the ESC module terminals to the L/F damper solenoid terminals, each wire had very low resistance. Next, the technician connected the ohmmeter leads from each terminal in the L/F damper solenoid connector to the ground. When connected to one of the damper solenoid terminals, the ohmmeter displayed an infinite reading indicating the wire was not grounded. However, when the ohmmeter leads were connected from the other damper solenoid terminal to the ground, the ohmmeter indicated a very low reading indicating contact between this wire and the chassis.

A closer examination of the L/F damper solenoid wires indicated these wires were jammed against the chassis about 2 ft. from the damper solenoid. The insulation was worn on the wires, and one wire was contacting the chassis. The technician repaired the wiring insulation and repositioned the harness so the wires were not jammed against the chassis. The technician used the scan tool to clear the DTC from the ESC module, and road-tested the car to be sure the DTC did not reset and illuminate the SERVICE RIDE CONTROL light.

DRIVER ASSIST SYSTEMS

The driver assist systems include, but are not limited to, adaptive cruise control, lane departure, park assist systems, dynamic steering systems and stability control systems. Many of these systems require distance sensor calibration after:

Front or rear bumper cover replacement
Front or rear bumper support replacement

Significant structural collision repair
Distance sensor replacement
Rear suspension repair
Steering column repair or replacement
Replacement of steering position or steering torque sensor angle sensor

Note: The signal transmitted or received by the distance sensors can be affected by coverage of dirt, snow, and ice. Additionally, improper application of paint covering the distance sensor during a front or rear bumper repair could affect sensor operation.

All driver assist systems rely on a properly set base vehicle wheel alignment and special tools and calibration targets may be required for some sensor calibration procedures. All driver assist modules are calibrated to the thrust angle of the vehicle. After the vehicle has received a proper four-wheel alignment, a post-alignment calibration of sensors can be performed. The alignment software will guide you through the process of recalibration and instruct you to attach an interface connector to the vehicle 16-pin OBD-II DLC. On Hunter alignment equipment, this interface is called Codelink (**Figure 11-35**). To calibrate the steering angle sensor on the alignment terminal, select "Begin Dynamic Steering Calibration" for calibration of the steering position and torque sensor angle sensor

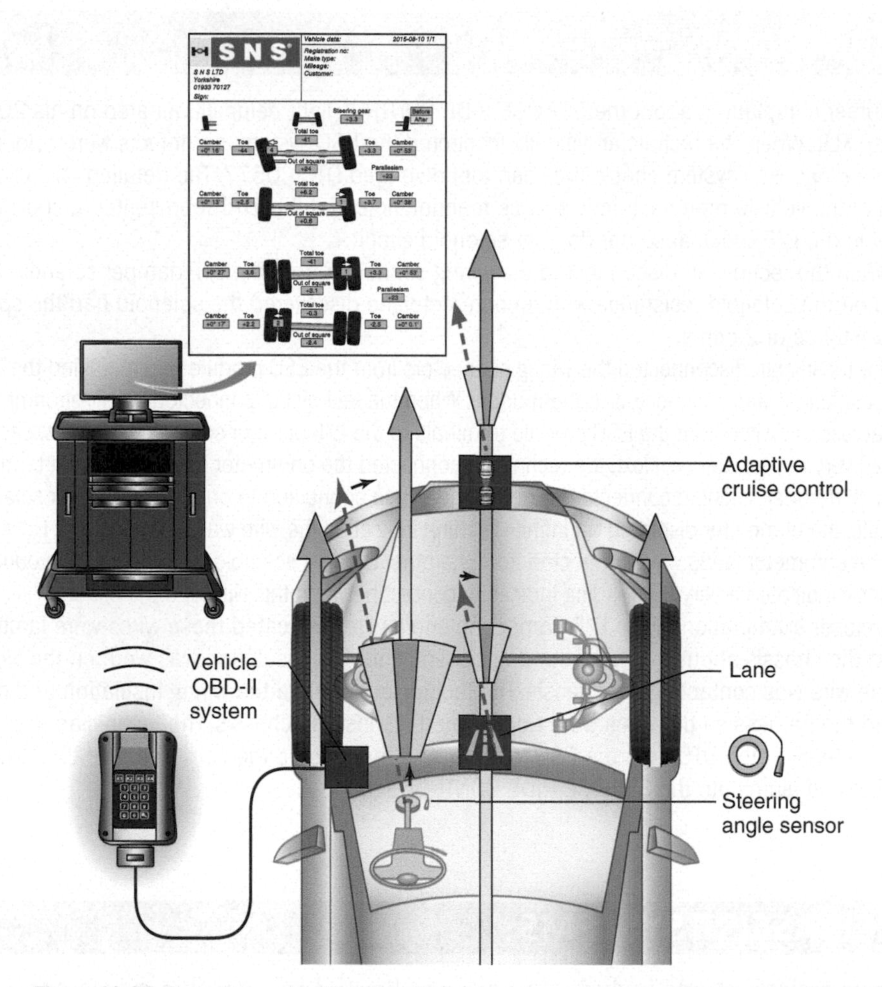

Figure 11-35 Post-alignment calibration of driver assistance system sensors is performed with the guidance of the alignment equipment OBD-II interface. On Hunter alignment equipment, this interface adapter is called Codelink.

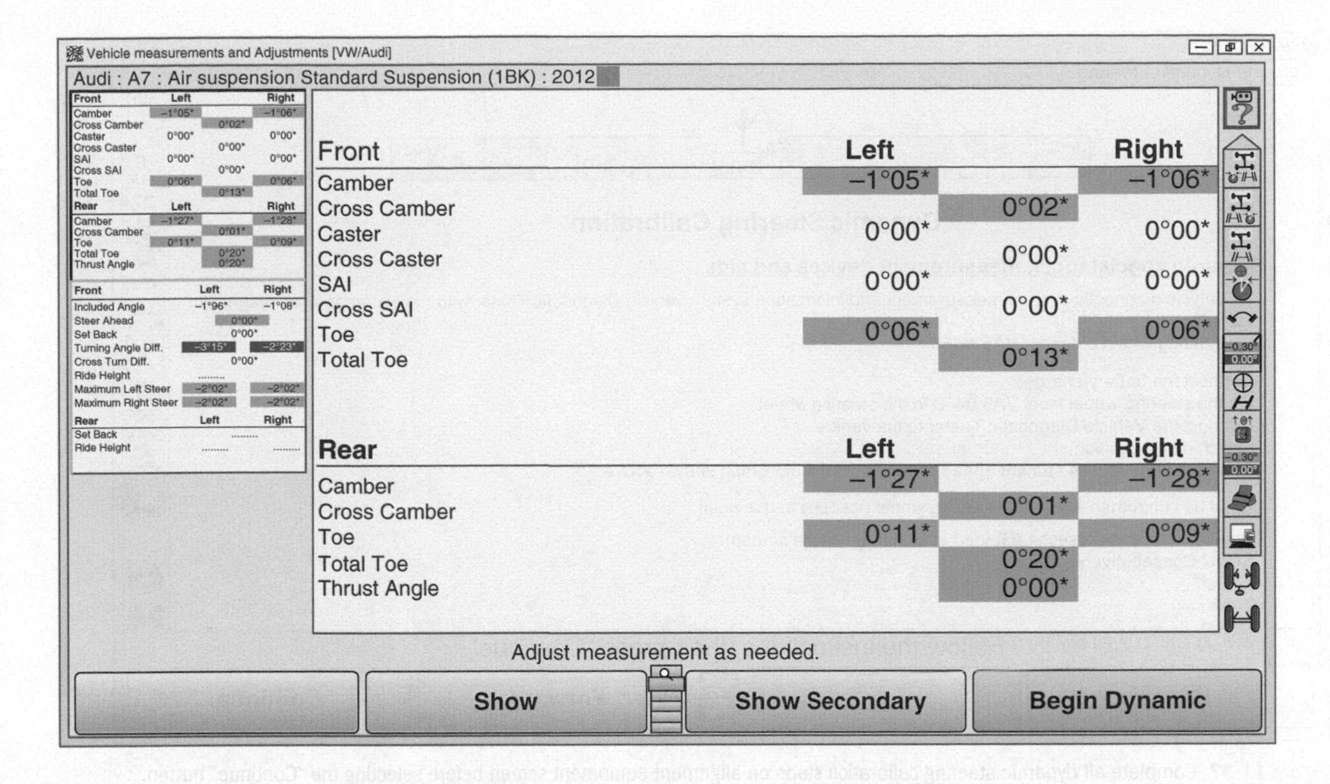

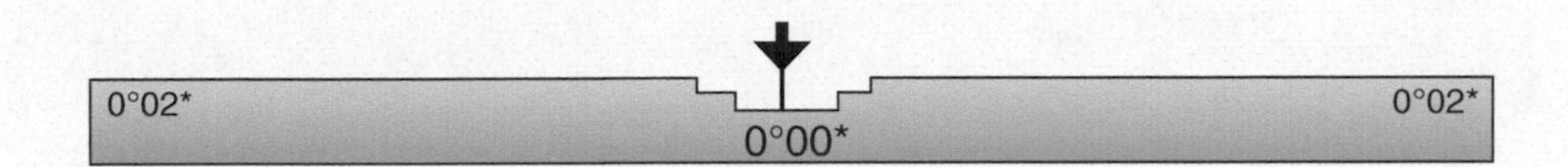

Dynamic Steering Calibration

Figure 11-36 Select "Begin Dynamic Steering Calibration" on alignment equipment screen.

(**Figure 11-36**). A bar graph will appear on the screen showing steer ahead position. Instructions will be displayed on the screen to walk you through the relearn steps. Follow the steps outlined. Do Not Press "Continue" until the calibration process is finished (**Figure 11-37**). The exact steps for driver assist system and sensor calibration will vary depending on the year, make, and model of the vehicle. The alignment software program will guide you through the procedure and identify any special equipment or tools required. Due to the high cost of some of the specialty equipment and targeting systems some calibrations may only be available at the manufacturer's dealer service department.

On a vehicle equipped with adaptive cruise control, the alignment software will provide an option to enter the calibration process (**Figure 11-38**) at the end of the alignment procedure for the radar sensor. Specialized targeting equipment is required to be placed in front of vehicle and used with alignment equipment (**Figure 11-39**). The adaptive cruise control (ACC) radar sensors allow for both horizontal and vertical adjustment compensation (**Figure 11-40**).

As an example, on a Volkswagen with either a lane departure warning system or night vision, a VAS 6430 targeting system (**Figure 11-41**) is required to be used with the alignment program to calibrate the camera system.

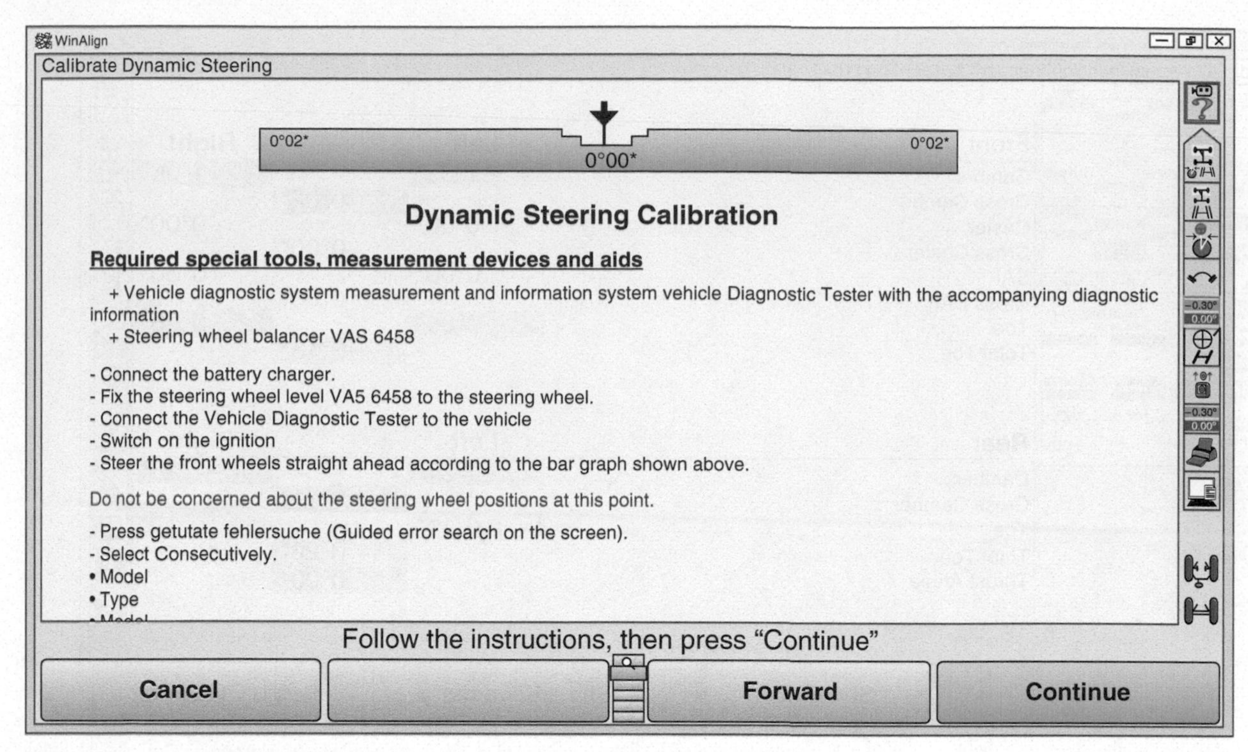

Figure 11-37 Complete all dynamic steering calibration steps on alignment equipment screen before selecting the "Continue" button.

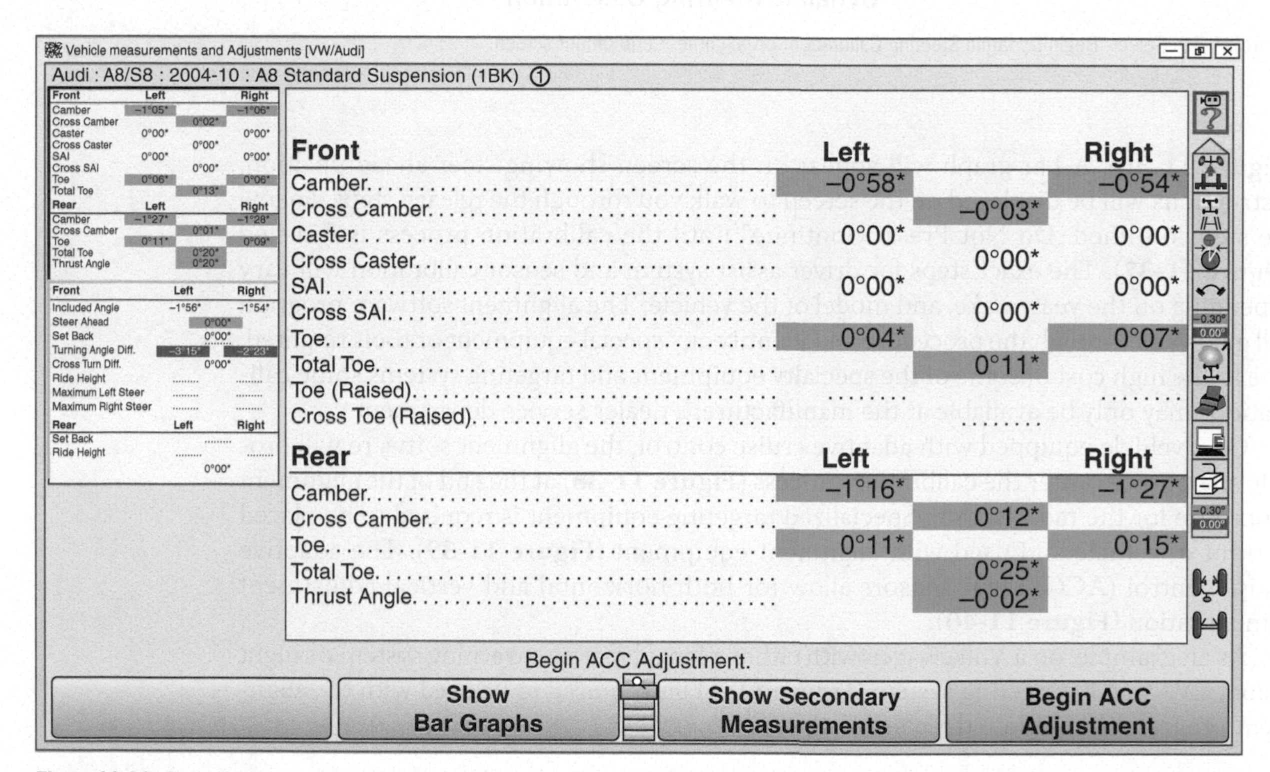

Figure 11-38 Post-alignment calibration of adaptive cruise control (ACC) is performed by following the information on the alignment equipment screen.

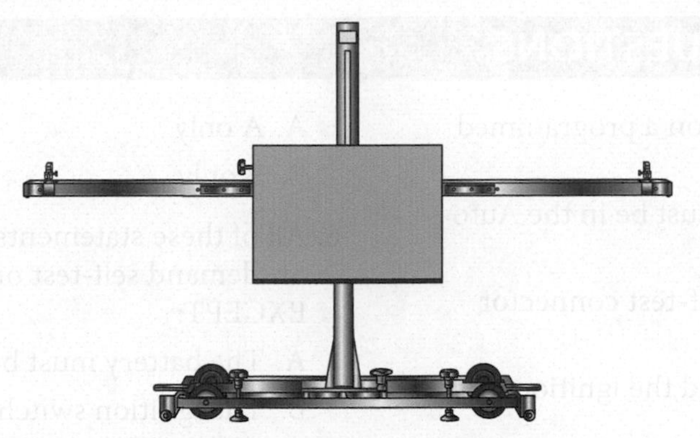

Figure 11-39 Post-alignment calibration of adaptive cruise control (ACC) radar sensor requires the use of specialized targeting equipment to be used in conjunction with alignment program.

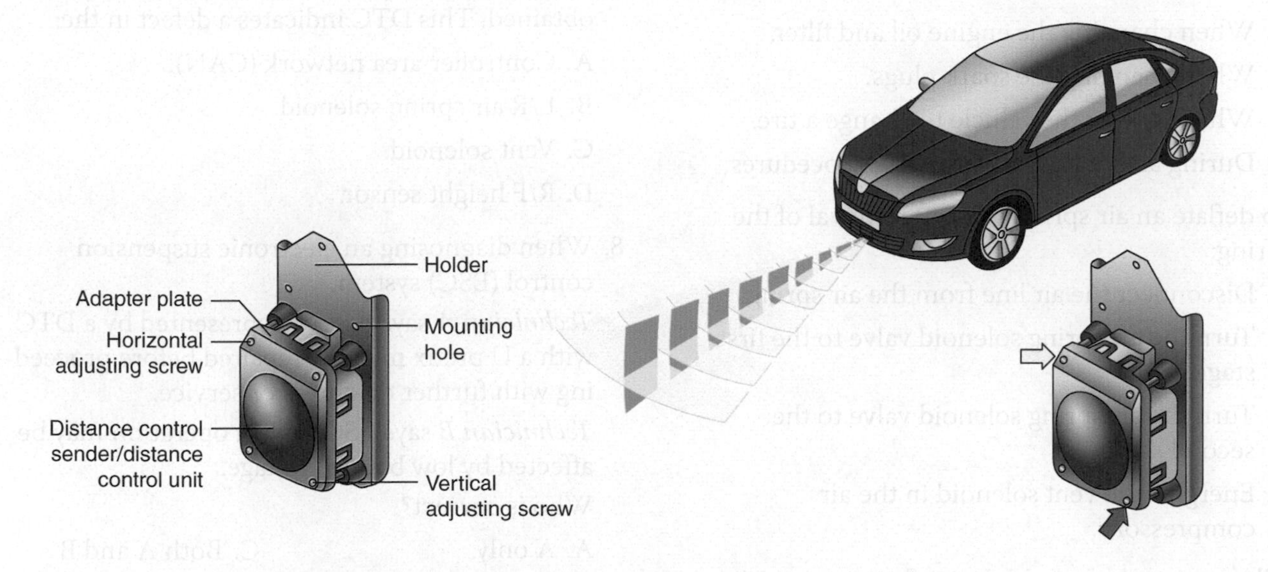

Figure 11-40 Adaptive cruise control (ACC) radar sensor has both horizontal and vertical adjustment compensation.

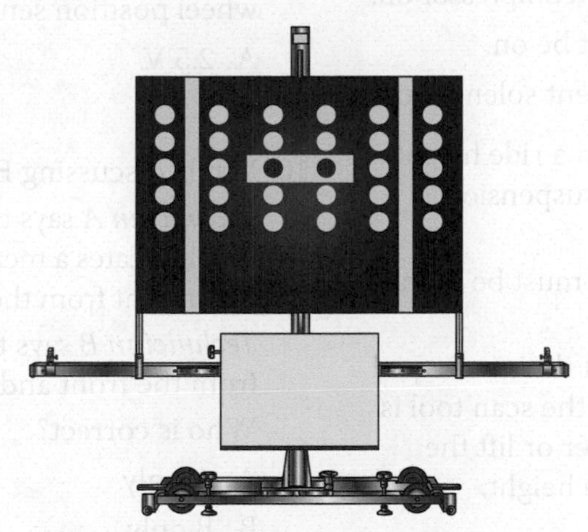

Figure 11-41 The Volkswagen VAS 6430 targeting system is required to be used with the alignment program to calibrate the camera system for the lane departure warning system and the night vision camera.

ASE-STYLE REVIEW QUESTIONS

1. When performing a self-test on a programmed ride control system:
   A. The mode select switch must be in the Auto position.
   B. One of the wires in the self-test connector must be grounded.
   C. The engine must be off and the ignition switch turned on.
   D. The headlights must be on during the self-test.
2. When servicing a vehicle with an air suspension system, the air suspension switch must be turned off:
   A. When changing the engine oil and filter.
   B. When changing the spark plugs.
   C. When jacking the vehicle to change a tire.
   D. During any of the above service procedures.
3. To deflate an air spring prior to removal of the spring:
   A. Disconnect the air line from the air spring.
   B. Turn the air spring solenoid valve to the first stage.
   C. Turn the air spring solenoid valve to the second stage.
   D. Energize the vent solenoid in the air compressor.
4. When using the scan tool to inflate an air spring:
   A. The vehicle must be raised on a tire-contact lift.
   B. The scan tool commands the air compressor on.
   C. The air suspension switch must be on.
   D. The scan tool commands the vent solenoid on.
5. When using a scan tool to perform a ride height adjustment on a vehicle dynamic suspension (VDS),
   Technician A says the VDS switch must be in the Off position.
   Technician B says if the ride height does not equal the manufacturer's specifications, the scan tool is used to command the VDS to lower or lift the vehicle to obtain the specified ride height.
   Who is correct?
   A. A only
   B. B only
   C. Both A and B
   D. Neither A nor B
6. All of these statements about performing an on-demand self-test on a VDS system are true EXCEPT:
   A. The battery must be fully charged.
   B. The ignition switch must be on.
   C. The vehicle must be raised on a lift.
   D. The 4L mode must not be selected on four-wheel-drive vehicles.
7. When diagnosing a VDS system, a U1900 DTC is obtained. This DTC indicates a defect in the:
   A. Controller area network (CAN).
   B. L/R air spring solenoid.
   C. Vent solenoid.
   D. R/F height sensor.
8. When diagnosing an electronic suspension control (ESC) system,
   Technician A says defects represented by a DTC with a U prefix must be repaired before proceeding with further diagnosis or service.
   Technician B says ESC system operation may be affected by low battery voltage.
   Who is correct?
   A. A only
   B. B only
   C. Both A and B
   D. Neither A nor B
9. On an ESC system a normal voltage signal from a wheel position sensor is:
   A. 2.5 V.
   B. 5.5 V.
   C. 6.1 V.
   D. 6.8 V.
10. While discussing ESC system diagnosis,
    Technician A says the normal force signal on a scan tool indicates a measured level of road surface condition sent from the ESC module to the EBCM.
    Technician B says the normal force data is sent from the front and rear wheels.
    Who is correct?
    A. A only
    B. B only
    C. Both A and B
    D. Neither A nor B

ASE CHALLENGE QUESTIONS

1. *Technician A* says during air spring inflation the vehicle weight must be applied to the suspension system.
   Technician B says during air spring inflation the vehicle should be positioned on a lift so the wheels are dropped downward.
   Who is correct?
   A. A only
   B. B only
   C. Both A and B
   D. Neither A nor B
2. All of these statements about a programmed ride control system are true EXCEPT:
   A. An air spring is mounted at each corner of the vehicle.
   B. An electric actuator is located in each strut.
   C. A mode indicator light is positioned in the tachometer.
   D. The PRC module provides a firm ride during severe braking.
3. *Technician A* says the suspension switch must be turned off before raising any corner of a car with an electronic air suspension.
   Technician B says the ignition switch must not be turned on while any corner of a car with electronic air suspension is raised.
   Who is correct?
   A. A only
   B. B only
   C. Both A and B
   D. Neither A nor B
4. A vehicle with an electronic air suspension system with mechanical trim height adjustment requires front and rear trim height adjustment.
   Technician A says to adjust the front trim height, rotate the threaded mounting bolt in the upper end of the height sensor.
   Technician B says to adjust the rear trim height, loosen the attaching bolt(s) on the upper height sensor bracket and move the bracket upward or downward.
   Who is correct?
   A. A only
   B. B only
   C. Both A and B
   D. Neither A nor B
5. When removing and replacing an air spring and shock absorber assembly on a VDS:
   A. The self-locking nuts on the upper strut mount can be reused.
   B. Retainer tabs on the lower end of the spring must be depressed to separate the spring and the shock absorber.
   C. The spring must be vented by loosening the spring solenoid valve to the first stage.
   D. The VDS switch and the ignition switch must be in the On position.

Name ______________________________ Date ______________

JOB SHEET 53

ELECTRONIC SUSPENSION CONTROL SUSPENSION SYSTEM

Upon completion you should be able to describe the function of suspension and steering control systems and components (i.e., active suspension and stability control). Determine the automatic suspension system type and design.

ASE Education Foundation Task Correlation

The following job sheet addresses the following **AST/MAST** task:

D.3. Describe the function of suspension and steering control systems and components, (i.e., active suspension and stability control). **(P-3)**

We Support ASE Education Foundation

Tools and Materials

Modern vehicle with a vehicle dynamic suspension (VDS) system.

Scan tool

Describe the vehicle being worked on:

Year ______________ Make ______________ Model ______________

VIN ______________ Engine type and size ______________

Procedure

1. Determine the type of automatic ride control system.

__________ Automatic air ride control
__________ Rear electronic level control system
__________ Magneto-rheological fluid ride control

2. Is the vehicle also equipped with any of the following systems? Check all that apply:

☐ Stability control system
☐ Antilock brake system
☐ Traction control system
☐ Lane departure warning
☐ Park assist system
☐ Adaptive cruise control
☐ Collision-mitigation systems

3. Describe the function of suspension and steering control systems and components (i.e., active suspension and stability control).

____

____

____

4. Where are the ESC system height sensors located and what type of sensors are they?

____

____

____

5. What are the advantages of an ESC system with magneto-rheological fluid in the shock absorbers compared to other computer-controlled suspension systems?

____

____

____

6. In the space provided below list all Input and Output devices for the electronic suspension control (ESC) system:

Inputs	Outputs
______	______
______	______
______	______
______	______
______	______
______	______
______	______
______	______
______	______

7. What components are shared by the active suspension and stability control system? Include sensors, control modules, and output devices.

____

____

____

8. What diagnostic trouble code would be set if the left front suspension height sensor developed and open circuit (signal circuit open)?

Code # ______________________________

9. Describe the diagnostic trouble code retrieval procedure. (Look in the service manual or electronic service information.)

____

____

10. Describe the diagnostic trouble code clearing procedure. (Look in the service manual or electronic service information.)

____

____

Instructor's Response

____

____

____

Name ______________________ Date ____________

JOB SHEET 54

DIAGNOSTIC PROCEDURES AND STEPS TO FOLLOW

Upon completion of this job sheet you should be able to follow the four principal steps to every repair procedure. Any repair procedure requires a systematic approach to diagnose and repair a vehicle.

ASE Education Foundation Task Correlation:

This job sheet addresses the following **MLR/AST/MAST** task:

A.1. Research vehicle service information including fluid type, vehicle service history, service precautions, and technical service bulletins. **(P-1)**

This job sheet addresses the following **MLR** task:

A.3. Identify suspension and steering system components and configurations. **(P-1)**

This job sheet addresses the following **AST** task:

A.2. Identify and interpret suspension and steering system concerns; determine needed action. **(P-2)**

This job sheet addresses the following **MAST** task:

A.2. Identify and interpret suspension and steering system concerns; determine needed action. **(P-1)**

We Support ASE | Education Foundation

Tools and Materials

Late-model vehicle

Service manual or information system

Safety glasses or goggles

Hand tools, as required

Describe the vehicle being worked on:

Year ____________ Make ____________ Model ____________

VIN ____________ Engine type and size ____________

Procedure

Always follow the procedures outlined in the service manual. Give a brief description of your procedure following each step. Ensure that the engine is cold and wear eye protection before you begin.

The four principal steps to every repair procedure should always be followed.

1. Verify the customer concern (complainant) and the involved system:

a. What is the original customer concern with the vehicle?

____

____

b. Road-test the vehicle. Was the customer's concern present and verifiable?

____

____

2. Isolate the problem:

a. To which system is the concern related to?

____

____

b. Are there any *Technical Service Bulletins* (TSBs) that may be related to the problem area?

____

____

c. Are there manufacturer published diagnostic procedures available?

____

____

3. Repair the failure area (component or circuit):

a. *Erase any stored DTC* after all information has been recorded. Remember, freeze frame data will be erased and major monitors will be reset.

4. Recheck (verify) repair performed:

a. Road-test the vehicle *under the same conditions* in which the original complaint was detected to verify that repair has been successfully completed.

b. *Perform any required drive cycle procedures,* which may be necessary for OBD-II System Monitor to run to completion. Failure to perform proper drive cycle could cause MIL to be illuminated after the customer picks up the vehicle and system monitor detects the same or another system failure.

b. *Was the customer's concern corrected?* ______________________

____

Instructor's Response

____

____

____

Name ______________________ **Date** ____________

JOB SHEET
55

INSPECTION AND PRELIMINARY DIAGNOSIS OF COMPUTER-CONTROLLED SUSPENSION SYSTEM

Upon completion of this job sheet, you should be able to inspect computer-controlled suspension systems.

ASE Education Foundation Task Correlation

This job sheet addresses the following **MLR** task:

B.9. Inspect upper and lower control arms, bushings, and shafts. **(P-1)**

This job sheet addresses the following **MAST** task:

B.18. Inspect, test, and diagnose electrically assisted power steering systems (including using a scan tool); determine needed action. **(P-2)**

This job sheet addresses the following **AST/MAST** tasks:

C.1. Diagnose short-and-long arm suspension system noises, body sway, and uneven ride height concerns; determine needed action. **(P-1)**

C.2. Diagnose strut suspension system noises, body sway, and uneven ride height concerns; determine needed action. **(P-1)**

We Support
ASE | Education Foundation

Tools and Materials

Modern vehicle with electronically controlled suspension system.

Describe the vehicle being worked on:

Year ____________ Make ____________ Model ____________

VIN ________________ Engine type and size ________________

Procedure

1. Discuss the vehicle complaint with the customer.

   Describe the customer complaint(s). ____________________

   ____

   ____

2. Road-test the vehicle and verify the vehicle complaint.

   Describe the vehicle complaint experienced during the road test.

   ____

   ____

   ____

3. Inspect the vehicle for collision damage or other damage that could affect the computer-controlled suspension system.

   Describe any collision or other vehicle damage that could affect the computer-controlled suspension system ____________________

   ____

   ____

7. Use a tape measure to calculate the ride height at the locations specified by the vehicle manufacturer.

   Specified ride height, Front ______________ Rear ______________

   Actual ride height, L/F ______________ R/F ______________

   L/R ______________ R/R ______________

8. If the ride height does not equal the manufacturer's specifications, open the left front door, and connect the scan tool to the DLC.

   Scan tool properly connected to the DLC ☐ Yes ☐ No

9. Select the correct vehicle year, model, and engine type on the scan tool, and then select *Vehicle Dynamic Module* (VDM).

   Scan tool selections, Vehicle year ______________

   Model ______________

   Engine Size ______________

   Vehicle Dynamic Module (VDM) selected on scan tool ☐ Yes ☐ No

10. Use the active command mode in the scan tool to vent or lift the vehicle to achieve the specified trim height.

    Active command mode on scan tool used to lift or vent VDS system to specified trim height ☐ Yes ☐ No

11. When a tape measure indicates the specified front and rear ride height, select *Save Calibration Values* on the scan tool to calibrate the VDM to the specified trim height.

    Actual ride height L/F ______________

    R/F ______________

    L/R ______________

    R/R ______________

Instructor's Response

____

____

____

Name ______________________________ **Date** ____________

JOB SHEET
57

PERFORM AN ON-DEMAND SELF-TEST ON A VEHICLE DYNAMIC SUSPENSION (VDS) SYSTEM

Upon completion of this job sheet, you should be able to perform an on-demand self-test on a vehicle dynamic suspension (VDS) system.

ASE Education Foundation Task Correlation

This job sheet addresses the following **MAST** task:

B.18. Inspect, test, and diagnose electrically-assisted power steering systems (including using a scan tool); determine needed action. **(P-2)**

This job sheet addresses the following **AST** task:

B.19. Inspect electric power steering assist system. **(P-3)**

We Support
ASE | Education Foundation

Tools and Materials

Modern vehicle with a vehicle dynamic suspension (VDS) system
Scan tool

Describe the vehicle being worked on:

Year ____________ Make ____________ Model ____________

VIN ____________ Engine type and size ____________

Procedure

1. Be sure the transmission is in Park.

Transmission in Park ☐ Yes ☐ No

2. Be sure the battery is fully charged.

Battery fully charged ☐ Yes ☐ No

3. Connect the scan tool to the DLC.

Scan tool properly connected to the DLC ☐ Yes ☐ No

4. Lower the driver's window and close all doors, liftgate, and liftgate glass.

Driver's window down ☐ Yes ☐ No

All doors, liftgate, and liftglass closed ☐ Yes ☐ No

5. Be sure the 4L mode is not selected on four-wheel-drive models.

4L switch Off ☐ Yes ☐ No

6. Turn on the ignition switch.

Ignition switch On ☐ Yes ☐ No

7. Select *Vehicle Dynamic Module* and *On-Demand Self-Test* on the scan tool.

Vehicle Dynamic Module selected on scan tool ☐ Yes ☐ No

On-Demand Self-Test selected on scan tool ☐ Yes ☐ No

____

8. Record all the DTCs displayed on the scan tool.

____

____

____

9. Use the vehicle manufacturer's service information to interpret the meaning of each DTC, and list the meaning of each DTC in the following spaces.

____

____

____

10. Turn off the ignition switch, and disconnect the scan tool.

Ignition switch Off ☐ Yes ☐ No

Scan tool disconnected ☐ Yes ☐ No

Instructor's Response

____

____

____

CHAPTER 12
POWER STEERING PUMP DIAGNOSIS AND SERVICE

Upon completion and review of this chapter, you should be able to:

- Check power steering belt condition and adjust belt tension.
- Diagnose power steering belt problems.
- Check power steering fluid level and add fluid as required.
- Drain and flush power steering system.
- Bleed air from power steering system.
- Perform power steering pump pressure test.
- Check power steering pump fluid leaks.
- Remove and replace power steering pumps, and inspect pump mounts.
- Remove and replace power steering pump pulleys.
- Remove and replace power steering pump integral reservoirs.
- Remove, replace, and check flow control valve and pressure relief valve.
- Remove, replace, and check power steering pump rotating components.
- Remove and replace power steering pump seals and O-rings.
- Check, remove, and replace power steering lines.
- Inspect, test, diagnose, and service hybrid electric vehicles (HEVs) and electrohydraulic power steering (EHPS) systems.

Basic Tools

Basic technician's tool set
Service manual
Pry bar
Floor jack
Safety stands
Oil drain pan
Crocus cloth

Terms To Know

Belt tension gauge
Crocus cloth
Flow control valve
Integral reservoir
Multi-ribbed V-belt
Pressure gauge
Pressure relief valve
Remote reservoir
Serpentine belt
Steering effort
Woodruff key

Power steering pump diagnosis and service is very important to maintain vehicle safety and driver convenience. If a badly worn power steering pump belt is undetected, the belt may suddenly break while the driver is completing a turn. Under this condition, steering effort is greatly increased, and this may result in a collision. Therefore, the power steering belt should be inspected at regular intervals. A loose power steering belt may cause hard steering intermittently, and this condition requires increased driver steering effort, which reduces driver comfort and convenience. Therefore, one of the first checks when diagnosing a power steering system is to inspect the belt condition and test the belt tension.

POWER STEERING PUMP BELT SERVICE

Checking Belt Condition and Tension

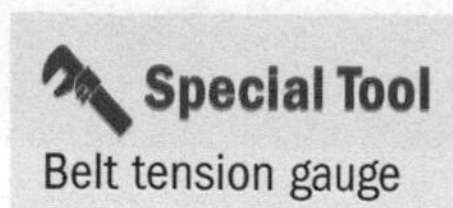

Belt tension gauge

Classroom Manual
Chapter 12, page 351

Do not pry on the pump reservoir with a pry bar. This action may damage the reservoir.

Power steering belt condition and tension are extremely important for satisfactory power steering pump operation. A loose belt causes low pump pressure and hard steering. The steering wheel may jerk and surge during a turn if the power steering pump belt is loose. A loose, dry, or worn belt may cause squealing and chirping noises, especially during engine acceleration and cornering.

The power steering pump belt should be checked for tension, cracks, oil-soaking, worn or glazed edges, tears, and splits (**Figure 12-1**). If any of these conditions are present, belt replacement is necessary.

> **SERVICE TIP** During rainy or wet conditions, a loose power steering belt, which causes low power steering pump pressure and increased steering effort, may be more noticeable because the belt becomes wet and slips more easily.

Most power steering pumps have a **multi-ribbed V-belt** (or **serpentine belt**). Many multi-ribbed V-belts have an automatic tensioning pulley; therefore, a tension adjustment is not required. The multi-ribbed V-belt should be checked to make sure it is installed properly on each pulley in the belt drive system (**Figure 12-2**). The tension on a multi-ribbed V-belt may be checked with a belt tension gauge. Prior to removal of the serpentine belt locate the routing diagram similar to the one shown in **Figure 12-3** (pulley alignment). This diagram is often located under the hood in the engine bay, often on the radiator support cover. The diagram may also be located in the vehicle service information. Some technicians choose to draw a sketch of the belt routing prior to removal.

Many multi-ribbed V-belts have a spring-loaded tensioner pulley that automatically maintains belt tension. As the belt wears or stretches, the spring moves the tensioner pulley to maintain the belt tension. Some of these tensioners have a belt length scale that indicates the new belt range and used belt range (**Figure 12-4**). If the indicator on the tensioner is out of the used belt length range, belt replacement is required. Many belt tensioners have a 0.5-in. drive opening in which a ratchet or flex handle may be installed to move the tensioner pulley off the belt during belt replacement (**Figure 12-5**).

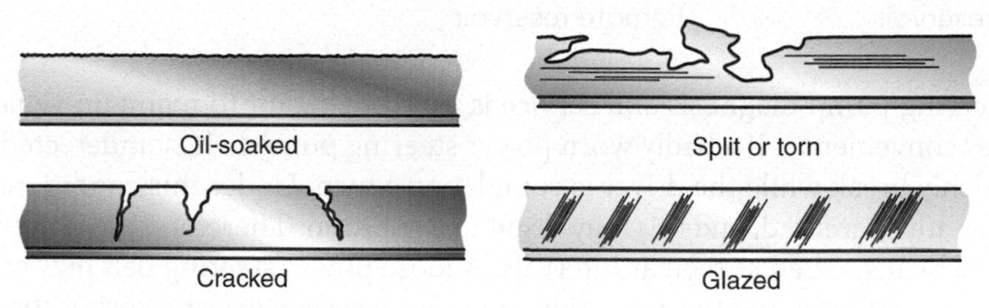

Figure 12-1 Defective belt conditions.

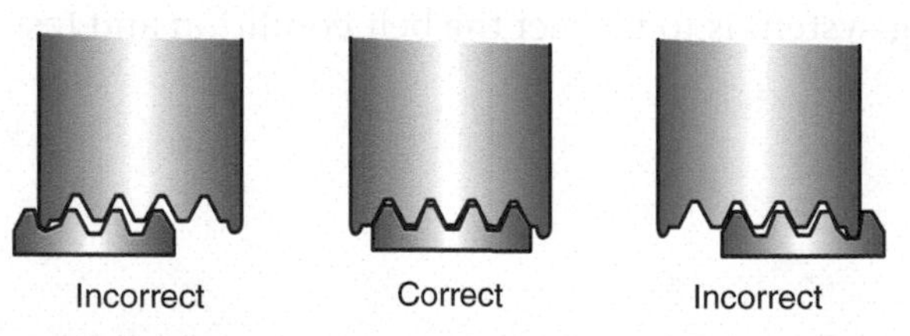

Figure 12-2 Proper and improper installation of multi-ribbed V-belt on a pulley.

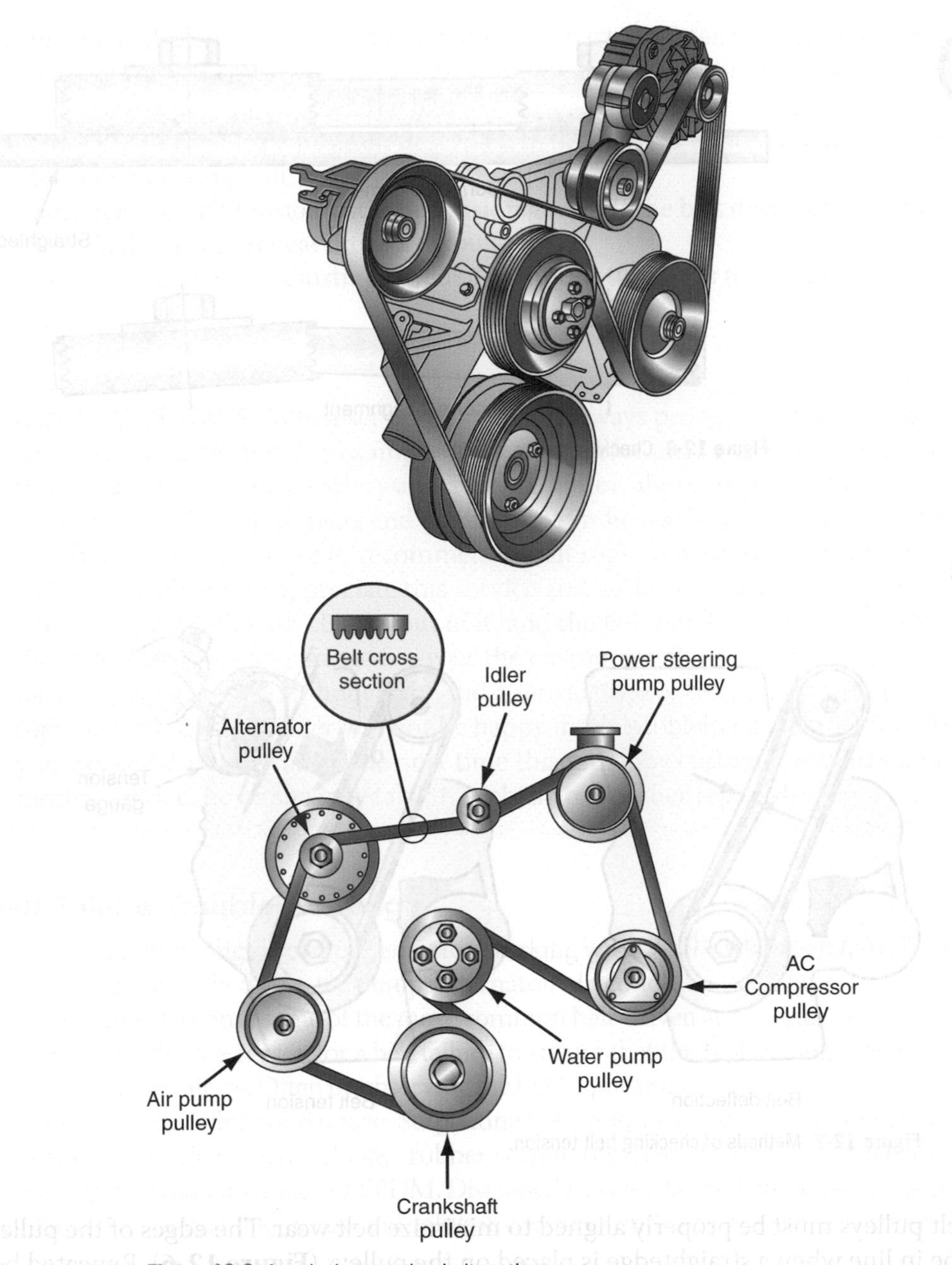

Figure 12-3 A typical serpentine belt routing.

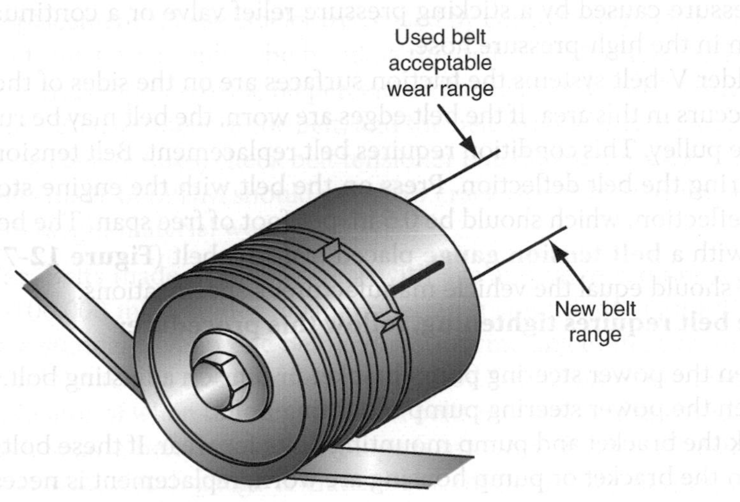

Figure 12-4 Belt tension scale.

Figure 12-8 Multi-ribbed belt cracks are caused by continuous flexing and bending of the material in high heat conditions, and are the most common reason for belt replacement.

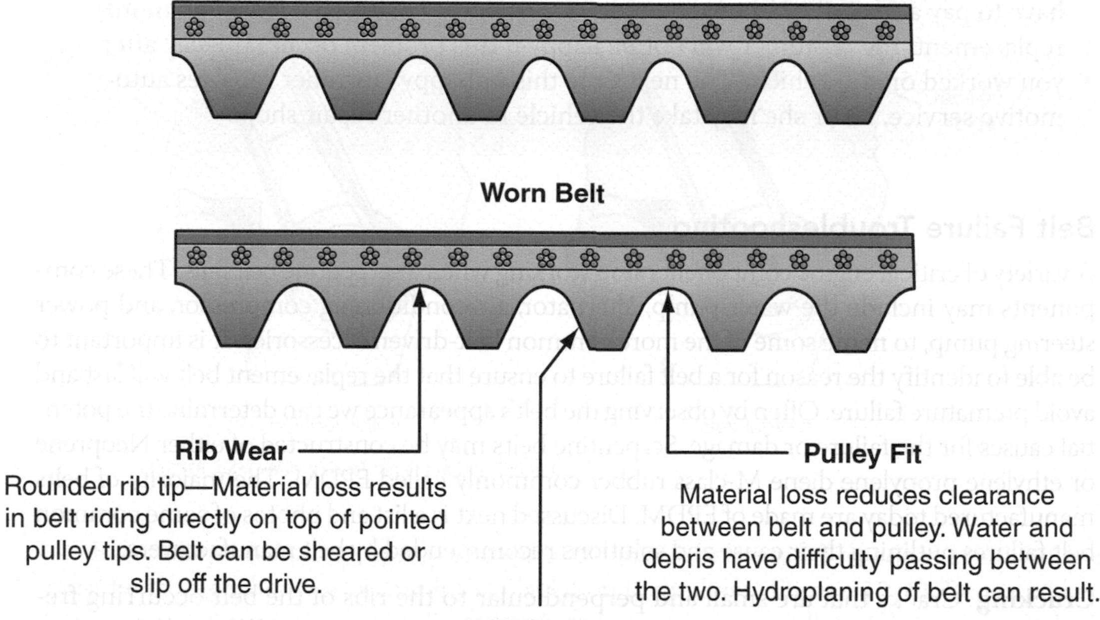
New Belt
Worn Belt
Rib Wear
Rounded rib tip—Material loss results in belt riding directly on top of pointed pulley tips. Belt can be sheared or slip off the drive.
Pulley Fit
Material loss reduces clearance between belt and pulley. Water and debris have difficulty passing between the two. Hydroplaning of belt can result.
Belt Seating
Material loss results in belt seating further down in pulley. This reduces wedging force necessary to transmit power.

Figure 12-9 Belts made of EPDM are designed to last 100,000 miles, but over time the material on the sides and valleys will wear. A 10 percent loss of material is enough to cause performance issues.

beltwear). As little as 5 to 10 percent wear can effect operation if belt wear indicates replacement is required.

Abrasion Abrasion may cause a shiny or glazed appearance on one or both sides of the belt. In severe cases cord fibers may be visible (**Figure 12-11**). It is caused by belt making contact with an object, such as a flange bolt or a foreign object. It may also be caused by improper belt tension, pulley surface wear, or a binding pulley bearing. Inspect belt tensioner for proper operation, replace if necessary.

Figure 12-10 Gates produces a simple belt wear gauge that can be used to detect EPDM belt grove wear.

Figure 12-11 Belt abrasion is due to a belt making contact with an object such as a flange bolt or a foreign object.

Improper Installation A belt rib on either edge begins separating from main body plies (**Figure 12-12**). If left unnoticed the outer covering may separate causing the belt to unravel. It is caused by improper replacement procedure when one of the belt ribs is placed outside of one of the pulley grooves during installation. Verify that the belt has the same number of ribs as its companion pulley grooves. Once damage has occurred the belt must be replaced. Ensure that all belt ribs fit properly in all pulley groves when replacing.

Pilling Pilling occurs as belt material is sheared off from the ribs building up in the grooves (**Figure 12-13**). It is generally caused by insufficient belt tension, pulley misalignment, or wear. It is more common on diesel engines, but can happen on any powertrain. Pilling can lead to belt noise and vibration; pulley grooves should be inspected for buildup and alignment should be checked to see if belt needs to be replaced.

It is wise to replace the drive belt along as well as the automatic belt tensioner when any pulley-driven component is replaced. By replacing only one system component vibration is increased dramatically (**Figure 12-14**). This vibration may be both felt and heard, leading to customer complaints, and the performance and longevity of the new component may be compromised. The green line on the graph indicates vibration minimization with a new belt, tensioner, and component. The red graph line indicates vibration caused by an aging drive belt, tensioner, and belt-driven components at 150,000 miles. The blue line indicates the vibration results of just replacing one belt-driven component and reusing the old belt and tensioner. The old belt and tensioner do not dampen new component vibration effectively. Excessive vibration leads to excessive noise and component bearing wear. Today all belt drive

Figure 12-12 A separation of a belt rib on either edge from the main body plies caused by improper replacement procedure.

Figure 12-13 A belt pilling occurs as belt material is sheared off from the ribs building up in the grooves.

components are closely integrated and each depends on the belt and tensioner to keep the entire system at optimum operating performance. It is critical that you as a technician check every component in the system; this includes idler pulleys if used.

POWER STEERING PUMP FLUID SERVICE

Fluid Level Inspection

Steering effort is the amount of effort required by the driver to turn the steering wheel.

Most car manufacturers recommend power steering fluid or automatic transmission fluid in power steering systems. Always use the type of power steering fluid recommended in the vehicle manufacturer's service manual. If the power steering fluid level is low, **steering effort** is increased and may be erratic. A low fluid level may cause a growling noise in the power steering pump. Some car manufacturers now recommend checking the power steering pump fluid level with the fluid at an ambient temperature of 176°F (80°C).

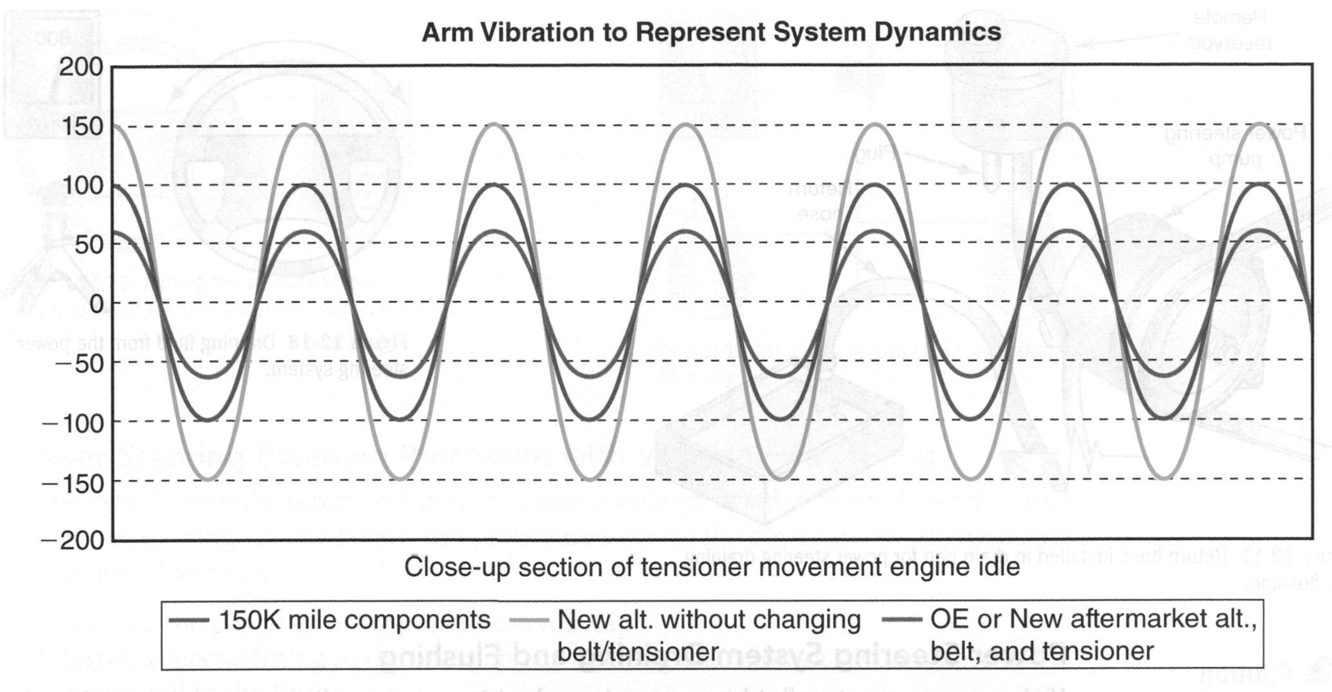

Figure 12-14 When a drive belt component is replaced it is advisable to also replace the belt and automatic tensioner to limit system vibration.

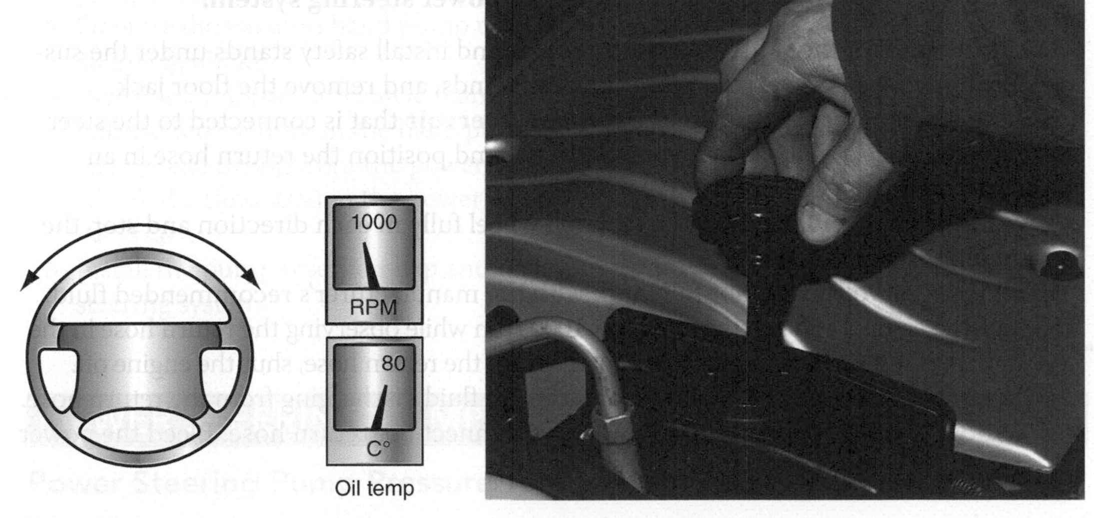

Figure 12-15 Boosting fluid temperature.

Figure 12-16 Power steering pump dipstick.

Follow these steps to check the power steering fluid level:

1. With the engine idling at 1000 rpm or less, turn the steering wheel slowly and completely in each direction several times to boost the fluid temperature (**Figure 12-15**).
2. If the vehicle has a remote power steering fluid reservoir, check for foaming in the reservoir, which indicates low fluid level or air in the system.
3. Observe the fluid level in the remote reservoir. This level should be at the hot full mark. Shut off the engine, and remove dirt from the neck of the reservoir with a shop towel. If the power steering pump has an integral reservoir, the level should be at the hot level on the dipstick. When an external reservoir is used, the dipstick is located in the external reservoir (**Figure 12-16**).
4. Pour the required amount of the car manufacturer's recommended power steering fluid into the reservoir to bring the fluid level to the hot full mark on the reservoir or dipstick with the engine idling.

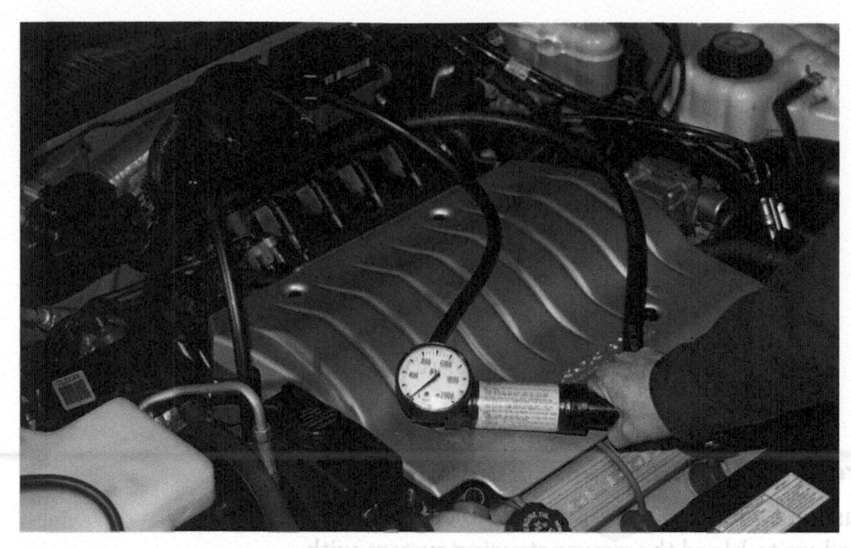

Figure 12-21 Pressure gauge connection to power steering pump.

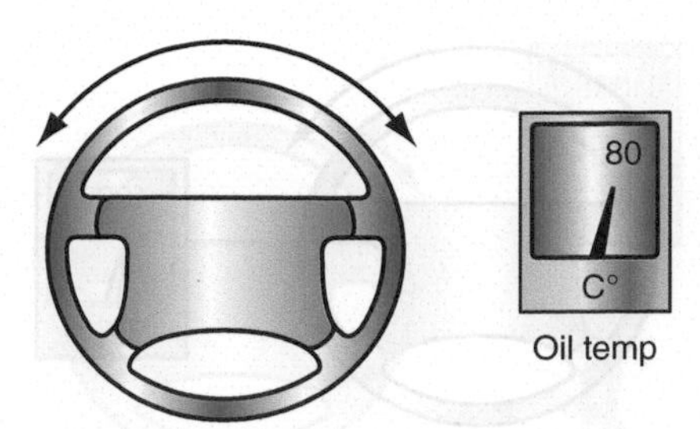

Figure 12-22 Bleeding air from the power steering pump system and checking fluid temperature.

WARNING During the power steering pump pressure test with the engine idling, close the pressure gauge valve for no more than 10 seconds; excessive pump pressure may cause power steering hoses to rupture, resulting in personal injury.

WARNING Do not allow the fluid to become too hot during the power steering pump pressure test. Excessively high fluid temperature reduces pump pressure. Wear protective gloves, and always shut the engine off before disconnecting gauge fittings, because the hot fluid may cause burns.

3. With the engine idling, close the pressure gauge valve for no more than 10 seconds, and observe the pressure gauge reading (**Figure 12-23**). Turn the pressure gauge valve to the fully open position. If the pressure gauge reading does not equal the vehicle manufacturer's specifications, repair or replace the power steering pump.
4. Inspect the power steering pump pressure with the engine running at 1000 rpm and 3000 rpm, and record the pressure difference between the two readings (**Figure 12-24**). If the pressure difference between the pressure readings at 1000 rpm and 3000 rpm does not equal the vehicle manufacturer's specifications, repair or replace the **flow control valve** in the power steering pump.
5. With the engine running, turn the steering wheel fully in one direction and observe the steering pump pressure while holding the steering wheel in this position (**Figure 12-25**). If the pump pressure is less than the vehicle manufacturer's specifications, the steering gear housing has an internal leak and should be repaired or replaced.
6. Be sure the front tire pressure is correct and center the steering wheel with the engine idling. Connect a spring scale to the steering wheel and measure the steering effort in both directions (**Figure 12-26**). If the power steering pump pressure is satisfactory and the steering effort is more than the vehicle manufacturer's specifications, the power steering gear should be repaired. **Photo Sequence 24** shows a typical procedure for pressure testing a power steering pump.

The **flow control valve** controls power steering fluid flow from the pump to the steering gear.

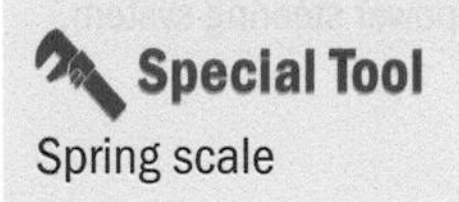

Special Tool

Spring scale

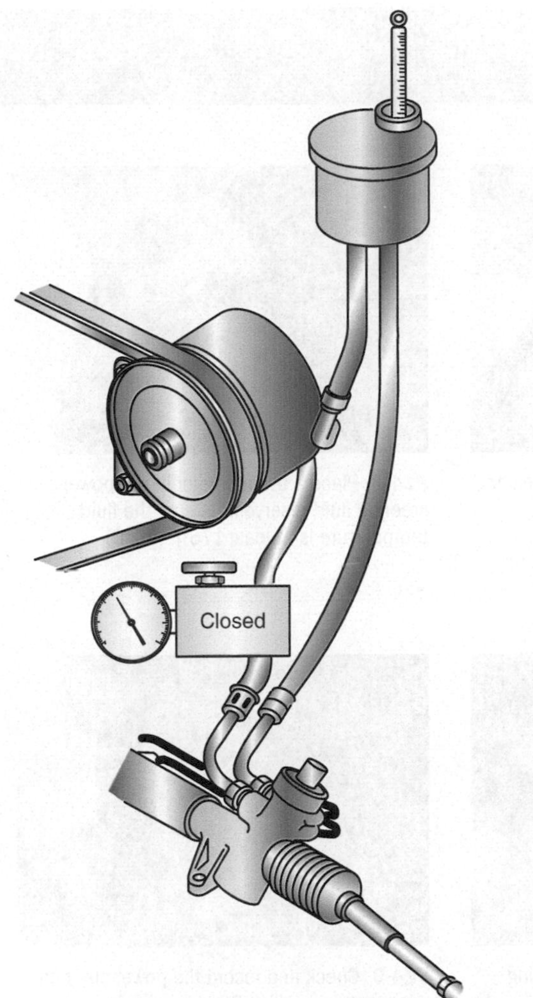

Figure 12-23 Power steering pump pressure test with pressure gauge valve closed.

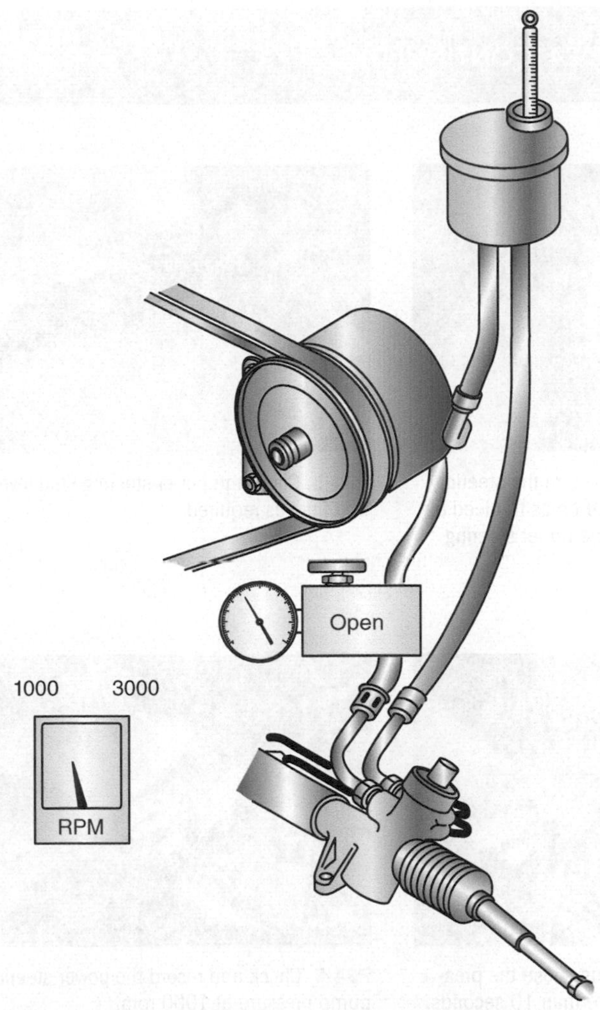

Figure 12-24 Power steering pump pressure test at 1000 and 3000 rpm.

PHOTO SEQUENCE 24
Typical Procedure for Pressure Testing a Power Steering Pump

P24-1 Connect the pressure gauge to the power steering pump.

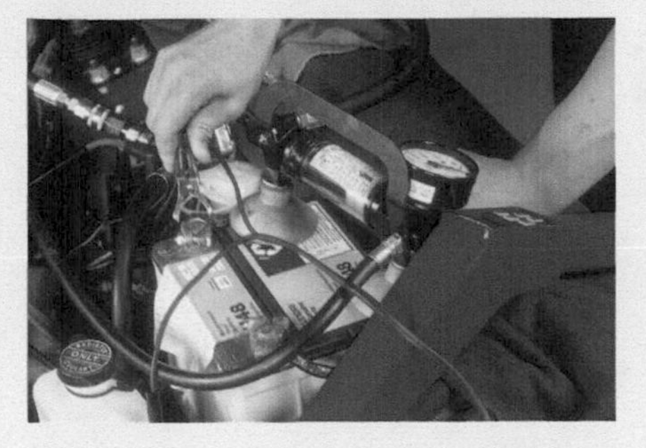

P24-2 Connect a tachometer to the ignition system.

P24-3 Look up the vehicle manufacturer's specified power steering pump pressure in the service manual.

PHOTO SEQUENCE 24 (CONTINUED)

P24-4 Start the engine, and turn the steering wheel from lock to lock three times to bleed air from the system and heat the power steering fluid.

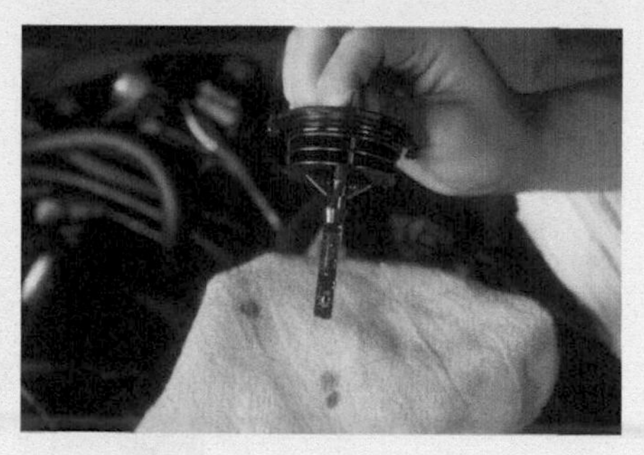

P24-5 Check the power steering fluid level, and add fluid as required.

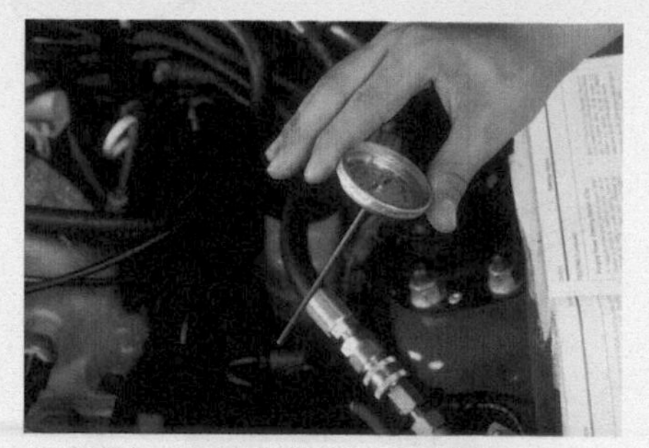

P24-6 Place a thermometer in the power steering fluid reservoir. Be sure the fluid temperature is at least 176°F (80°C).

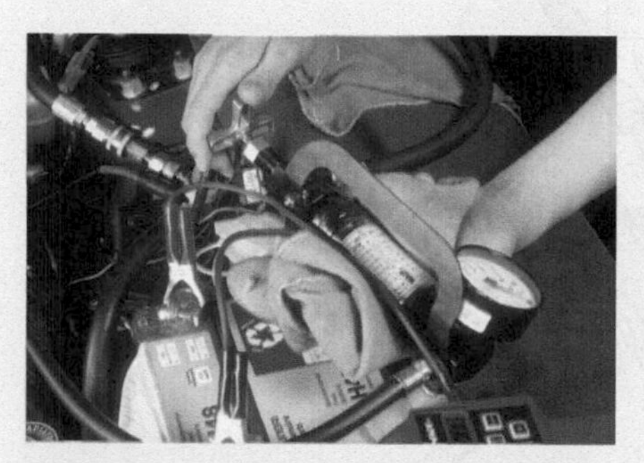

P24-7 With the engine idling, close the pressure gauge valve for no more than 10 seconds. Observe and record the pressure gauge reading.

P24-8 Check and record the power steering pump pressure at 1000 rpm.

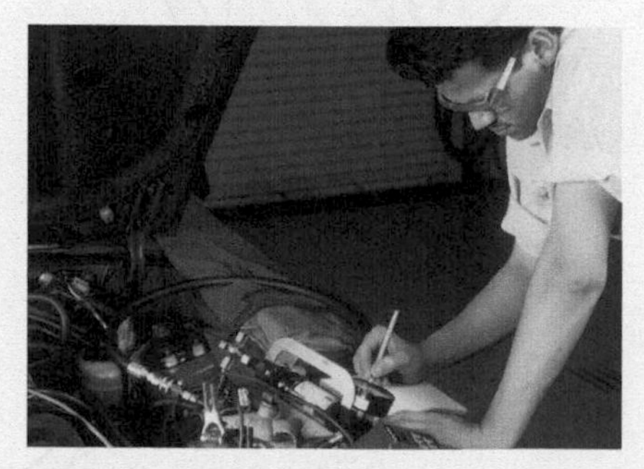

P24-9 Check and record the power steering pump pressure at 3000 rpm.

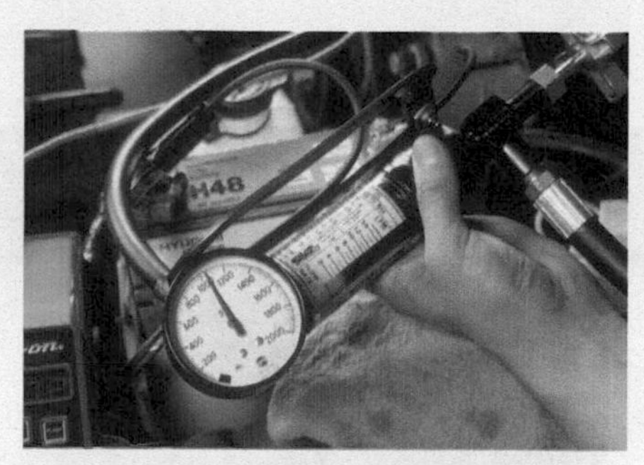

P24-10 Check and record the power steering pump pressure with the steering wheel turned fully in one direction and the engine idling.

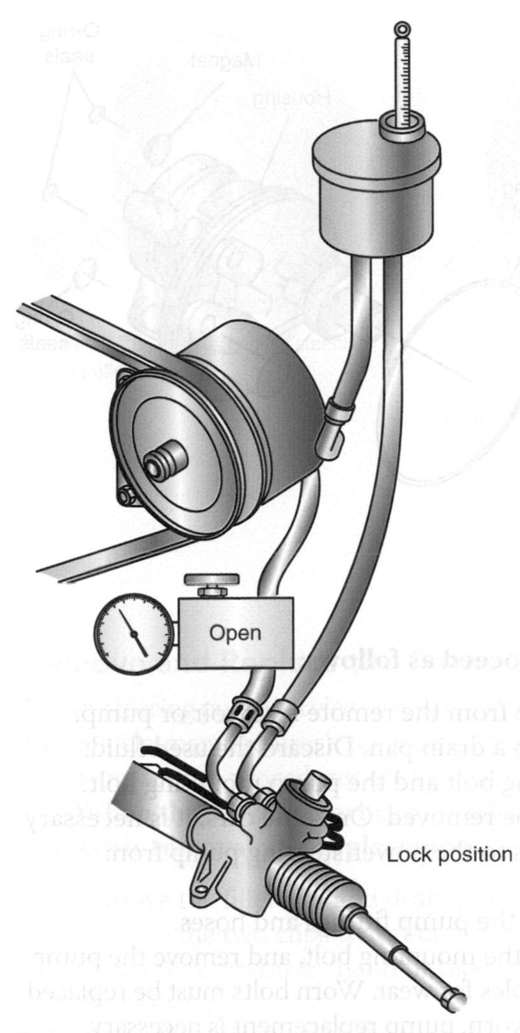

Figure 12-25 Power steering pump pressure test with the front wheels turned fully in one direction.

Figure 12-26 Steering effort measurement.

Power Steering Pump Oil Leak Diagnosis

The possible sources of power steering pump oil leaks are the drive shaft seal, reservoir O-ring seal, high-pressure outlet fitting, and the dipstick cap. If leaks occur at any of the seal locations, seal replacement is necessary. When a leak is present at the high-pressure outlet fitting, tighten this fitting to the specified torque (**Figure 12-27**). If the leak still occurs, replace the O-ring seal on the fitting and retighten the fitting.

POWER STEERING PUMP SERVICE

Power Steering Pump Replacement

WARNING If the vehicle has been driven recently, the pump, hoses, and fluid could be extremely hot. Use caution when handling components to avoid burns.

If a growling noise is present in the power steering pump after the fluid level is checked and air has been bled from the system, the pump bearings or other components are defective and pump replacement or repair is required. When the power steering pump pressure is lower than specified, pump replacement or repair is necessary.

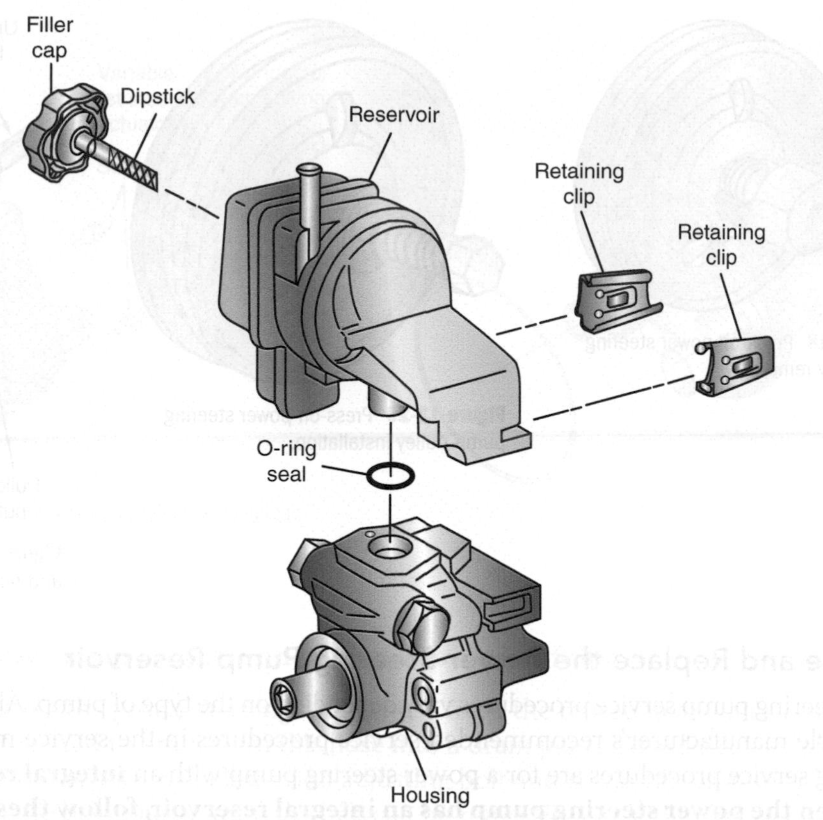

Figure 12-31 Integral power steering pump reservoir seal and O-ring.

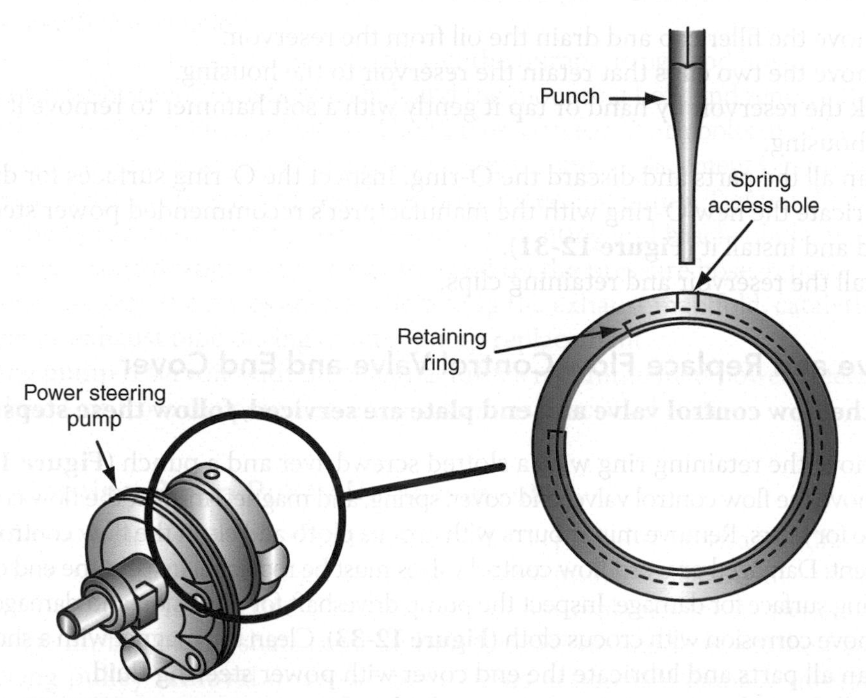

Figure 12-32 Retaining ring removal.

INSPECTING AND SERVICING POWER STEERING LINES AND HOSES

Power steering lines should be inspected for leaks, dents, sharp bends, cracks, and contact with other components. Lines and hoses must not rub against other components. This action could wear a hole in the line or hose. Many high-pressure power steering lines are made from high-pressure steel-braided hose with molded steel fittings on each end (**Figure 12-35**).

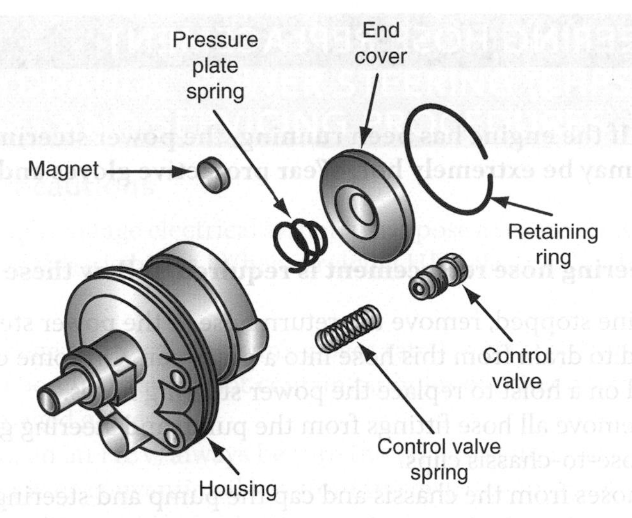

Figure 12-33 End cover and related component service.

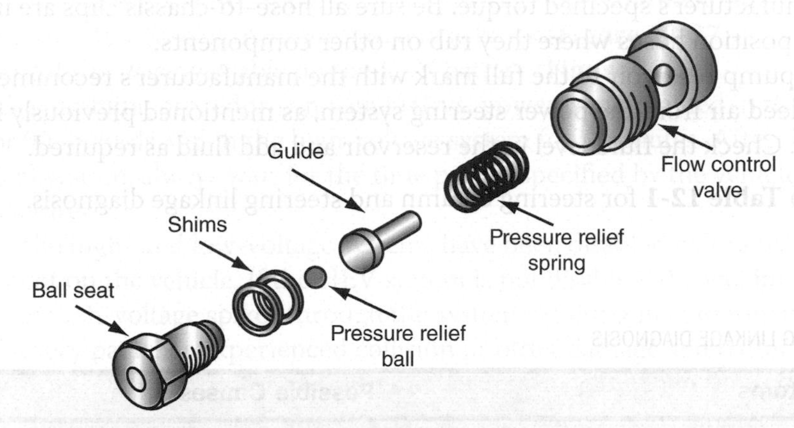

Figure 12-34 Pressure relief valve removal.

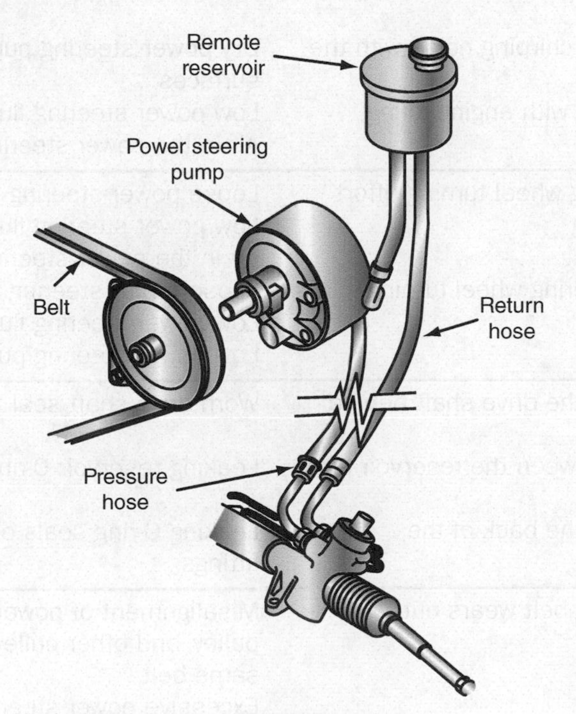

Figure 12-35 Power steering hoses and lines.

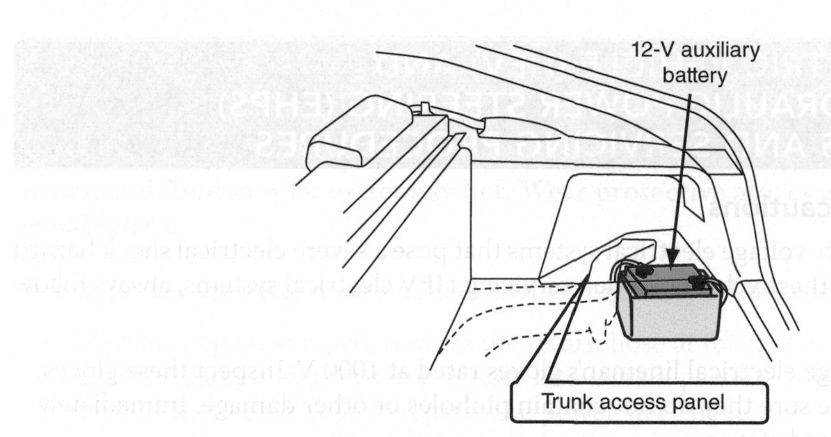

Figure 12-38 Disconnecting the 12-V battery.

8. Inspect high-voltage (orange) cables for frayed insulation and damage, which could result in electrical shocks, sparks, and fires. Frayed, damaged orange cables must be replaced.
9. If HEVs are towed with the drive wheels on the road surface, the generator may begin producing voltage and current even if the high-voltage electrical system is disabled. This may result in sparks and a fire.
10. The battery packs in many HEVs contain nickel–metal hydride type batteries. These batteries self-discharge quickly compared with a conventional lead acid battery. A nickel–metal hydride battery may lose 30 percent of its charge per month. Do not leave an HEV sitting for 2 or 3 months without starting the engine. If an HEV is not being driven, the engine should be started and allowed to run for 30 minutes every two or three weeks to maintain the battery pack's state of charge. Some test equipment manufacturers are designing battery chargers for HEV battery packs, but this test equipment is not available at the present time. The engine must be operated to allow the generator to recharge the battery pack. Some HEVs have an auxiliary starter to start the engine if the battery pack is discharged.
11. When working on an HEV with a smart key, always be sure the smart key is removed from the instrument panel, and the disable button under the steering column is pushed to the Off position. This action prevents the HEV system from being activated by pressing the smart key when it is located in close proximity to the vehicle.
12. Use caution when hoisting or lifting HEVs to avoid damage to the high-voltage cable compartments and cables.
13. On HEVs the refrigerant oil specified by the vehicle manufacturer, such as SE 10-Y, ND-OIL, or 11 must be used in the AC system. Many HEVs have an electric-drive AC compressor that remains operational during the engine stop mode to maintain AC system operation. In many of these compressors, 300 V from the inverter is supplied to the compressor drive motor. This higher voltage requires lower amperes of current through the compressor clutch coil. The refrigerant oil in these compressors must have special insulating qualities or the high-voltage insulation inside the compressor may be damaged.

Starting and Stopping Procedures and Warning Indicators

Technicians must be familiar with the proper starting and stopping procedures and warning indicators on HEVs. The engine starting and stopping procedures and warning indicators may vary on hybrid vehicles depending on the vehicle and model year. Always refer to the vehicle manufacturer's information. The following procedures and indicators are

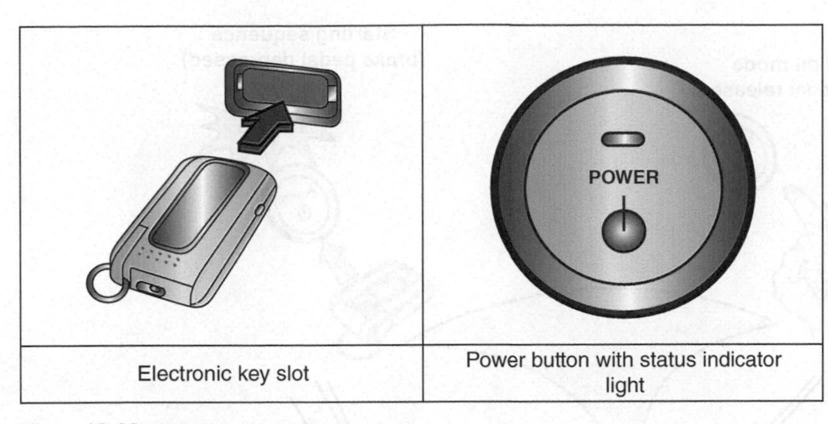

Figure 12-39 Electronic key and power button.

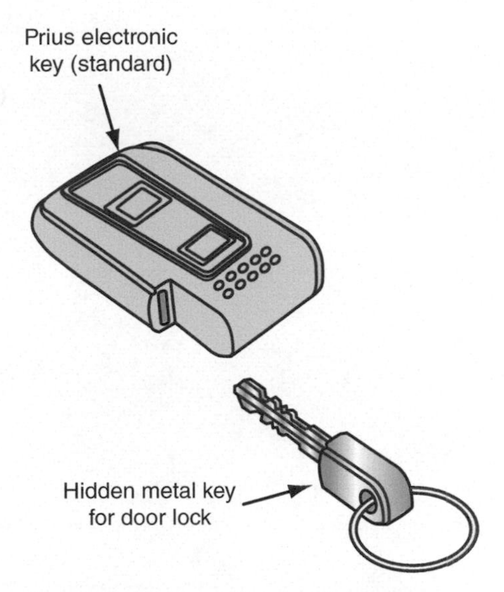

Figure 12-40 Electronic key with remote keyless entry button and hidden metal key.

typical. Some hybrid vehicles have an electronic key that fits into a key slot in the dash. A power button and an indicator light are located above the electronic key slot in the dash (**Figure 12-39**). The power button is pressed to cycle through the ignition modes rather than rotating the electronic key in the slot. The electronic key also contains the remote keyless entry buttons and a hidden metal key (**Figure 12-40**).

The doors are unlocked and locked by pressing the appropriate buttons on the electronic key. Another option for opening the doors is to remove the hidden metal key from the electronic key, and insert the metal key into the key cylinder in the driver's door. Rotating the metal key once unlocks the driver's door, and rotating this key a second time unlocks the other doors. There are no lock cylinders in the other doors.

To activate the HEV system and other electronic systems, insert the electronic key into the slot in the dash. With the brake pedal released, pressing the power button once turns on the accessory mode, and pushing the power button again turns on the ignition mode. Pressing the power button a third time turns the ignition off. When the ignition mode is off, the power button indicator light is also off. In the accessory mode the power button indicator light is green, and in the ignition mode this light is amber. When a defect occurs in the electrical system, the amber power button light starts blinking.

With the brake pedal depressed, pressing the power button once activates the HEV and other electronic systems. Starting the vehicle takes priority over all other modes. When the HEV system is activated, the power button indicator light goes off and the Ready light in the instrument panel is illuminated (**Figure 12-41**). When starting off under light load conditions, the engine does not start because the vehicle is only driven by the propulsion motor. Never work on an HEV with the electronic key installed in the dash and the HEV system activated. Under this condition, the engine may start and cause personal injury and/or vehicle component damage if the battery pack voltage reaches a specific discharged state. To shut off the vehicle, bring the vehicle to a complete stop and press the power button once and remove the electronic key. When the Ready light is illuminated, the electronic key cannot be removed from the key slot.

The vehicle may have an optional smart entry system with a smart electronic key. This key contains a transceiver that communicates bidirectionally and allows the vehicle to recognize the smart key in close proximity to the vehicle. The smart entry and smart key can unlock and lock the doors without pushing the buttons on the smart key. This system can also start the hybrid system without inserting the smart key into the dash slot. If the

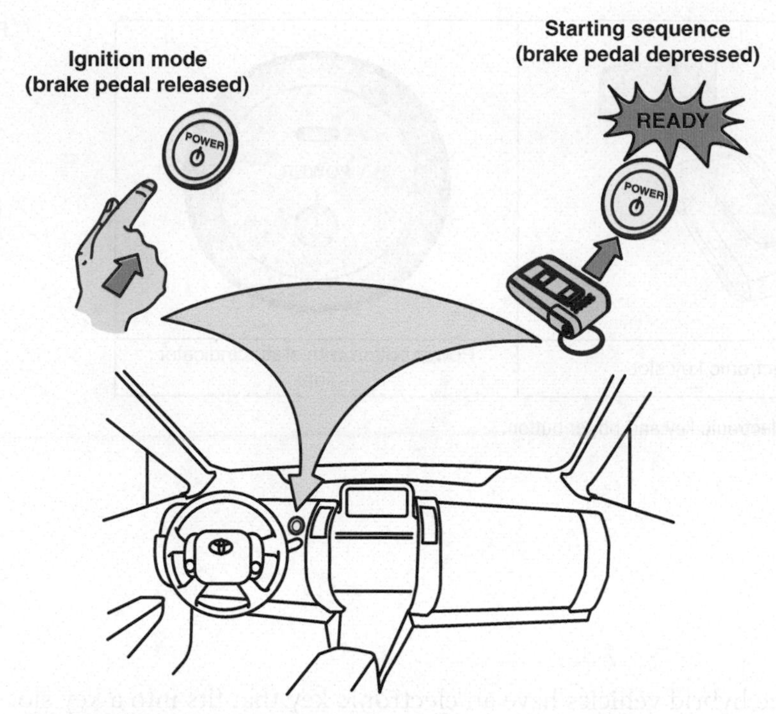

Figure 12-41 Ignition modes and starting sequence.

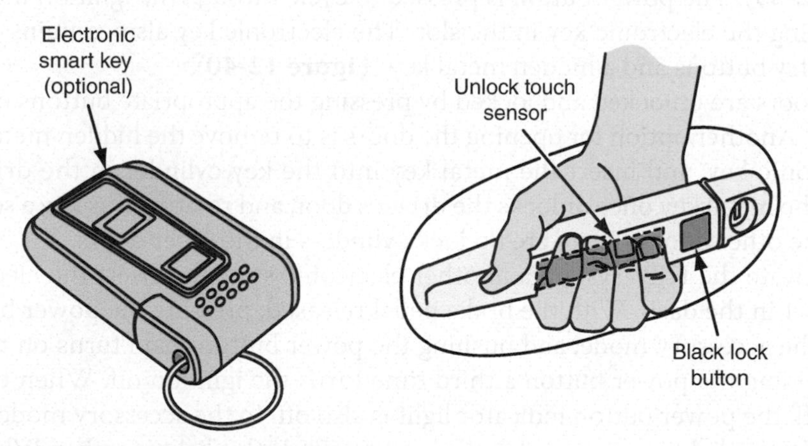

Figure 12-42 Smart key and smart entry system.

smart key is in close proximity to the vehicle, the doors may be unlocked by touching the sensor on the back of either exterior front door handle (**Figure 12-42**). Press the black lock button on either exterior front door handle to lock the doors. If the smart key is in close proximity to the vehicle, the normal ignition modes and start mode are operational. The smart key also contains a hidden metal key. Vehicles equipped with a smart key and smart entry system have a disabling button located in the dash under the steering column. If the disable button is pressed, the smart key must be inserted into the key slot to activate the ignition modes and start the vehicle (**Figure 12-43**).

The speedometer, gear shift indicator, Ready light, fuel gauge, and warning lights are located in a digital display in the center of the dash near the base of the windshield. An LCD monitor located at the top, center of the dash displays fuel consumption, A/C controls, and energy monitor (**Figure 12-44**). The electronic gearshift selector is a momentary select, shift-by-wire system that engages the transaxle in Reverse, Neutral, Drive, or engine Brake modes. There are no mechanical connections between the dash indictors and the transaxle. The dash indicator also informs the driver regarding the gear selected

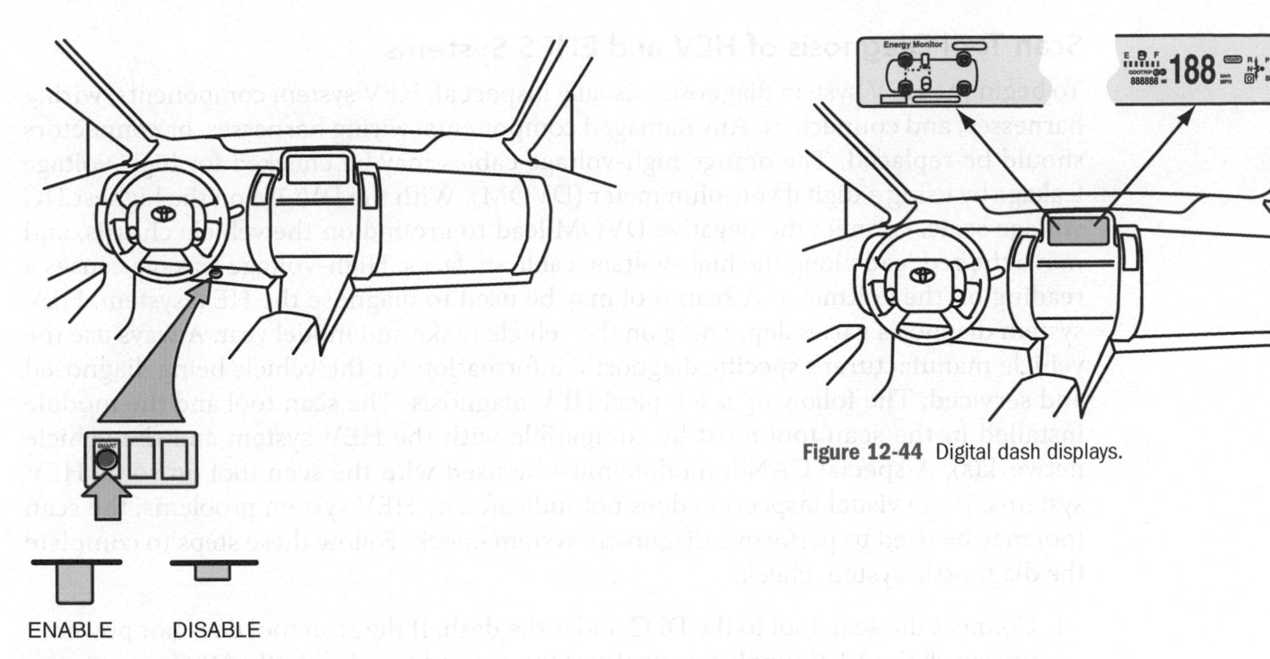

Figure 12-44 Digital dash displays.

Figure 12-43 Disable button on smart key systems.

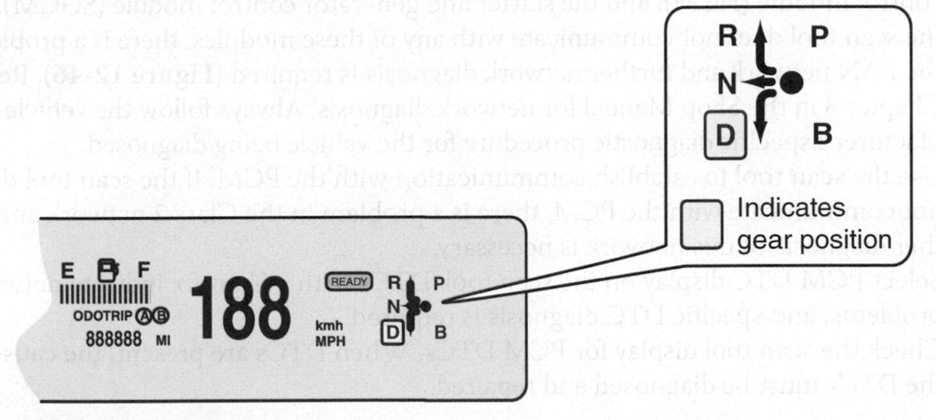

Figure 12-45 Gearshift indicator.

(**Figure 12-45**). After selecting the desired gear position, the transaxle remains in that position as indicated by the dash display, but the shift selector returns to a default mode. When the vehicle is stopped, Park may be selected by pressing the P switch in the dash display. If the power button is pressed to shut off the vehicle, Park is automatically engaged. Voltage is supplied to the gearshift selector and the electromechanical parking pawl by the 12-V battery. If the 12-V battery is discharged, or disconnected, the vehicle cannot be started or shifted out of park.

When the Ready light is illuminated, the vehicle is operational, even though the engine is not running. The driver simply selects Drive (D) or Reverse (R) on the gearshift selector, releases the brake pedal, and steps on the accelerator pedal to begin driving the vehicle. The vehicle computer decides when to start and stop the engine depending on many factors such as vehicle load and speed, and battery pack state of charge. A hybrid vehicle does not have the usual noise associated with starting motor operation, because the battery simply passes current through the generator windings to rotate and start the engine. The vehicle computer stops the engine during deceleration below a specific speed, and during idle operation, to improve fuel economy.

Scan Tool Diagnosis of HEV and EHPS Systems

To begin the HEV system diagnosis, visually inspect all HEV system components, wiring harnesses, and connectors. Any damaged components, wiring harnesses, or connectors should be replaced. The orange high-voltage cables may be checked for high-voltage leakage by using a digital volt-ohm meter (DVOM). With the DVOM on the highest DC voltage scale, connect the negative DVOM lead to ground on the vehicle chassis, and move the red lead along the high-voltage cable surfaces. High-voltage leakage causes a reading on the voltmeter. A scan tool may be used to diagnose the HEV system. HEV system diagnosis varies depending on the vehicle make and model year. Always use the vehicle manufacturer's specific diagnostic information for the vehicle being diagnosed and serviced. The following is a typical HEV diagnosis. The scan tool and the module installed in the scan tool must be compatible with the HEV system and the vehicle network(s). A special CANdi module must be used with the scan tool on some HEV systems. If the visual inspection does not indicate any HEV system problems, the scan tool may be used to perform a diagnostic system check. Follow these steps to complete the diagnostic system check:

1. Connect the scan tool to the DLC under the dash. If the scan tool does not power up, check the 12-V supply terminal and the ground terminal in the DLC.
2. Turn the ignition switch on with the engine off. Using the scan tool, attempt to establish communication with the energy storage control module (ESCM), hybrid control module (HCM), and the starter and generator control module (SGCM). If the scan tool does not communicate with any of these modules, there is a problem in the LAN network and further network diagnosis is required (**Figure 12-46**). Refer to Chapter 8 in the Shop Manual for network diagnosis. Always follow the vehicle manufacturer's specific diagnostic procedure for the vehicle being diagnosed.
3. Use the scan tool to establish communication with the PCM. If the scan tool does not communicate with the PCM, there is a problem in the Class 2 network and further diagnosis of this network is necessary.
4. Select PCM DTC display on the scan tool. DTCs with a U prefix indicate network problems, and specific DTC diagnosis is required.
5. Check the scan tool display for PCM DTCs. When DTCs are present, the cause of the DTCs must be diagnosed and repaired.

Diagnostic Example for a U0131 Network DTC

Current vehicles have hundreds of DTCs, and thus it is impossible to include an interpretation and diagnostic procedure for all DTCs in this publication. A U0131 DTC is set when the communication is lost between the EHPS module and the PCM. Under this condition, the EHPS operates in a default mode. The possible causes of this DTC are as follows:

1. An open circuit in the EHPS module connector.
2. An open circuit in the LAN network.
3. An open circuit in the voltage supply or ground circuits to the EHPS module.
4. An internal module malfunction. Many modules have specific DTCs for internal module defects.

To locate the root cause of the U0131 DTC, the technician must perform tests with a DVOM or lab scope on the suspected components.

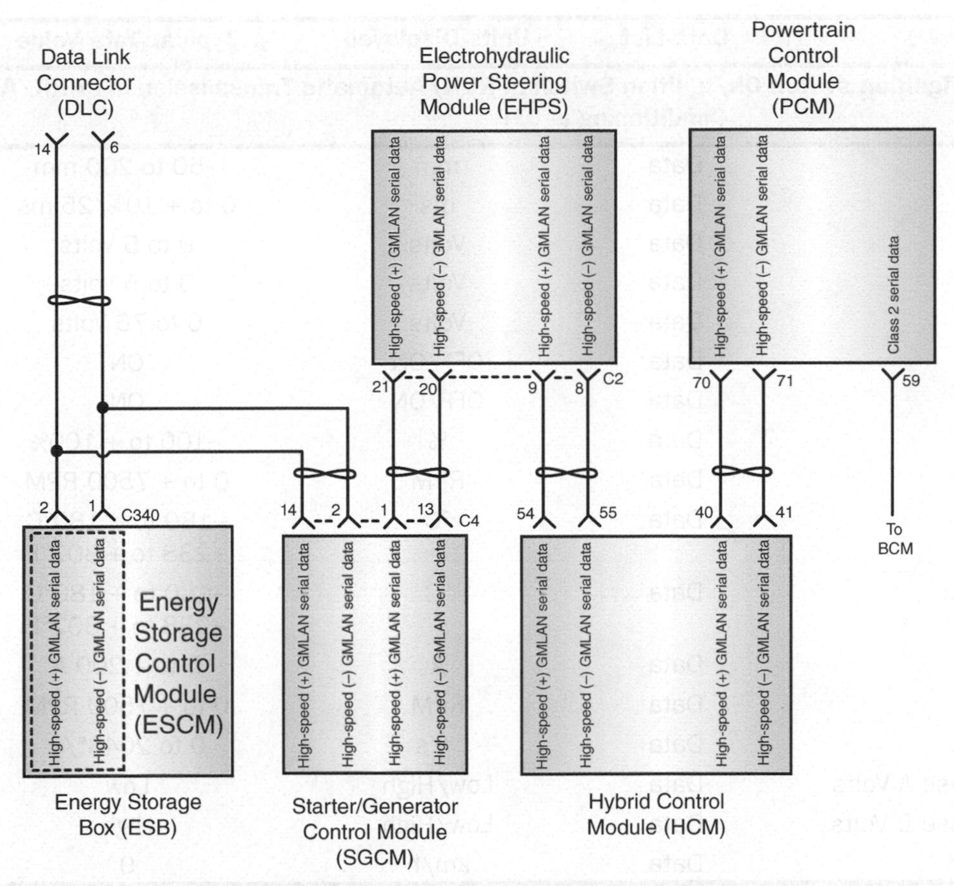

Figure 12-46 The LAN bus interconnects the SGCM, PCM, EHPS modules, ESB, and DLC.

Diagnostic Example for a P0562 DTC

A P0562 DTC is set when the PCM detects an improper voltage below 11 V for 5 seconds. When this DTC is set in the PCM memory, the PCM takes these actions:

1. The PCM stores faulty circuit conditions that were present when the DTC set.
2. The PCM disables many outputs.
3. The transmission defaults to a predetermined gear.
4. Torque converter clutch lockup is inhibited.
5. The instrument panel cluster (IPC) displays a warning message.
6. The malfunction indicator light (MIL) will not illuminate.

When this DTC is present in the PCM memory, the technician must use an amp-volt tester to test the charging circuit and use a DVOM to test all the electrical circuits between the charging circuit, battery, and PCM to locate the root cause of the problem. DTCs may be erased with the scan tool.

A scan tool data display may be useful when diagnosing HEV and EHPS systems especially when there is a system problem with no DTCs. Before using the data display, the diagnostic system check should be performed and all DTCs corrected. Available scan tool data from a HEV with EHPS is illustrated in **Figure 12-47**. **Photo Sequence 25** illustrates the safety precautions that must be observed when working on hybrid vehicles.

Scan Tool Parameter	Data List	Units Displayed	Typical Data Value
Park brake set with the ignition switch ON/Ignition Switch in RUN/Automatic Transmission in PARK/Air Conditioning is OFF			
Brake Pedal Travel	Data	mm	−50 to 200 mm
BPP Sensor Rate	Data	ms	0 to + 10%/25 ms
BPP Sensor Input Volts	Data	Volts	0 to 5 Volts
BPP Sensor Output Volts	Data	Volts	0 to 5 Volts
EHPS 42 V Bus Voltage	Data	Volts	0 to 76 Volts
EHPS Ignition 0	Data	OFF/ON	ON
EHPS Ignition 1	Data	OFF/ON	ON
EHPS Demand	Data	%	−100 to + 100%
EHPS Demand Speed	Data	RPM	0 to + 7500 RPM
EHPS ECU Temperature	Data	°C	−180 to + 180°C (−238 to + 302°F)
EHPS Manifold Temperature	Data	°C	−180 to + 180°C (−238 to + 302°F)
EHPS Motor Current	Data	A	0 to + 200 A
EHPS Motor RPM	Data	RPM	0 to + 7500 RPM
Steering Wheel Rate	Data	°/s	0 to 2048°/s
Steering Wheel Sensor Phase A Volts	Data	Low/High	Low
Steering Wheel Sensor Phase B Volts	Data	Low/High	High
Vehicle Speed	Data	km/h	0

Figure 12-47 EHPS data.

PHOTO SEQUENCE 25
Observing Safety Precautions When Servicing Hybrid Electric Vehicles (HEVs)

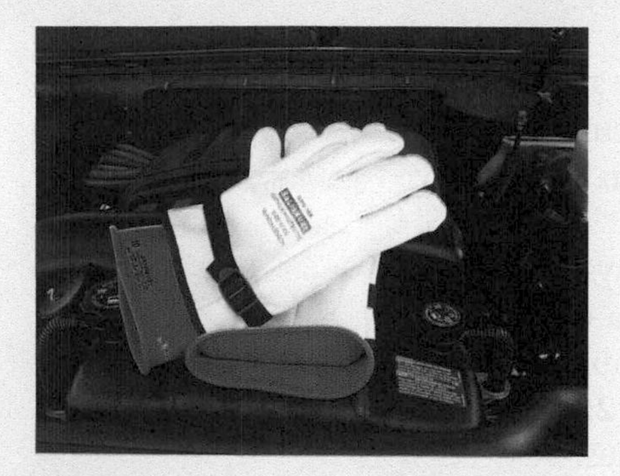

P25-1 Always wear high-voltage lineman's gloves when servicing HEV electrical systems.

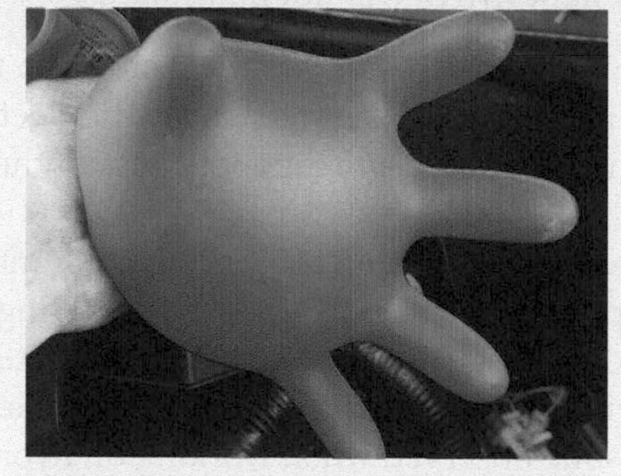

P25-2 Periodically inspect the high-voltage gloves for pinholes or damage.

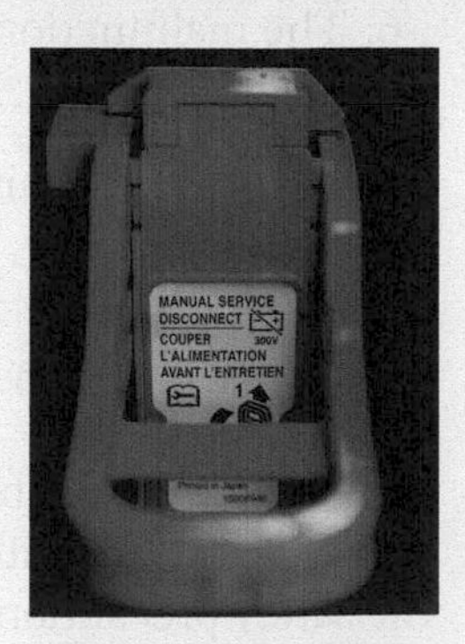

P25-3 Shut off the high-voltage circuit switch to isolate the battery pack from the electrical system.

PHOTO SEQUENCE 25 (CONTINUED)

P25-4 Disconnect the negative cable on the 12V battery.

P25-5 After the high- and low-voltage systems are disabled, wait for the time specified by the vehicle manufacturer to allow these systems to completely power down.

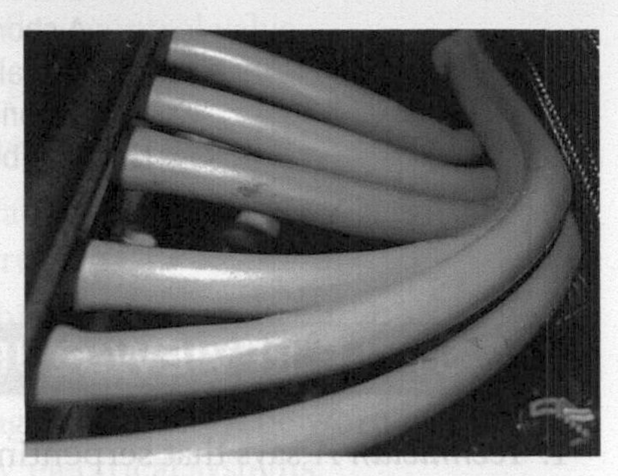

P25-6 Inspect the high-voltage (orange) cables for frayed insulation and damage.

P25-7 If the HEV is not driven for a few weeks, the engine should be started and run for 30 minutes every two weeks to maintain the state of charge in the high-voltage battery pack.

P25-8 Always remove the electronic or conventional key from the instrument panel to prevent the engine from starting when working on an HEV electrical system.

CASE STUDY

The owner of a Dodge Magnum with a 5.7-L turbocharged engine requested a price on a power steering gear replacement. The service writer questioned the owner regarding the reason for the steering gear replacement, and discovered that the problem was intermittent hard steering. Further discussion with the owner also revealed that the car had been taken to another automotive service center, and the owner was informed that his or her car required a steering gear replacement. The owner was now looking for a second opinion.

The technician road-tested the vehicle and found no indication of hard steering. The owner was asked about the exact driving conditions when the hard steering was experienced, and the answer to this question provided the solution to the problem. The owner indicated that the hard steering was usually experienced while cornering and always on a rainy day. Once this information was revealed, the technician suspected a problem with the power steering belt, because a wet belt slips more easily. A check of the power steering belt indicated the belt was oil soaked

Task Completed

7. Fill the reservoir to the hot full mark with the manufacturer's recommended fluid.

   Specified type of power steering fluid ______________________

   Power steering reservoir filled to the specified level with the proper power steering fluid? ☐ Yes ☐ No

   Instructor check ______________________

8. Start the engine, and run the engine at 1000 rpm while observing the return hose in the drain pan. When fluid begins to discharge from the return hose, shut the engine off. ☐

9. Repeat Steps 4, 5, and 6 until there is no air in the fluid discharging from the return hose.

   Have Steps 4, 5, and 6 been repeated? ☐ Yes ☐ No

   Instructor check ______________________

10. Remove the plug from the reservoir and reconnect the return hose.

    Is the return hose connected and tightened? ☐ Yes ☐ No

    Instructor check ______________________

11. Fill the power steering pump reservoir to the specified level. ☐

12. With the engine running at 1000 rpm, turn the steering wheel fully in each direction three or four times. Each time the steering wheel is turned fully to the right or left, hold it there for 2 to 3 seconds before turning it in the other direction.

    Number of times the steering wheel was turned fully in each direction

    ____

13. Check for foaming of the fluid in the reservoir. When foaming is present, repeat Steps 8 and 9.

    Is foaming present in the power steering pump reservoir? ☐ Yes ☐ No

    Have Steps 8 and 9 been repeated? ☐ Yes ☐ No

    Instructor check ______________________

14. Check the fluid level and be sure it is at the hot full mark. ☐

15. Raise vehicle with a floor jack, remove safety stands, and lower vehicle. ☐

16. Explain how you know that all the air has been bled from the power steering system.

    ____

    ____

    ____

Instructor's Response

____

____

____

Name ______________________________ Date ______________

JOB SHEET 59

TESTING POWER STEERING PUMP PRESSURE

Upon completion of this job sheet, you should be able to test power steering pump pressure as well as diagnose power steering gear binding, uneven turning effort, looseness, hard steering, and noise concerns, and determine necessary action.

ASE Education Foundation Task Correlation

This job sheet addresses the following **AST/MAST** tasks:

B.4. Diagnose power steering gear (non-rack and pinion) binding, uneven turning effort, looseness, hard steering, and noise concerns; determine needed action. **(P-2)**

B.5. Diagnose power steering gear (rack and pinion) binding, uneven turning effort, looseness, hard steering, and noise concerns; determine needed action. **(P-2)**

This job sheet addresses the following **MAST** task:

B.20. Test power steering system pressure; determine needed action. **(P-2)**

We Support ASE Education Foundation

Tools and Materials

Specified type of power steering fluid

Power steering pressure test gauge

Thermometer

Describe the Vehicle Being Worked On:

Year ______________ Make ______________ Model ______________

VIN ______________ Engine type and size ______________

Procedure **Task Completed**

1. Start vehicle and turn steering wheel to both the left and right bump stops. Note any power steering gear binding, uneven turning effort, looseness, hard steering, and noise concerns.

____

____

____

2. With the engine stopped, disconnect the pressure line from the power steering pump. ☐

3. Connect the gauge side of the pressure gauge to the pump outlet fitting. Connect the valve side of the gauge to the pressure line.

   Is the power steering pressure gauge properly connected? ☐ Yes ☐ No

   Are all power steering gauge fittings tightened to the specified torque? ☐ Yes ☐ No

4. Start the engine and turn the steering wheel fully in each direction two or three times to bleed air from the system. ☐

5. Install a thermometer in the pump reservoir fluid to measure the fluid temperature. Be sure the fluid level is correct and the fluid temperature is at least 176°F (80°C)

   Is the power steering fluid at the specified level? ☐ Yes ☐ No

   Is the power steering fluid at the proper temperature? ☐ Yes ☐ No

   Instructor check ______________________________

WARNING During the power steering pump pressure test with the engine idling, close the pressure gauge valve for no more than 10 seconds because excessive pump pressure may cause power steering hoses to rupture, resulting in personal injury.

WARNING Do not allow the fluid to become too hot during the power steering pump pressure test. Excessively high fluid temperature reduces pump pressure. Wear protective gloves, and always shut the engine off before disconnecting gauge fittings, because the hot fluid may cause burns.

6. With the engine idling, close the pressure gauge valve for no more than 10 seconds, and observe the pressure gauge reading. Turn the pressure gauge valve to the fully open position. If the pressure gauge reading does not equal the vehicle manufacturer's specifications, repair or replace the power steering pump.

   Specified power steering pump pressure with pressure gauge valve closed

   ____

   Actual power steering pump pressure with the pressure gauge valve closed

   ____

   Recommended power steering pump service

   ____

7. Check the power steering pump pressure with the engine running at 1000 rpm and 3000 rpm. Record the pressure difference between the two readings.

   Specified power steering pump pressure at 1000 rpm ______________

   Specified power steering pump pressure at 3000 rpm ______________

   Actual power steering pump pressure at 1000 rpm ______________

   Actual power steering pump pressure at 3000 rpm ______________

   Recommended power steering pump service ______________________________

8. With the engine running, turn the steering wheel fully in one direction. Observe the steering pump pressure while holding the steering wheel in this position.

   Specified power steering pump pressure with the steering wheel turned fully in one direction ______________________________

   Actual power steering pump pressure with the steering wheel turned fully in one direction ______________________________

   Recommended power steering service ______________________________

9. Be sure the front tire pressure is correct, and center the steering wheel with the engine idling. Connect a spring scale to the steering wheel, and measure the steering effort in both directions.

   Specified force required to turn the steering wheel to the right ______________

   Specified force required to turn the steering wheel to the left ______________

Actual force required to turn the steering wheel to the right _______________

Actual force required to turn the steering wheel to the left _______________

List all the required power steering service and explain the reasons why this service is necessary. ____

____

____

Instructor's Response

____

____

____

Name ______________________________ Date ____________

JOB SHEET 60

MEASURE AND ADJUST POWER STEERING BELT TENSION AND ALIGNMENT

Upon completion of this job sheet, you should be able to measure and adjust power steering belt tension and alignment.

ASE Education Foundation Task Correlation

This job sheet addresses the following **MLR** task:

B.5. Remove, inspect, replace, and/or adjust power steering pump drive belt. **(P-1)**

This job sheet addresses the following **AST/MAST** task:

B.12. Remove, inspect, replace, and/or adjust power steering pump drive belt. **(P-1)**

We Support ASE Education Foundation

Tools and Materials

Belt tension gauge

Belt automatic tensioner tool

Describe the Vehicle Being Worked On:

Year ____________ Make ____________ Model ____________

VIN ____________ Engine type and size ____________

Procedure **Task Completed**

1. Describe the power steering problems and noises that occur when the power steering pump belt is loose.

____

____

____

2. Inspect the power steering belt for fraying, oil soaking, wear on friction surfaces, cracks, glazing, and splits.

Belt condition: ☐ Satisfactory ☐ Unsatisfactory

3. With the engine stopped, press on the belt at the longest belt span to measure the belt deflection, which should be 0.5 in. per foot of free span.

Length of belt span where belt deflection is measured

Amount of belt deflection ______________________________

Belt tension: ☐ Satisfactory ☐ Unsatisfactory

4. Install a belt tension gauge over the belt in the center of the longest span to measure the belt tension.

Specified belt tension ______________________________

Actual belt tension ______________________________

5. Following manufacturer's service procedures, move automatic belt tensioner to release tension on belt using belt tensioner tool. ☐

Task Completed

a. If vehicle is equipped with manual belt tension adjustment, follow manufacturer's service procedures and loosen the power steering pump bracket or tension adjusting bolt.

6. Remove the accessory drive belt. ☐

7. Release automatic tensioner slowly. ☐

8. Check the bracket and pump mounting bolts for wear.

Power steering pump mounting bolt holes and bolt condition:

☐ Satisfactory ☐ Unsatisfactory

Power steering pump bracket and bracket bolt condition:

☐ Satisfactory ☐ Unsatisfactory

List the required power steering pump and bracket service and explain the reasons for your diagnosis.

____

____

____

____

9. Route the new drive belt over all the pulleys except the idler pulley. ☐

10. Using belt tensioner tool rotate automatic belt tensioner in order to relieve the tension of the tensioner and install belt on idler pulley. ☐

a. If equipped with a manual tensioner pry against the pump ear and hub with a pry bar to tighten the belt. Some pump brackets have a 0.5-in. square opening in which a breaker bar may be installed to move the pump and tighten the belt.

b. Hold the pump in the position described in step a) and tighten the bracket or tension adjusting bolt to specified torque. ☐

11. Slowly release the tension on the belt tensioner.

12. Recheck the belt tension with the tension gauge.

Specified power steering belt tension _______________

Actual power steering pump belt tension _______________

13. Check alignment of the power steering pump pulley in relation to the other pulleys surrounded by the power steering belt.

Power steering pump pulley alignment: ☐ Satisfactory ☐ Unsatisfactory

Explain the service required to align the power steering pump pulley with other related pulleys and give the reasons for your diagnosis. _______________________

____

Instructor's Response

____

____

____

Name ______________________________ Date ______________

JOB SHEET
61

REMOVE AND REPLACE POWER STEERING PUMP AND PULLEY

Upon completion of this job sheet, you should be able to inspect, remove, and/or replace power steering pump and pulley assembly.

ASE Education Foundation Task Correlation

This job sheet addresses the following **AST/MAST** tasks:

B.13. Remove and reinstall power steering pump. **(P-2)**

B.14. Remove and reinstall press fit power steering pump pulley; check pulley and belt alignment. **(P-2)**

We Support ASE Education Foundation

Tools and Materials

Hand tool

Power steering pump

Pulley puller

Describe the Vehicle Being Worked On:

Year ______________ Make ______________ Model ______________

VIN ______________ Engine type and size ______________

Procedure **Task Completed**

1. Remove power steering pump pulley. ☐
   - Remove the accessory drive belt.
   - Install power steering pump pulley removal tool and remove pulley
2. Remove power steering fluid from reservoir using a vacuum siphon. ☐
3. Place drain pan under the vehicle as needed. ☐
4. Following manufacturer's service procedures, remove both the pressure and return hose from power steering pump. ☐
5. Following manufacturer's service procedures, remove power steering pump assembly. ☐
6. Transfer any parts to new pump as needed. ☐
7. Install new power steering pump following manufacturer's service procedures.
   - Torque specifications for mounting bolts:______________________
8. Connect power steering pressure and return hoses to pump assembly.
   - Torque specifications for fittings: ______________________
9. Clean any excess fluid spilled from vehicle and remove drain pan. ☐
10. Install power steering pump pulley following manufacturer service procedures. ☐
    - Place the power steering pump pulley on the end of the power steering pump shaft.
    - Install power steering pump pulley using pulley installation tool.
    - Ensure that the pulley is flush against the power steering pump shaft, general allowable variance is 0.010 in. (0.25 mm).
    - Install accessory drive belt and check pulley alignment.
11. Fill and bleed power steering system. ☐

Name ______________________ Date ______

JOB SHEET 62

REMOVE AND REPLACE POWER STEERING HOSE

Upon completion of this job sheet, you should be able to inspect, remove, and/or replace power steering hose(s) and fittings.

ASE Education Foundation Task Correlation

This job sheet addresses the following AST/MAST task:

B.15. Inspect, remove, and/or replace power steering hoses and fittings. (P-2)

We Support ASE Education Foundation

Tools and Materials

Hand tool

Flare nut wrenches

Power steering fluid

Describe the Vehicle Being Worked On:

Year ______ Make ______ Model ______

VIN ______ Engine type and size ______

Procedure **Task Completed**

1. Remove power steering fluid from reservoir using a vacuum siphon. ☐
2. Place drain pan under the vehicle as needed. ☐
3. Following manufacturer's service procedures, remove both the pressure and/or return hose from power steering pump. ☐
4. Following manufacturer's service procedures, remove both the pressure and/or return hose from the steering gear. ☐
5. Connect power steering pressure and return hoses to pump assembly.
   a. Torque specifications for fittings: ______
6. Connect power steering pressure and return hoses to steering gear.
   a. Torque specifications for fittings: ______
7. Clean any excess fluid spilled from vehicle and remove drain pan. ☐
8. Fill and bleed power steering system. ☐

Instructor's Response

______

______

______

CHAPTER 13

RACK AND PINION STEERING GEAR AND FOUR-WHEEL STEERING DIAGNOSIS AND SERVICE

Upon completion and review of this chapter, you should be able to:

- Perform a manual or power rack and pinion steering gear inspection.
- Remove and replace manual or power rack and pinion steering gears.
- Disassemble, inspect, repair, and reassemble manual rack and pinion steering gears.
- Adjust manual rack and pinion steering gears.
- Diagnose manual rack and pinion steering systems.
- Diagnose oil leaks in power rack and pinion steering gears.
- Disassemble, inspect, and repair power rack and pinion steering gears.
- Adjust power rack and pinion steering gears.
- Diagnose variable effort steering.
- Diagnose column-driven electronic power steering systems.
- Diagnose active steering system.
- Perform a preliminary inspection on a four-wheel steering (4WS) system.
- Perform rear wheel steering and rear wheel alignment procedure.
- Diagnose four-wheel active steering systems (4WAS).

Basic Tools

Basic technician's tool set

Service manual

Floor jack

Safety stands

Machinist's rule

Terms To Know

Active roll stabilization (ARS) ECU

Active steering ECU

Bellows boots

Bump steer

Claw washers

Column-drive EPS

Dynamic stability control (DSC) ECU

Electronically controlled orifice (ECO) valve

On-board diagnostic-II (OBD-II)

Rack-drive electronic power steering

Safety and gateway module (SGM)

Servotronic valve

Soft-jaw vise

Steering effort imbalance

Steering wheel free play

Proper rack and pinion steering gear operation is essential to maintain vehicle safety and reduce driver fatigue. Such steering gear conditions as looseness and excessive steering effort may contribute to a loss of steering control, resulting in a vehicle collision. Worn steering gear mountings may cause improper wheel alignment and **bump steer**. Improper wheel alignment increases tire tread wear, and bump steer may increase driver fatigue. Excessive steering gear looseness or high steering effort also contributes to driver fatigue. Therefore, in the interest of vehicle safety and driver alertness, rack and pinion steering gear diagnosis and service are extremely important.

Bump steer occurs when one of the front wheels strikes a road irregularity while driving straight ahead, and the steering suddenly veers to the right or left.

WARNING Bent steering components must be replaced. Never straighten steering components, because this action may weaken the metal and result in sudden component failure, serious personal injury, and vehicle damage.

8. Grasp each outer tie rod end and check for vertical movement. While an assistant turns the steering wheel one-quarter turn in each direction, watch for looseness in the outer tie rod ends. If any looseness or vertical movement is present, replace the tie rod end. Check the outer tie rod end seals for cracks and proper installation of the nuts and cotter pins. Cracked seals must be replaced. Inspect the tie rods for a bent condition. Bent tie rods or other steering components must be replaced. Do not attempt to straighten these components.

MANUAL OR POWER RACK AND PINION STEERING GEAR REMOVAL AND REPLACEMENT

The replacement procedure is similar for manual or power rack and pinion steering gears. This removal and replacement procedure varies depending on the vehicle. On some vehicles, the front crossmember or engine support cradle must be lowered to remove the rack and pinion steering gear. Always follow the vehicle manufacturer's recommended procedure in the service manual.

The following is a typical rack and pinion steering gear removal and replacement procedure:

1. Place the front wheels in the straight-ahead position and remove the ignition key from the ignition switch to lock the steering wheel. Place the driver's seat belt through the steering wheel to prevent wheel rotation if the ignition switch is turned on (**Figure 13-3**). This action maintains the clock spring electrical connector or spiral cable in the centered position on air bag–equipped vehicles.
2. Lift the front end with a floor jack, and place safety stands under the vehicle chassis. Lower the vehicle onto the safety stands. Remove the left and right fender apron seals (**Figure 13-4**).

Figure 13-3 Driver's seat belt wrapped around the steering wheel to prevent wheel rotation.

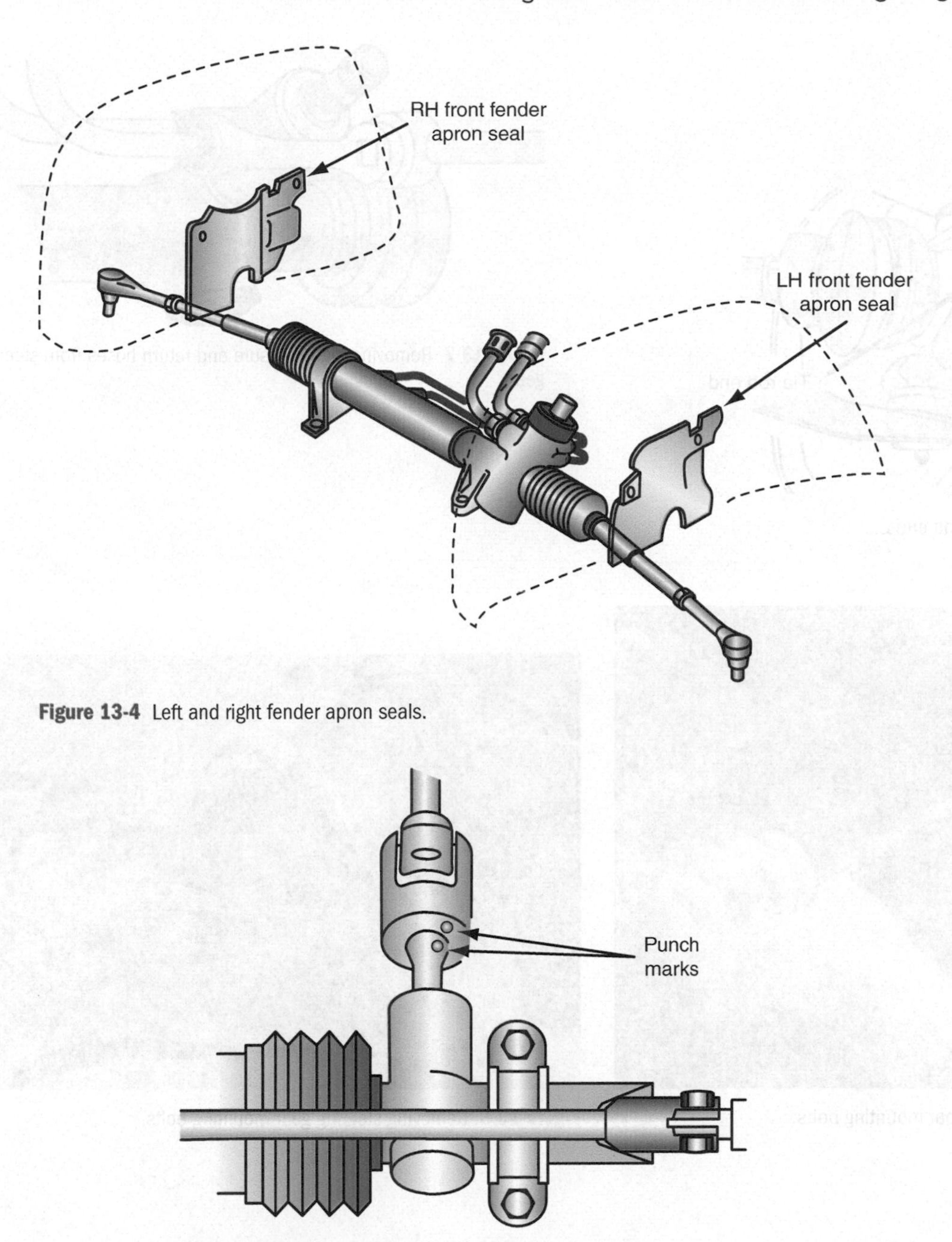

Figure 13-4 Left and right fender apron seals.

Figure 13-5 Punch marks on universal joint and pinion shaft.

3. Place punch marks on the lower universal joint and the steering gear pinion shaft so they may be reassembled in the same position (**Figure 13-5**). Loosen the upper universal joint bolt, remove the lower universal joint bolt, and disconnect this joint.
4. Remove the cotter pins from the outer tie rod ends. Loosen, but do not remove, the tie rod end nuts. Use a tie rod end puller to loosen the outer tie rod ends in the steering arms (**Figure 13-6**). Remove the tie rod end nuts and remove the tie rod ends from the arms.

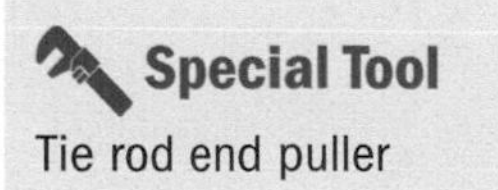

5. Use the proper wrenches to disconnect the high-pressure hose and the return hose from the steering gear (**Figure 13-7**). This step is not required on a manual steering gear. The removal procedure for a power rack and pinion steering gear is shown in **Photo Sequence 26**.
6. Remove the four stabilizer bar mounting bolts (**Figure 13-8**).
7. Remove the steering gear mounting bolts (**Figure 13-9**).

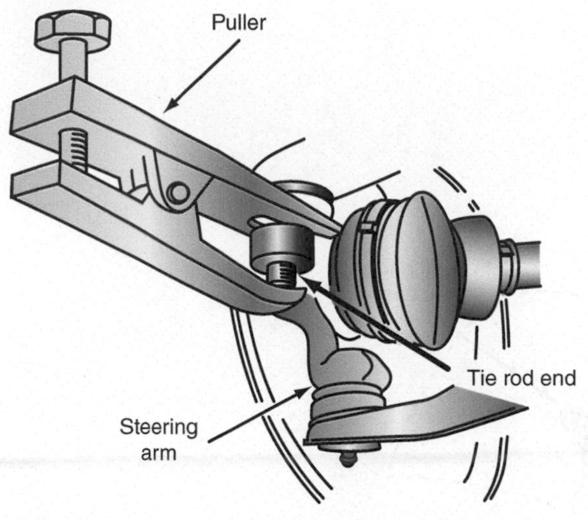

Figure 13-6 Removing outer tie rod ends.

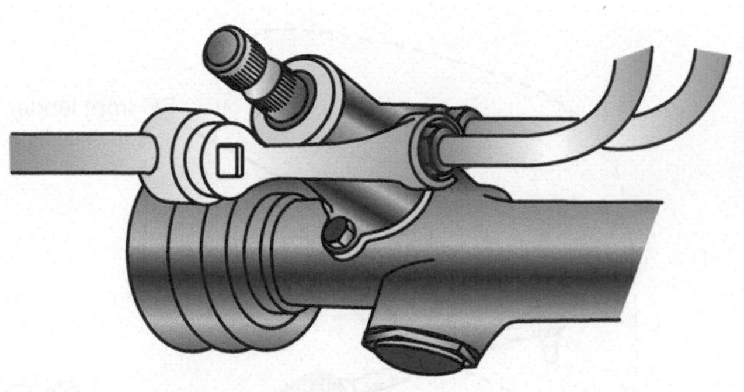

Figure 13-7 Removing high-pressure and return hoses from steering gear.

Figure 13-8 Removing stabilizer bar mounting bolts.

Figure 13-9 Removing steering gear mounting bolts.

8. Remove the steering gear assembly from the right side of the car (**Figure 13-10**).
9. Position the right and left tie rods the specified distance from the steering gear housing (**Figure 13-11**). Install the steering gear through the right fender apron.
10. Install the pinion shaft in the universal joint with the punch marks aligned. Tighten the upper and lower universal joint bolts to the specified torque.
11. Install the steering gear mounting bolts and tighten these bolts to the specified torque.
12. Install the stabilizer bar mounting bolts and torque these bolts to specifications.
13. Install and tighten the high-pressure and return hoses to the specified torque. (This step is not required on a manual rack and pinion steering gear.)
14. Install the outer tie rod ends in the steering knuckles, and tighten the nuts to the specified torque. Install the cotter pins in the nuts.
15. Check the front wheel toe and adjust as necessary. Tighten the outer tie rod end jam nuts to the specified torque, and tighten the outer bellows boot clamps.
16. Install the left and right fender apron seals, and lower the vehicle with a floor jack.

PHOTO SEQUENCE 26
Typical Procedure for Removing and Replacing a Power Rack and Pinion Steering Gear

P26-1 Position the vehicle on a tire-contact lift, and place the front wheels in the straight-ahead position.

P26-2 Remove the key from the ignition switch to lock the steering column, and place the driver's seat belt through the steering wheel opening to prevent wheel rotation if the ignition switch is turned on.

P26-3 Raise the vehicle a short distance off the floor, and remove the left and right fender apron seals.

P26-4 Place punch or chalk marks on the lower universal joint and steering gear pinion shaft so they may be reassembled in the same position.

P26-5 Loosen the upper universal joint bolt, remove the lower universal joint bolt, and disconnect the lower universal joint.

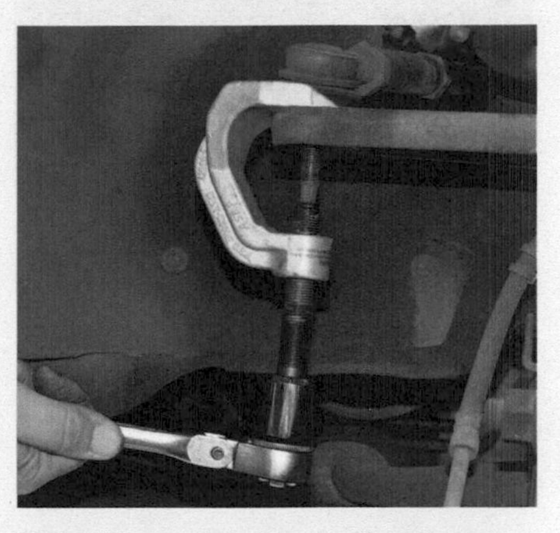

P26-6 Remove the cotter pins from the outer tie rod ends, and loosen but do not remove the tie rod end nuts. Use a tie rod end puller to loosen the outer tie rod ends in the steering arms. Remove the tie rod end nuts, and remove the tie rod ends from the steering arms.

P26-7 Use the proper-sized wrenches to remove the high-pressure hose and return hose from the steering gear.

P26-8 Remove the four stabilizer bar mounting bolts.

P26-9 Remove the steering gear mounting bolts.

PHOTO SEQUENCE 26 (CONTINUED)

P26-10 Remove the steering gear through one of the fender apron openings.

P26-11 Measure the distance from the outer end of the steering gear housing to the outer edge of the outer tie rod end on both sides of the steering gear, and record these measurements.

P26-12 Rotate the pinion shaft to obtain the measurements obtained in Step 11, and install the steering gear through one of the fender apron openings.

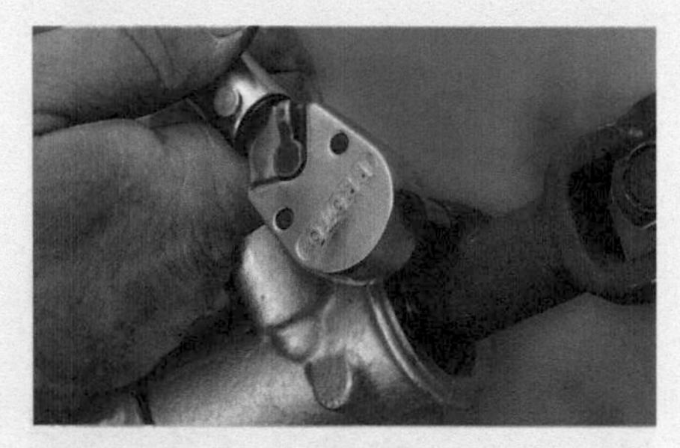

P26-13 Install the pinion shaft into the universal joint with the punch marks aligned. Tighten the upper and lower universal joint bolts to the specified torque.

P26-14 Install the steering gear mounting bolts and tighten these bolts to the specified torque.

P26-15 Install the stabilizer mounting bolts and tighten these bolts to the specified torque.

P26-16 Install and tighten the high-pressure and return line fittings to the specified torque.

P26-17 Install the outer tie rod ends in the steering arms, and tighten the retaining nuts to the specified torque. Install the cotter pins in the nuts.

P26-18 Lower the vehicle to a convenient working height, and install the fender apron seals.

PHOTO SEQUENCE 26 (CONTINUED)

P26-19 Fill the power steering pump reservoir to the proper level with the specified power steering fluid. Start the engine for a short time, then shut the engine off, and add power steering fluid as required. Run the engine for several minutes, turn the steering wheel fully in both directions, and then add power steering fluid as necessary.

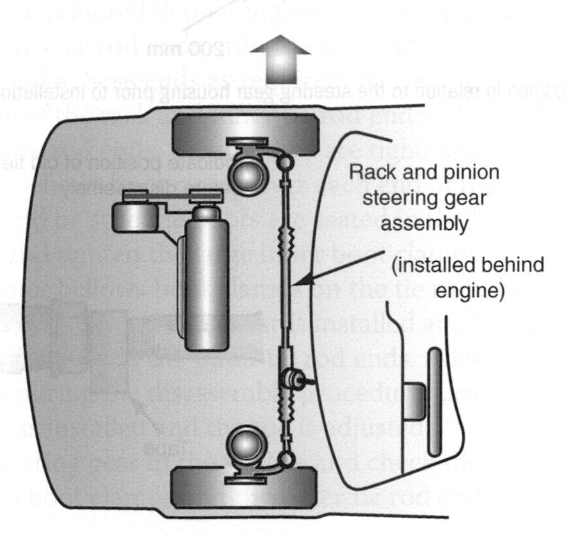

Figure 13-10 Removing steering gear from the car.

17. Fill the power steering pump reservoir with the vehicle manufacturer's recommended power steering fluid and bleed the air from the power steering system. (Refer to Chapter 12 for these tasks. This step is not required on a manual rack and pinion steering gear.)
18. Road-test the vehicle and check for proper steering gear operation and steering control.

Caution

Do not loosen the tie rod end nuts to align cotter pin holes. This action causes improper torquing of these nuts.

MANUAL RACK AND PINION STEERING GEAR DIAGNOSIS AND SERVICE

Manual Rack and Pinion Steering Gear Tie Rod Service

Follow these steps for manual rack and pinion steering gear disassembly:

1. Clamp the center of the steering gear housing in a **soft-jaw vise**. Do not apply excessive force to the vise.

A **soft-jaw vise** has brass jaws to prevent marking components.

TABLE 13-1 *(Continued)*

Condition	Possible Cause	Correction
	4. Intermediate shaft-to-steering-column-shaft attaching bolt loose	4. Tighten the bolt to specifications.
	5. Flex coupling damaged or worn	5. Replace as required.
	6. Front and/or rear suspension components loose, damaged, or worn	6. Tighten or replace as required.
	7. Steering rack adjustment set too low	7. Check the steering rack adjustment and readjust as required.
Noise: Knock, clunk, rapping, squeaking noise when turning	1. Steering wheel-to-steering-column shroud interference	1. Adjust or replace as required.
	2. Lack of lubrication where speed control brush contacts steering wheel pad	2. Lubricate as required.
	3. Steering column mounting bolts loose	3. Tighten the bolt to specifications.
	4. Intermediate shaft-to-steering-column-shaft attaching bolt loose	4. Tighten the bolt to specifications.
	5. Flex coupling-to-steering-rack attaching bolt loose	5. Tighten the bolt to specifications.
	6. Steering rack mounting bolts loose	6. Tighten the bolt to specifications.
	7. Tires rubbing or grounding out against body or chassis	7. Adjust/replace as required.
	8. Front suspension components loose, worn, or damaged	8. Tighten or replace as required.
	9. Steering linkage ball stud rubber deteriorated	9. Replace as required.
	10. Normal lash between pinion gear teeth and rack gear teeth when the steering gear is in the off-center position permits gear tooth chuckle noise while turning on rough road surfaces	10. This is a normal condition and cannot be eliminated.
Vehicle pulls/drifts to one side Note: This condition cannot be caused by the manual steering gear.	1. Vehicle overloaded or unevenly loaded	1. Correct as required.
	2. Improper tire pressure	2. Adjust air pressure as required.
	3. Mismatched tires and wheels	3. Install correct tire and wheel combination.
	4. Unevenly worn tires	4. Replace as required (check for cause).

(continued)

TABLE 13-1 *(Continued)*

Condition	Possible Cause	Correction
	5. Loose, worn, or damaged steering linkage	5. Tighten or replace as required.
	6. Steering linkage stud not centered within the socket	6. Repair or replace as required.
	7. Bent spindle or spindle arm	7. Repair or replace as required.
	8. Broken and/or sagging front and/or rear suspension springs	8. Replace as required.
	9. Loose, bent, or damaged suspension components	9. Tighten or replace as required.
	10. Bent rear axle housing	10. Replace as required.
	11. Excessive camber/caster split (excessive side-to-side variance in the caster/camber settings)	11. Adjust camber/caster split as required.
	12. Improper toe setting	12. Adjust as required.
	13. Front wheel bearings out of adjustment	13. Adjust as required.
	14. Investigate tire variance (conicity for radial tires and unequal circumference for bias-ply and belted bias-ply tires)	14. Repair or replace as required.

POWER RACK AND PINION STEERING GEAR DIAGNOSIS AND SERVICE

Fluid, Filter, and Fluid Leak Diagnosis

Power steering fluid should be inspected for discoloration and contamination with metal particles, dirt, or water. A suction gun may be used to remove some fluid from the power steering reservoir for inspection. If the fluid is contaminated or discolored, the power steering system must be flushed and then filled to the specified level with the vehicle manufacturer's specified fluid. After refilling the power steering system, an air bleeding procedure is usually necessary. Some power steering systems have a filter that is usually located in the remote power steering reservoir (**Figure 13-16**). On some systems, the filter and reservoir must be replaced as an assembly. Filter/reservoir replacement is necessary if the power steering system has been opened for repairs or if the fluid is contaminated with water, metal, or dirt particles. The filter/reservoir and fluid should also be changed at the vehicle manufacturer's specified service intervals. A restricted power steering filter may cause erratic steering operation.

WARNING **If the engine has been running for a length of time, power steering gears, pumps, lines, and fluid may be very hot. Wear eye protection and protective gloves when servicing these components.**

If power steering fluid leaks from the cylinder end of the power steering gear, the outer rack seal is leaking (**Figure 13-17**).

The inner rack seal is defective if oil leaks from the pinion end of the housing when the rack reaches the left internal stop (**Figure 13-18**). An oil leak at one rack seal may result in oil leaks from both boots because the oil may travel through the breather tube between the boots.

If an oil leak occurs at the pinion end of the housing and this leak is not influenced when the steering wheel is turned, the pinion seal is defective (**Figure 13-19**).

Turning Imbalance Diagnosis

Steering effort imbalance is the difference in steering effort when making right and left turns.

The same amount of effort should be required to turn the steering wheel in either direction. A pressure gauge connected to the high-pressure hose should indicate the same pressure when the steering wheel is turned in each direction. **Steering effort imbalance** or lower power assist in each direction may be caused by defective rack seals (**Figure 13-22**). Steering effort imbalance or low power assist in both directions may also be caused by defective rotary valve rings and seals or restricted hoses and lines (**Figure 13-23**).

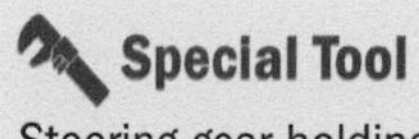
Special Tool

Steering gear holding tool

In a power rack and pinion steering gear, a condition that causes excessive steering effort when the vehicle is first started may be called "morning sickness."

SERVICE TIP A common problem with power rack and pinion steering gears is excessive steering effort when the car is first started after sitting overnight. The steering effort becomes normal, however, after the steering wheel is turned several times. This problem usually indicates severe scoring where the pinion sealing rings contact the pinion housing. Steering gear replacement is required to correct this problem.

Rotary Valve and Spool Valve Operation

The spool valve and rotary valve are mounted over the upper end of the pinion gear. During a turn, torsion bar twisting moves the spool valve inside the rotary valve and this action directs fluid to the appropriate side of the rack piston to provide steering assist. The pinion gear spool valve directs fluid pressure to the power rack piston right or left chamber for power assist (**Figure 13-24**) during a turn.

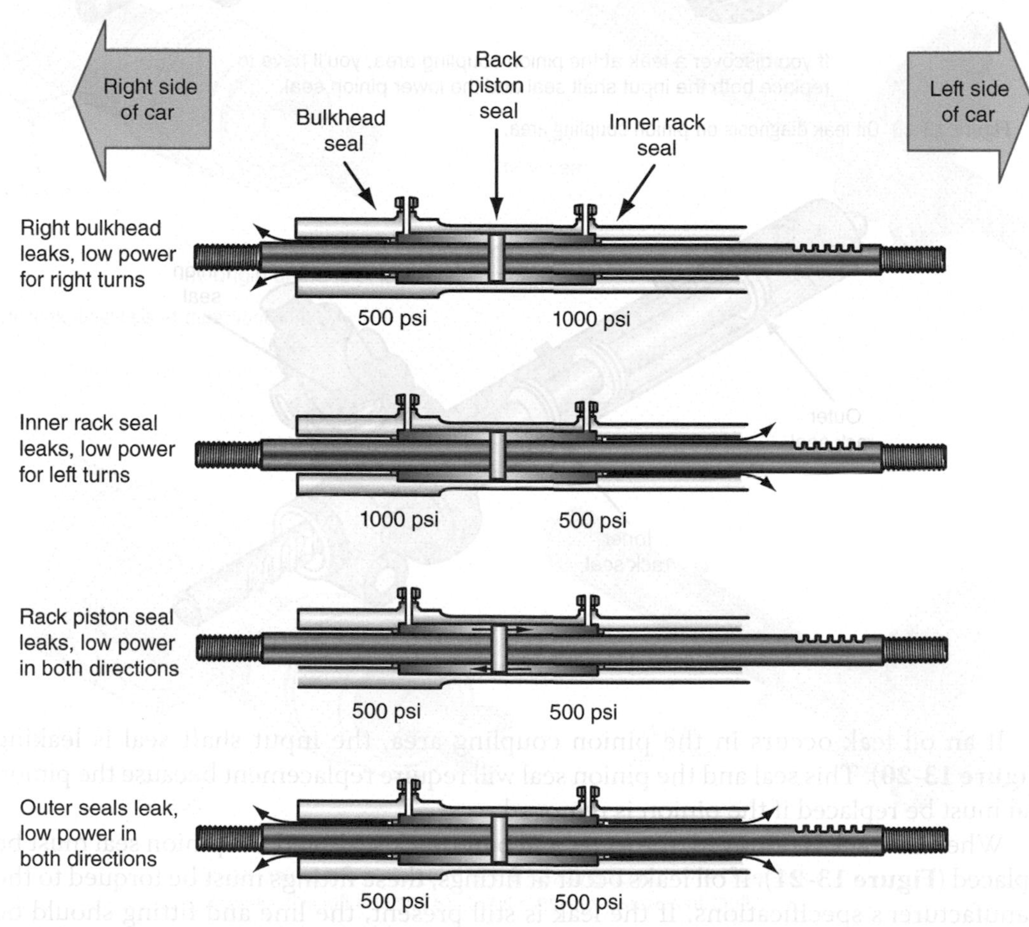

Figure 13-22 Effect of defective rack seals on steering effort imbalance and low power assist.

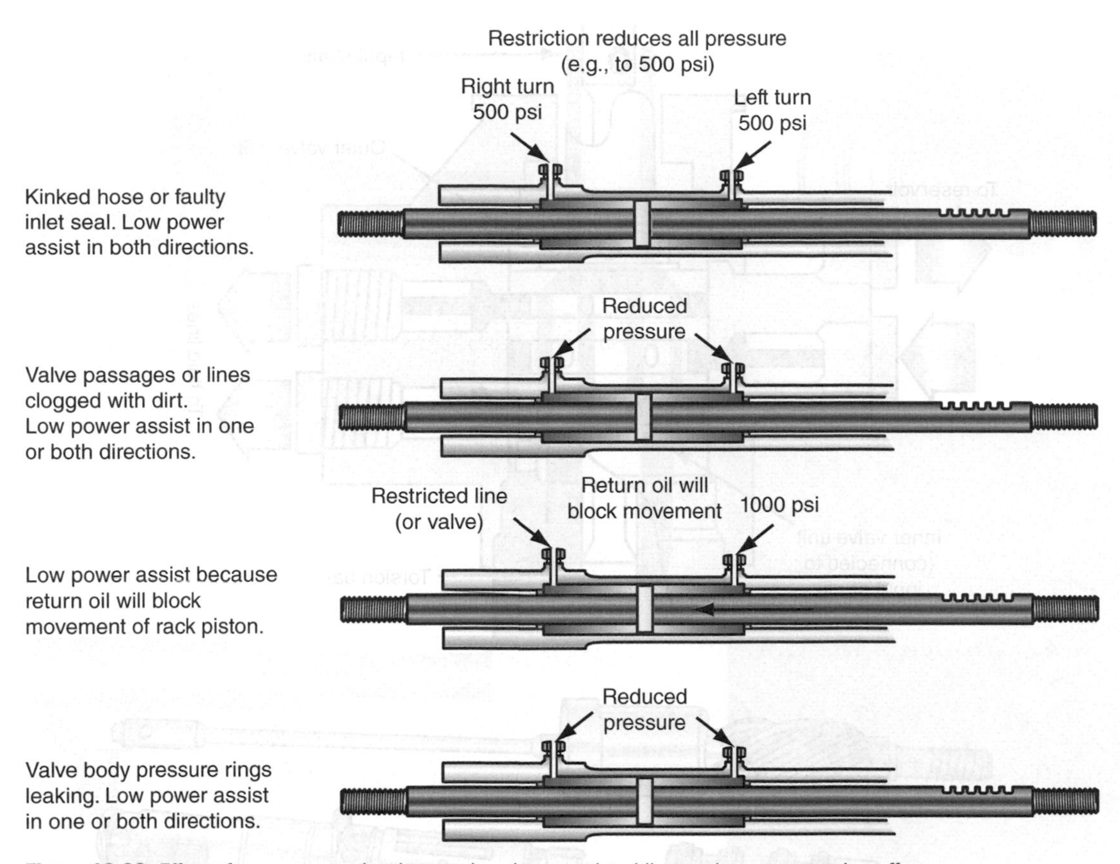
Restriction reduces all pressure
(e.g., to 500 psi)
Right turn
500 psi
Left turn
500 psi
Kinked hose or faulty inlet seal. Low power assist in both directions.
Reduced pressure
Valve passages or lines clogged with dirt.
Low power assist in one or both directions.
Restricted line (or valve)
Return oil will block movement
1000 psi
Low power assist because return oil will block movement of rack piston.
Reduced pressure
Valve body pressure rings leaking. Low power assist in one or both directions.

Figure 13-23 Effect of worn rotary valve rings and seals or restricted lines or hoses on steering effort.

When driving with the front wheels straight ahead, most of the fluid is directed through the rotary valve to the return hose under normal operation. During a left turn, spool valve movement directs fluid to the left side of the rack piston. During a right turn, spool valve movement directs fluid to the right side of the rack piston.

SERVICE TIP To provide equal turning effort in both directions, the rack must be centered with the front wheels straight ahead.

The pinion gear spool valve can bind or stick causing fluid to be directed to the power rack piston chamber causing a drift or pull to one side (**Figure 13-25**). The symptom for this is a vehicle drift or pull to one side. Alignment angle will be within vehicle specifications but a drift is still present. With both front wheels of the vehicle jacked off the ground, start the vehicle while paying attention to steering wheel position. If the steering wheel moves slightly to the left or right suspect a binding spool valve directing pressure to one side of the rack piston. Be aware that a front tire with excessive conicity can also cause the vehicle to pull or drift to one side. This is often referred to as a radial tire pull.

Power Rack and Pinion Steering Gear Tie Rod and Rack Bearing Service

1. Install a holding tool on the steering gear housing, and clamp this tool in a vise (**Figure 13-26**).
2. Mark the outer tie rod ends, jam nuts, and tie rods (**Figure 13-27**). Loosen the jam nuts and remove the tie rod ends and jam nuts.
3. Loosen the inner and outer clamps on both bellows boots, and remove these boots (**Figure 13-28**).

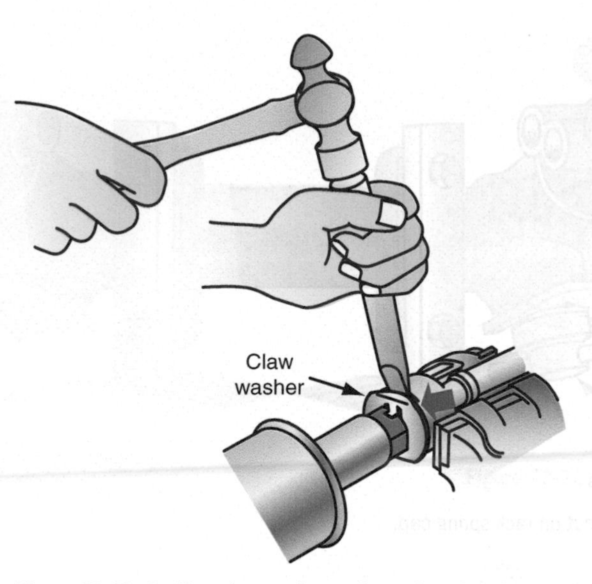

Figure 13-39 Staking claw washer on inner tie rod end.

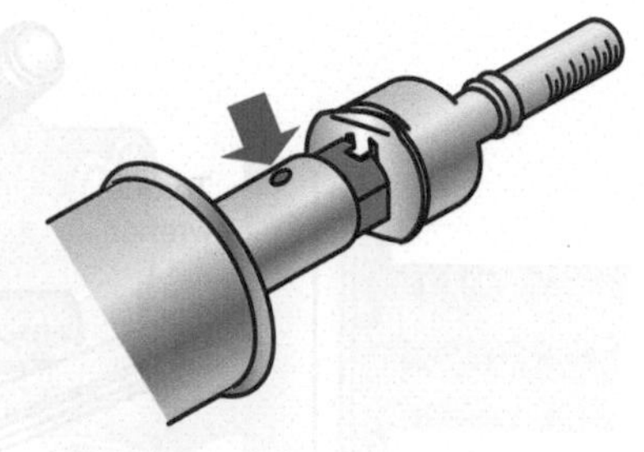

Figure 13-40 Rack vent holes must be open.

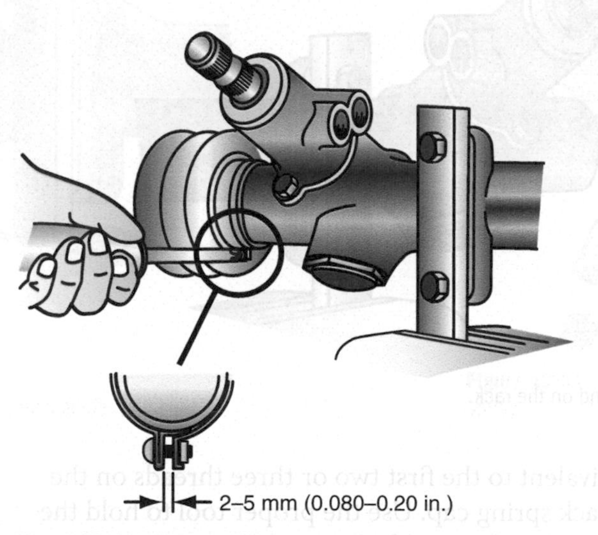

Figure 13-41 Minimum clearance between boot clamp ends.

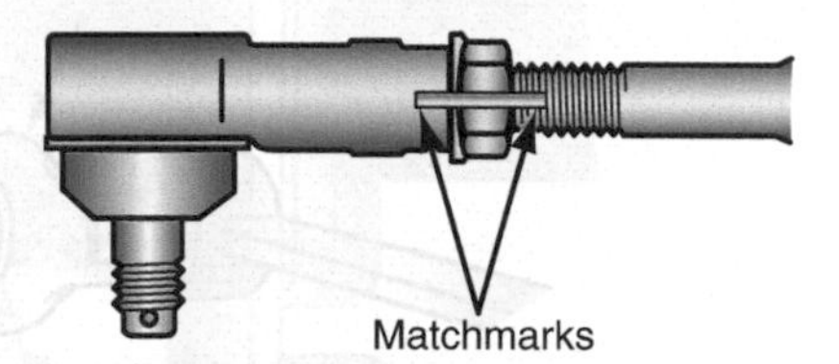

Figure 13-42 Aligning marks on the outer tie rod ends, jam nuts, and tie rods.

16. Install the outer tie rod ends and jam nuts. Align the marks placed on the tie rod ends, jam nuts, and tie rods during disassembly (**Figure 13-42**). Leave the jam nuts loose until the steering gear is installed and the front wheel toe is adjusted.
17. Follow the steering gear installation procedure described earlier in this chapter to install the steering gear in the vehicle. Tighten all fasteners to the specified torque. (See Chapter 12 for power steering pump filling and bleeding procedure.) Check the front wheel toe and tighten the outer tie rod end jam nuts to the specified torque. Tighten the outer bellows boot clamps. Road-test the vehicle and check for proper steering operation and control.

Photo Sequence 27 shows a typical procedure for removing and replacing an inner tie rod end on a power rack and pinion steering gear.

PHOTO SEQUENCE 27
Typical Procedure for Removing and Replacing Inner Tie Rod End, Power Rack, and Pinion Steering Gear

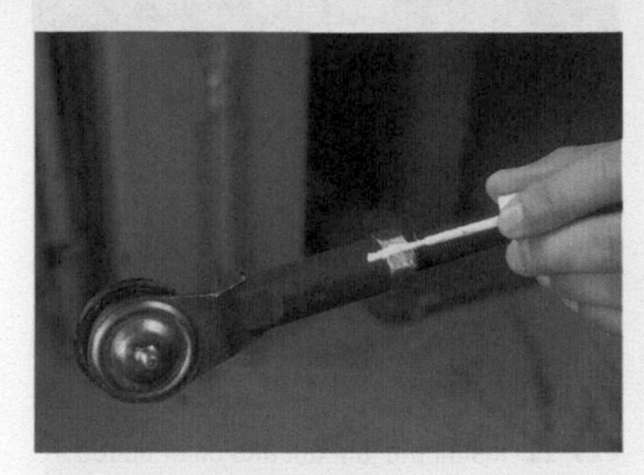

P27-1 Place an index mark on the outer tie rod end, jam nut, and tie rod.

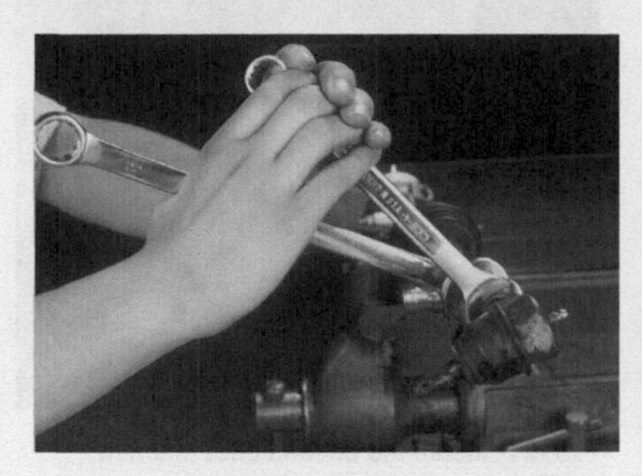

P27-2 Loosen the jam nut and remove the outer tie rod end and jam nut.

P27-3 Remove the inner and outer boot clamps.

P27-4 Remove the bellows boot from the tie rod.

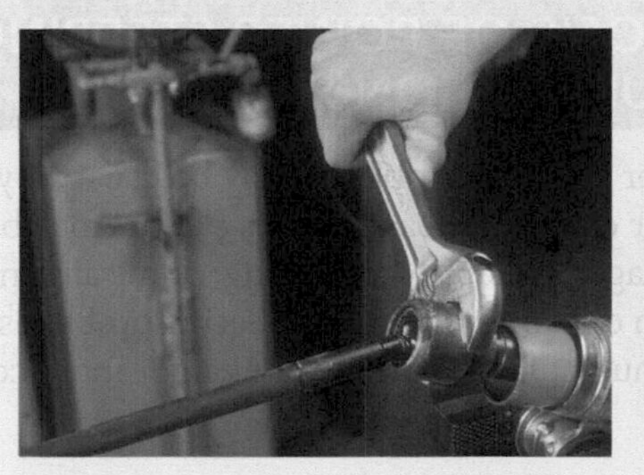

P27-5 Hold the rack with the proper size wrench and loosen the inner tie rod end with the proper wrench.

P27-6 Remove the inner tie rod from the rack.

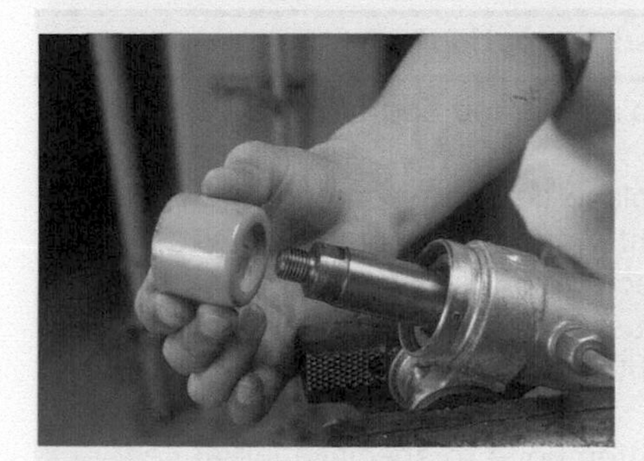

P27-7 Be sure the shock damper ring is in place on the rack.

P27-8 Install the inner tie rod on the rack, and tighten the tie rod to the specified torque while holding the rack with the proper size wrench.

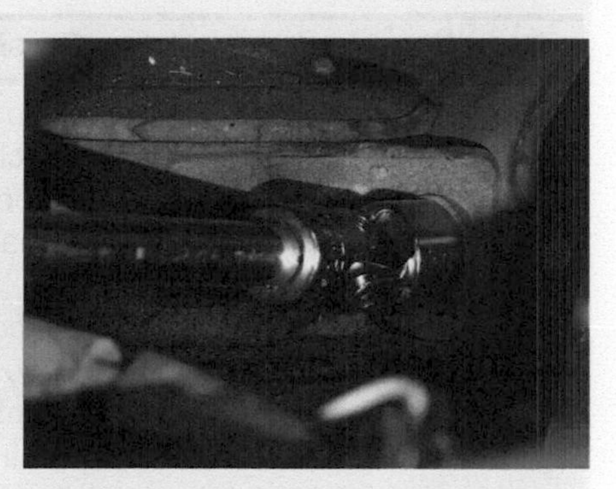

P27-9 Support the inner tie rod on a vise, and stake both sides of the inner tie rod joint with a hammer and punch.

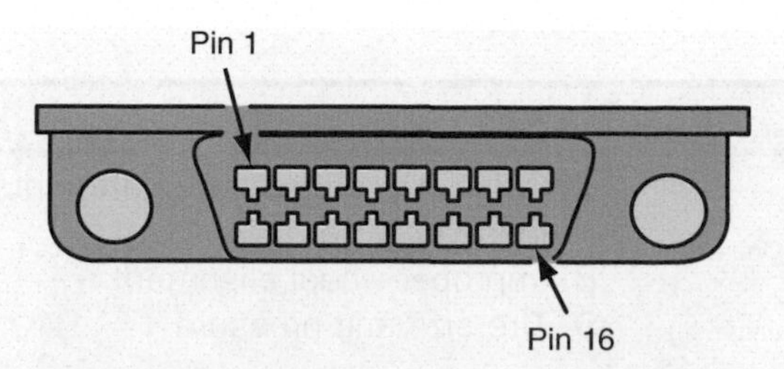

Figure 13-43 Data link connector (DLC) on-board diagnostic-II (OBD-II) system.

DIAGNOSIS OF VARIABLE EFFORT STEERING SYSTEMS

Caution

Always be sure the ignition switch is off when connecting or disconnecting the scan tool. Computer system and/or scan tool damage may occur if the ignition switch is on when connecting or disconnecting the scan tool from the data link connector (DLC).

The most common customer complaints on variable effort steering systems are continually heavy or light steering effort. When diagnosing a variable effort steering system, the first step is to visually inspect the system. Be sure the power steering fluid level is correct, and check the power steering belt tension. Inspect the wiring harness connected to the actuator solenoid in the steering gear.

If the visual inspection does not locate any defects, connect a scan tool to the DLC near the steering column under the dash (**Figure 13-43**). Turn the ignition switch on and select the Electronic Brake and Traction Control Module (EBTCM) on the scan tool menu. If there is an electrical defect in the actuator solenoid, connecting wires, or EBTCM, Diagnostic Trouble Code (DTC) C1241 will be displayed on the scan tool. To locate the exact cause of this DTC, perform voltmeter tests on the actuator solenoid and connecting wires. A detailed diagnostic chart is included in the vehicle manufacturer's service manual.

SERVICE TIP Some generic scan tools do not have the capability to reprogram computers.

If the customer complains about continually heavy or light steering and there are no defects in the variable effort steering system, reprogramming of the system may be required.

To reprogram the system, follow these steps:

1. Connect the scan tool to the DLC.
2. Select Variable Effort Steering and Recalibration on the scan tool.
3. Respond to the VIN query on the scan tool. If the displayed VIN is incorrect, enter the correct VIN.
4. When "Factory Standard Calibration Will Be Used For This VIN" appears on the scan tool, press Enter on the scan tool to install factory calibration.
5. When calibration is complete, disconnect the scan tool.

DIAGNOSIS OF RACK-DRIVE ELECTRONIC POWER STEERING

When the ignition switch is turned on in a vehicle with **rack-drive electronic power steering**, the electronic power steering (EPS) light in the instrument panel should be illuminated (**Figure 13-44**). This EPS light action proves the bulb and related circuit are functioning normally. When the engine starts, the EPS light is turned off by the EPS control unit if there are no electrical defects in the EPS system. The EPS light should remain off while the engine is running.

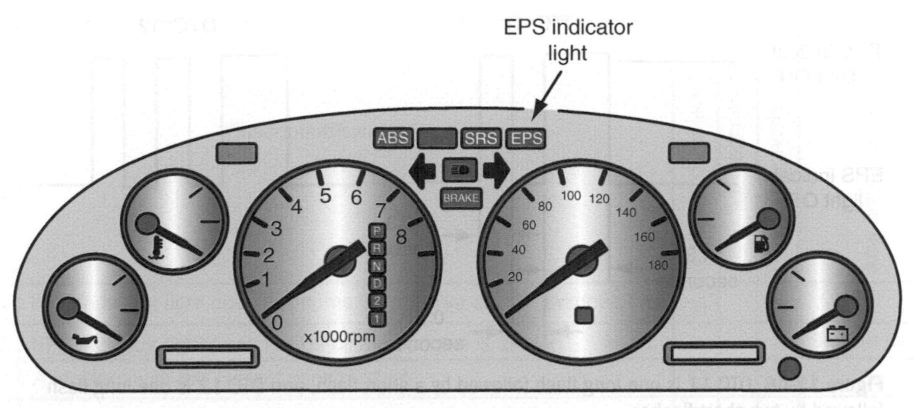

Figure 13-44 Electronic power steering (EPS) warning light.

If an electrical defect occurs in the EPS system, the EPS control unit turns on the EPS light to alert the driver that a defect exists in the EPS system. Under this condition, the EPS control unit shuts down the EPS system, and electric assist is no longer available. Manual steering with increased steering effort is provided in this mode. When the EPS control unit senses an electrical defect in the EPS system, a diagnostic trouble code (DTC) is stored in the control unit memory. The EPS control unit memory is erased if battery voltage is disconnected from the control unit. The 7.5-ampere clock fuse may be disconnected for 10 seconds to disconnect battery voltage from the EPS control unit and erase the DTCs from the control unit memory.

SERVICE TIP If the special jumper wire is left in the EPS service connector and the engine is started, the malfunction indicator light (MIL) for the engine computer system is illuminated.

If the EPS light is on with the engine running, shut the ignition switch off and disconnect the clock fuse for 10 seconds; then road-test the car to determine if the EPS light comes back on. If the EPS light comes on again with the engine running, proceed with the diagnosis. Locate the two-wire EPS service check connector under the glove box. With the ignition switch off, connect a special jumper wire to the service check connector (**Figure 13-45**). This connects the two wires together in the service check connector.

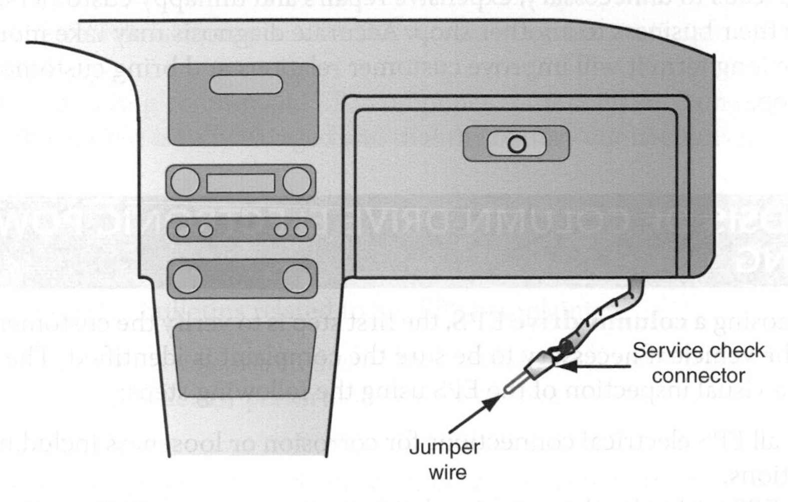

Figure 13-45 EPS system service check connector and special jumper wire.

2. Turn on the ignition switch, and select Vehicle DTC Information on the scan tool. If the scan tool displays "No Comm. From Any Computer," repair the data link communication problem by referring to the appropriate diagnostic procedure in the vehicle manufacturer's service manual.
3. Be sure the engine cranks and starts properly.
4. Turn on the ignition switch with the engine stopped, and select List All DTCs on the scan tool.
5. Record all DTCs displayed on the scan tool. Check for DTCs from other computer systems that may affect the EPS system. For example, a DTC representing the VSS is displayed as a PCM DTC, but this sensor signal is transmitted on the data links to the power steering control module (PSCM) and affects EPS operation.

Repair the causes of the DTCs displayed on the scan tool. Some typical EPS DTCs are:

1. C0460 steering position sensor
2. C0475 electric steering motor circuit
3. C0545 steering wheel torque input sensor
4. C0870 PSCM voltage range performance

On-Board Diagnostic-II (OBD-II) systems are a type of computer system mandated on cars and light trucks since 1996. OBD-II systems have a number of mandated standardized features, including several monitoring systems in the PCM.

In these **On-Board Diagnostic-II (OBD-II)** DTCs, the first letter C indicates the DTC is related to a chassis problem. If the second digit in the DTC is a 0, the DTC is a universal Society of Automotive Engineers (SAE) code. When the second digit in the DTC is a 1, a vehicle manufacturer's code is indicated. The third digit in the DTC indicates the subsystem to which the DTC belongs, and the last two digits indicate the specific area where the fault exists. A complete list of DTCs is provided in the vehicle manufacturer's service manual. Some scan tools provide an explanation of each DTC indicating the area where the problem is located. A DTC indicates only the area in which a problem exists. Voltmeter and ohmmeter tests may be required to test the circuit and locate the exact cause of the problem. The vehicle manufacturer's service manual also provides a detailed diagnostic procedure for each DTC. Always follow these diagnostic procedures when diagnosing the cause of a DTC.

SERVICE TIP When connecting certain scan tools to controller area network (CAN) or local area network (LAN) data link systems, a CANdi module must be connected in series between the scan tool and the DLC. Always refer to the vehicle manufacturer's recommended procedure.

Scan Tool Data Display

The scan tool also displays the following EPS data, which are very helpful when diagnosing EPS defects:

1. Battery voltage—The scan tool displays 6.5 V to 16.5 V as sensed at the power steering control module.
2. Calculated system temperature—The ambient temperature up to 275°F (135°C) as sensed by an internal temperature sensor in the PSCM.
3. Motor command—The amount of EPS assist motor current commanded by the PSCM.
4. Steering shaft torque—The amount of torque being applied to the steering column shaft. A positive value indicates a right turn, and a negative number indicates a left turn.

5. Steering position sensor voltage 1—The amount of steering position sensor voltage from the sensor signal 1 circuit. The voltage range is 0 V to 5 V as the steering wheel is turned.
6. Steering position sensor voltage 2—The amount of steering position sensor voltage from the sensor signal 2 circuit. The voltage range is 0 V to 5 V as the steering wheel is turned.
7. Steering tuning—The scan tool displays 1 or 2 depending on the vehicle. Steering tuning 1 is for 4-cylinder engines, and steering tuning 2 is for 6-cylinder engines. Steering tuning 3 is for the extended sedan.
8. Steering wheel position—The scan tool displays from 562° to −562° indicating the number of degrees the steering wheel has been turned for the center position. A positive number indicates a right turn, and a negative number indicates a left turn.
9. Torque sensor signal 1—The scan tool displays 0.25 V to 4.75 V indicating the voltage from the steering shaft torque sensor 1 circuit.
10. Torque sensor signal 2—The scan tool displays 0.25 V to 4.75 V indicating the voltage from the steering shaft torque sensor 2 circuit.
11. Vehicle speed—The actual vehicle speed from 0 to 158 mph (255 km/h).

SERVICE TIP After the ignition switch is turned off, wait 25 seconds before performing any procedures that require disconnecting the vehicle battery. If this precaution is not observed, the PSCM memory may be erased. This precaution also applies to torque sensor calibration and steering tuning selection.

Steering Position Sensor Calibration

After completing a wheel alignment on some system, or if the power steering control module (PSCM) or steering column is replaced, use the scan tool to calibrate the center position of the steering wheel position sensor as follows:

1. Be sure the ignition switch is off, and connect the scan tool to the DLC.
2. Turn on the ignition switch, but do not start the engine.
3. Turn the steering wheel from lock-to-lock and count the number of turns. From the fully left or fully right position, turn the steering wheel back toward the center position one-half the total number of turns.
4. Select Special Functions on the scan tool.
5. Select Steering Position Sensor Calibration on the scan tool and press the enter key. The scan tool screen indicates Calibration in Progress during the calibration procedure.
6. When the calibration process is completed, and the scan tool displays Calibration Completed, press the exit key.
7. Use the scan tool to clear any DTCs.
8. Turn off the ignition switch and disconnect the scan tool.

Torque Sensor Calibration

If the PSCM or steering column is replaced, a scan tool must be used to perform a center position calibration for the steering shaft torque sensor as described in the following steps:

1. Be sure the ignition switch is off, and connect the scan tool to the DLC.
2. Turn on the ignition switch, but do not start the engine.
3. Turn the steering wheel from lock-to-lock and count the number of turns. From the fully left or fully right position, turn the steering wheel back toward the center position one-half the total number of turns.

4. After the steering wheel is centered, remove your hands or any other objects from the steering wheel. Be sure the suspension is relaxed, and that no uneven force is applied to the steering system.
5. Select Special Functions on the scan tool.
6. Select Torque Sensor Calibration on the scan tool and press the enter key. The scan tool screen indicates Calibration in Progress during the calibration procedure.
7. When the calibration process is completed, and the scan tool displays Calibration Completed, press the exit key.
8. Use the scan tool to clear any DTCs.
9. Turn off the ignition switch and disconnect the scan tool.

If the PSCM and motor assembly is removed from the steering column, inspect the steering column assist mechanism input shaft for the rotor isolator bumper (**Figure 13-48**). If present, the isolator bumper should be removed from the input shaft, and this bumper must be installed in the rotor isolator in the PSCM and motor assembly. Remove any debris from the steering column assist mechanism housing, but do not remove the remaining grease on the steering column assist mechanism input shaft.

Steering Tuning Selection

If the PSCM is replaced, these steps must be followed to perform a steering tuning procedure:

1. Be sure the ignition switch is off, and connect the scan tool to the DLC.
2. Turn on the ignition switch, but do not start the engine.
3. Select Special Functions on the scan tool.
4. Select Steering Tuning Selection on the scan tool and press the enter key. The scan tool screen indicates Selection in Progress during the selection procedure.
5. When the selection process is completed, and the scan tool displays Selection Completed, press the exit key.
6. Use the scan tool to clear any DTCs.
7. Turn off the ignition switch and disconnect the scan tool.

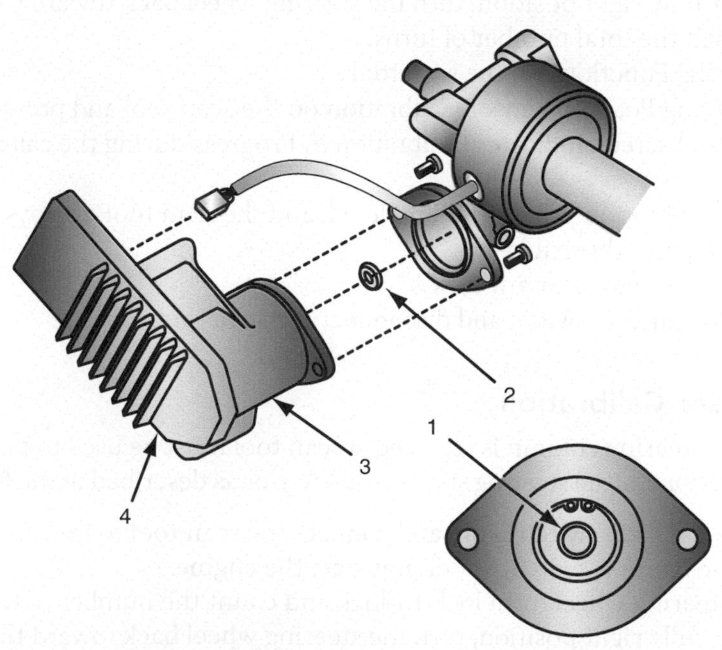

Figure 13-48 Removing and replacing the PSCM and motor assembly.

ACTIVE STEERING SYSTEM PRELIMINARY DIAGNOSIS

The active steering system has a warning light and a service light in the instrument panel (**Figure 13-49**). When the active steering system is operating normally, the active steering warning light and service light go out after the engine starts. If an electronic defect occurs in the active steering system, the warning and service lights are illuminated. The service light indicates that a service message is available in the central information display screen in the instrument panel. When "service" is selected in the central information display screen, the following messages may be displayed:

Classroom Manual
Chapter 13, page 401

1. Active steering! Exercise care when steering.
   Active steering fault.
   Steering behavior altered. Steering wheel might be at an angle. Possible to continue journey with caution. Exercise care when steering. Have the problem checked by the nearest BMW service.
2. Active steering inactive.
   Active steering.
   Active steering inactive.
   Steering behavior altered. Steering wheel might be at an angle. Possible to continue journey with caution. Exercise care when steering.
3. Servotronic failure!
   Servotronic failure.
   Possible to continue journey with caution.
   Important: Power steering assistance is no longer automatically adapted to the vehicle's speed.
   Have problem checked by the nearest BMW service.

Preliminary Inspection

If the active steering warning lights are illuminated indicating a defect in the system, the first step in the diagnostic procedure is to perform a visual inspection of all system components, wiring, and fiber-optic cables. During the preliminary inspection, check these items:

1. Check the power steering fluid level and condition.
2. Inspect all wiring harness and electrical connectors in the system for damage, corrosion, and frayed insulation.

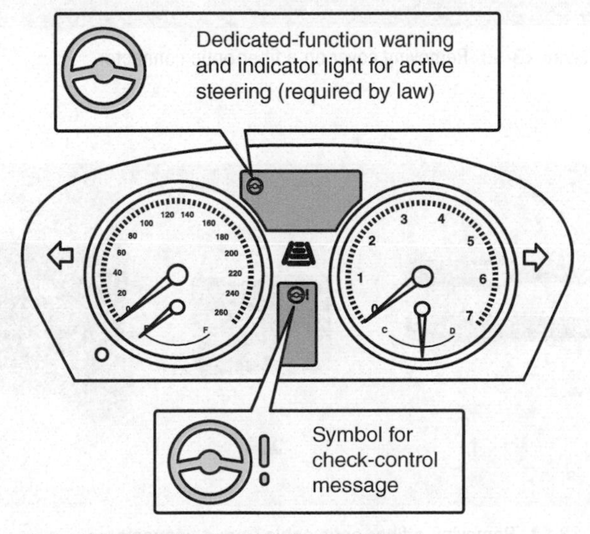

Figure 13-49 Active Steering System light.

3. Inspect all fiber-optic cables for sharp bends, heat damage, punctures, cuts, or looseness.
4. Check all the hydraulic lines and fittings for damage or leaks.
5. Inspect all system mounting bolts for looseness.

Fiber-Optic Cable Service

Many vehicles with active steering have fiber-optic cable networks. Media oriented systems transport (MOST) networks have green fiber-optic cables, and intelligent safety and integration systems (ISIS) networks have yellow fiber-optic cables. Some manufacturers provide orange fiber-optic cables for repair purposes. Fiber-optic cables should always be inspected for sharp bends, heat damage, punctures or cracks, and loose connectors. Repair or replace fiber-optic cables as necessary. To remove a fiber-optic cable from a connector, follow this procedure:

1. To remove the connector, use a small screwdriver in the connector openings (**Figure 13-50**), and expand the cover catches. Remove the cover.
2. To remove the fiber-optic cable from the connector lift the lock (**Figure 13-51**) and carefully feed the fiber-optic cable out of the connector.
3. To install the fiber-optic cable in the connector, push the cable all the way into the connector and be sure the lock secures the cable.
4. If it is necessary to remove the fiber-optic cable from its mounting position, press to release the lock in the bore (**Figure 13-52**) and remove the cable connector.

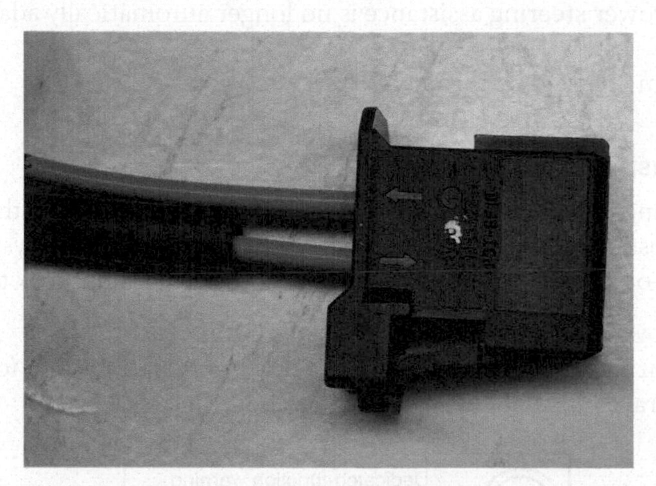

Figure 13-50 Removing cover on a fiber-optic connector.

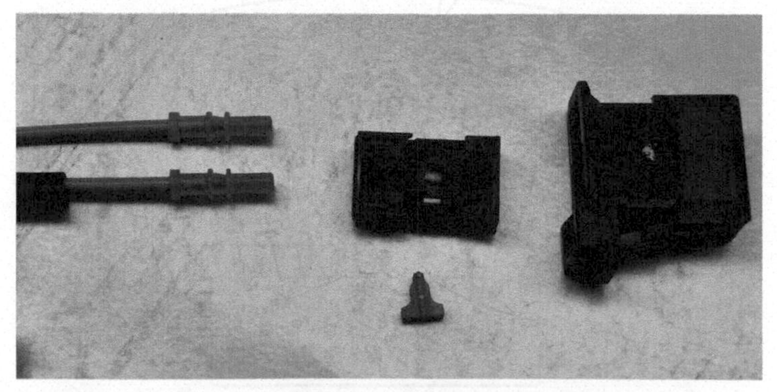

Figure 13-51 Removing a fiber-optic cable from a connector.

Figure 13-52 Removing a fiber-optic connector from its mounting surface.

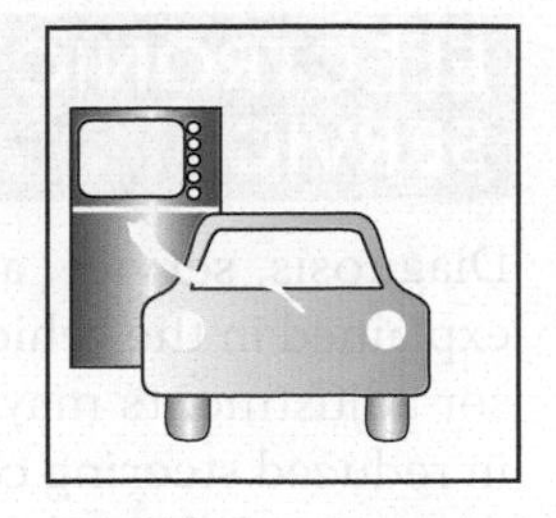

Figure 13-53 Diagnosis system.

Active Steering Diagnosis System

A diagnosis system is supplied by the vehicle manufacturer (**Figure 13-53**). The diagnosis system is connected to the vehicle DLC, and this system is used for all coding, programming, and diagnosis of the active steering system and other systems. ECU and module replacement nearly always require coding/programming of the replacement ECU or module. After the diagnosis system is connected to the DLC, to access the active steering system, select "Service functions," "Suspension," and "Active steering" on the diagnosis system menus. Always follow the vehicle manufacturer's recommended service, diagnosis, or programming procedure when using the diagnosis system.

If an electronic defect occurs in the active steering system, the diagnosis system displays DTCs. As in other electronic systems, the DTCs represent defects in a certain area, and voltmeter or ohmmeter tests may have to be performed to locate the root cause of the problem. The technician must select the appropriate ECU to access the DTCs. For example, the **servotronic valve** and the **electronically controlled orifice (ECO) valve** are controlled by the **safety and gateway module (SGM)** on some models. Therefore, DTCs representing electronic faults in these valves are stored in the SGM, and the technician must access the SGM in the diagnosis system to obtain these DTCs. On other models, the servotronic valve and ECO valve are controlled by the **active steering ECU**, and DTCs representing defects in these components are stored in this ECU.

Certain service procedures and specific adjustments must be performed using the diagnosis system. For example, a steering angle sensor adjustment must be performed after any of the following procedures:

1. Adjustment procedures on the front steering or suspension.
2. Disconnecting the battery.
3. Any replacement of steering components.
4. Replacement of steering column sensors or switches.
5. Replacement of **dynamic stability control (DSC) ECU**, **active roll stabilization (ARS) ECU**, or active steering ECU.

When the active steering system is accessed in the diagnosis system, the steering angle sensor adjustment procedure is entered by selecting "Initial operation/adjustment for active front steering." Follow the vehicle manufacturer's recommended procedure in the diagnosis system to complete the steering angle sensor adjustment. Service and adjustment procedures may vary depending on the model and year of vehicle.

The need for a wheel alignment is indicated by any of the following conditions:

1. Inclination of the steering wheel from the straight-ahead position with no faults stored in the active steering ECU memory.
2. Fault(s) stored in the active steering ECU memory and diagnosis system troubleshooting provides "Check wheel alignment" message.
3. Replacement of suspension and/or steering components that requires wheel alignment.

ELECTRONIC FOUR-WHEEL STEERING DIAGNOSIS AND SERVICE

Diagnosis, service, and adjustments on 4WS systems must be performed precisely as explained in the vehicle manufacturer's service manual. On some systems inaccurate sensor adjustments may cause improper rear wheel steering operation, and this may result in reduced steering control. Always follow the diagnostic, service, and adjustment procedures carefully and accurately!

PRELIMINARY INSPECTION

Prior to any 4WS diagnosis, the following concerns should be considered:

1. Have any suspension modifications been made that would affect steering?
2. Are the tire sizes the same as specified by the vehicle manufacturer?
3. Are the tires inflated to the pressure specified by the vehicle manufacturer?
4. Is the power steering belt adjusted to the vehicle manufacturer's specified tension?
5. Is the power steering pump reservoir filled to the proper level with the type of fluid specified by the vehicle manufacturer?
6. Is the engine idling at the speed specified by the vehicle manufacturer? Is the idle speed steady?
7. Is the steering wheel original equipment?
8. Is the battery fully charged?
9. Are all electrical connections clean and tight?
10. Is there any damaged electrical wiring in the system?
11. Are the rear wheel steering fuses in satisfactory condition?
12. Are there any damaged or worn rear steering gear or rear axle components?
13. Is the Service 4 Wheel Steer indicator illuminated in the instrument panel cluster?

A fail-safe mode may be called a backup mode.

The SAE J1930 terminology is an attempt to standardize electronics terminology in the automotive industry.

Photo Sequence 28 illustrates a preliminary four-wheel steering inspection.

PHOTO SEQUENCE 28
Preliminary Inspection, Four-Wheel Steering Diagnosis

P28-1 Inspect the front and rear suspension and steering for modifications and damage that could affect steering.

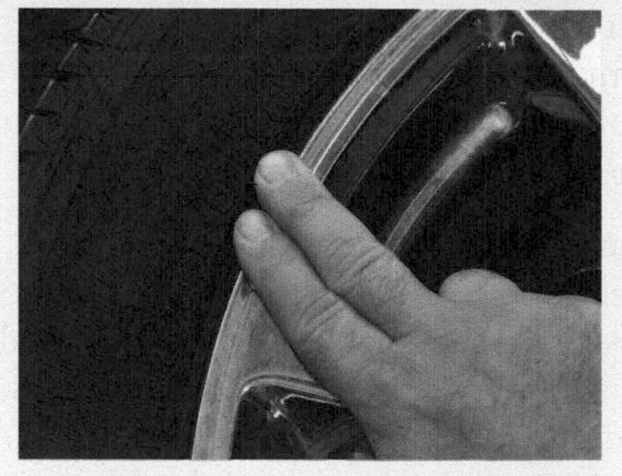

P28-2 Check the tire sizes to be sure they are the size specified by the vehicle manufacturer.

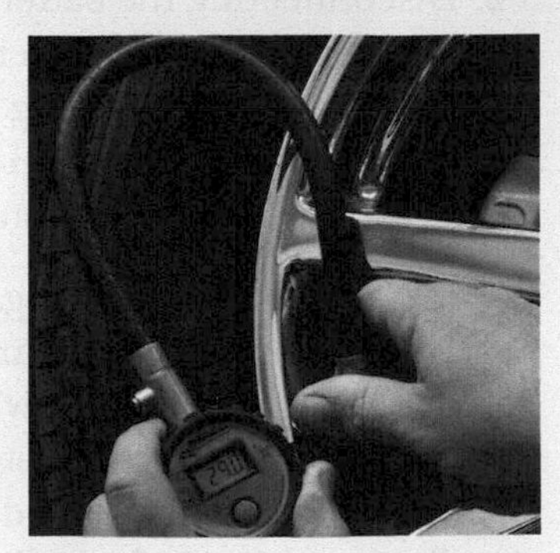

P28-3 Inflate the tires to the specified pressure.

PHOTO SEQUENCE 28 (CONTINUED)

P28-4 Check the power steering belt tension.

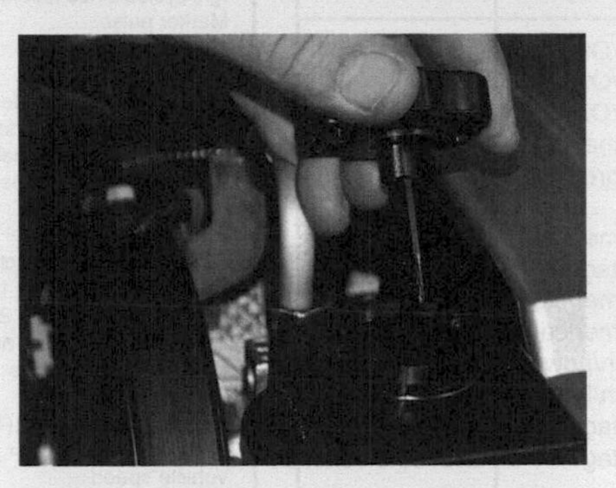
P28-5 Check the fluid level in the power steering pump reservoir.

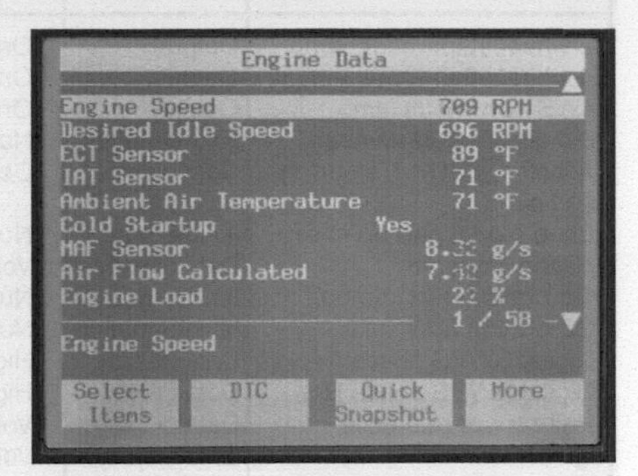

P28-6 Be sure the engine idle speed is correct.

P28-7 Be sure the steering wheel is original equipment.

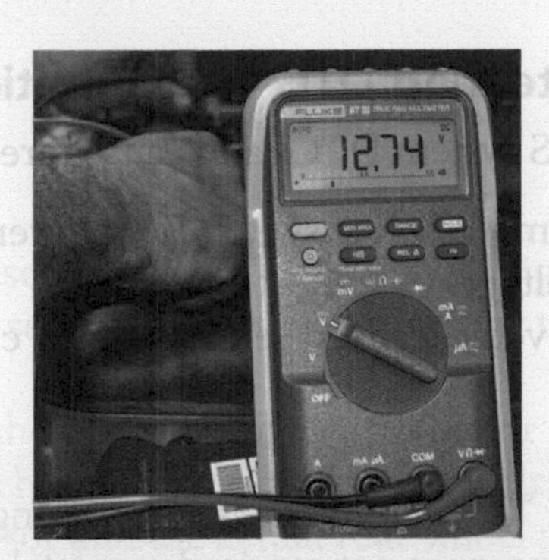

P28-8 Be sure the battery is fully charged.

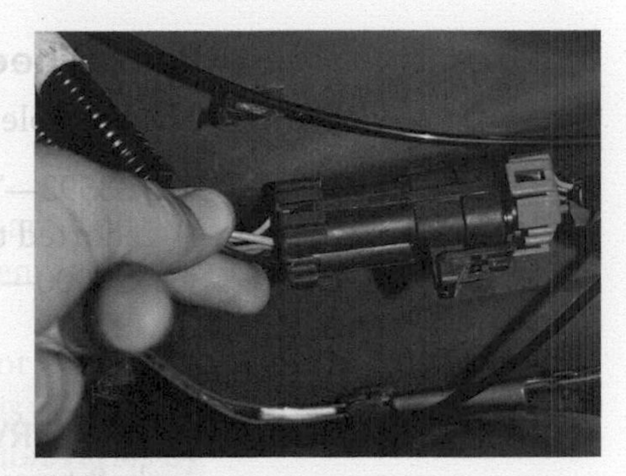
P28-9 Inspect all electrical connections and harness in the 4WS system to be sure they are in satisfactory condition.

P28-10 Check the rear wheel steering system fuses.

P28-11 Check the 4WS warning light in the instrument panel for proper operation.

Rear Wheel Steering Data Display

Select Rear Wheel Steering and Data Display on the scan tool. Observe the data displayed on the scan tool and illustrated in **Figure 13-54** and **Figure 13-55**. Compare the displayed data to the vehicle manufacturer's specified data. Repair the cause of any incorrect data.

Compare the incorrect data to the DTCs recorded previously. If any DTCs and incorrect data are from the same component, repair this component or the related circuit.

In the SAEJ1930 terminology, the term malfunction indicator light (MIL) replaces other terms for computer system indicator lights.

CASE STUDY

A customer complains about excessive steering effort on a 2012 Buick Enclave. The technician road-tested the car and found no evidence of hard steering. Further questioning of the customer revealed that the problem only occurred when, with the engine idling, a 90° parking maneuver was attempted in either forward or reverse; it occurred during both right and left turns.

Prior to starting the engine, the technician connected a power steering pressure test gauge in series with the high-pressure hose. When the engine was started, the technician discovered the steering effort seemed normal when the vehicle was driven and only lane change maneuvers were attempted. The line pressure was at the lower end of the normal range compared to specifications when the wheels were not turned. But when a 90° turn was attempted at idle speed, the line pressure dropped off significantly, and the steering felt much like manual steering while the steering wheel was turned. A deadhead pressure test was performed and line pressure was below specifications.

Since the power steering pump pressure was lower than specified, the technician concluded the power steering pump was faulty. The technician also reasoned that other causes of excessive steering effort, such as binding in the steering column, flexible coupling, or power steering rack spool valve would be constant.

The customer was advised that the power steering pump required replacement. A replacement power steering pump was installed, and the system was flushed and refilled with power steering fluid. When the engine was started, the steering effort and the power steering pump pressure were normal.

ASE-STYLE REVIEW QUESTIONS

1. While discussing steering wheel free play:
   Technician A says excessive steering wheel free play may be caused by a worn outer tie rod end.
   Technician B says excessive steering wheel free play may be caused by loose steering gear mounting bushings.
   Who is correct?
   A. A only C. Both A and B
   B. B only D. Neither A nor B

2. While discussing steering effort:
   Technician A says a tight rack bearing adjustment may cause excessive steering effort.
   Technician B says a loose inner tie rod end may cause excessive steering effort.
   Who is correct?
   A. A only C. Both A and B
   B. B only D. Neither A nor B

3. While discussing steering gear removal and replacement on air bag–equipped vehicles:
   Technician A says the steering column should be locked in the centered position to prevent air bag clock spring damage.
   Technician B says that an index mark should be placed on the lower universal joint and steering gear pinion shaft.
   Who is correct?
   A. A only C. Both A and B
   B. B only D. Neither A nor B

4. While being test driven a vehicle equipped with power rack and pinion steering has a clunking noise, which can also be felt in the steering wheel as it is driven straight ahead on rough pavement. What is the most likely cause?
   A. Loose center link
   B. Loose wheel bearing
   C. Worn flexible steering coupler
   D. Warped brake rotors

5. A steering imbalance or low power assist in both directions may be caused by all of the following EXCEPT:
   A. Defective rack seals
   B. Defective rotary valve rings or seals
   C. Worn lower control arm bushings
   D. Restricted power steering lines or hoses
6. While discussing power rack and pinion steering gears:
   Technician A says power steering fluid should be inspected for discoloration and contamination with metal particles.
   Technician B says after flushing and refilling the power steering system, it is not necessary to perform an air bleeding procedure.
   Who is correct?
   A. A only C. Both A and B
   B. B only D. Neither A nor B
7. When performing a Learn Rear Wheel Alignment procedure:
   Technician A says the steering wheel must be centered.
   Technician B says during this procedure the front wheels must be lifted off the shop floor.
   Who is correct?
   A. A only C. Both A and B
   B. B only D. Neither A nor B
8. *Technician A* says the pinion gear spool valve can bind or stick causing fluid to be directed to the power rack piston chamber causing a drift or pull to one side.
   Technician B says that a front tire with excessive conicity can cause the vehicle to pull or drift to one side and is often referred to as a radial tire pull.
   Who is correct?
   A. A only C. Both A and B
   B. B only D. Neither A nor B
9. When diagnosing a 4WAS with a scan tool:
   Technician A says the CAN diagnostic monitor indicates if transmit/receive communication is satisfactory.
   Technician B says that during the active tests the scan tool activates specific system components.
   Who is correct?
   A. A only C. Both A and B
   B. B only D. Neither A nor B
10. When diagnosing an active steering system with a scan tool:
    Technician A says if the DTC's first letter is a C, it indicates it is related to a chassis problem.
    Technician B says that when the DTC's second digit is a 1, the DTC is a universal code.
    Who is correct?
    A. A only C. Both A and B
    B. B only D. Neither A nor B

ASE CHALLENGE QUESTIONS

1. While discussing power rack and pinion steering gears:
   Technician A says inner rack seal leaks can cause power rack and pinion steering to turn more easily in one direction than another.
   Technician B says a hissing noise is normal when the steering is turned to the limit on either side.
   Who is correct?
   A. A only C. Both A and B
   B. B only D. Neither A nor B
2. A pull, wandering, or poor straight-ahead tracking may be caused by all of the following EXCEPT:
   A. Worn or loose rack mounting bushings.
   B. Rack piston seal leaks.
   C. Binding spool valve.
   D. Loose or worn tie rod ends.
3. A customer says his front-wheel-drive car with power rack and pinion steering is hard to steer for several minutes the first time the car is driven during the day.
   Technician A says the cause of the problem could be a plugged spool valve spring.
   Technician B says the cause of the problem could be a scored steering gear cylinder.
   Who is correct?
   A. A only C. Both A and B
   B. B only D. Neither A nor B

Instructor's Response

Name ______________________________ Date ______________

JOB SHEET
64

DIAGNOSE POWER RACK AND PINION STEERING GEAR OIL LEAKAGE PROBLEMS

Upon completion of this job sheet, you should be able to diagnose power rack and pinion steering gear oil leakage problems.

ASE Education Foundation Task Correlation

This job sheet addresses the following **MLR** task:

B.4. Inspect for power steering fluid leakage. **(P-1)**

This job sheet addresses the following **AST/MAST** tasks:

B.5. Diagnose power steering gear (rack and pinion) binding, uneven turning effort, looseness, hard steering, and noise concerns; determine needed action. **(P-2)**

B.11. Inspect for power steering fluid leakage; determine needed action. **(P-1)**

We Support ASE Education Foundation

Tools and Materials

Power steering fluid

Describe the vehicle being worked on:

Year ______________ Make ______________ Model ______________

VIN ______________ Engine type and size ______________

Procedure **Task Completed**

WARNING If the engine has been running for a length of time, power steering gears, pumps, lines, and fluid may be very hot. Wear eye protection and protective gloves when servicing these components.

1. Be sure the power steering reservoir is filled to the specified level with the proper steering fluid. ☐
2. Be sure the power steering fluid is at normal operating temperature. If necessary, rotate the steering wheel several times from lock-to-lock to bring the fluid to normal operating temperature.

   Is the power steering reservoir filled to the specified level? ☐ Yes ☐ No

   Is the power steering fluid at normal temperature? ☐ Yes ☐ No

   Instructor check ______________________________
3. Inspect the cylinder end of the steering gear for oil leaks.

   Oil leaks at cylinder end of the power steering gear:

   ☐ Satisfactory ☐ Unsatisfactory

   If there are oil leaks in this area, state the necessary repairs and explain the reasons for your diagnosis.

   ____

   ____

   ____

4. Inspect the steering [illegible] running and the steering wheel turned so the rack is against the left internal stop.

SERVICE TIP An oil leak [illegible] boots, because the oil may travel through the breather tube between the boots.

Oil leaks at the pinion end of the steering gear with the rack against the left inner stop:

☐ Satisfactory ☐ Unsatisfactory

If there are oil leaks in this area, state the necessary repairs and explain the reasons for your diagnosis.

5. Inspect the pinion end of the power steering gear housing for oil leaks with the steering wheel turned in either direction.

Oil leak at the pinion end of the housing with the steering wheel turned in either direction: ☐ Satisfactory ☐ Unsatisfactory

If there are oil leaks in the area, state the necessary repairs and explain the reasons for your diagnosis.

6. Inspect the pinion coupling area for oil leaks.

Oil leaks in the pinion coupling area: ☐ Satisfactory ☐ Unsatisfactory

If there are oil leaks in the area, state the necessary repairs and explain the reasons for your diagnosis.

7. Inspect all the lines and fittings on the steering gear for oil leaks [illegible] the wheels with the engine running.

Oil leaks at steering gear fittings: ☐ Satisfactory ☐ Unsatisfactory

Oil leaks in steering gear hoses and lines: ☐ Satisfactory ☐ Unsatisfactory

If there are oil leaks in the power steering [illegible] lines, or hoses, state the necessary repairs and explain the reasons for your diagnosis.

Name ______________________________ Date ______________

JOB SHEET 65

PERFORM A LEARN REAR WHEEL ALIGNMENT PROCEDURE ON A 4WS SYSTEM

Upon completion of this job sheet, you should be able to perform a Learn Rear Wheel Alignment procedure on a four-wheel steering system.

ASE Education Foundation Task Correlation

This job sheet addresses the following **MAST** task:

B.18. Inspect, test, and diagnose electrically assisted power steering systems (including using a scan tool); determine needed action. **(P-2)**

This job sheet addresses the following **AST** task:

B.19. Inspect electric power steering assist system. **(P-3)**

We Support ASE Education Foundation

Tools and Materials

Vehicle with a 4WS system

Compatible scan tool

Describe the vehicle being worked on:

Year ______________ Make ______________ Model ______________

VIN ______________ Engine type and size ______________

Procedure **Task Completed**

1. Turn on the ignition switch and start the engine.

   Engine running: ☐ Yes ☐ No

2. Connect the scan tool to the DLC under the dash.

   Scan tool properly connected to the DLC: ☐ Yes ☐ No

3. Center the steering wheel.

   Steering wheel properly centered: ☐ Yes ☐ No

4. Lift the rear of the vehicle so the rear tires are a few inches off the shop floor. Be sure the chassis is securely supported on safety stands, and the rear wheels are centered.

   Rear wheels raised off the shop floor: ☐ Yes ☐ No

   Chassis properly supported on safety stands: ☐ Yes ☐ No

   Steering wheel centered: ☐ Yes ☐ No

5. Select the Learn Alignment menu on the scan tool.

   Learn Alignment displayed on the scan tool: ☐ Yes ☐ No

6. Follow the directions on the scan tool. When directed, the front wheels must be turned 90° to the left and 90° to the right, and then returned to the center position.

   Steering wheel rotated 90° in each direction: ☐ Yes ☐ No

   Steering wheel centered: ☐ Yes ☐ No

6. Does the scan tool display DTC B1000? This DTC indicates an internal defect in the body control module (BCM) and causes the BCM to refuse all additional inputs.

DTC B1000 displayed: ☐ Yes ☐ No

Instructor's Response

____

____

____

CHAPTER 14

RECIRCULATING BALL STEERING GEAR DIAGNOSIS AND SERVICE

Upon completion and review of this chapter, you should be able to:

- Diagnose power recirculating ball steering gear problems.
- Remove and replace power recirculating ball steering gears.
- Adjust worm shaft thrust bearing preload in power recirculating ball steering gears.
- Adjust sector lash in power recirculating ball steering gears.
- Diagnose and repair oil leaks in power recirculating ball steering gears.
- Disassemble, repair, and reassemble power recirculating ball steering gears.

Basic Tools

Basic technician's tool set

Service manual

Terms To Know

Bearing preload	Sector shaft lash	Vehicle wander
Road feel	Steering wheel free play	

Steering gear and linkage inspection and service is extremely important to maintain vehicle safety! For example, if a vehicle has excessive steering wheel free play, this problem should be diagnosed and corrected as soon as possible. If the excessive steering wheel free play is caused by a loose sector shaft lash or worm bearing preload adjustments, follow the proper adjustment procedure to correct the problem. However, if this problem is caused by worn tie rod ends or loose steering gear mounting bolts, these conditions create a safety hazard. If a worn tie rod end becomes disconnected, or the steering gear becomes disconnected from the frame, the result is a complete loss of steering control. This condition may cause a collision resulting in vehicle damage and/or personal injury. Diagnostic **Table 14-1** on page 619 contains symptoms and causes of common steering gear issues.

Classroom Manual
Chapter 14, page 430

POWER RECIRCULATING BALL STEERING GEAR DIAGNOSIS

If the steering gear is noisy, check these items:

1. Loose pitman shaft lash adjustment—may cause a rattling noise when the steering wheel is turned.
2. Cut or worn dampener O-ring on the valve spool—when this defect is present, a squawking noise is heard during a turn.
3. Loose steering gear mounting bolts.
4. Loose or worn flexible coupling or steering shaft U-joints.

PHOTO SEQUENCE 29
Typical Procedure for Performing a Worm Shaft Bearing Preload Adjustment

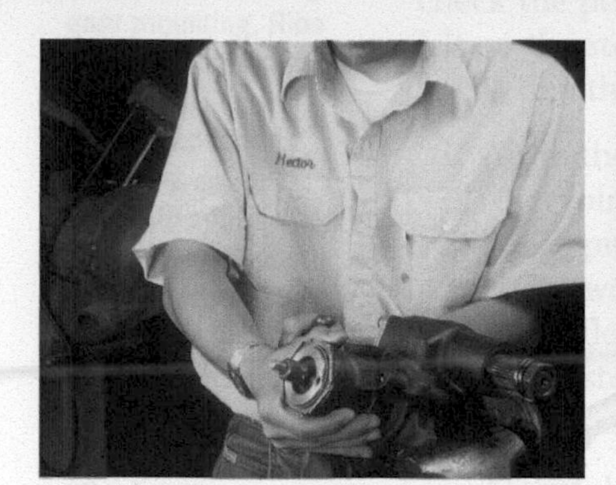

P29-1 Retain the steering gear in a vise.

P29-2 Remove the worm shaft bearing locknut with a hammer and brass punch.

P29-3 Use the proper tool to turn the worm shaft adjuster plug clockwise until it bottoms, and tighten this plug to 20 ft-lb. (27 Nm).

P29-4 Place an index mark on the steering gear housing next to one of the adjuster plug holes.

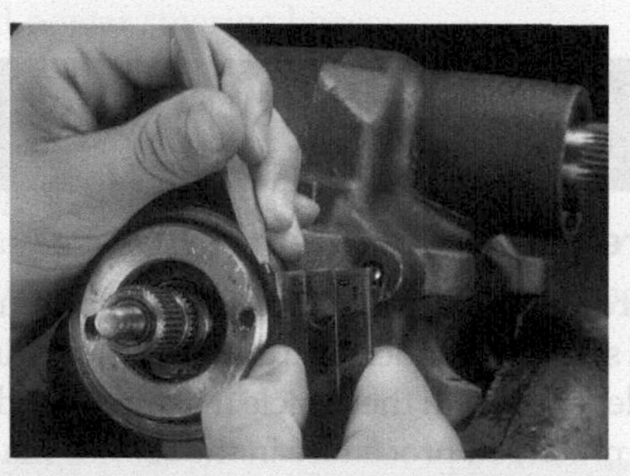

P29-5 Measure 0.50 in. (13 mm) counterclockwise from the mark placed on the housing in Step P11-4, and place a second mark on the housing.

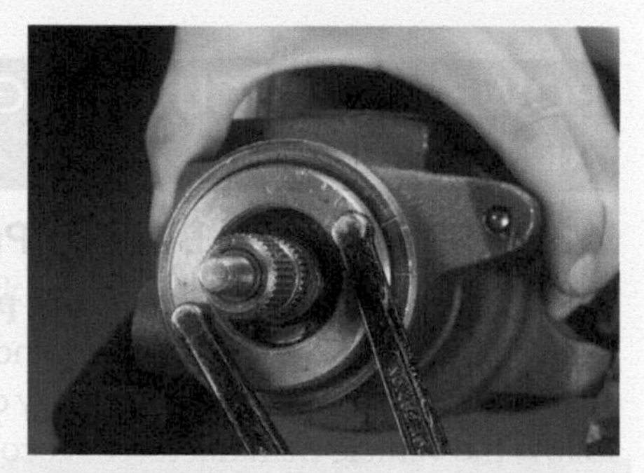

P29-6 Rotate the worm shaft adjuster plug counterclockwise until the adjuster plug hole is aligned with the second mark placed on the housing.

P29-7 Install and tighten the worm shaft adjuster plug locknut.

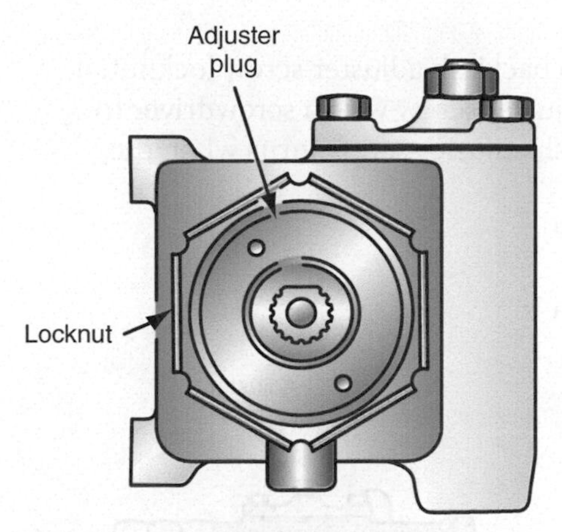

Figure 14-2 Removing worm shaft thrust bearing adjuster plug locknut.

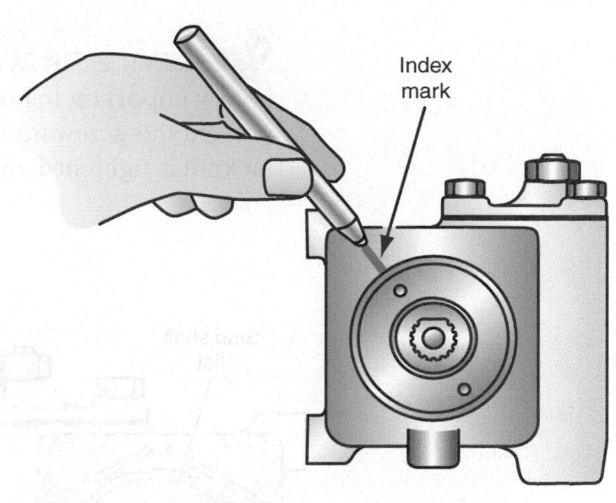

Figure 14-3 Placing index mark on steering gear housing opposite one of the adjuster plug holes.

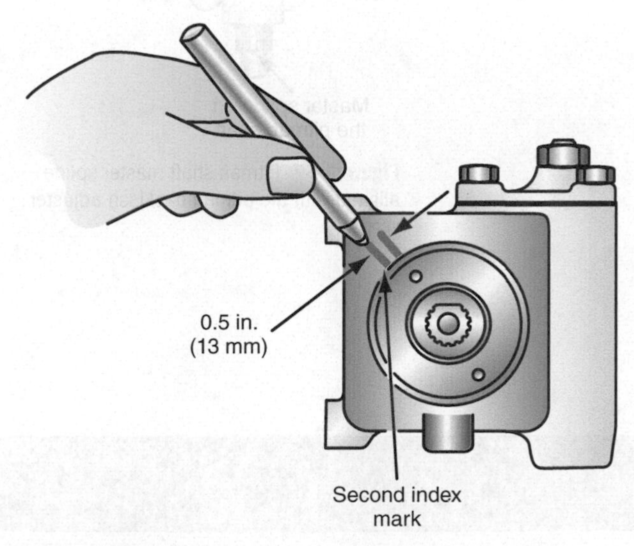

Figure 14-4 Measuring 0.50 in. (13 mm) counterclockwise from the index mark on the steering gear housing.

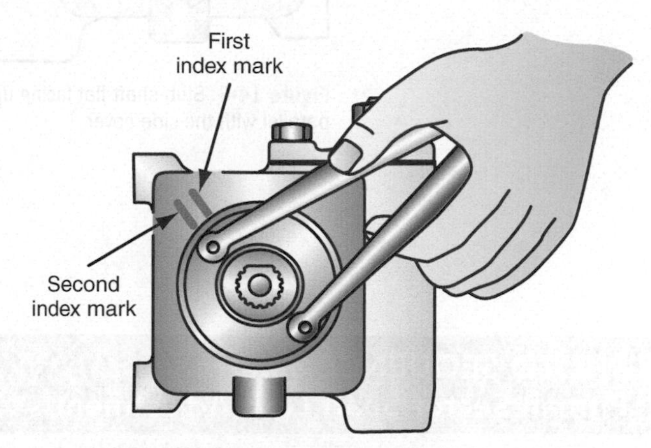

Figure 14-5 Aligning adjuster plug hole with the second index mark placed on steering gear housing.

When the pitman sector shaft lash adjustment is performed, proceed as follows:

1. Rotate the stub shaft from stop to stop and count the number of turns.
2. Starting at either stop, turn the stub shaft back one-half of the total number of turns. In this position, the flat on the stub shaft should be facing upward (**Figure 14-6**), and the master spline on the pitman shaft should be aligned with the pitman shaft backlash adjuster screw (**Figure 14-7**).
3. Loosen the locknut and turn the pitman shaft backlash adjuster screw fully counterclockwise, and then turn it clockwise one turn.
4. Use an inch-pound torque wrench to turn the stub shaft through a 45° arc on each side of the position in Step 2. Read the over-center torque as the stub shaft turns through the center position (**Figure 14-8**).
5. Continue to adjust the pitman shaft adjuster screw until the torque is 6 to 10 in.-lb. (0.6 to 1.2 Nm) more than the torque in Step 4.
6. Hold the pitman shaft adjuster screw in this position, and tighten the locknut to the specified torque as illustrated in **Photo Sequence 30**.

SERVICE TIP When tightening the pitman backlash adjuster screw locknut, it is very important to hold the pitman arm lash adjuster screw with a screwdriver to prevent this screw from turning. If the pitman lash adjuster screw turns when the locknut is tightened, the adjustment is changed.

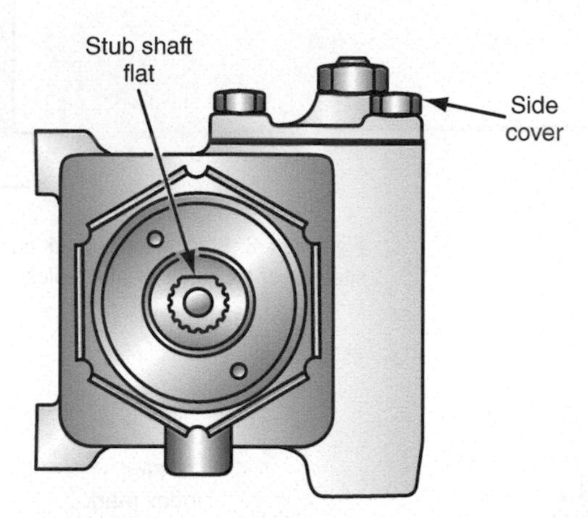

Figure 14-6 Stub shaft flat facing upward and parallel with the side cover.

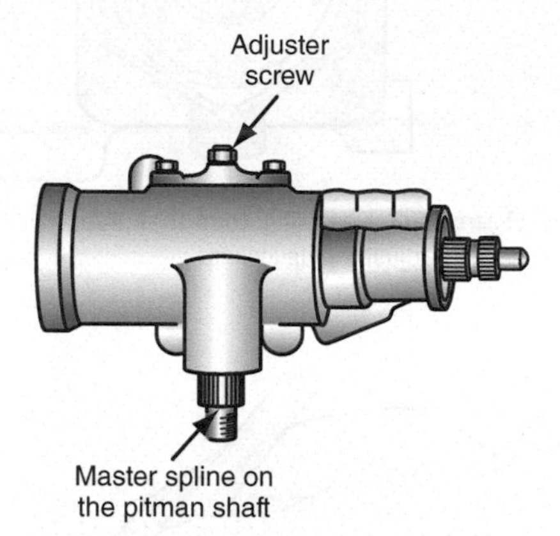

Figure 14-7 Pitman shaft master spline aligned with the pitman backlash adjuster screw.

PHOTO SEQUENCE 30
Pitman Sector Shaft Lash Adjustment

P30-1 Clamp the steering gear securely in a vise.

P30-2 Rotate the stub shaft from stop to stop and count the number of turns.

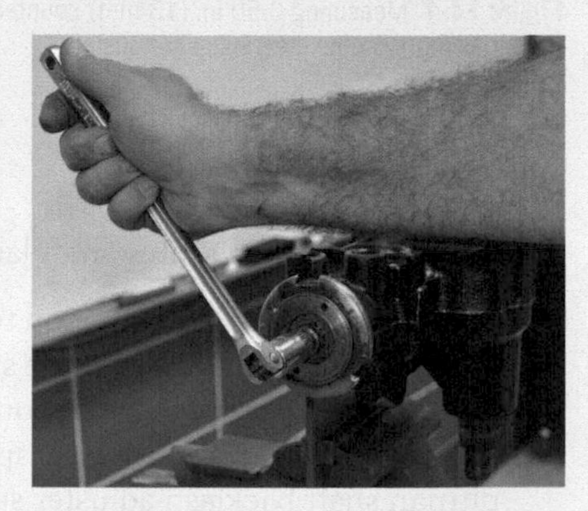

P30-3 Position the stub shaft at either stop and rotate the shaft back one-half the number of total turns.

PHOTO SEQUENCE 30 (CONTINUED)

P30-4 Be sure the master spline on the stub shaft is aligned with the pitman shaft backlash adjuster screw.

P30-5 Loosen the locknut and rotate the backlash adjuster screw fully counterclockwise, and then turn this screw clockwise one turn.

P30-6 Use an inch-pound torque wrench and the proper size socket to rotate the stub shaft through a 45° on each side of the stub shaft center position.

P30-7 Read the over-center torque reading on the torque wrench as the stub shaft is rotated through the center position.

P30-8 While rotating the stub shaft through the center position, turn the sector backlash adjuster screw until the torque wrench reading is 6–10 in.-lb. (0.6–1.2 Nm) more than the torque reading in Step 7.

P30-9 Hold the sector backlash adjuster screw in this position and tighten the locknut to the specified torque.

POWER RECIRCULATING BALL STEERING GEAR OIL LEAK DIAGNOSIS

Five locations where oil leaks may occur in a power steering gear are the following:

1. Side cover O-ring seal (**Figure 14-9**)
2. Adjuster plug seal
3. Pressure line fitting
4. Pitman shaft oil seals
5. End cover seal

If an oil leak is present at any of these areas, complete or partial steering gear disassembly and seal or O-ring replacement is necessary.

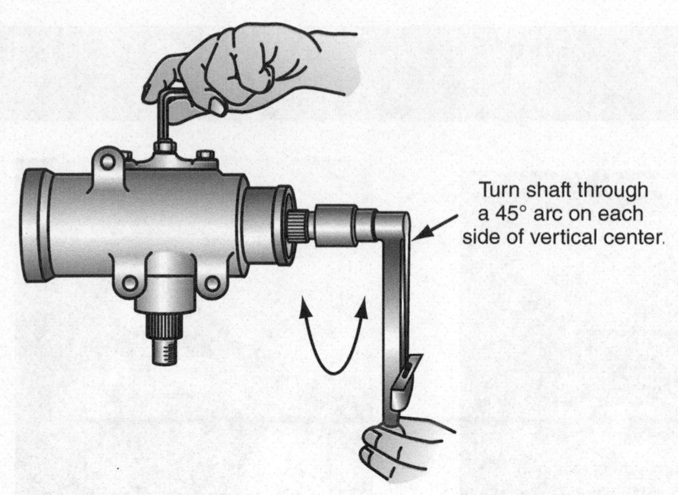

Figure 14-8 Measuring worm shaft turning torque to adjust pitman backlash adjuster screw.

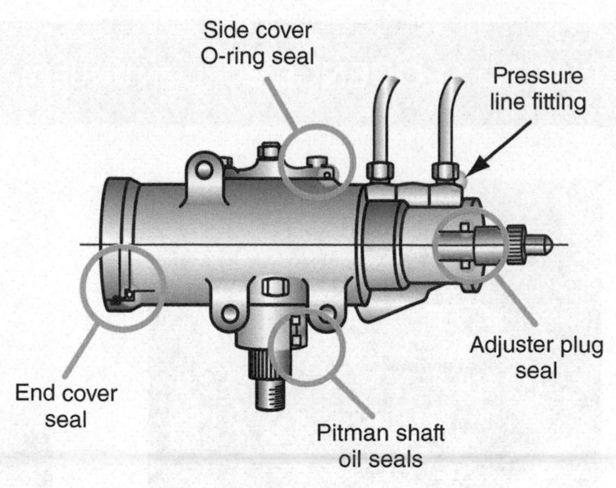

Figure 14-9 Power recirculating ball steering gear oil leak locations.

POWER RECIRCULATING BALL STEERING GEAR SEAL REPLACEMENT

Side Cover O-Ring Replacement

Prior to any disassembly procedure, clean the steering gear with solvent or in a parts washer. The steering gear service procedures vary depending on the make of gear. Always follow the vehicle manufacturer's recommended procedure in the service manual.

Following is a typical side cover O-ring replacement procedure:

1. Loosen the pitman backlash adjuster screw locknut and remove the side cover bolts. Rotate the pitman backlash adjuster screw clockwise to remove the cover from the screw (**Figure 14-10**).
2. Discard the O-ring and inspect the side cover matching surfaces for metal burrs and scratches.
3. Lubricate a new O-ring with the vehicle manufacturer's recommended power steering fluid and install the O-ring.
4. Rotate the pitman backlash adjuster screw counterclockwise into the side cover until the side cover is properly positioned on the gear housing. Turn this adjuster screw fully counterclockwise and then one turn clockwise. Install and tighten the side cover bolts to the specified torque. Adjust the pitman sector shaft lash as explained earlier.

Caution

The bearing identification number must face the driving tool to prevent bearing damage during installation.

End Plug Seal Replacement

Follow these steps for end plug seal replacement:

1. Insert a punch into the access hole in the steering gear housing to unseat the retaining ring, and remove the ring (**Figure 14-11**).
2. Remove the end plug and seal.
3. Clean the end plug and seal contact area in the housing with a shop towel.
4. Lubricate a new seal with the vehicle manufacturer's recommended power steering fluid, and install the seal.
5. Install the end plug and retaining ring.

Worm Shaft Bearing Adjuster Plug Seal and Bearing Replacement

Follow these steps for worm shaft bearing adjuster plug seal and bearing service:

1. Remove the adjuster plug locknut, and use a special tool to remove the adjuster plug.
2. Use a screwdriver to pry at the raised area of the bearing retainer to remove this retainer from the adjuster plug (**Figure 14-12**).
3. Place the adjuster plug face down on a suitable support, and use the proper driver to remove the needle bearing, dust seal, and lip seal.
4. Place the adjuster plug face up on a suitable support, and use the proper driver to install the needle bearing dust seal and lip seal.
5. Install the bearing retainer in the adjuster plug, and lubricate the bearing and seal with the vehicle manufacturer's recommended power steering fluid.
6. Install the adjuster plug and locknut, and adjust the worm shaft bearing preload as discussed previously.

TABLE 14-1 STEERING GEAR DIAGNOSIS

Problem	Symptoms	Possible Causes
Excessive steering wheel free play	Steering wander when driving straight ahead	Loose worm shaft preload adjustment Loose sector lash adjustment Loose steering gear mounting bolts Worn steering linkage components
Excessive steering effort	Excessive steering wheel turning effort when turning a corner or parking	Worn flexible coupling or universal joint in steering shaft
Underhood noise	Rattling noise when driving over road irregularities Squawking noise while turning a corner	Lack of steering gear lubricant Tight worm shaft bearing preload adjustment Tight sector lash adjustment Worn flexible coupling or universal joint in steering shaft Loose steering gear mounting bolts Cut or worn dampener O-ring on spool valve
Erratic steering effort	Erratic steering effort when turning a corner	Low power steering fluid level Air in power steering system Worn, damaged worm shaft, ball nut, or sector teeth

CUSTOMER CARE As an automotive technician, you should be familiar with the maintenance schedules recommended by various vehicle manufacturers. Of course, it is impossible to memorize all the maintenance schedules on different makes of vehicles, but maintenance schedule books are available. This maintenance schedule information is available in the owner's manual, but the vehicle owner may not take time to read this manual. If you advise the customer that his or her vehicle requires some service, such as a cooling system flush, according to the vehicle manufacturer's maintenance schedule, the customer will often have the service performed. The customer will usually appreciate your interest in his or her vehicle, and the shop will benefit from the increased service work.

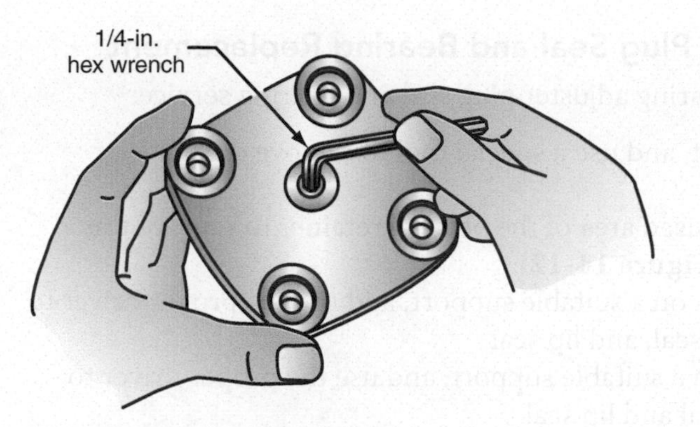

Figure 14-10 Removing steering gear side cover.

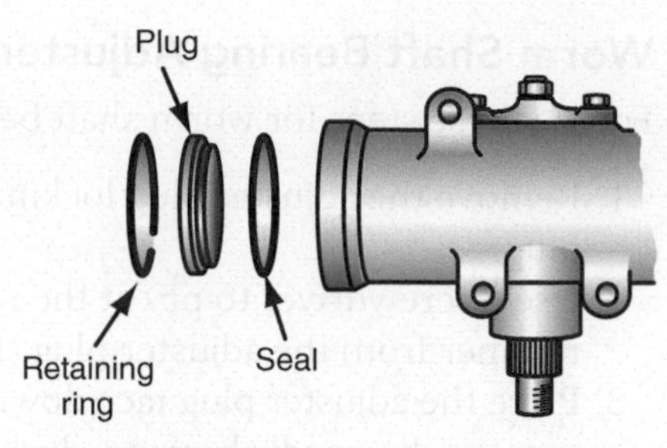

Figure 14-11 Removing steering gear end plug, retaining ring, and seal.

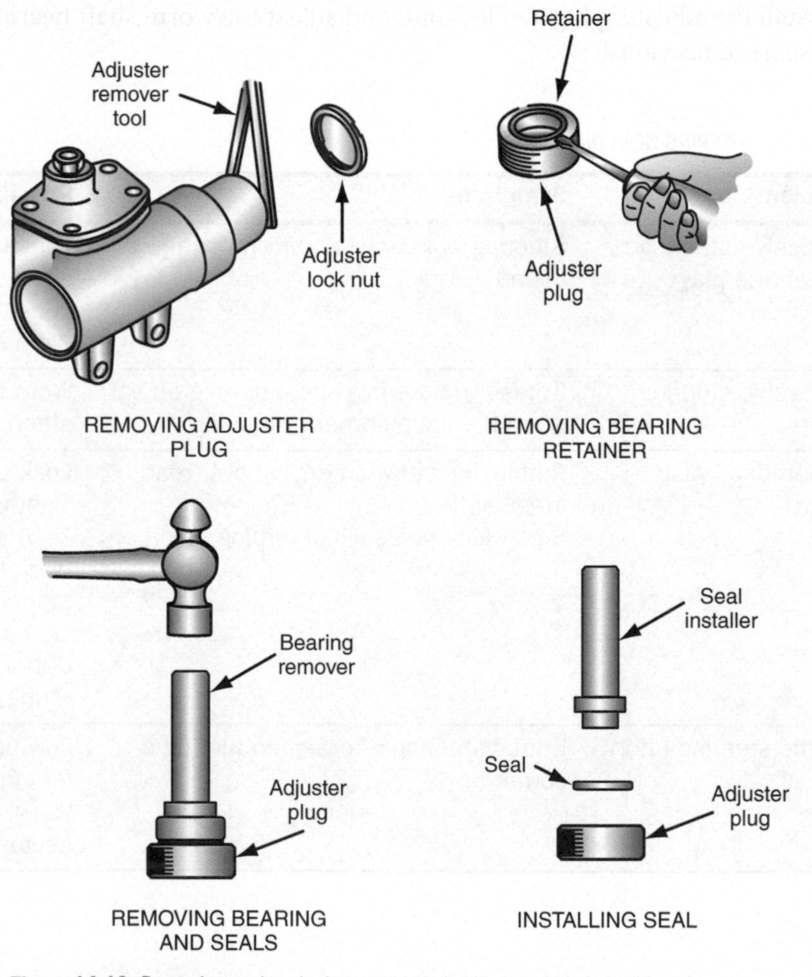

Figure 14-12 Removing and replacing worm shaft adjuster plug, bearing, and seal.

CASE STUDY

The owner of a 2005 Chevrolet Silverado truck complained about increased and somewhat erratic steering effort. The service writer asked the customer about the conditions when this problem occurred, and the customer said that the condition was always present. During a road test, the technician discovered that the customer's description of the problem was accurate.

The technician checked the power steering fluid level and made a careful check of the belt tension and condition without finding any problems. Next, the technician checked the power steering pump pressure and found it to be normal. A check of the power steering hoses did not reveal any hose restrictions.

The technician removed and disassembled the steering gear and found a severely scored cylinder bore in the gear housing. The ball nut piston ring was also worn and scored. A replacement steering gear was installed, and the system filled with the manufacturer's recommended power steering fluid. A road test indicated the new gear worked satisfactorily.

ASE-STYLE REVIEW QUESTIONS

1. While discussing power recirculating ball steering gear diagnosis:
   Technician A says excessive steering wheel free play may be caused by a worm shaft bearing preload adjustment that is tighter than normal.
   Technician B says excessive steering wheel free play may be caused by loose steering gear mounting bolts.
   Who is correct?
   A. A only C. Both A and B
   B. B only D. Neither A nor B
2. While discussing power recirculating ball steering gear service:
   Technician A says steering wander may be caused by a loose sector lash adjustment.
   Technician B says a loose sector lash adjustment may cause reduced feel of the road.
   Who is correct?
   A. A only C. Both A and B
   B. B only D. Neither A nor B
3. While discussing power recirculating steering gear diagnosis:
   Technician A says the sector lash adjustment is performed with the worm shaft halfway between the centered position and the full-right position.
   Technician B says the worm shaft bearing preload adjustment is performed before the sector lash adjustment.
   Who is correct?
   A. A only C. Both A and B
   B. B only D. Neither A nor B
4. A power recirculating ball steering gear has a rattling noise while driving the vehicle.
   The most likely cause of this problem is:
   A. Low power steering fluid level.
   B. Worn flexible coupling in the steering shaft.
   C. Loose worm bearing preload adjustment.
   D. Leaking pitman shaft seal.
5. A power recirculating ball steering gear has a fluid leakage problem, and fluid appears on the top of the gear.
   Technician A says the side cover O-ring seal may be leaking.
   Technician B says the high-pressure line fitting on the gear may be leaking.
   Who is correct?
   A. A only C. Both A and B
   B. B only D. Neither A nor B
6. A power recirculating ball steering gear has excessive kickback on the steering wheel.
   The most likely cause of this problem is:
   A. A loose sector lash adjustment.
   B. A loose worm shaft bearing preload adjustment.
   C. A defective poppet valve in the steering gear.
   D. A worn pitman shaft bearing.
7. All of these statements about power recirculating ball steering gear defects are true EXCEPT:
   A. A worn steering shaft U-joint may cause a growling noise when turning the front wheels.
   B. Excessive steering effort may be caused by a defective power steering pump.

C. Excessive steering effort may be caused by a scored steering gear cylinder.

D. Steering wheel jerking when turning may be caused by a slipping power steering belt.

8. When adjusting on a power recirculating ball steering gear:

   A. The worm shaft bearing adjuster plug should be bottomed and then backed off until the specified worm shaft turning torque is obtained.

   B. The sector lash adjustment screw should be tightened one turn before the worm shaft bearing preload is adjusted.

   C. The pitman shaft backlash adjuster screw is turned fully counterclockwise and then one turn clockwise prior to the backlash adjustment.

   D. The pitman shaft is positioned one-half turn from the fully right or left position prior to the pitman shaft backlash adjustment.

9. A power recirculating ball steering gear experiences excessive steering effort while parking. The most likely cause of this problem is:

   A. A restricted high-pressure steering hose.

   B. A misaligned power steering belt.

   C. A leaking pitman shaft seal.

   D. Excessive torque on the worm shaft adjuster plug locknut.

10. A power recirculating ball steering gear experiences hard steering for a short time after the vehicle sits overnight. The most likely cause of this problem is:

    A. Excessively tight worm bearing preload adjustment.

    B. A scored cylinder and worn piston ring in the steering gear.

    C. A restricted power steering return hose.

    D. A binding U-joint in the steering shaft.

ASE CHALLENGE QUESTIONS

1. A car with a power recirculating ball steering gear has excessive steering kickback, and a preliminary inspection shows no abnormal wear in the linkage. Which of the following could be the cause of the problem?

   A. Worn gear piston or bore.

   B. Slipping pump belt.

   C. Worn pump poppet valve.

   D. Sticking valve spool.

2. The complaint is loss of power assist, but there is no mention of any associated noise. All of the following could cause this problem EXCEPT:

   A. Low fluid.

   B. Improperly inflated tires.

   C. Broken pump belt.

   D. Steering column misalignment.

3. While discussing steering problems:

   Technician A says a "jerky" steering wheel and a "clunking" noise could indicate worn steering column U-joints.

   Technician B says lack of assist and a "growling" noise in a fluid-filled steering pump could indicate a hose or pump internal restriction.

   Who is correct?

   A. A only C. Both A and B

   B. B only D. Neither A nor B

4. A vehicle with a power recirculating ball steering system requires much higher than normal steering effort, especially in the parking lot.

   Technician A says the cause of the problem is a worn pitman shaft seal.

   Technician B says the cause of the problem is a worn worm shaft thrust bearing.

   Who is correct?

   A. A only C. Both A and B

   B. B only D. Neither A nor B

5. The complaint is excessive steering wheel free play. All of the following could cause this problem EXCEPT:

   A. Loose worm shaft bearing preload.

   B. Worn steering gears.

   C. Steering gear column misalignment.

   D. Worn flex coupling or U-joint.

Name ______________________ Date ____________

JOB SHEET
67

POWER RECIRCULATING BALL STEERING GEAR OIL LEAK DIAGNOSIS

Upon completion of this job sheet, you should be able to diagnose oil leaks in a power recirculating ball steering gear.

ASE Education Foundation Task Correlation

This job sheet addresses the following **AST/MAST** task:

B.4. Diagnose power steering gear (non-rack and pinion) binding, uneven turning effort, looseness, hard steering, and noise concerns; determine needed action. **(P-2)**

We Support ASE Education Foundation

Tools and Materials

Modern vehicle with a power recirculating ball steering gear.

Describe the Vehicle Being Worked On:

Year ____________ Make ____________ Model ____________

VIN ____________ Engine type and size ____________

Procedure **Task Completed**

1. Place fender, seat, and floor mat covers on the vehicle. ☐
2. Clean the outside of the steering gear. ☐
3. Be sure the power steering pump reservoir is filled to the proper level with the specified fluid. ☐
4. Start the engine and observe the power steering gear while an assistant turns the steering wheel fully in each direction several times. Turn off the ignition switch. ☐
5. If the power steering gear is leaking fluid on the lower side, raise the front of the vehicle with a floor jack and support the front suspension on safety stands positioned at the specified vehicle lift points. Repeat Steps 2–4. ☐
6. List the fluid leak locations on the power steering gear.

   ____

   ____

7. State the necessary repairs to correct the power steering gear leaks.

   ____

   ____

Instructor's Response

____

____

____

6. Install and tighten the adjuster plug locknut to the specified torque.

   Specified adjuster plug locknut torque ____________________

   Actual adjuster plug locknut torque ____________________

Instructor's Response

___

___

___

Name ______________________ Date ______________

JOB SHEET 69

ADJUST POWER RECIRCULATING BALL STEERING GEAR SECTOR LASH, STEERING GEAR REMOVED

Upon completion of this job sheet, you should be able to adjust sector lash on power recirculating ball steering gears.

ASE Education Foundation Task Correlation

This job sheet addresses the following **AST/MAST** task:

B.4. Diagnose power steering gear (non-rack and pinion) binding, uneven turning effort, looseness, hard steering, and noise concerns; determine needed action. **(P-2)**

We Support ASE Education Foundation

Tools and Materials

Torque wrench, in-lb.

Describe the Vehicle Being Worked On:

Year ______________ Make ______________ Model ______________

VIN ______________ Engine type and size ______________

Make of steering gear ______________

Procedure **Task Completed**

1. Rotate the worm shaft from stop to stop and count the number of turns.

   Total number of worm shaft turns from stop to stop ______________.

2. Starting at either stop, turn the worm shaft back two-thirds of the total number of turns.

   Number of worm shaft turns from fully right or fully left to position the worm shaft properly prior to sector lash adjustment ______________.

   Explain the reason for placing the worm shaft in this position prior to the sector lash adjustment.

   ____

   ____

   ____

   ____

3. In this position, the flat on the worm shaft should be facing upward, and the master spline on the pitman shaft should be aligned with the pitman shaft backlash adjuster screw.

   Is the worm shaft flat properly positioned? ☐ Yes ☐ No

   Is the master spline on the pitman shaft properly positioned? ☐ Yes ☐ No

   If the answer is no to either of the above questions, state the necessary corrective action.

   ____

   ____

   ____

4. Turn the pitman shaft backlash adjuster screw fully counterclockwise, and then turn it clockwise one turn.

Is the pitman shaft backlash adjuster screw properly positioned? ☐ Yes ☐ No

Instructor check ____________________

5. Use an inch-pound torque wrench to turn the worm shaft through a 45° arc on each side of the position in Step 2. Read the over-center torque as the worm shaft turns through the center position.

Stub shaft turning torque ____________________

6. Continue to adjust the pitman shaft adjuster screw until the torque is 6 to 10 in.-lb. (0.6 to 1.2 Nm) more than the torque in Step 5.

Final worm shaft turning torque after adjustment ____________________

7. Hold the pitman shaft adjuster screw in this position and tighten the locknut to the specified torque.

Specified pitman shaft adjuster screw locknut torque ____________________

Actual pitman shaft adjuster screw locknut torque ____________________

Instructor's Response

____

____

____

APPENDIX A
ASE PRACTICE EXAMINATION

1. After new tires and new alloy rims are installed on a sports car, the owner complains about steering wander and steering pull in either direction while braking.
   Technician A says there may be brake fluid on the front brake linings.
   Technician B says the replacement rims may have a different offset than the original rims.
   Who is correct?
   A. Technician A C. Both A and B
   B. Technician B D. Neither A nor B
2. *Technician A* says when a vehicle pulls to one side; the problem may be caused by a pinion gear binding on the rack gear of a power rack and pinion steering gear.
   Technician B says when a power steering rack and pinion steering gear spool valve binds it may cause a vehicle to pull to one side.
   Who is correct?
   A. Technician A C. Both A and B
   B. Technician B D. Neither A nor B
3. The outside edge of the left front tire on a front-wheel-drive car is badly worn.
   Technician A says the cause could be worn ball joints.
   Technician B says the cause could be that the camber setting is excessively negative on the left front suspension.
   Who is correct?
   A. Technician A C. Both A and B
   B. Technician B D. Neither A nor B
4. The owner of a 2-wheel drive pick-up says the front tires squeal loudly during low-speed turns. The most probable cause of this condition is:
   A. Excessive positive camber.
   B. Negative caster adjustment.
   C. Improper steering axis inclination (SAI).
   D. Improper turning angle.
5. A 2-wheel drive pickup has a severe shudder when the vehicle is started from a stop with a load in the bed.
   Technician A says the problem may be worn spring eyes.
   Technician B says the problem may be axle torque wrap-up.
   Who is correct?
   A. Technician A C. Both A and B
   B. Technician B D. Neither A nor B
6. Steering pull to the right occurs only when braking. The most likely cause of this problem is:
   A. Excessive positive caster on the right-front wheel.
   B. Excessive positive camber on the left-front wheel.
   C. A loose strut rod bushing on the right-front suspension.
   D. Excessive toe-out on the right-front wheel.
7. *Technician A* says hard steering may be caused by low hydraulic pressure due to a stuck flow control valve in the pump.
   Technician B says hard steering may be caused by low hydraulic pressure due to a worn steering gear spool valve piston ring or housing bore.
   Who is correct?
   A. Technician A C. Both A and B
   B. Technician B D. Neither A nor B
8. *Technician A* says incorrect wheel alignment angles will reduce tire tread life.
   Technician B says wheel alignment is necessary after ball joint replacement.
   Who is correct?
   A. Technician A C. Both A and B
   B. Technician B D. Neither A nor B

9. While discussing tire tread wear:
   Technician A says a scalloped pattern of tire wear indicates an out-of-round wheel or tire.
   Technician B says uneven wear on one side of a tire may indicate radial force variation.
   Who is correct?
   A. Technician A C. Both A and B
   B. Technician B D. Neither A nor B

10. *Technician A* says excessive front wheel bearing endplay may cause steering wander.
    Technician B says a defective front wheel bearing may cause a growling noise while cornering.
    Who is correct?
    A. Technician A C. Both A and B
    B. Technician B D. Neither A nor B

11. A vehicle pulls to the right when driving straight ahead.
    Technician A says the right-front tire may have excessive radial runout.
    Technician B says the right-front tire may have a conicity defect.
    Who is correct?
    A. Technician A C. Both A and B
    B. Technician B D. Neither A nor B

12. All of the following could cause shock absorber noise EXCEPT:
    A. A bent piston rod.
    B. Worn shock absorbers.
    C. Shock fluid leaks.
    D. Extreme temperatures.

13. When a vehicle pulls to one side, any of the following problems may be the cause EXCEPT:
    A. Worn ball joints.
    B. Reduced curb height.
    C. Bent strut rod.
    D. Improper turning angle.

14. A customer says when he applied the brakes hard on his front-wheel-drive car, "the whole car shook."
    Technician A says the problem could be worn sway bar bushings.
    Technician B says the problem could be worn or loose strut rod bushings.
    Who is correct?
    A. Technician A C. Both A and B
    B. Technician B D. Neither A nor B

15. A customer with a sport utility vehicle says after an off-road outing over the weekend, his vehicle pulls to the left on acceleration. Which of the following could cause this problem?
    A. Broken leaf spring center bolt.
    B. Loose steering unit.
    C. Stuck brake pad.
    D. Stuck in 4WD.

16. While discussing electronic air suspension:
    Technician A says the compressor must be running when a corner of the vehicle is lifted from the ground.
    Technician B says on a car with electronic air suspension, the switch should be in the "on" position.
    Who is correct?
    A. Technician A C. Both A and B
    B. Technician B D. Neither A nor B

17. Front wheel "shimmy"—a side-to-side movement of the front wheels that is felt in the steering wheel—may be caused by any of the following EXCEPT:
    A. Worn tie rod ends.
    B. Tire/wheel imbalance.
    C. Rack bushings and rack alignment.
    D. Tight sector shaft adjustment.

18. A customer says her front-wheel-drive car is hard to steer because the steering wheel no longer returns to center. After the turn, she has to bring it back to center.
    Technician A says a corroded or stuck strut bearing plate could be the cause of the problem.
    Technician B says a bent tension rod could be the cause of the problem.
    Who is correct?
    A. Technician A C. Both A and B
    B. Technician B D. Neither A nor B

19. The tire pressure monitoring system is being discussed.
    Technician A says that the tire pressure monitoring system must be reinitialized after a wheel pressure sensor has been replaced.
    Technician B says that the tire pressure monitoring system must be reinitialized after the tires have been rotated on some vehicles.
    Who is correct?
    A. Technician A C. Both A and B
    B. Technician B D. Neither A nor B

20. When one of the front wheels strikes a road irregularity, the steering suddenly veers to the right or left on a vehicle with a parallelogram steering linkage. The most likely cause of this problem is:

A. A weak stabilizer bar.

B. Worn-out shock absorbers.

C. Sagged coil springs.

D. Center link that is not level.

21. While discussing suspension height:

Technician A says raising the suspension height in the rear of a vehicle will affect the front suspension geometry.

Technician B says lowering the suspension height of a vehicle will affect the suspension geometry.

Who is correct?

A. Technician A C. Both A and B

B. Technician B D. Neither A nor B

22. A vehicle with recirculating ball steering has excessive steering wheel free play.

Technician A says a loose worm bearing preload adjustment may cause the problem.

Technician B says loose column U-joints may cause the problem.

Who is correct?

A. Technician A C. Both A and B

B. Technician B D. Neither A nor B

23. The engine has recently been replaced in a front-wheel-drive car with power rack and pinion steering The customer now complains about excessive steering effort. A preliminary check of the steering revealed the fluid level was OK and the belts were not slipping.

Technician A says perhaps the rack was knocked out of alignment when the engine was installed.

Technician B says perhaps the pressure line was bent or pinched when the engine was installed.

Who is correct?

A. Technician A C. Both A and B

B. Technician B D. Neither A nor B

24. While discussing rear suspension systems:

Technician A says semi-independent rear suspension systems momentarily have a slight negative camber and toe-in when the wheel goes over a bump.

Technician B says constantly carrying a lot of weight in the rear of a car with semi-independent rear suspension may cause the rear tires to wear on the inside edge.

Who is correct?

A. Technician A C. Both A and B

B. Technician B D. Neither A nor B

25. The right front wheel setback is excessive on a Macpherson strut front suspension system. The most likely cause of this problem is:

A. Broken insulators on the right coil spring.

B. A binding right upper strut mount.

C. A broken stabilizer bar.

D. A bent engine cradle.

26. During a wheel alignment, a technician suspects that the front cradle on a front-wheel-drive vehicle may have shifted. If the cradle was shifted from the right to the left, what effect would it have?

A. Left caster will become more positive and right caster will become more negative evenly.

B. Right caster will become more positive and left caster will become more negative evenly.

C. Left camber will become more positive and right camber will become more negative evenly.

D. Right camber will become more positive and left camber will become more negative evenly.

27. The customer says his vehicle has a rapid "thumping" noise and a vibration in the steering wheel that is most noticeable when the vehicle is at a steady speed in a long curve.

Technician A says the problem could be caused by a tire defect.

Technician B says the problem could be improper dynamic wheel balance.

Who is correct?

A. Technician A C. Both A and B

B. Technician B D. Neither A nor B

28. While discussing power steering fluids:

Technician A says a foamy, milky power steering fluid is caused by mixing automatic transmission fluid with hydraulic fluid intended for power steering use.

Technician B says using automatic transmission fluid instead of power steering hydraulic fluids will lower pump pressure and increase steering effort.

Who is correct?

A. Technician A C. Both A and B

B. Technician B D. Neither A nor B

29. On a car with power rack and pinion steering, the steering suddenly swerves to the right or left when a front wheel strikes a road irregularity. The most likely cause of the problem is:
   A. A loose rack bearing adjustment.
   B. Worn front struts.
   C. Loose steering gear mounting bushings.
   D. Bent steering arm.

30. A customer complains of a steering wheel shake on a rear wheel drive vehicle that occurs at speeds between 50 and 60 mph. What is the most likely cause of this complaint?
   A. Worn control arm bushings.
   B. An out of balance front tire and wheel assembly.
   C. Steering gear is adjusted too tightly.
   D. Worn shocks/struts.

31. When diagnosing an HEV with a scan tool a U0131 is obtained. This DTC indicates a defect in the:
   A. Data network.
   B. Inverter.
   C. Propulsion motor and related circuit.
   D. Generator and related circuit.

32. *Technician A* says front curb riding height is reduced by worn control arm bushings.
   Technician B says front curb riding height is reduced by weak coil springs.
   Who is correct?
   A. Technician A C. Both A and B
   B. Technician B D. Neither A nor B

33. In a four-wheel active steering system (4WAS), when changing lanes at high speed the:
   A. Rear wheels are steering a few degrees in the same direction as the front wheels.
   B. Rear wheels are steered 5 degrees in the opposite direction to the front wheels.
   C. Camber angle is changed on the rear wheels.
   D. The rear wheels remain in the straight ahead position.

34. While discussing unibody and frame problems:
   Technician A says an indication of possible frame damage is excessive tire wear when the alignment angles are correct.
   Technician B says a worn strut lower ball joint is often an indicator of unibody torsional damage.
   Who is correct?
   A. Technician A C. Both A and B
   B. Technician B D. Neither A nor B

35. While discussing frame damage:
   Technician A says wrinkles in the upper flange of a truck frame indicate sag.
   Technician B says diamond-shaped distortion of a 4WD sport utility vehicle frame was possibly caused by towing or being towed from a frame corner.
   Who is correct?
   A. Technician A C. Both A and B
   B. Technician B D. Neither A nor B

36. Power steering fluid is leaking from the high-pressure outlet fitting on a nonsubmerged power steering pump. To correct this problem, you should:
   A. Replace the cover seal.
   B. Replace the O-ring.
   C. Replace the fitting.
   D. Replace the check valve.

37. During a road test on a front-wheel-drive vehicle, it is determined that the vehicle pulls right after a right turn, and pulls left after a left turn. The most likely cause of this condition is:
   A. A misaligned subframe/cradle assembly.
   B. Excessive positive caster.
   C. Excessive lower ball joint play.
   D. A binding upper strut bearing.

38. When the projected steering axis of a front wheel viewed from the side is tilted rearward from the true vertical center of the wheel, this alignment angle is referred to as:

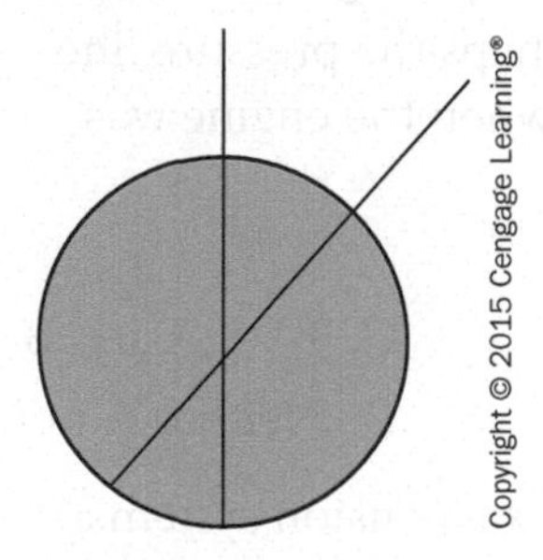

   A. Negative camber
   B. Positive camber
   C. Positive scrub radius
   D. Positive caster

39. While discussing unibody and frame damage:
Technician A says a vehicle that has been rear-ended could have later problems with steering and tracking.
Technician B says towing a 2,000 pound trailer on a class II hitch bolted to the frame of a minivan will cause diamond distortion of the frame.
Who is correct?
A. Technician A C. Both A and B
B. Technician B D. Neither A nor B

40. *Technician A* says in some short arm long arm (SLA) front suspension systems the springs are mounted between the lower control arm and the chassis.
Technician B says in an SLA type of suspension system the upper ball joints are the load-carrying ball joints.
Who is correct?
A. Technician A C. Both A and B
B. Technician B D. Neither A nor B

41. A customer says the steering wheel of her front-wheel-drive car does not return to the center position after a turn. A test drive reveals that the steering wheel is stiff and only returns to within approximately 90° of center after a 180° turn.
Technician A says the problem is memory steer and could be caused by binding in the steering shaft universal joints.
Technician B says memory steer in a front-wheel-drive car may be also caused by binding upper strut mounts.
Who is correct?
A. Technician A C. Both A and B
B. Technician B D. Neither A nor B

42. A customer says his front-wheel-drive car shakes and moans when decelerating at low speeds. The most likely cause of the problem is:
A. Worn upper strut bearings.
B. Worn inner drive axle joints.
C. Loose wheel bearings.
D. Excessive tire inflation pressure.

43. A sport utility vehicle with a power recirculating ball steering gear makes a rattling noise when the steering wheel is turned. This noise is noticeable whether the car is in motion or standing still.
Technician A says the cause of the problem could be a loose pitman arm or shaft.
Technician B says the cause of the problem could be the high-pressure power steering hose touching some part of the vehicle.
Who is correct?
A. Technician A C. Both A and B
B. Technician B D. Neither A nor B

44. While a vehicle is taken for a test drive, a loud clunking noise is heard from the right front of the vehicle only during hard brake maneuvers. What is the most likely cause?
A. A loose right front shock absorber
B. A worn rack and pinion mounting bushing
C. A loose sway bar link
D. A worn strut rod bushing

45. On a vehicle equipped with rear parallel leaf springs and a solid rear axle, the customer is complaining that the vehicle reacts erratically (it darts) during turns. What is the most likely cause of this complaint?
A. Loose rear axle U-bolts
B. Missing jounce/rebound bumpers
C. Incorrect ride height
D. Incorrect drive line angle

46. Excessive play in the steering wheel is often because of loose steering components. A check of these components should be:
A. Done with vehicle weight on the tires.
B. Done with a computerized alignment system.
C. Completed after a front wheel alignment.
D. Performed with the vehicle raised and the wheels unsupported.

47. While discussing shock absorber diagnosis:
Technician A says the "bounce test" is the sure way to quickly pinpoint bad shocks and struts.
Technician B says oil film on the lower chamber of a shock or strut indicates leakage and requires replacement.
Who is correct?
A. Technician A C. Both A and B
B. Technician B D. Neither A nor B

48. While discussing front suspension angles:
Technician A says turning angle is fixed by the steering and suspension system design.

Technician B says toe-out on turns may be adjusted with the steering stopper bolts.

Who is correct?

A. Technician A C. Both A and B

B. Technician B D. Neither A nor B

49. While discussing rear suspension alignment: *Technician A* says in straight-ahead driving, the rear wheels must track exactly behind the front wheels or the vehicle will not handle correctly.

Technician B says a dynamic tracking rear suspension that lets the rear wheels turn in the same direction as the front wheels makes curve negotiation and lane changes much quicker and safer.

Who is correct?

A. Technician A C. Both A and B

B. Technician B D. Neither A nor B

50. The most likely cause of excessive body lean and sway while cornering is:

A. Worn upper control arm bushings.

B. A broken stabilizer link

C. Sagged coil springs

D. Worn strut rod bushings

APPENDIX B
METRIC CONVERSIONS

TO CONVERT THESE	TO THESE	MULTIPLY BY
TEMPERATURE		
Centigrade Degrees	Fahrenheit Degrees	C degrees $\times$ 1.8 + 32 = F degrees
Fahrenheit Degrees	Centigrade Degrees	F degrees = 32 $\times$ .556 = C degrees
LENGTH		
Millimeters	Inches	0.03937
Inches	Millimeters	25.4
Meters	Feet	3.28084
Feet	Meters	0.3048
Kilometers	Miles	0.62137
Miles	Kilometers	1.60935
AREA		
Square Centimeters	Square Inches	0.155
Square Inches	Square Centimeters	6.45159
VOLUME		
Cubic Centimeters	Cubic Inches	0.06103
Cubic Inches	Cubic Centimeters	16.38703
Cubic Centimeters	Liters	0.001
Liters	Cubic Centimeters	1000
Liters	Cubic Inches	61.025
Cubic Inches	Liters	0.01639

TO CONVERT THESE	TO THESE	MULTIPLY BY
Liters	Quarts	1.05672
Quarts	Liters	0.94633
Liters	Pints	2.11344
Pints	Liters	0.47317
Liters	Ounces	33.81497
Ounces	Liters	0.02957
WEIGHT		
Grams	Ounces	0.03527
Ounces	Grams	28.34953
Kilograms	Pounds	2.20462
Pounds	Kilograms	0.45359
WORK		
Centimeter-Kilograms	Inch-Pounds	0.8676
Inch-Pounds	Centimeter-Kilograms	1.15262
Newton-Meters	Foot-Pounds	0.7376
Foot-Pounds	Newton-Meters	1.3558
PRESSURE		
Kilograms/Sq. Cm	Pounds/Sq. Inch	14.22334
Pounds/Sq. Inch	Kilograms/Sq. Cm	0.07031
Bar	Pounds/Sq. Inch	14.504
Pounds/Sq. Inch	Bar	0.06895

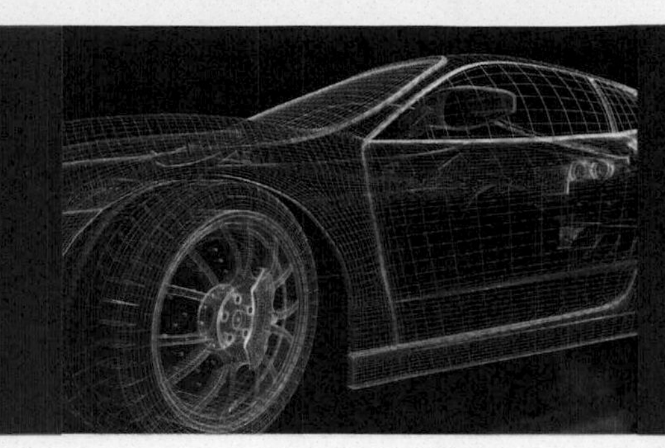

APPENDIX C
SPECIAL TOOL SUPPLIERS

Automotive Group
Kent-Moore Division, SPX Corporation Roseville, MI

Fluke Corporation
Everett, WA

Easco/KD Tools
Lancaster, PA

Hennessy Industries, Inc.
LaVerge, TN

Hunter Engineering Company
Bridgeton, MO

Mac Tools, Inc.
Washington Courthouse, OH

OTC Division, SPX Corporation
Owatonna, MN

Sears Industrial Sales
Cincinnati, OH

Snap-On Tools Corporation
Kenosha, WI

Specialty Products Company
Longmont, CO

Federal Mogul Corporation
Ann Arbor, MI

APPENDIX D
MANUFACTURER WEBSITES

Audi diagnostic tool and service information
http://www.ebahn.com/audi

Chrysler service information
http://www.techauthority.com

General Motors tool and service information
http://www.gmgoodwrench.com
http://www.acdelco.com

Ford Motor Company vehicle information
http://www.ford.com

Helm Incorporation, distributors of service manuals and training publications for many original equipment manufacturers
http://www.helm.com

Honda service information
http://www.serviceexpress.honda.com

Training materials, tools, and equipment
https://gmspecialservicetools.service-solutions.com/

Special service tools
http://www.otctools.com/product-landing

KIA general information
http://www.kia.com

Mac Tools, tool and equipment information
http://www.mactools.com

Nissan service manuals, 2000 and later models
http://www.nissan-techinfo.com

OTC Division of SPX Corporation, tools and equipment information
http://www.otctools.com

Snap-on Tools Corporation, tools and service equipment
http://www.snapon.com

Toyota Motor Corporation, vehicles and parts information
http://www.toyota.com

Volkswagen service information
http://www.ebahn.com/vw

Volvo service information, 1999 and later models
https://www.volvotechinfo.com/index.cfm

Service information pre-1999
http://www.volvotechinfo.com

APPENDIX E
SUSPENSION AND STEERING PROFESSIONAL ASSOCIATIONS

Automotive Service Association (ASA)
www.asashop.org
P.O. Box 929
Bedford Texas 76021

International Automotive Technicians Network (IATN)
www.iatn.net

Society of Automotive Engineers (SAE)
www.sae.org
400 Commonwealth Dr.
Warrendale PA 15096

Moog Service Link
www.moogauto.com
Cooper Moog Automotive
P.O. Box 7224
6565 Wells Ave.
St. Louis MO 63177

GLOSSARY

Note: **Terms are highlighted in color**, followed by **Spanish translation in bold**.

Accidental air bag deployment An unintended air bag deployment caused by improper service procedures.

Despliegue accidental del Airbag Despliegue imprevisto del Airbag ocasionado por procedimientos de reparación inadecuados.

Active roll stabilization (ARS) ECU A computer that receives input signals and controls output functions in an ARS system.

ECU de estabilización de rodamiento activo (ARS, en inglés) Una computadora que recibe señales de entraday controla las funciones de salida en un sistema ARS.

Active steering ECU A computer that receives input signals and controls output functions in an active steering system.

ECU de dirección activa Una computadora que recibe señales de entrada y controla las funciones de salida en un sistema de dirección activa.

Advanced vehicle handling (AVH) Computer software on CD that allows the technician to locate the cause of hidden suspension, body, and chassis problems.

Manejo avanzado de vehículos (AVH, en inglés) Programa de software en CD que permite que el técnico localice la causa de problemas ocultos de suspensión, carrocería y chasis.

Air bag deployment module The air bag and deployment canister assembly that is mounted in the steering wheel for the driver's side air bag, or in the dash panel for the passenger's side air bag.

Unidad de despliegue del air bag El conjunto del Airbag y elemento de despliegue montado en el volante de dirección para proteger al conductor, o en el tablero de instrumentos para proteger al pasajero.

Air bag system Is designed to protect the driver and/or passengers in a vehicle collision.

Sistema de bolsa de aire Se diseñó para proteger al conductor y/o a los pasajeros en caso de un choque.

Alignment ramp A metal ramp positioned on the shop floor on which vehicles are placed during wheel alignment procedures.

Rampa de alineación Rampa de metal ubicada en el suelo del taller de reparación de automóviles sobre la que se colocan los vehículos durante los procedimientos de alineación de ruedas.

Antilock brake system (ABS) A computer-controlled system that prevents wheel lockup during brake applications.

Sistema de frenado antibloqueo (ABS) Un sistema controlado por computadora que previene el bloqueo de las ruedas durante la aplicación de los frenos.

Antitheft locking wheel covers Locking wheel covers that help to prevent wheel theft.

Cubrerruedas anti-robo Cubrerruedas autobloqueantes que ayudan a evitar el robo de las ruedas.

Antitheft wheel nuts Special wheel nuts designed to prevent wheel and tire theft.

Tuercas antirrobo de fijación de las ruedas Tuercas especiales de fijación de las ruedas diseñadas para evitar el robo de las ruedas y los neumáticos.

Approved gasoline storage can A special gasoline container that meets safety requirements.

Bidón aprobado para almacenamiento de gasolina Un recipiente especial para gasolina que cumple los requisitos de seguridad.

Aqueous parts cleaning tank Uses a water-based, environmentally friendly cleaning solution, such as Greasoff® 2, rather than traditional solvents.

Tanque de limpieza de partes acuosas Usa una solución a base de agua y compatible con el medio ambiente, tal como Greasoff® 2, en lugar delos solventes tradicionales.

ASE blue seal of excellence An ASE logo displayed by automotive service shops that employ ASE-certified technicians.

Sello azul de excelencia de la ASE Logotipo exhibido en talleres de reparación de automóviles donde se emplean mecánicos certificados por la ASE.

Axle offset A condition where the complete rear axle assembly has turned so one rear wheel has moved forward, and the opposite rear wheel has moved rearward.

Eje descentrado Una condición en la cual la asamblea total del eje trasero se ha girado para que una rueda se ha movido hacia afrente, y la rueda opuesta trasera se ha movido hacia atrás.

Axle pullers Are usually slide-hammer type. These pullers attach to the axle studs to remove the axle.

Extractores del eje Son generalmente de tipo martillo de corredera. Estos extractores se vinculan a los espárragos roscados del eje para quitarlo.

Backup power supply A voltage source, usually located in the air bag computer, that is used to deploy an air bag if the battery cables are disconnected in a collision.

Alimentación de reserva Fuente de tensión, por lo general localizada en la computadora del Airbag, que se utiliza para desplegar el Airbag si se desconectan los cables de la batería a consecuencia de una colisión.

Ball joint radial measurement Vertical movement in a ball joint because of internal joint wear.

Medida de radial de la junta esférica Movimiento horizontal en una junta esférica ocasionado por el desgaste interno de la misma.

Ball joint removal and installation tools Special tools required for ball joint removal and replacement.

Herramientas para la remoción y el ajuste de la junta esférica Herramientas especiales requeridas para la remoción y el reemplazo de la junta esférica.

Ball joint vertical movement Vertical movement in a ball joint because of internal joint wear.

Movimiento vertical de la junta esférica Un movimiento vertical en la junta esférica debido al desgaste interior de la junta.

Ball joint wear indicator A visual method of checking ball joint wear.

Indicador de desgaste de la junta esférica Método visual de revisar el desgaste interior de una junta esférica.

Dial indicators A precision-measuring device with a stem and a rotary pointer.

Indicador de cuadrante Dispositivo para medidas precisas con vástago y aguja giratoria.

Digital adjustment photos Photos that are available in some computer-controlled wheel aligners to inform the technician regarding suspension adjustment procedures.

Fotos digitales de ajuste Las fotos que son disponibles en algunos alineadores de ruedas controlados por computadoras para informar a los técnicos en cuanto a los procedimientos de ajustes de la suspensión.

Digital signal processor (DSP) An electronic device in some wheel sensors on computer-controlled wheel aligners.

Procesor de señales digitales Un dipositivo electrónico en algunos sensores de ruedas en los alineadores de ruedas controlados por computadora.

Directional stability The tendency of a vehicle steering to remain in the straight-ahead position when driven straight ahead on a smooth, level road surface.

Estabilidad direccional Tendencia de la dirección del vehículo a permanecer en línea recta al ser así conducido en un camino cuya superficie es lisa y nivelada.

Downshift test A test for chassis vibrations performed with the transmission downshifted.

Prueba de cambio descendente Una prueba de las vibraciones del chasis que se efectúan durante un cambio de carrera de transmisión descendente.

Dynamic stability control (DSC) electronic control unit A computer that receives input signals and controls output functions in a DSC system.

ECU de control de estabilidad dinámico (DSC, en inglés) Una computadora que recibe señales de entrada y controla las funciones de salida en un sistema DSC.

Dynamic wheel balance Refers to proper balance of the tire-and-wheel assembly during tire-and-wheel rotation.

Equilibrado dinámico de la rueda Se refiere al equilibrado apropiado del ensamblaje de la llanta y la rueda durante su rotación.

Eccentric camber bolt A bolt with an out-of-round metal cam on the bolt head that may be used to adjust camber.

Perno de combadura excéntrica Perno con una leva metálica con defecto de circularidad en su cabeza, que puede utilizarse para ajustar la combadura.

Eccentric cams Out-of-round metal cams mounted on a retaining bolt with the shoulder of the cam positioned against a component. When the cam is rotated, the component position is changed. Delete eccentric cams or bushings.

Levas excéntricas Las levas de metal ovaladas montadas en un perno retenedor con lo saliente de la leva posicionado contra un componente. Cuando se gira la leva cambia la posición del componente.

Electronic brake and traction control module (EBTCM) A module that controls antilock brake, traction control, and vehicle stability functions.

Módulo electrónico de control del frenado y tracción (EBTCM) Un módulo que controla las funciones del freno antibloqueo, control de tracción, y control de estabilidad del vehículo.

Electronically controlled orifice (ECO) valve A computer-controlled solenoid valve in the power steering pump that controls fluid flow from the pump to the servotronic valve.

Válvula de orificio controlado electrónicamente (ECO, en inglés) Una válvula solenoide controlada por computadora que se encuentra en la bomba de dirección de potencia que controla el flujo de fluidos desde la bomba hasta la válvula servotrónica.

Electronic suspension control (ESC) An electronically controlled suspension system in which the computer controls strut firmness.

Control electrónico de la suspensión (CES) Los vehículos que se equipan con este sistema pueden tener el tablero de control análogo o digital.

Electronic vibration analyzer (EVA) A tester that measures component vibration.

Analizador electrónico de vibraciones Un detector que mide la vibración de un componente.

Electronic wheel balancer A computer-controlled balancer that provides static and dynamic wheel balance.

Equilibrador de ruedas electrónico Un equilibrador controlado por computadora que provee la equilibración estática y dinámico.

Environmental Protection Agency (EPA) A federal government agency that is responsible for air and water quality in the United States.

Agencia de Protección Ambiental (EPA, en inglés) Una agencia del Gobierno Federal responsable de la calidad del aire y del agua en los Estados Unidos.

Fail-safe mode A mode entered by a computer if the computer detects a fault in the system.

Modo de seguridad Un modo iniciado por la computadora si ésta descubre un fallo en el sistema.

Floor jack A hydraulically operated lifting device for vehicle lifting.

Gato de pie Dispositivo activado hidráulicamente que se utiliza para levantar un vehículo.

Flow control valve A special valve that controls fluid movement in relation to system demands.

Válvula de control de flujo Válvula especial que controla el movimiento del fluido de acuerdo a las exigencias del sistema.

Fluorescent trouble lights Contain fluorescent-type bulbs.

Luces de peligro fluorescentes Incluyen bombillas de tipo fluorescente.

Frame flange The upper or lower horizontal edge on a vehicle frame.

Brida del armazón Borde horizontal superior o inferior en el armazón del vehículo.

Frame web The vertical side of a vehicle frame.

Malla del armazón Lado vertical del armazón del vehículo.

Front and rear wheel alignment angle screen A display on a computer wheel aligner that provides readings of the front and rear wheel alignment angles.

Pantalla para la visualización del ángulo de alineación de las ruedas delanteras y traseras Representación visual en una computadora para alineación de ruedas que provee lecturas de los ángulos de alineación de las ruedas delanteras y traseras.

Geometric centerline An imaginary line through the exact center of the front and rear wheels.

Línea central geométrica Línea imaginaria a través del centro exacto de las ruedas delanteras y traseras.

Hazard Communication Standard A forerunner to the right-to-know laws published by the Occupational Health and Safety Administration (OSHA).

Estándar de comunicación de riesgos Un precursor de las leyes del derecho a saber publicado por la Administración de Seguridad y Salud Ocupacional (OSHA, en inglés).

Heavy spot Is a location in a tire with excessive weight.

Zona Dura es un lugar en la llanta con peso excesivo.

High-frequency transmitter An electronic device that sends high-frequency voltage signals to a receiver. Some wheel sensors on computer-controlled wheel aligners send high-frequency signals to a receiver in the wheel aligner.

Transmisor de alta frecuencia Un dispositivo electrónico que manda las señales de voltaje de alta frecuencia a un receptor. Algunos sensores de ruedas en los alineadores controlados por computadora mandan señales de alta frecuencia a un receptor en el alineador de ruedas.

History code A fault code in a computer that represents an intermittent defect.

Código histórico Código de fallo en una computadora que representa un defecto intermitente.

Hot cleaning tank Uses a heated solution to clean metal parts.

Tanque de limpieza en caliente Usa una solución caliente para limpiar las partes de metal.

Hybrid vehicle A hybrid vehicle has a power train with two power sources. The most common type of hybrid vehicle has a gasoline engine and an electric drive motor(s).

Vehículo híbrido Un vehículo híbrido posee un tren de potencia con dos fuentes de energía. El tipo más común de vehículo híbrido posee un motor de gasolina y uno o varios motores eléctricos.

Hydraulic press A hydraulically operated device for disassembling and assembling components that have a tight press-fit.

Prensa hidráulica Dispositivo activado hidráulicamente que se utiliza para desmontar y montar componentes con un fuerte ajuste en prensa.

Ignitable A liquid, a solid, or a gas that can be set on fire spontaneously or by an ignition source.

Inflamable Un líquido, sólido o gas que se puede encender espontáneamente o mediante una fuente de ignición.

Incandescent bulb-type trouble lights Contain incandescent-type bulbs.

Luces de peligro con bombilla de tipo incandescente Incluyen bombillas de tipo incandescente.

Included angle The sum of the camber and steering axis inclination (SAI) angles.

Ángulo incluido La suma del ángulo de inclinación de la comba y de ángulo de inclinación del eje de dirección (SAI).

Integral reservoir A reservoir that is joined with another component such as a power steering pump.

Tanque integral Tanque unido a otro componente, como por ejemplo una bomba de la dirección hidráulica.

International system (SI) A system of weights and measures.

Sistema internacional Sistema de pesos y medidas.

Lateral axle sideset The amount that the rear axle is moved straight sideways in relation to the front axle.

Desplazamiento lateral del eje La cantidad que el eje trasero está desplazado directamente hacia un lado en relación al eje delantero.

Lateral chassis movement Movement from side to side.

Movimiento lateral del chasis Movimiento de un lado a otro.

Lateral tire runout The variation in side-to-side movement.

Desviación lateral del neumático Variación del movimiento de un lado a otro.

Machinist's rule A steel ruler used for measuring short distances. These rulers are available in USC or metric measurements.

Regla para mecánicos Regla de acero que se utiliza para medir distancias cortas. Dichas reglas están disponibles en medidas del USC o en medidas métricas.

Main menu A display on a computer wheel aligner from which the technician selects various test procedures.

Menú principal Representación visual en una computadora para alineación de ruedas de la que el mecánico puede eligir varios procedimientos de prueba.

Magnetic wheel alignment gauge A gauge that may be attached magnetically to the wheel hub to indicate some wheel alignment angles.

Indicador de alineación de rueda magnética Un indicador que se puede adjuntar magnéticamente al buje de la rueda para indicar algunos ángulos de alineación de la rueda.

Manual test A shock absorber test in which the lower end of the shock is disconnected from the suspension.

Prueba manual Una prueba del amortiguador en la cual el extremo inferior del amortiguador se disconecta de la suspensión.

Memory steer The tendency of the vehicle steering not to return to the straight-ahead position after a turn, but to keep steering in the direction of the turn.

Dirección de memoria Tendencia de la dirección del vehículo a no regresar a la posición de línea recta después de un viraje, sino a continuar girando en el sentido del viraje.

Molybdenum disulfide lithium-based grease A special lubricant containing molybdenum disulphide and lithium that may be used on some steering components.

Grasa de bisulfuro de molibdeno con base de litio Lubricante especial compuesto de bisulfuro de molibdeno y litio que puede utilizarse en algunos componentes de la dirección.

Multipull unitized body straightening system A hydraulically operated system that pulls in more than one location when straightening unitized bodies.

Sistema de tiro múltiple para enderezar la carrocería unitaria Sistema activado hidráulicamente que tira hacia más de una dirección al enderezar carrocerías unitarias.

Multipurpose dry chemical fire extinguisher Contains a dry chemical powder used to extinguish various types of fires.

Extintor de incendios multipropósito de polvo químico seco Contiene un polvo químico seco que se utiliza para apagar diversos tipos de incendios.

Multi-ribbed V-belt A belt containing a series of small "V" grooves on the underside of the belt.

Correa trapezoidal de ranuras múltiples Una correa que posee varias ranuras pequeñas en forma de "V" en la cara interna.

National Institute for Automotive Service Excellence (ASE) An organization that provides voluntary automotive technician certification in eight areas of expertise.

Instituto Nacional para la excelencia en la reparación de automóviles Organización que provee una certificación voluntaria de mecánico de automóviles en ocho áreas diferentes de especialización.

Neutral coast-down test A chassis vibration test performed while the vehicle is coasting in neutral.

Tire thump A pounding noise as the tire and wheel rotate, usually caused by improper wheel balance.

Ruido sordo del neumático Ruido similar al de un golpe pesado que se produce mientras el neumático y la rueda están girando. Por lo general este ruido lo ocasiona el desequilibrio de la rueda.

Tire tread depth gauge A special tool required to measure tire tread depth.

Calibrador de la profundidad de la huella del neumático Herramienta especial requerida para medir la profundidad de la huella del neumático.

Tire vibration Vertical or sideways tire oscillations.

Vibración del neumático Oscilaciones verticales o laterales del neumático.

Toe gauge A special tool used to measure front or rear wheel toe.

Calibrador del tope Herramienta especial que se utiliza para medir el tope de las ruedas delanteras o traseras.

Toe-in A condition in which the distance between the front edges of the tires is less than the distance between the rear edges of the tires.

Convergencia Condición que ocurre cuando la distancia entre los bordes frontales de los neumáticos es menor que la distancia entre los bordes traseros de los mismos.

Toe-out A condition in which the distance between the front edges of the tires is more than the distance between the rear edges of the tires.

Divergencia Condición que ocurre cuando la distancia entre los bordes delanteros de los neumáticos es mayor que la distancia entre los bordes traseros de los mismos.

Torque steer The tendency of the steering to pull to one side during hard acceleration on front-wheel-drive vehicles with unequal length front drive axles.

Dirección de torsión Tendencia de la dirección a desviarse hacia un lado durante una aceleración rápida en un vehículo de tracción delantera con ejes de mando desiguales.

Total toe The sum of the toe angles on both wheels.

Tope total Suma de los ángulos del tope en ambas ruedas.

Track gauge A long straight bar with adjustable pointers used to measure rear wheel tracking in relation to the front wheels.

Calibrador del encarrilamiento Barra larga y recta con agujas ajustables que se utiliza para medir el encarrilamiento de las ruedas traseras con relación a las ruedas delanteras.

Tram gauge A long, straight bar with adjustable pointers used to measure unitized bodies.

Calibrador del tram Barra larga y recta con agujas ajustables que se utiliza para medir carrocerías unitarias.

Tread wear indicators Raised portions near the bottom of the tire tread that are exposed at a specific tread wear.

Indicadores del desgaste del neumático Secciones elevadas cerca de la parte inferior de la huella del neumático que quedan expuestas cuando la huella alcanza una cantidad de desgaste específica.

Trim height The normal chassis riding height on a computer-controlled air suspension system.

Altura equilibrada Altura normal de viaje del chasis en un sistema de suspensión controlado por computadora.

Turning radius gauge A gauge with a degree scale mounted on turntables under the front wheels.

Calibrador del radio de giro Calibrador con una escala de grados montado sobre plataformas giratorias debajo de las ruedas delanteras.

Turntable A mechanical device placed under each front wheel during a wheel alignment to allow the front wheels to be turned in each direction to measure various front suspension angles.

Plataforma giratoria Un dispositivo mecánico colocado debajo de cada rueda delantera durante la alineación de ruedas para permitir girar las ruedas delanteras en cada dirección para medir varios ángulos de la suspensión delantera.

Understeer The tendency of a vehicle not to turn as much as desired by the driver.

Dirección pobre Tendencia de un vehículo a no girar tanto como lo desea el conductor.

U.S. Customary (USC) system A system of weights and measures patterned after the British system.

Sistema usual estadounidense (USC) Sistema de pesos y medidas desarrollado según el modelo del Sistema Imperial Británico.

Vacuum hand pump A mechanical pump with a vacuum gauge and hose used for testing vacuum-operated components.

Bomba de vacío manual Bomba mecánica con un calibrador de vacío y una manguera que se utiliza para probar componentes a depresión.

Vehicle Dynamic Suspension (VDS) A computer-controlled air suspension system used on some current model vehicles.

Suspensión dinámica del vehículo (SDV) Es un sistema de suspensión de aire que se usa en algunos vehículos de modelo reciente.

Vehicle lift A hydraulically or air-operated mechanism for lifting vehicles.

Levantamiento del vehículo Mecanismo activado hidráulicamente o por aire que se utiliza para levantar vehículos.

Vehicle wander The tendency of the steering to pull to the right or left when the vehicle is driven straight ahead on a straight road.

Desviación de la marcha del vehículo Tendencia de la dirección a desviarse hacia la derecha o hacia la izquierda cuando se conduce el vehículo en línea recta en un camino cuya superficie es lisa.

V-type belt A drive belt with a V shape.

Correa en V Una correa de transmisión en forma de V.

Wear indicators Wear indicators show ball joint wear by the position of the grease fitting in the ball joint.

Indicadores de desgaste Los indicadores de desgaste muestran el desgaste de las rótulas de la suspensión por la posición de la válvula engrasadora en la rótula de suspensión.

Woodruff key A half-moon-shaped metal key used to retain a component, such as a pulley, on a shaft.

Chaveta woodruff Chaveta de metal en forma de media luna que se utiliza para sujetar un componente, como por ejemplo una roldana, en un árbol.

Workplace hazardous materials information systems (WHMIS) Are similar to SDSs and they provide information regarding the handling of hazardous waste materials.

Sistemas de información acerca de materiales peligrosos en el lugar de trabajo (WHMIS, en inglés) Son similares a las hojas de datos MSDS y brindan información acerca del manejo de materiales de desecho peligrosos.

INDEX

S

W

Y